Acta Physica Austriaca

Supplementum XIII

Proceedings of the
XIII. Internationale Universitätswochen für Kernphysik 1974
der Karl-Franzens-Universität Graz
at Schladming (Steiermark, Austria)
4th February—15th February 1974

Sponsored by
Bundesministerium für Wissenschaft und Forschung
Steiermärkische Landesregierung
Sektion Industrie der Kammer der
gewerblichen Wirtschaft für Steiermark
International Centre for Theoretical Physics, Triest

1974

Springer-Verlag
Wien New York

Progress in Particle Physics

Edited by Paul Urban, Graz

With 175 Figures

1974

Springer-Verlag
Wien New York

Organizing Committee

Chairman

Prof. Dr. Paul Urban
Vorstand des Institutes für Theoretische Physik
der Universität Graz

Committee Members

Dr. C. B. Lang
Dr. H. Latal
Dr. A. Mas-Parareda
Dr. L. Pittner
Dr. W. Plessas

Secretary

M. Krautilik

Softcover reprint of the hardcover 1st edition 1974

Library of Congress Cataloging in Publication Data

Internationale Universitätswochen für Kernphysik der Karl-Franzens-Universität Graz, 13th, Schladming, Austria, 1974.
Progress in particle physics.

(Acta physica Austriaca. Supplementum 13)
"Sponsored by Bundesministerium für Wissenschaft und Forschung [and others]."
1. Particles (Nuclear physics)--Congresses.
I. Urban, Paul Oskar, 1905- ed. II. Austria. Bundesministerium für Wissenschaft und Forschung.
III. Title. IV. Series.
QC793.I57 1974 539.7'21 74-13975

ISBN-13:978-3-7091-8377-9 e-ISBN-13:978-3-7091-8375-5
DOI: 10.1007/978-3-7091-8375-5

CONTENTS

Acta Physica Austriaca, Suppl. XIII, 1–3 (1974)

OPENING ADDRESS

by

P. URBAN
Institut für Theoretische Physik
Universität Graz

With great pleasure I welcome you most cordially at our "XIII. Internationale Universitätswochen für Kernphysik der Universität Graz" here in Schladming und thank you for your coming.

It is always a difficult problem to choose the general topic of a meeting, especially if this meeting, as in our case, takes place almost a year after these decisions. Since at the last conference we dealt with mathematical physics and there the problems were of very abstract nature, we decided this year to give a cross-section of the current status of elementary particle physics under the title "Progress in Particle Physics", an attempt therefore to review our knowledge about the four basic interactions and to present the most recent results in the lectures. In September last year, however, experimental results were published which may allow the speculation that actually we have to deal with only three interactions; also this will be discussed here.

In strong interaction physics the main efforts of theoretical and experimental research lie in the field of multiparticle production at highest energies, where especially theoreticians have much to do in order to explain the wealth of existing data in a consistent manner. Therefore the phenomenology and some of the current models for these processes will be discussed in three different lectures. In the same way as one tries to obtain simple physical laws at the highest energies, one also hopes to have a better understanding of the processes at low energies. Here meson-meson interactions are at the focus of the investigations and are of course included in our program. On the more abstract side we will be hearing about gauge theories in strong interactions and about recent attempts to include relativity into the concept of higher symmetries.

The transition to the next field is provided by lectures about electromagnetic interactions of hadrons. But even well known quantumelectrodynamics still isn't understood clearly and completely, therefore an attempt to achieve this will also be discussed. As I already mentioned, recent experiments, which may provide evidence for the existence of neutral currents, gave new impetus to the theories of a unified description for weak and electromagnetic interactions. The problems connected with these and some consequences of the models have been discussed already yesterday and today. These experiments were made with neutrinos at the big accelerators, but, as you will be hearing extensively, the largest sources of neutrinos lie outside our laboratories and give rise to plenty of interesting problems. This year also a series of lectures in the second week is devoted to a topic, which usually is treated rather perfunctory in elementary physics, namely gravitation as expressed through the theory of general relativity. And finally an interesting development is presented which deals with the evaluation by a computer of complex expressions in physics by means of symbolic representation and the application of these methods to elementary particle physics. As usual some seminars will round off the scientific program.

I hope that through these lectures we may succeed in extending and deepening our knowledge in the various fields of elementary particle physics and I wish to all of you a pleasant and fruitful stay in Schladming.

The material presented in the three lectures

EMPIRICAL AND MATHEMATICAL FOUNDATIONS
OF GENERAL RELATIVITY

by

J. EHLERS
Max Planck-Institut für Physik und Astrophysik
Munich, Germany

is essentially contained in the following publications:

J. Ehlers, Survey of General Relativity Theory, in Relativity, Astrophysics and Cosmology, edited by W. Israel, Reidel Publ. Comp., Dordrecht-Holland, p. 1-125, particularly chapters 1 and 2, 1973.

J. Ehlers, The Nature and Structure of Spacetime, in The Physicist's Conception of Nature, edited by J. Mehra, Reidel Publ. Comp., p. 71-91, 1973.

S. W. Hawking, S. F. R. Ellis, The Large-Scale Structure of Spacetime, Cambridge University Press, 1973.

An extended version of the material will be presented under the title"Spacetime Structures"in the University of Pittsburgh Series in the Philosophy of Science, vol.6, to appear early in 1975.

Acta Physica Austriaca, Suppl. XIII, 5–56 (1974)

NEW TRENDS IN MULTIPERIPHERAL DYNAMICS*) +)

by

A. BASSETTO
Istituto di Fisica dell'Università
Padova, Italy
Istituto Nazionale di Fisica Nucleare
Sez. Padova, Italy

INTRODUCTION

The original idea [1] underlying multiperipheral dynamics was to describe a high energy many particle production event with a chain of many low energy amplitudes of peripheral nature, which are, in a dispersive language, dominated by few nearby singularities (the pion pole in the simplest case). The main use of these production amplitudes was not in the detailed calculation of exclusive differential cross-sections, but as an approximate input to the unitarity equation for the absorptive part of the elastic amplitude and for the inclusive distributions.

---

*) Lecture given at the XIII. Internationale Universitätswochen für Kernphysik, Schladming, February 4 - 15, 1974.

+) Dedicated to the memory of Ziro Koba.

In these lectures we shall confine our study to the multiperipheral equation in the forward direction. As is well known, the forward absorptive amplitude is related to the total cross-section through the optical theorem and is also the main ingredient in evaluating inclusive distributions.

We shall first examine with pedagogical purposes the ABFST model; then we shall discuss the physical motivations and the ways to generalize it in a new formulation of multiperipheral dynamics which has been recently proposed [2]. In this scheme important physical effects, like diffractive and absorptive contributions, can be properly included in a way which is largely model independent, still retaining the basic multiperipheral features. More important, this scheme will provide us with experimentally testable consequences: it can therefore be supported or disproved by data.

While the data we presently know agree, we must admit the scheme is, in a sense, incomplete and leaves room for further dynamical input. In fact neither it is a complete picture of hadron dynamics nor it gives detailed information on the Pomeranchuk singularity. At present it can be regarded, not as a theory, but as a phenomenological picture, which seems however to possess some germ of truth.

1. DEFINITIONS AND KINEMATICS

We shall consider for simplicity the scattering of two spinless particles a and b with masses μ_a and μ_b, which produces in the final state n identical spinless particles of mass m larger than μ_a and μ_b [1]. We choose an arbitrary ordering for the outgoing particles and indicate the various four-momenta as in fig. 1, where the four-momentum transfers Q_i related to that ordering are also defined. At this stage fig. 1 does not possess any dynamical meaning.

The vectors P_a, P_b and P_i obey the mass shell constraints

$$P_a^2 = \mu_a^2, \quad P_b^2 = \mu_b^2 , \tag{1.1}$$

$$P_j^2 = m^2, \quad j = 1, \ldots, n . \tag{1.2}$$

They lie inside the future light cone. The vectors Q_i are spacelike; in fact one can easily realize that their time component changes sign going from a rest frame of b to a rest frame of a.

Therefore we have:

$$Q_i^2 = t_i < 0, \quad i = 1, \ldots, n-1 . \tag{1.3}$$

It is also useful to introduce spacelike unit vectors:

$$x_i = \frac{Q_i}{\sqrt{-t_i}} , \quad i = 1, \ldots, n-1 . \tag{1.4}$$

They span the spacelike hyperboloid P $(x^2 = -1)$ and can be parametrized as follows:

$$x_i = (\mathrm{sh}\, \sigma_i, \underline{x}_i \cosh \sigma_i), \; i = 1,\ldots, n-1, \tag{1.5}$$

where σ_i is a real quantity and $\underline{x}_i$ is a vector on the three-dimensional unit sphere $\underline{x}^2 = 1$.

The invariant measure on the hyperboloid is:

$$dP = 2\delta(x^2+1)d^4x = (\cosh\sigma)^2 \, d\sigma \, d(\cos\Theta) \, d\phi \, , \tag{1.6}$$

whereas

$$d^4Q = \frac{|t|dt}{2} dP \quad . \tag{1.7}$$

We need 3n-4 independent kinematical variables to describe a n-particle production event. If we take as variables the vectors Q_i, they are not independent as they must obey the mass-shell constraints of eqs.(1.1) and (1.2). A complete set of independent variables is provided by Bali-Chew-Pignotti (BCP variables) [3].

For each vertex in fig. 1 we introduce a frame of reference. These frames are connected with a given arbitrary frame by means of the Lorentz transformations $a_1,\ldots,a_n$. We define these group elements a_i implicitly through the formulae:

$$\begin{aligned}
P_b &= L(a_n)(\mu_b,0,0,0) = L(q_b)(\mu_b,0,0,0) \, , \\
Q_i &= L(a_i)(0,0,0,\sqrt{-t_i}) \, , \\
Q_i &= L(a_{i+1}\, a_z(\chi_{i+1}))(0,0,0,\sqrt{-t_i}) \, , \\
P_a &= L(a_1 a_z(\chi_1)) \, (\mu_a,0,0,0) = L(q_a)(\mu_a,0,0,0) \, ,
\end{aligned} \tag{1.8}$$

$$i = 1,\ldots, n-1 \, .$$

The Lorentz transformation a_z is a boost along the z-axis. The quantities χ_i can be evaluated from the mass-shell constraints in eqs. (1.1), (1.2). We have:

$$P_1^2 = (P_a - Q_1)^2 = m^2 \quad ,$$

$$P_j^2 = (Q_{j-1} - Q_j)^2 = m^2, \quad j = 2,\ldots,n-1 \quad , \qquad (1.9)$$

$$P_n^2 = (Q_{n-1} + P_b)^2 = m^2.$$

Introducing eqs. (1.8) we get:

$$\text{sh}\ \chi_1 = \frac{m^2-\mu_a^2-t_1}{2\mu_a\sqrt{-t_1}} \quad ,$$

$$\cosh \chi_i = \frac{m^2-t_{i-1}-t_i}{2\sqrt{t_i t_{i-1}}} \ , \quad \chi_i > 0, \quad i = 2,\ldots,n-1,$$

$$\text{sh}\ \chi_n = \frac{m^2-t_{n-1}-\mu_b^2}{2\mu_b\sqrt{-t_{n-1}}} \quad . \qquad (1.10)$$

If we define the Lorentz transformations

$$g_i = a_z(-\chi_{i+1})\ a_{i+1}^{-1}\ a_i, \ i = 1,\ldots,n-1, \qquad (1.11)$$

from eqs. (1.8) we see that they leave invariant the z component of a four vector. Therefore they belong to the subgroup SU(1,1) of SL(2,C) and can be parametrized as follows:

$$g_i = u_z(\rho_i)\ a_x(\xi_i)\ u_z(\nu_i),\ i = 1,\ldots,n-1, \qquad (1.12)$$

where u_z represents a rotation around the z-axis and a_x a boost along the x-axis. Actually the variables ρ_i and ν_i are not independent, as eqs. (1.8) are invariant under the substitution

$$a_i \rightarrow a_i\ u_z\ (\gamma) \quad , \qquad (1.13)$$

that is

$$g_i \rightarrow g_i\ u_z\ (\gamma)\ ,$$

$$g_{i-1} \rightarrow u_z(-\gamma)\ g_{i-1}\ . \qquad (1.14)$$

It follows that the independent variables are:

$$t_i, \quad i = 1,\ldots,n-1, \qquad -\infty < t_i < 0\ ,$$

$$\xi_i, \quad i = 1,\ldots,n-1, \qquad 0 \leq \xi_i < +\infty \quad , \qquad (1.15)$$

$$\omega_i = \nu_i + \rho_{i-1}\ , \quad i = 2,\ldots,n-1, \qquad 0 \leq \omega_i < 2\pi.$$

In the sequel however, we shall refer to the redundant variables t_i and g_i as BCP variables.

Other useful kinematical quantities are the sub-energies s_i, defined as

$$s_i = (P_i + P_{i+1})^2\ , \quad i = 1,\ldots,n-1\ . \qquad (1.16)$$

They can be expressed through the BCP variables as follows:

$$s_i = t_{i-1} + t_{i+1} - \frac{1}{2t_i} \{[T(m^2, t_{i-1}, t_i) \cdot$$

$$\cdot T(m^2, t_i, t_{i+1})]^{1/2} \cosh \xi_i + \tag{1.17}$$

$$+ (m^2 - t_{i-1} - t_i)(m^2 - t_i - t_{i+1})\} \quad ,$$

$$i = 1,\dots,n-1,$$

where:

$$T(a,b,c) = a^2+b^2+c^2-2ab-2ac-2bc \quad , \tag{1.18}$$

and the convention

$$t_o = \mu_a^2, \; t_n = \mu_b^2 \tag{1.19}$$

has been used.

The cross section for n particle production is:

$$\sigma_n(s) = \frac{1}{2} [T(s,\mu_a^2,\mu_b^2)]^{-1/2} \; (2\pi)^{4-3n}$$

$$\frac{1}{n!} \int |M_n|^2 \delta^4 (P_a+P_b - \sum_{j=1}^{n} P_j) \prod_{j=1}^{n} \frac{d^3 \underline{P}_j}{2P_j^o} \; , \tag{1.2o}$$

where M_n is the invariant amplitude.

We define

$$\sigma(s) = \sum_{n=3} \sigma_n(s) \tag{1.21}$$

as the total inelastic cross-section [2].

We define the average multiplicity as:

$$F_1(s) = \langle n \rangle = [\sigma(s)]^{-1} \sum_{n=3} n\, \sigma_n(s) \, , \tag{1.22}$$

and, more generally, the binomial multiplicity moments:

$$F_k(s) = [\sigma(s)]^{-1} \sum_{n=3} \frac{n!}{(n-k)!} \sigma_n(s) \, , \tag{1.23}$$

$$k = 1,2,\ldots \quad .$$

2. THE ABFST MULTIPERIPHERAL MODEL

From Bose statistics the amplitude M_n has to be invariant under permutations of the final particles. In the ABFST model the amplitude is constructed by taking the lowest order Feynman graphs. Therefore the correct recipe for realizing Bose symmetry in this case is to sum over the n! permutations of the final lines:

$$M_n = \sum_{\pi} F_n(P_a, P_b, P_{\pi(1)}, \ldots, P_{\pi(n)}) \, . \tag{2.1}$$

F_n refers to the contribution of a single Feynman graph, characterized by a given ordering of the final lines (see fig. 2).

In evaluating $|M_n|^2$ mixed terms obviously appear. Only if they are neglected we can write:

$$|M_n|^2 = \sum_{\pi} \left| F_n(P_a, P_b, P_{\pi(1)}, \ldots, P_{\pi(n)}) \right|^2 . \qquad (2.2)$$

The approximation of neglecting mixed terms is often justified by making recourse to the peripherality idea: the Feynman graphs are damped if the four-momentum transfers are large owing to the propagators.

Diagrams with a different order in the final particles are relevant in different phase space regions; therefore mixed terms are depressed.

We must remember however that the number of terms one drops is very large; therefore this assumption must be regarded at the best as an approximation. We would like to emphasize that this procedure is not at all required by the multiperipheral dynamics. We shall see later in detail how the permutation problem can be solved in a way which is fully satisfactory.

In the ABFST approach eq. (2.2) is accepted and the following factorized structure is considered

$$F_n = c\, g^n \prod_{i=1}^{n-1} (\mu^2 - t_i)^{-1}, \qquad n \geq 2 , \qquad (2.3)$$

g being a coupling constant. Introducing eqs. (2.2) and (2.3) into eq. (1.20) and choosing the vectors Q_i as integration variables, we get:

$$\sigma_n(s) = \pi c^2 g^2 \left[T(s, \mu_a^2, \mu_b^2) \right]^{-1/2} \int d^4Q \, .$$

$$\cdot (t-\mu^2)^{-2} \, \delta[(P_a - Q)^2 - m^2] \, \Theta(P_a^0 - Q^0) \, A_{n-1}(Q, P_b) , \qquad (2.4)$$

where

$$A_{n-1}(Q_1,P_b) = [(2\pi)^{-3}g^2]^{n-1} \int \prod_{i=2}^{n-1} [d^4Q_i \cdot$$

$$\cdot(t_i-\mu^2)^{-2}] \prod_{i=1}^{n-2} \{\delta[(Q_i-Q_{i+1})^2-m^2]\ \Theta(Q_i^o-Q_{i+1}^o)\}\cdot$$

$$\cdot\delta[(P_b+Q_{n-1})^2-m^2]\ \Theta(P_b^o + Q_{n-1}^o) \quad , \tag{2.5}$$

$$n \geq 2 .$$

From this equation a recursive relation can be easily obtained:

$$A_{n-1}(Q,P_b) = (2\pi)^{-3}\ g^2 \int d^4Q'\ (t'-\mu^2)^{-2}\cdot$$

$$\cdot\delta[(Q-Q')^2-m^2]\ \Theta(Q^o-Q'^o)\ A_{n-2}(Q',P_b), \tag{2.6}$$

$$n \geq 3 .$$

We define:

$$A(Q,P_b) = \sum_{n=1} A_n\ (Q,P_b) \tag{2.7}$$

Summing both sides of eq. (2.6) over n we get the multi-peripheral integral equation:

$$A(Q,P_b) = A_1(Q,P_b) + (2\pi)^{-3}\ g^2 \int d^4Q'\ (t'-\mu^2)^{-2}\cdot$$

$$\delta[(Q-Q')^2-m^2]\ \Theta(Q^o-Q'^o)\ A\ (Q',P_b) \quad . \tag{2.8}$$

Eq. (2.8) is completely equivalent to eqs. (2.6) and (2.7), as an iterative procedure of solving eq. (2.8) stops, at a fixed value of $\bar{s} = (Q + P_b)^2$, when

$$n > \frac{\sqrt{\bar{s}}}{m} \quad . \tag{2.9}$$

In other words, at fixed $\bar{s}$, the sum in eq. (2.7) is a finite sum and eq. (2.8) is clever enough as to take into account the threshold conditions.

Eq. (2.8) could be solved by iteration; however, when $\bar{s}$ is large, the number of iterations is also large and this way of proceeding is not at all convenient. We shall now describe an alternative way of solving eq. (2.8). We choose a frame of reference in which:

$$P_b = (\mu_b,0,0,0) \tag{2.1o}$$

and parametrize the vectors Q and Q' as in eqs. (1.4) and (1.5). From spherical symmetry the amplitudes $A(Q,P_b)$ and $A_1(Q,P_b)$ can only depend on σ and t:

$$A(\sigma,t) = A_1(\sigma,t)+(2\pi)^{-3}g^2\int\frac{|t'|dt'}{2}(\cosh\sigma')^2\,d\sigma'\cdot$$

$$\cdot d^2\underline{x}'(t'-\mu^2)^{-2}\ A(\sigma',t')\ \Theta(\sqrt{-t}\ \mathrm{sh}\ \sigma-\sqrt{-t'}\ \mathrm{sh}\ \sigma')\cdot$$

$$\cdot\delta[t+t'-m^2-2\sqrt{tt'}\ (\mathrm{sh}\ \sigma\cdot\mathrm{sh}\ \sigma' - \underline{x}\cdot\underline{x}'\ \cosh\sigma\cdot\cosh\sigma')] \quad . \tag{2.11}$$

We perform the angular integration:

$$A(\sigma,t) = A_1(\sigma,t)+(2\pi)^{-2}\, g^2\int\sqrt{\frac{t'}{t}}\,\frac{dt'}{4}\,\frac{\cosh\sigma'}{\cosh\sigma}\,d\sigma'\cdot$$

$$\cdot(t'-\mu^2)^{-2}\,A(\sigma',t')\,\Theta(\sigma-\sigma'-\chi)\ , \tag{2.12}$$

where:

$$\cosh\chi\ (t,t') = \frac{m^2-t-t'}{2\sqrt{tt'}}\ , \qquad \chi > 0\ . \tag{2.13}$$

Eq. (2.12) can be partially diagonalized by means of a Laplace transform:

$$A(\lambda,t) = \int_0^\infty d\sigma\ \ e^{-\lambda\sigma}\cosh\sigma\cdot A(\sigma,t)\ . \tag{2.14}$$

We get:

$$A(\lambda,t) = A_1(\lambda,t) + (2\pi)^{-2}\, g^2\ (4\lambda)^{-1}\cdot$$
$$\cdot\int_{-\infty}^{0}\sqrt{\frac{t'}{t}}\ \frac{dt'}{(t'-\mu^2)^2}\ \exp[-\lambda\chi(t,t')]\ A(\lambda,t')\ . \tag{2.15}$$

Eq. (2.15) is the "radial" equation. It involves only one variable and cannot be solved by relying on symmetry properties any more. It is model dependent as its kernel depends on the structure of the propagators and of the vertices. Changing these structures, one can obtain different detailed solutions. As we are not interested in them, since they are unrealistic in any case, we continue with a discussion of the general features of the equation. From these features important concepts can be learnt.

Eq. (2.15) is a "bona fide" integral equation; if

we try to solve it by iteration, the procedure does not stop after a finite number of cycles (of course this happens because in the projection formula (2.14) we have integrated up to an infinite value of the energy). However threshold properties are not lost; they are converted into asymptotic properties of $A(\lambda,t)$ when $\mathrm{Re}\lambda\to+\infty$.

To see that, let us first invert eq. (2.14):

$$A(\sigma,t) = \frac{1}{2\pi i}\int_{L-i\infty}^{L+i\infty} d\lambda\, \frac{e^{\lambda\sigma}}{\cosh\sigma}\, A(\lambda,t)\ , \tag{2.16}$$

where L is to the right of any singularity of $A(\lambda,t)$ [3]. From eqs. (2.5) and (2.14) we get:

$$A_1(\lambda,t) = \frac{(2\pi)^{-3}g^2}{2\mu_b\sqrt{-t}}\exp[-\lambda\chi_b(t)]\ , \tag{2.17}$$

where:

$$\mathrm{sh}\,\chi_b(t) = \frac{m^2-t-\mu_b^2}{2\mu_b\sqrt{-t}}\ . \tag{2.18}$$

The first iteration of eq. (2.15) gives:

$$A_2(\lambda,t) = \frac{(2\pi)^{-5}g^4}{8\mu_b\lambda\sqrt{-t}}\int_{-\infty}^{0}\frac{dt'}{(t'-\mu^2)^2}\ \exp[-\lambda\chi_b(t')-\lambda\chi(t,t')]\ . \tag{2.19}$$

The asymptotic behaviour of $A_2(\lambda,t)$ as $\mathrm{Re}\lambda\to+\infty$ is determined by searching the minimum of the quantity $\chi_b(t')+\chi(t,t')$ with respect to the variable t', in the integration region.

A simple algebraic calculation shows that the minimum is attained at the value:

$$\bar{t}' = \frac{t + \mu_b^2 - 2m^2}{2} \quad . \qquad (2.2o)$$

Inserting eq. (2.19) into eq. (2.16), we see that $A_2(\sigma_2, t)$ vanishes unless

$$\sigma_2 \geq \chi_b(\bar{t}') + \chi(t, \bar{t}') \quad , \qquad (2.21)$$

as the integral could be evaluated by closing the integration contour to the right. Eq. (2.21) can be written as:

$$\text{sh}\ \sigma_2 \geq \text{sh}\ \chi_b(\bar{t}')\ \cosh \chi(t, \bar{t}') + \cosh \chi_b(\bar{t}') \cdot$$

$$\cdot \text{sh}\ \chi(t, \bar{t}') = \frac{4m^2 - t - \mu_b^2}{2\mu_b\sqrt{-t}} \quad , \qquad (2.22)$$

where, in the last step, eqs. (2.13), (2.18) and (2.2o) have been used.

Proceeding by induction with the same technique, one could show that iterations of eq. (2.15) lead to functions $A_n(\lambda, t)$ decreasing in the limit $\text{Re}\lambda \to +\infty$ in such a way that $A_n(\sigma_n, t)$, evaluated from eq. (2.16), is zero, unless

$$\text{sh}\ \sigma_n \geq \frac{(nm)^2 - t - \mu_b^2}{2\mu_b\sqrt{-t}} \quad . \qquad (2.23)$$

In other words, although iteration of eq. (2.15) does

not stop, it leads to functions which are faster and faster decreasing in the limit $\mathrm{Re}\lambda\to+\infty$. When introduced in eq. (2.16), only a finite number of them contributes for a given value of $\bar{s}$, namely we recover the trheshold condition (2.9).

When $\bar{s}$ is large, it is convenient to evaluate the integral in eq. (2.16) by shifting the integration contour to the left. In this way we meet the singularities of the function $A(\lambda,t)$ in the complex λ-plane. These singularities depend on the structure of the "radial" integral equation (2.15). In the ABFST model we are considering, the Kernel is a positive Fredholm Kernel for real positive λ, analytic in the half-plane $\mathrm{Re}\lambda>0$. It can be made symmetric by defining:

$$A_s(\lambda,t) = -\frac{\sqrt{-t}}{t-\mu^2}\, A(\lambda,t) \tag{2.24}$$

and, replacing $A(\lambda,t)$ in eq. (2.15)

$$A_s(\lambda,t) = \frac{\sqrt{-t}}{\mu^2-t}\, A_1(\lambda,t) + {} + g^2 \int_{-\infty}^{0} K_s(\lambda,t,t')\, A_s(\lambda,t')\, dt' \,. \tag{2.25}$$

The Kernel K_s has a pole at $\lambda = 0$ and is a decreasing function of $\mathrm{Re}\lambda>0$ in all the integration region. In the limit $\mathrm{Re}\lambda\to+\infty$ it vanishes exponentially.

From all these properties it follows that the resolvent Kernel $R(\lambda,g^2,t,t')$ is meromorphic in the half-plane $\mathrm{Re}\lambda>0$, its singularity farthest to the right in the λ-plane being a simple pole at the positive value $\lambda = \alpha_p$. The residue factorizes with respect to t and t':

$$\lim_{\lambda=\alpha_p} (\lambda-\alpha_p) R(\lambda,t,t') = \phi(\alpha_p,t)\phi(\alpha_p,t') \ . \tag{2.26}$$

The dependence on g^2 is understood.

The function ϕ can be chosen to be positive; the position of the pole α_p is an increasing function of the coupling constant g^2, starting from the value $\alpha_p = 0$ in the limit $g^2 \to 0$.

This pole controls the asymptotic behaviour of $A(\sigma,t)$ in the large $\bar{s}$ limit, as we see from eq. (2.16), by summing over the many production channels; in this sense it represents a collective effect. If we approximate $A(\lambda,t)$ by retaining only this pole contribution, we get:

$$A(\sigma,t) \simeq \frac{\mu^2-t}{\mu_b\sqrt{-t}} \ \beta_b \frac{e^{\alpha_p\sigma}}{\cosh\sigma} \ \phi(\alpha_p,t) \ , \tag{2.27}$$

where

$$\beta_b = \mu_b \int_{-\infty}^{0} \frac{\sqrt{-t'}}{\mu^2-t'} \, dt' \, \phi(\alpha_p,t') \ A_1(\alpha_p,t',\mu_b) . \tag{2.28}$$

We explicitly notice that the dependence on the particle b, affecting the known term, is factorized in eq. (2.27)

We can now evaluate the total inelastic cross-section summing eq. (2.4) over n, starting from n = 3. If for $A(\sigma,t)$ we consider the approximation (2.27), we get, in the high energy limit:

$$\sigma(s) \simeq \pi \ c^2 \ g^2[T(s,\mu_a^2,\mu_b^2)]^{-1/2} \ \ \beta_b(\mu_b)^{-1}.$$

$$\cdot\int d^4Q[(\mu^2-t)\sqrt{-t}]^{-1}\ \delta[(P_a-Q)^2-m^2]$$ (2.29)

$$\Theta(P_a^o-Q^o)\ \frac{e^{\alpha_p\sigma}}{\cosh\sigma}\ \phi(\alpha_p,t)\ .$$

We parametrize the vector P_a as follows:

$$P_a = \mu_a\ (\cosh\eta, 0, 0, \text{sh}\ \eta)$$ (2.3o)

and, from eq. (2.1o), we have:

$$s = \mu_a^2 + \mu_b^2 + 2\mu_a\mu_b \cosh\eta \quad .$$ (2.31)

We integrate eq. (2.29) with the same techniques we used before:

$$\sigma(s) \simeq \frac{\pi^2c^2g^2\beta_b}{4\mu_a^2\mu_b^2(\text{sh}\ \eta)^2}\int_{-\infty}^{o}\frac{dt}{\mu^2-t}\ \phi(\alpha_p,t)\int^{\eta-\chi_a(t)} d\sigma \quad e^{\alpha_p\sigma}$$

$$\simeq \frac{4\pi^5c^2\beta_a\beta_b}{\mu_a^2\mu_b^2\alpha_p}\ \frac{e^{\alpha_p\eta}}{(\text{sh}\ \eta)^2} = \gamma_a\gamma_b\ \frac{e^{\alpha_p\eta}}{(\text{sh}\ \eta)^2}\ .$$ (2.32)

The quantities γ_a and γ_b are dimensional constants related to the incoming particles a and b respectively. We have recovered the usual factorized expression for the total cross-section, given by the Regge pole theory.

Factorization is a consequence of the leading output singularity being a simple pole in the complex λ-plane. We shall refer to it as to inclusive factorization. It is indeed a quite different concept from the exclusive factorization we assumed for the input production amplitudes (eq. (2.3)); as in the sequel these concepts will play a very important role, a clear distinction should be kept in mind.

Another important remark concerns the dependence of α_p on the coupling constant g^2. If we require that the total inelastic cross-section is constant for large s, we must impose the condition:

$$\alpha_p = 2 \quad . \tag{2.33}$$

This in turn requires a particular value for the coupling constant g^2. Saturation of the Froissart bound (apart from logarithmic terms) is not included in this model and must be imposed as an independent condition, by a peculiar choice of the coupling constant. A larger value of it would entail a violation of the Froissart bound. Of course this feature is highly unsatisfactory; it is not hard to trace back this shortcoming of the ABFST model to its lack of s-channel unitarity [1]. In fact, if we consider the t-channel singularities the model takes into account, by cutting diagrams according to the Cutkoski rules (see fig. 3), we realize that only states with two particles of mass μ and 0,1,... particles of mass m contribute. Even in the original papers [1] iterations of ladder diagrams in the direct channel were considered, giving rise to branch point singularities in the angular momentum plane.

Before examining more general multiperipheral dynamics, we would like to review further simple consequences of the model we are considering.

In these lectures we shall not study inclusive

distributions, as their physics has been largely discussed in the literature, whereas a rigorous mathematical treatment should involve harmonic analysis results [4]. We shall instead discuss average multiplicity and binomial multiplicity moments.

From the definition (1.22) and eqs. (1.21), (2.4) and (2.5) we get:

$$\langle n\rangle = [\sigma(s)]^{-1} g^2 \frac{\partial}{\partial g^2} \sigma(s) . \qquad (2.34)$$

There is no problem in changing the order of summation and derivation, as, for a given s, the sum is over a finite number of terms. An approximate expression could be immediately derived by inserting eq. (2.32) into eq. (2.34). We find it instructive however to get first an exact representation for the quantity in eq. (2.34).

From eqs. (1.21), (2.4), (2.7), (2.8), (2.14), (2.3o) we obtain:

$$\sigma(s) = \frac{\pi^2 c^2 g^2}{4\mu_b \mu_a^2 (\mathrm{sh}\ \eta)^2} \frac{1}{2\pi i} \int_{L-i\infty}^{L+i\infty} \frac{d\lambda}{\lambda} \int_{-\infty}^{0} \frac{\sqrt{-t}\, dt}{(t-\mu^2)^2} \cdot$$

$$\cdot \exp [\lambda(\eta-\chi_a(t))] \ \hat{A}(\lambda,t) , \qquad (2.35)$$

where

$$\hat{A}(\lambda,t) = A(\lambda,t) - A_1(\lambda,t) . \qquad (2.36)$$

Actually, for any given value of η, the integration domain over t is compact. We have understood in $\sigma(s)$ and in $A(\lambda,t)$ the variable g^2. We perform the derivative with respect to g^2:

$$g^2 \frac{\partial}{\partial g^2} \sigma(s) = \sigma(s) + \frac{\pi^2 c^2 g^2}{4\mu_a^2 \mu_b (\text{sh}\ \eta)^2} \frac{1}{2\pi i} \int_{L-i\infty}^{L+i\infty} \frac{d\lambda}{\lambda} \cdot$$

$$\cdot \int_{-\infty}^{0} \frac{\sqrt{-t}\ dt}{(t-\mu^2)^2} \exp\ [\lambda(\eta - \chi_a(t))]\ g^2 \frac{\partial}{\partial g^2} \hat{A}(\lambda, t) \ . \qquad (2.37)$$

From eqs. (2.15) and (2.17) we get:

$$g^2 \frac{\partial}{\partial g^2} \sigma(s) = 2\sigma(s) + \frac{\pi^3 c^2 g^2}{4\mu_a \mu_b (\text{sh}\ \eta)^2} \frac{1}{2\pi i} \int_{L-i\infty}^{L+i\infty} \frac{d\lambda}{\lambda^2} \cdot$$

$$\cdot \iint_{-\infty}^{0} \frac{\sqrt{tt'}\ dt\ dt'}{(t-\mu^2)^2 (t'-\mu^2)^2} \exp [\lambda(\eta - \chi(t,t'))]\ A_1(\lambda, t, \mu_a) ,$$

$$g^2 \frac{\partial}{\partial g^2} A(\lambda, t', \mu_b) = 2\sigma(s) + \frac{4\pi^5 c^2}{\mu_a \mu_b (\text{sh}\ \eta)^2} \frac{1}{2\pi i} \int_{L-i\infty}^{L+i\infty} \frac{d\lambda}{\lambda} \cdot$$

$$\cdot e^{\lambda\eta} \int_{-\infty}^{0} \frac{|t|\ dt}{(t-\mu^2)^2} A_1(\lambda, t, \mu_a)[g^2 \frac{\partial}{\partial g^2} A(\lambda, t, \mu_b) - A(\lambda, t, \mu_b)] , \qquad (2.38)$$

where we have made explicit the dependence on the incoming particles a and b.

Now, from eq. (2.25) and:

$$g^2 \frac{\partial}{\partial g^2} A_s(\lambda, t) = A_s(\lambda, t) + g^2 \int_{-\infty}^{0} dt'\ K_s(\lambda, t, t') g^2 \frac{\partial}{\partial g^2} A_s(\lambda, t') , \qquad (2.39)$$

remembering that the resolvent Kernel is also symmetric, we get:

$$g^2 \frac{\partial}{\partial g^2} \sigma(s) = 2\sigma(s) + \frac{\pi^3 c^2 g^2}{4\mu_a \mu_b (\mathrm{sh}\ \eta)^2} \frac{1}{2\pi i} \int_{L-i\infty}^{L+i\infty} \frac{d\lambda}{\lambda^2}$$

$$\iint_{-\infty}^{0} \frac{dt\, dt' \sqrt{tt'}}{(t-\mu^2)^2 (t'-\mu^2)^2} \exp[\lambda(\eta-\chi(t,t')] A(\lambda,t,\mu_a) A(\lambda,t,\mu_b) \ . \tag{2.4o}$$

Eq. (2.4o) is our basic result.

We introduce in eq. (2.4o) the pole approximation we have already considered for $A(\lambda,t)$; by keeping only the leading term in the high energy limit, we obtain:

$$g^2 \frac{\partial}{\partial g^2} \sigma(s) \simeq \frac{4\pi^5 c^2}{\mu_a^2\ \mu_b^2} \frac{\beta_a \beta_b}{\alpha_p} \frac{\eta e^{\alpha_p \eta}}{(\mathrm{sh}\ \eta)^2} \ \Phi^2 , \tag{2.41}$$

where

$$\Phi^2 = \int_{-\infty}^{0} dt\ \phi^2\ (\alpha_p, t) \ . \tag{2.42}$$

Then, from eqs. (2.32) and (2.34) we have:

$$\langle n \rangle \ \simeq \ \Phi^2 \eta \qquad . \tag{2.43}$$

We notice that a factorized pole as leading singularity in the λ-plane implies logarithmic increase with energy of the average multiplicity, the coefficient of η being independent of the incoming particles a and b. We also notice that the position of the pole α_p affects this coefficient, but not the functional dependence on the energy.

With the same technique, we can evaluate the binomial multiplicity moments; for instance

$$F_2(s) = \langle n(n-1)\rangle = [\sigma(s)]^{-1} g^2 \frac{\partial}{\partial g^2}[(g^2 \frac{\partial}{\partial g^2}\sigma(s)) - \sigma(s)] \quad . \tag{2.44}$$

It is useful to put eq. (2.4o) in the form:

$$g^2 \frac{\partial}{\partial g^2}\sigma(s) = 2\sigma(s) + \frac{4\pi^5 c^2 g^2}{\mu_a \mu_b (\text{sh } \eta)^2} \frac{1}{2\pi i} \int_{L-i\infty}^{L+i\infty} \frac{d\lambda}{\lambda} e^{\lambda\eta} .$$

$$\cdot \iint_{-\infty}^{0} dt\, dt' \, A_s(\lambda,t,\mu_a) K_s(\lambda,t,t') A_s(\lambda,t',\mu_b) \quad . \tag{2.45}$$

Introducing this expression in eq. (2.44) and taking eqs. (2.39) and (2.4o) into account, we get:

$$F_2(s)\cdot\sigma(s) = 2\sigma(s) + \frac{8\pi^5 c^2 g^2}{\mu_a \mu_b (\text{sh } \eta)^2} \frac{1}{2\pi i} \int_{L-i\infty}^{L+i\infty} \frac{d\lambda}{\lambda} e^{\lambda\eta} .$$

$$\cdot \iint_{-\infty}^{0} dt\, dt' \, A_s(\lambda,t,\mu_a) K_s(\lambda,t,t') A_s(\lambda,t',\mu_b) + \frac{8\pi^5 c^2}{\mu_a \mu_b (\text{sh } \eta)^2} \cdot$$

$$\frac{1}{2\pi i} \int_{L-i\infty}^{L+i\infty} \frac{d\lambda}{\lambda} e^{\lambda\eta} \iint_{-\infty}^{0} dt\, dt' \, A_s(\lambda,t,\mu_a) R(\lambda,t,t') A_s(\lambda,t',\mu_b) , \tag{2.46}$$

$R(\lambda,t,t')$ being the resolvent Kernel of eq. (2.25).

Again we introduce the pole approximation we have already considered and keep only the leading term in the high energy limit. We get:

$$F_2(s)\sigma(s) \simeq \frac{4\pi^5 c^2}{\mu_a^2\,\mu_b^2}\,\frac{\beta_a\beta_b}{\alpha_p}\,\frac{\eta^2 e^{\alpha_p\eta}}{(\text{sh}\,\eta)^2}\,\Phi^4 \,, \qquad (2.47)$$

that is

$$F_2(s) = \Phi^4\,\eta^2 + O(\eta) \quad . \qquad (2.48)$$

Eq. (2.48) is obviously generalized into:

$$F_k(s) = \Phi^{2k}\,\eta^k + O(\eta^{k-1}) \quad ,$$

$$k = 1\,,\ldots.\,, \qquad (2.49)$$

which can be proved with the same technique.

We notice that a simple factorized pole as leading singularity entails a cancellation between F_2 and $\langle n\rangle^2$; precisely:

$$F_2(s) - \langle n\rangle^2 = f_2(s) = O(\eta) \quad . \qquad (2.5o)$$

The function $f_2(s)$ represents the integrated inclusive two particle correlation function [4]; if the produced particles were emitted according to a Poisson distribution, $f_2(s)$ should be identically zero.

Of course this cannot happen, as at least correlations due to conservation laws must be present. We see however that in the ABFST model we are considering (and more generally in any model having a simple pole as leading output singularity), $f_2(s)$ increases with energy more slowly than $F_2(s)$.

Starting from eq. (2.32) one could prove that $f_k(s)$, namely the fully integrated k-particle inclusive correlation, increases with energy only as log s [5] in this type of models. For this reason they are called weakly correlated models [6]. They seem to be over-simplified as data indicate a relation

$$f_2(s) \simeq \eta^2 \tag{2.51}$$

rather than eq. (2.5o).

3. GENERALIZING THE MULTIPERIPHERAL MODEL

In its origin the Multiperipheral Model was formulated on a phenomenological basis. It was recognized that further important effects were not properly treated. Two features were particularly relevant:

i) the damping of the momentum transfers introduced by the pion pole was weaker than the observed one,

ii) unitarity in the direct channel was poor; in particular diffractive exchanges and rescattering effects were neglected.

As far as the point ii) is concerned, iteration of ladder diagrams in the direct channel was considered [1], giving rise to branch points (AFS cuts) in the angular momentum plane.

A different approach is proposed in the Multi-Regge model [7]. The physical idea underlying the Multi-Regge model is that the simple pion exchange is injustified when the subenergies become large and can be better replaced with a Regge pole exchange. This suggestion can be made mathematically precise by performing a complete angular momentum analysis of the many particle production amplitudes.

A Regge pole exchange provides a stronger damping of the momentum transfers, according to the point i). At the same time it also improves unitarity in the direct channel, as discontinuities with four particles of mass μ and 0,1,... particles of mass m in the multi-ladder diagram are taken into account, according to fig. 4. Moreover it is an example of a more general expression for the production amplitudes, which is not completely factorized with respect to the momentum transfers:

$$F_n = G_a(P_a,Q_1,Q_2) \prod_{i=2}^{n-2} G(Q_{i-1},Q_i,Q_{i+1}) \cdot G_b(Q_{n-2},Q_{n-1},P_b) , \quad n \geq 3 . \tag{3.1}$$

(Compare with eq. 2.3). In the multi-Regge model we have:

$$G(Q_{i-1}, Q_i, Q_{i+1}) \approx \gamma(t_{i-1},t_i)\, \gamma(t_i,t_{i+1}) \cdot [(Q_{i-1} - Q_{i+1})^2]^{\alpha(t_i)} , \quad i = 1,\ldots,n-1 . \tag{3.2}$$

One could generalize further eq. (3.1) by considering factors depending on a larger, but finite, number of adjacent variables. This property is called "short range order" in ref. [7]. It is related to the correlations between produced particles in exclusive measures; therefore it has not to be confused with the inclusive factorization, which controls the correlations between produced particles in inclusive measures. The exclusive factorization is governed by the input singularities for the exclusive production amplitudes (Feynman or Regge poles); the inclusive factorization depends on the output singularities of the multiperipheral equation. There is of course a dynamical connection between the two patterns of singularities, but they are not the same.

The two patterns of singularities would be equal if a complete bootstrap program would be successful. However even a partial bootstrap (namely the one requiring self-consistency only for the leading singularity) encounters serious difficulties in the multi-Regge model. Presumably the leading output singularity (Pomeron) is mainly due to low-lying input singularities (ρ-f,π,...). Moreover one must realize that eq. (3.2) is a good approximation in eq. (3.1) only for a small part of the available phase-space, namely when all the subenergies are large. From data one can see that the mean subenergy is not large; therefore the expression (3.2) is unjustified for the largest number of actual events. Of course this criticism mainly applies to the approximation (3.2) and not to the factorized structure (3.1). To this regard one could improve eq. (3.1) by considering a function G depending on a larger number of neighbouring momenta (longer range correlations in the language of ref. [7]). However this cannot be pushed to a definite limit, since to take all possible exclusive correlations into account with this technique would simply amount to introduce different functions for production amplitudes with a different number of final particles and this in turn would prevent the possibility of any recursive relation.

A different possibility is to consider several Regge exchanges as input singularities. In this approach the function G of eq. (3.1) will become a matrix with the dimensions given by the number of Regge poles we are considering, while the "external" functions G_a and G_b will be vectors of the corresponding linear space. This way of generalizing the multi-Regge model is still unsatisfactory, as the true n-particle production amplitude cannot factorize as in eq. (3.1), if we consider finite dimension matrices, as for instance thresholds do not factorize.

Rather surprisingly however, this second approach is fruitful, if we allow our linear space to be infinite dimensional. Before doing that, in order to understand the physical meaning of such a procedure, it is worth discussing the general assumptions, which are the basis of any form of multiperipheral dynamics.

In our opinion these basic ingredients are [2]:

i) the decrease of the production amplitudes for a suitable ordering of the secondaries as the momentum transfers become large,

ii) the existence of a recursive relation connecting the n-particle production amplitude M_n with M_{n-1}.

The first requirement can be stated in a mathematically clear way as an upper bound on $|M_n|$. The second requirement is related to the possibility of factorization; we have already seen that numerical factorization and factorization in a linear finite dimensional space are impossible for the true amplitudes. We shall therefore require factorization in an infinite dimensional linear space (actually a Banach space). The functions G, G_a and G_b of few neighbouring momenta as in eq. (3.1), will be considered as operator and vectors respectively, in such a space.

Numerical factorization can be retained as a property of the upper bound; in ref. (2) in fact the following important theorem has been proved:
"A necessary and sufficient condition for the existence of a factorized representation in the operator sense for the production amplitudes is the existence of a numerically factorized upper bound".

We realize that both the basic concepts under i) and ii) are contained in the requirement that the amplitudes have upper bounds which are decreasing with momentum transfers and numerically factorized. This assumption is sufficient in order to develop the whole mathematical formalism without any approximation.

This upper bound provides a mathematically clear definition of multiperipheral dynamics, which does not rely on any diagram approach. As we have said, it is general enough as to take diffractive and absorptive effects into account, and it leads, as we shall see, to testable consequences on multiplicity distributions, which can be supported or disproved by data. To this regard we must anticipate that the bound we shall assume contains an arbitrary numerical factor and, therefore, the amplitude being continuous, it does not impose any restriction when we consider the amplitude in a compact region of the space of the kinematical variables (including the multiplicity). In other words, our condition has an asymptotic character. It follows that all the consequences of this condition on measurable quantities have an asymptotic nature as well and their comparison with experiments implies necessarily some extrapolation procedure.

In order to treat carefully the concept of decrease with the momentum transfers, we have first to examine how to realize the invariance of the production amplitude under permutation of the final particles.

We shall refer to Bali-Chew-Pignotti varialbes; we call Ω_n the manifold spanned by these variables and x a point in this manifold. Therefore x stands for 3n-4 parameters. Let us now discuss the permutation problem. As we are considering identical bosons, the amplitude M_n is invariant under permutation of the final lines. Let us call P_π the operator that generates the permutation π. P_π generates a mapping of Ω_n onto itself and we have:

$$M_n(P_\pi x) = M_n(x) \quad , \qquad x \in \Omega_n \; . \tag{3.3}$$

In the case of Feynman diagrams, Bose symmetry is realized by the equation:

$$M_n(x) = \sum_\pi F_n \, (P_\pi x) \quad , \tag{3.4}$$

F_n being the contribution of a given Feynman graph. Eq. (3.4) is not however the correct recipe for realizing Bose symmetry in other models (for instance in the multi-Regge model) as it can lead to double counting, if we choose F_n as in eqs. (3.1), (3.2). The ambiguity in many models lies in the very definition of the function F_n.

Let us therefore for the moment impose our upper bound on the complete amplitude M_n:

$$|M_n(x)| \leq \sup_{\pi} \hat{f}_n(P_\pi x) \quad , \tag{3.5}$$

where the function $\hat{f}_n$ is strongly decreasing in its arguments $t_1,\ldots,t_{n-1}$.

The meaning of eq. (3.5) is the following: the production amplitude M_n is very small unless x belongs to the one of the n! regions Δ_n^π, defined, for each fixed permutation π, by the condition that the n-1 invariants

$$t_i^\pi = (P_a - \sum_{j=1}^{i} P_{\pi(j)})^2, \quad i = 1,\ldots, n-1, \tag{3.6}$$

are small.

In order to justify this assumption experimentally, it is useful to decompose the various four-vectors into longitudinal and transverse parts. We consider here the region Δ_n corresponding to the identical permutation. We choose the z-axis along the direction of the incoming particles and write:

$$- t_i = [(Q_i^z)^2-(Q_i^o)^2] + [(Q_i^x)^2+(Q_i^y)^2] \quad . \tag{3.7}$$

One can show that both the quantities in the brackets are non-negative and, therefore, they are both small if t_i is small. We get easily the inequality:

$$[(P_i^x)^2 + (P_i^y)^2]^{1/2} \leq \sqrt{-t_i} + \sqrt{-t_{i-1}} \; , \tag{3.8}$$

from which we realize that small momentum transfers imply small transverse momenta, as experiments require.

However the assumption we are considering is stronger than the decrease with transverse momenta at least for two reasons. First we see that, in order to have quantities

$$Q_k^x = - \sum_{i=1}^{k} P_i^x \tag{3.9}$$

small, it is not sufficient to require that the quantities P_i^x are small, but, if k is large, a negative correlation among these quantities must be present. Second, the requirement of small momentum transfers implies also conditions on the longitudinal momenta, as we see from eq. (3.7).

The condition (3.5) is rather complicated, due to the symmetry of the function $M_n(x)$. Is is therefore convenient to introduce a different function $F_n(x)$ which is not symmetric, but has simpler decrease properties in the variables $|t_i|$. With regard to the connection between the function M_n and F_n we adopt the following attitude [2]: for a given choice of the ordering of the final particles, the function F_n coincides with the function M_n in a given region Γ_n of the space Ω_n:

$$F_n(x) = M_n(x) \quad , \quad x \in \Gamma_n \; . \tag{3.1o}$$

Then if we indicate by Γ_n^π the region in Ω_n defined by:

$$x \in \Gamma_n^\pi \longleftrightarrow P_\pi \quad x \in \Gamma_n \tag{3.11}$$

and assume that:

$$\bigcup_{\pi} \Gamma_n^{\pi} = \Omega_n \tag{3.12}$$

$M_n(x)$ can be expressed by $F_n(x)$ in the whole Ω_n, using its symmetry property:

$$M_n(x) = F_n(P_\pi x) , \quad x \in \Gamma_n^{\pi} . \tag{3.13}$$

If some of the regions Γ_n^{π} overlap, we have the constraint:

$$F_n(x) = F_n(P_\pi x) , \quad x \in \Gamma_n \cap \Gamma_n^{\pi} . \tag{3.14}$$

We choose the region Γ_n to be open. Then we can find a continuous function $\eta_n(x)$ with support in Γ_n and with the property:

$$\sum_{\pi} \eta_n(P_\pi x) = 1 . \tag{3.15}$$

We shall give later an explicit expression for η_n. Then from eq. (3.13) we get:

$$M_n(x) = \sum_{\pi} F_n(P_\pi x)\, \eta_n(P_\pi x) \tag{3.16}$$

and

$$|M_n(x)|^2 = \sum_{\pi} |F_n(P_\pi x)|^2 \cdot \eta_n(P_\pi x) . \tag{3.17}$$

We see that in this formulàtion there is no problem with mixed terms.

Let us now explicitly construct the quantities we have used. We choose a continuous function $f_n(x)$ such that:

$$f_n(x) > \hat{f}_n(x) , \quad x \in \Omega_n , \tag{3.18}$$

and we define Γ_n as the region where the inequality:

$$f_n(x) > \sup_{\pi} \hat{f}_n (P_{\pi} x) \tag{3.19}$$

holds. In this region

$$|F_n(x)| = |M_n(x)| < f_n(x) . \tag{3.2o}$$

Outside this region $F_n(x)$ is arbitrary and can be chosen such as the following relation is satisfied:

$$|F_n(x)| \leq f_n(x) . \tag{3.21}$$

We see that the functions $F_n(x)$ are not symmetric but have simpler decrease properties.

In order to show that eq. (3.12) is satisfied, we remark that

$$\sup_{\pi} f_n (P_{\pi} x) > \sup_{\pi} \hat{f}_n (P_{\pi} x) . \tag{3.22}$$

Finally we can choose a simple expression for the function $\eta_n(x)$:

$$\eta_n(x) = g_n(x) \quad [\sum_{\pi} g_n (P_{\pi} x)]^{-1} , \tag{3.23}$$

where:

$$g_n(x) = [f_n(x) - \sup_{\pi} \hat{f}_n (P_{\pi} x)] .$$

$$\cdot\Theta[f_n(x) - \sup_{\pi} \hat{f}_n (P_{\pi} x)] \qquad (3.24)$$

Now we want to take into account only general features of multiperipherism, namely the decrease in the momentum transfers t_i and the polynomial boundedness of the production amplitudes in the subenergies. Therefore we choose for the upper bound the form:

$$f_n(x) = c \prod_{i=1}^{n-1} [h(t_i)(\frac{s_i}{4m^2})^{\alpha}], \qquad (3.25)$$

where h(t) is suitably decreasing with $|t|$ and α is an exponent of the order of 1.

We stress that our bound holds in the physical region; besides the decrease in $|t_i|$ and the polynomial boundedness in s_i, a very restrictive condition is implied on the behaviour of the amplitudes for increasing multiplicity n. This limitation will have very important consequences on multiplicity fluctuations, as we shall see in the last part of these lectures.

We also notice that the expression (3.25) is reminiscent of the multi-Regge model. The key point is that here it is not intended as an approximate expression for the amplitude, but as a rigorous upper bound in a suitable phase space region.

In ref. [2] the following important theorem is proved: "Given an arbitrary sequence of continuous functions

$$F_n (g_{n-1}, t_{n-1}, \ldots, g_1, t_1), \; n = 2, \ldots, \qquad (3.26)$$

which are invariant with respect to the transformations (1.14) and satisfy a bound of the kind (3.25), we can always find the vector valued functions $D(g_{n-1}, t_{n-1})$, $E(t_1)$ and the operator valued functions $K(t_{i+1}, g_i, t_i)$, $U(\gamma,t)$ satisfying the conditions:

$$U(\gamma,t_{i+1})K(t_{i+1},g_i,t_i)U(\gamma',t_i) = K(t_{i+1},u_z(\gamma)g_i u_z(\gamma'),t_i),$$

$$U(\gamma,t_1)\ E\ (t_1) = E(t_1)\ ,$$

$$U^T(\gamma,t_{n-1})D(g_{n-1},t_{n-1}) = D(g_{n-1}\ u_z(\gamma),t_{n-1})\ , \qquad (3.27)$$

and

$$||D(g_{n-1},t_{n-1})|| \leq d(\xi_{n-1},t_{n-1})\ ,$$

$$||K(t_{i+1},g_i,t_i)|| \leq k(t_{i+1},\xi_i,t_i)\ ,$$

$$||E(t_1)|| \leq e(t_1)\ ; \qquad (3.28)$$

here we have written the upper bound on F_n in the form:

$$|F_n(g_{n-1},t_{n-1},\dots,g_1,t_1)| \leq d(\xi_{n-1},t_{n-1}) \prod_{i=1}^{n-2} k(t_{i+1},\xi_i,t_i)e(t_1). \qquad (3.29)$$

In terms of these functions, the following representation holds:

$$F_n(g_{n-1},t_{n-1},\dots,g_1,t_1)=(D(g_{n-1},t_{n-1}), \prod_{i=n-2}^{1} \cdot K(t_{i+1}, g_i, t_i)\ E(t_1))\ . \qquad (3.3o)$$

$E(t)$ is an element of a Banach space B_t, $D(g,t)$ an element of the dual space B_t' and $K(t',g,t)$ a bounded operator from B_t to $B_{t'}$.

The proof will not be reported here. We only remark that it is based on an explicit construction of the operator K and of the vectors E and D [2].

We introduce eq. (3.17) into eqs. (1.2o), (1.21) and get:

$$\sigma(s) = \frac{1}{2}[T(s,\mu_a^2,\mu_b^2)]^{-1/2} \sum_{n=3} (2\pi)^{4-3n} \cdot$$

$$\cdot \int |F_n(x)|^2 \, \eta_n(x) \, \delta^4 \left(P_a+P_b-\sum_{j=1}^{n} P_j\right) \prod_{j=1}^{n} \frac{d^3 P_j}{2P_j^o} \quad . \tag{3.31}$$

It is convenient to introduce the auxiliary variables:

$$P_a' = L(a_1 a_z(\chi_1))(\mu_a',0,0,0) \ ,$$

$$P_b' = L(a_n)(\mu_b',0,0,0)$$

$$Q_i' = L(a_{i+1}\ a_z(\chi_{i+1}))(0,0,0,\sqrt{-t_i'}) \ ,$$

$$i = 1,\dots,n-1 \ . \tag{3.32}$$

We choose the momentum transfers as new variables:

$$\sigma(s) = \frac{1}{2}[T(s,\mu_a^2,\mu_b^2)]^{-1/2} \sum_{n=3} (2\pi)^{4-3n} \cdot$$

$$\cdot \int |F_n(x)|^2 \, \eta_n(x) \prod_{i=1}^{n-1} [d^4 Q_i d^4 Q_i' \delta^4 (Q_i - Q_i')] .$$

$$\cdot \delta[(P_a-Q_1)^2 - m^2] \, \Theta(P_a^o - Q_1^o) \prod_{i=1}^{n-2} [\delta[(Q_i-Q_{i+1})^2 - m^2] \cdot$$

$$\cdot\Theta(Q_i^o - Q_{i+1}^o)]\,\delta[(Q_{n-1}+P_b)^2 - m^2]\;\Theta(Q_{n-1}^o + P_b^o)\cdot$$

$$\cdot\delta^4(P_a - P_a')\; d^4P_a'\;\delta^4(P_b - P_b')\; d^4P_b' \;. \qquad (3.33)$$

In ref.[2] the following relations have been proved:

$$\delta^4(Q_i - Q_i') = \frac{2\pi^2}{-t_i}\;\delta(t-t_i')\;\delta_-^3(g_i)\;,$$

$$i = 1,\ldots,\, n-1\;, \qquad (3.34)$$

where δ_-^3 is a measure concentrated in the set $g\varepsilon SU(1,1)$, and (see eq.(1.8):

$$\delta^4(P_a - P_a') = \frac{2\pi^2}{\mu_a^2}\;\delta(\mu'^2_a - \mu_a^2)\;\delta_+^3(q_a^{-1}a_1 a_z(\chi_1))\;,$$

$$\delta^4(P_b - P_b') = \frac{2\pi^2}{\mu_b^2}\;\delta(\mu'^2_b - \mu_b^2)\;\delta_+^3(q_b^{-1}a_n)\;, \qquad (3.35)$$

where δ_+^3 is concentrated on the elements of SU(2). It has also been proved one can perform the following replacements in the integral (3.33) [2]:

$$\delta[(Q_i - Q_{i+1})^2 - m^2]\,\Theta(Q_i^o - Q_{i+1}^o)\,d^4Q_i'\,d^4Q_{i+1} \to$$

$$\to\; \frac{1}{4\pi}\,[T(m^2, t_i', t_{i+1})]^{1/2}\; d^6a_{i+1}\; dt_{i+1}\; dt_i'\;,$$

$$i = 1,\ldots,n-2, \qquad (3.36)$$

$$\delta[(P_a - Q_1)^2 - m^2]\,\Theta(P_a^o - Q_1^o)\; d^4Q_1\; d^4P_a' \to$$

$$\to\; \frac{1}{4\pi}\,[T(m^2, \mu'^2_a, t_1)]^{1/2}\; d^6a_1\; d\mu'^2_a\; dt_1\;, \qquad (3.37)$$

$$\delta[(P_b+Q_{n-1})^2-m^2]\Theta(P_b^o+Q_{n-1}^o)\ d^4Q'_{n-1}\ d^4P'_b \rightarrow$$

$$\rightarrow \frac{1}{4\pi}[T(m^2,\mu'^2_b,t'_{n-1})]^{1/2}\ d^6a_n\ d\mu'^2_b\ dt'_{n-1} \quad . \qquad (3.38)$$

By performing these substitutions we obtain:

$$\sigma(s) = \frac{1}{2}\ [T(s,\mu_a^2,\mu_b^2)]^{-1/2} \sum_{n=3} (2\pi)^{4-3n} \cdot$$

$$\cdot\int |F_n(x)|^2\ \eta_n(x)\ \frac{2\pi^2}{\mu_a^2}\ \delta_+^3(q_a^{-1}a_1a_z(\chi_1))\ \frac{2\pi^2}{\mu_b^2}\ \delta_+^3(q_b^{-1}a_n)\cdot$$

$$\cdot \prod_{i=1}^{n-1} [\frac{2\pi^2}{-t_i}\delta_-^3(g_i)dt_i] \prod_{j=1}^{n} [\frac{d^6a_j}{4\pi}[T(m^2,t_j,t_{j-1})]^{1/2}] \ , \qquad (3.39)$$

where the convention (1.19) has been used. Eq.(3.39) has the remarkable property that all the kinematical terms which appear in the integrand are factorized into sets of neighbouring variables.

We now apply to the function $|F_n(x)|^2\ \eta_n(x)$ the factorization theorem result:

$$|F_n(x)|^2\eta_n(x) = (D(g_{n-1},t_{n-1}),\ \prod_{i=n-2}^{1} H(t_{i+1},g_i,t_i)E(t_1)) \qquad (3.4o)$$

and define the new quantities:

$$\underline{H}(t_{i+1},a_{i+1}^{-1}a_i,t_i) = \frac{1}{16\pi^2}\ \frac{1}{|t_i|}\ [T(m^2,t_{i+1},t_i)]^{1/2}\cdot$$

$$\cdot H(t_{i+1},g_i,t_i)\ \delta_-^3(g_i), \qquad i = 1,\dots,\ n-2\ , \qquad (3.41)$$

and

$$\underline{D}(a_n^{-1}a_{n-1},t_{n-1}) = \frac{1}{8\pi|t_{n-1}|}[T(m^2,\mu_b^2,t_{n-1})]^{1/2} \cdot$$

$$\cdot\delta_-^3(g_{n-1})D(g_{n-1},t_{n-1}) , \tag{3.42}$$

$$\underline{E}(t_1) = \frac{1}{4\pi}[T(m^2,\mu_a^2,t_1)]^{1/2}\,E(t_1) . \tag{3.43}$$

Then eq. (3.39) takes the form:

$$\sigma(s) = \tfrac{1}{2}[T(s,\mu_a^2,\mu_b^2)]^{-1/2}\sum_{n=3}\int\frac{2\pi^2}{\mu_b^2}\,\delta_+^3(q_b^{-1}a_n)\cdot$$

$$\cdot(\underline{D}(a_n^{-1}a_{n-1},t_{n-1}),\prod_{i=n-2}^{1}\underline{H}(t_{i+1},a_{i+1}^{-1}a_i,t_i)\underline{E}(t_1))\cdot$$

$$\cdot\frac{2\pi^2}{\mu_a^2}\,\delta_+^3(a_z(-\chi_1)a_1^{-1}q_a)\prod_{i=1}^{n-1}dt_i\prod_{j=1}^{n}d^6a_j . \tag{3.44}$$

This expression can be simplified introducing the operator valued Kernel $\underline{R}$ $(t', a'^{-1}a, t)$, which is solution of the integral equation:

$$\underline{R}(t',a'^{-1}a,t) = \underline{H}(t',a'^{-1}a,t) +$$

$$+ \quad \underline{H}(t',a'^{-1}a'',t'')\underline{R}(t'',a''^{-1}a,t)dt''\,d^6a'' . \tag{3.45}$$

We get:

$$\sigma(s) = \frac{1}{2}[T(s,\mu_a^2,\mu_b^2)]^{-1/2}\int\frac{2\pi^2}{\mu_b^2}\,\delta_+^3(q_b^{-1}a')\cdot$$

$$\cdot\frac{2\pi^2}{\mu_a^2}\,\delta_+^3(a_z(-\chi_1)a_1^{-1}q_a)\,d^6a_1\,d^6a\,d^6a'\,dt_1\,dt\cdot$$

$$\cdot(\underline{D}(a'^{-1}a,t),\ \underline{R}(t,a^{-1}a_1,t_1)\ \underline{E}(t_1))\ . \tag{3.46}$$

The Lorentz invariance manifests itself in the fact that eq. (3.45) contains a convolution over SL(2C). As a consequence it can be diagonalized by projection on the irreducible representations of this group, according to the techniques developed in refs. [8]. As is well known, the projected equation determines the singularities in the λ-plane, which control the asymptotic behaviour of the total cross-section and of the inclusive distributions.

Eq. (3.45), complemented by the appropriate bounds, can be considered as an exact dynamical equation in the same sense as the Schroedinger equation. As the Schroedinger equation gives the scattering amplitude in terms of a potential which is unknown, but energy independent, eq. (3.45) gives the total cross-section (and the inclusive distributions) in terms of a Kernel $\underline{H}$ which is also unknown, but independent of the total energy and the multiplicity.

4. CONSEQUENCES ON MULTIPLICITY FLUCTUATIONS

Our purpose in this section is to derive from the upper bound (3.5) regarding the production amplitudes, an upper bound on multiplicity fluctuations. Our proof begins by deriving an upper bound for the total cross-sections at fixed multiplicity $\sigma_n(s)$. Actually, instead of eq. (3.17), we use the weaker condition:

$$|M_n(x)|^2 \leq \sum_{\pi} \quad [f_n(P_\pi x)]^2 \quad , \tag{4.1}$$

where f_n is given by eq. (3.25). From eq. (1.2o) we get:

$$\sigma_n(s) \leq (2\pi)^{4-3n} \frac{1}{2}[T(s,\mu_a^2,\mu_b^2)]^{-1/2} \cdot$$

$$\cdot \int \delta^4(P_a+P_b-\sum_{i=1}^{n} P_i)\ [f_n(P_a,P_b,P_1,\ldots,P_n)]^2 \prod_{i=1}^{n} \frac{d^3\underline{P}_i}{2P_i^o} \quad . \tag{4.2}$$

Now we use the kinematical inequality:

$$\prod_{i=1}^{n-1} (\frac{s_i}{2m^2}) \leq \frac{s}{m^2} \prod_{i=1}^{n-1} (1+\frac{\sqrt{-t_i}}{m})^4 \quad , \tag{4.3}$$

which can be obtained from the BCP variables.
By putting:

$$\hat{h}(t) = h(t)(1+\frac{\sqrt{-t}}{m})^{4\alpha}\ 2^{-\alpha} \quad , \tag{4.4}$$

eq. (4.2) can be written as:

$$\sigma_n(s) \leq (2\pi)^{4-3n} \frac{c^2}{2} [T(s,\mu_a^2,\mu_b^2)]^{-1/2} (\frac{s}{m^2})^{2\alpha} \cdot$$

$$\cdot \int \prod_{i=1}^{n-1} \{[\hat{h}(t_i)]^2\ d^4Q_i\}\ \delta[(P_a-Q_1)^2-m^2]\Theta(P_a^o-Q_1^o) \cdot$$

$$\cdot \prod_{i=1}^{n-2} [\delta[(Q_i-Q_{i+1})^2-m^2]\ \Theta(Q_i^o-Q_{i+1}^o)]\ \ \Theta(P_b^o+Q_{n-1}^o) \cdot$$

$$\cdot \delta[(P_b+Q_{n-1})^2-m^2] \quad . \tag{4.5}$$

We perform the integration in a rest system of the particle b, using eqs. (1.3)-(1.7) and (2.3o). Following the same procedure we have outlined in section 2, we perform the angular integrations using the mass-shell constraints:

$$\sigma_n(s) \leq (2\pi)^{3-2n} \frac{c^2}{2^{2n}} \frac{1}{2\mu_a^2\mu_b^2(\text{sh } \eta)^2} \left(\frac{s}{m^2}\right)^{2\alpha} .$$

$$\cdot \int \prod_{i=1}^{n-1} [h(t_i)]^2 dt_i \int \prod_{i=1}^{n-2} d\sigma_i \ \Theta(\eta-\sigma_1-\chi_1) \cdot$$

$$\cdot \prod_{i=2}^{n-2} \Theta(\sigma_{i-1}-\sigma_i-\chi_i) \quad \Theta(\sigma_{n-2}-\chi_{n-1}-\chi_n) \ . \tag{4.6}$$

As $\chi_i > 0$, we enlarge the integration region and, consequently, majorize the integral by putting $\chi_i = 0$, $i = 1,\dots,n$.

If we denote by H the integral:

$$H = \frac{1}{(4\pi)^2} \int_{-\infty}^{0} dt \ [h(t)]^2 \quad , \tag{4.7}$$

(we have assumed h(t) decreasing enough as to make the integral (4.7) converge), we get:

$$\sigma_n(s) \leq \hat{\sigma}_n(s) = \frac{\pi c^2 H}{4\mu_a^2\mu_b^2(\text{sh } \eta)^2} \left(\frac{s}{m^2}\right)^{2\alpha} \frac{(H\eta)^{n-2}}{(n-2)!} ,$$

$$n = 3,\dots \ . \tag{4.8}$$

Eq. (4.8) is our basic result. From eq. (4.8) and the uncontroversial assumption that the total inelastic cross-section $\sigma(s)$ does not decrease asymptotically faster than any inverse power of s, we shall derive the following upper bound on the binomial multiplicity moments (see eq. (1.23)):

$$F_k(s) \leq [A\eta]^k \quad , \quad k = 1,\ldots \quad , \tag{4.9}$$

which generalizes the case k = 1, obtained in ref. 9 . In eq. (4.9) A is a suitable positive constant.

For a given value of s, we define the quantity ν through the inequalities:

$$\sum_{n=\nu+1} \hat{\sigma}_n(s) < \sum_{n=3} \sigma_n(s) = \sigma(s) \leq \sum_{n=\nu} \hat{\sigma}_n(s) \quad . \tag{4.10}$$

Now, from the hypothesis that $\sigma(s)$ does not decrease for asymptotic s faster than any inverse power of s, we derive the bound:

$$\nu(s) < M\eta \quad , \tag{4.11}$$

M being a suitable positive constant. In fact we get:

$$\sigma(s) \leq \frac{\pi c^2 H}{4\mu_a^2\mu_b^2(\text{sh }\eta)^2} \left(\frac{s}{m^2}\right)^{2\alpha} \sum_{n=\nu} \frac{(H\eta)^{n-2}}{(n-2)!} <$$

$$< \frac{\pi c^2 H}{4\mu_a^2\mu_b^2(\text{sh }\eta)^2} \left(\frac{s}{m^2}\right)^{2\alpha} \left(\frac{eH\eta}{\nu-2}\right)^{\nu-2} \left[1 - \frac{eH\eta}{\nu-2}\right]^{-1} \quad . \tag{4.12}$$

From eq. (4.10) we obtain the following relation:

$$\sum_{n=3} \frac{n!}{(n-k)!} \sigma_n < \sum_{n=\nu} \frac{n!}{(n-k)!} \hat{\sigma}_n \ , \ k = 1, \ldots \quad . \tag{4.13}$$

Therefore it follows:

$$F_k(s) < [\sum_{n=\nu+1} \hat{\sigma}_n(s)]^{-1} \sum_{n=\nu} \frac{n!}{(n-k)!} \hat{\sigma}_n(s) =$$

$$= [\sum_{n=\nu+1} \frac{(H\eta)^{n-2}}{(n-2)!}]^{-1} \sum_{n=\nu} \frac{n(n-1)(H\eta)^{n-2}}{(n-k)!} \quad . \tag{4.14}$$

We use the identity:

$$n(n-1) = (n-k)(n-k-1)+2k(n-k)+k^2-k \ , \tag{4.15}$$

and obtain:

$$F_k(s) < \tilde{\eta}^k \sum_{n=0}^{k+1} (\frac{\nu-1}{\tilde{\eta}})^n + 2k\, \tilde{\eta}^{k-1} \sum_{n=0}^{k} (\frac{\nu-1}{\tilde{\eta}})^n +$$

$$+ (k^2-k)\, \tilde{\eta}^{k-2} \sum_{n=0}^{k-1} (\frac{\nu-1}{\tilde{\eta}})^n <$$

$$< \frac{1}{H} \frac{M^2}{M-H} (\eta M + 2)^k \ , \qquad \tilde{\eta} = H\eta \ , \tag{4.16}$$

from which eq. (4.9) follows.

A comparison of eq. (4.9) with eq. (2.49) shows that the bound we have obtained in this general formulation of multiperipherism is quite strong, having essentially the same form of the asymptotic behaviour of F_k in the ABFST model. The crucial feature is that eq. (4.9) is an upper bound; therefore from it one cannot conclude there is the cancellation between $F_2(s)$ and $[F_1(s)]^2$, which is characteristic of short range correlations. The most one can indeed obtain is:

$$f_2(s) \quad < \quad (B\eta)^2 \quad , \qquad (4.17)$$

where B is a suitable constant. There are actually models which obey our upper bound and exhibit long range correlations [1o].

Eq. (4.9) discriminates first models in which Feynman scaling is violated, in particular diffractive excitation models [11]. Such models, predicting a power increase with energy of the multiplicity moments for $k \geq 2$, are in conflict with recent high energy data.

Eq. (4.9) does also operate a much subtler distinction, as it imposes a bound on the growth with k of the constant multiplying the logarithmic term. In fact many models, based on geometrical or statistical considerations, which still have $F_k \sim \eta^k$, cannot be accepted in the multiperipheral scheme [12].

Actually from eq. (4.9) it is easy to derive:

$$\sum_{n=k} \frac{\sigma_n(s)}{\sigma(s)} < \frac{(A\eta)^k}{k!} \quad , \qquad k = 3,\ldots, \qquad (4.18)$$

and this equation implies that:

$$\lim_{\substack{s\to\infty \\ z \text{ fixed}}} \quad \eta \frac{\sigma_k(s)}{\sigma(s)} = 0, \qquad z > C \ , \qquad (4.19)$$

where C is a suitable constant and $Z = \frac{k}{\eta}$. Eq. (4.19) says that K.N.O. scaling [13] in this scheme is obtained with a scaling function $\psi(Z)$ which is zero for Z larger than a suitable constant.

Therefore all the models which lead to a KNO scaling function different from zero are again outside the multiperipheral dynamics. Eq. (4.19) was derived in a different way in ref. [14].

A comparison of the prediction (4.9) with data is delicate, because it is an asymptotic prediction as we have already explained, and data are still affected by large errors. The problem is at present under investigation; first results [15] indicate that data are actually compatible with the prediction (4.9), but to settle the matter, further and better data are probably required.

THREE GENERAL REMARKS SHOULD BE ADDED:

1. More general mass configurations can be treated without any essential change.
2. We do not include in our treatment the two body contribution as it requires different concepts from the ones we develop here.
3. The existence of a hal-plane of analyticity for $A(\lambda,t)$ is implicit in the use of the Laplace transform and can be checked a posteriori in the solution of eq. (2.15).

REFERENCES

1. L. Bertocchi, S. Fubini, M. Tonin, Nuovo Cimento 25, 626 (1962);
   D. Amati, S. Fubini, A. Stanghellini, Nuovo Cimento 26, 896 (1962);
   hereafter quoted as ABFST.
2. A. Bassetto, L. Sertorio, M. Toller, Nuovo Cimento 11A, 447 (1972).
3. N. F. Bali, G. F. Chew, A. Pignotti, Phys. Rev. Lett. 19, 614 (1967), and Phys. Rev. 163, 1572 (1967).
4. For a review paper, see: A. Bassetto, General features and new trends in multiperipheral dynamics; Fortschritte der Physik, to be published, where references to original papers can be found.
5. A. H. Mueller, Phys. Rev. D4, 15o (1971).
6. A. Bassetto, La Rivista del Nuovo Cimento 3, 119 (1973).
7. G. F. Chew, M. L. Goldberger and F. E. Low, Phys. Rev. Lett. 22, 2o8 (1969) and references quoted there.
8. M. Ciafaloni, C. de Tar, M. N. Misheloff, Phys. Rev. 188, 2522 (1969);
   M. Ciafaloni, C. de Tar, Phys. Rev. D1, 2917 (197o);
   A. H. Mueller, I. J. Muzinich, Ann. of Phys. (N. Y.) 57, 2o and 5oo (197o);
   S. Ferrara, G. Mattioli, G. Rossi, M. Toller, Nucl. Phys. B53, 366 (1973).
9. A. Bassetto, L. Sertorio, M. Toller, Lettere al Nuovo Cimento 4, 73 (1972).
1o. J. Finkelstein, F. Zachariasen, Phys. Lett. 34B, 631 (1971);

L. Caneschi, A. Schwimmer, Nucl. Phys. B44, 31 (1972).

11. R. C. Hwa, Phys. Rev. Lett. 26, 1143 (1971);
C. Quigg, J. N. Wang, C. N. Yang, Phys. Rev. Lett. 28, 129o (1972);
M. Jacob, R. Slansky, Phys. Rev. D5, 1847 (1972).

12. A. Giovannini, Nuovo Cimento loA, 713 (1972);
A. Bassetto, G. Ranft, J. Ranft, Lettere al Nuovo Cimento 5, 841 (1972);
A. J. Buras, Z. Koba, Lettere al Nuovo Cimento 6, 629 (1973).

13. Z. Koba, H. B. Nielsen, P. Olesen, Nucl. Phys. B4o, 317 (1972).

14. A. Bassetto, P. Pasti, Lettere al Nuovo Cimento 7, 269 (1973).

15. A. Bassetto, A. J. Buras, Lettere al Nuovo Cimento, to be published.

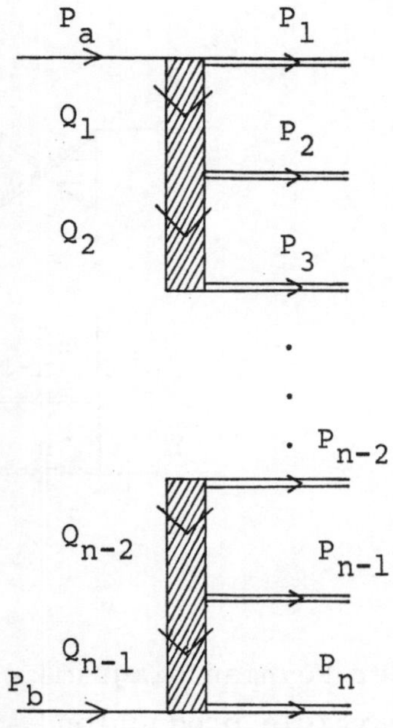

Fig. 1: Definitions of four-momenta for n-particle production

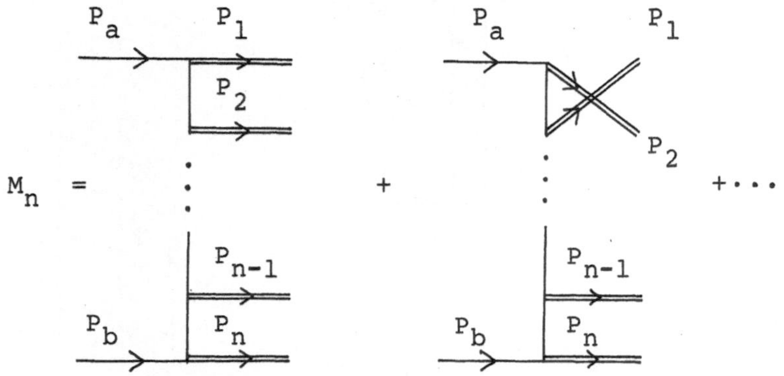

Fig. 2: Lowest order Feynman diagrams contributing to the n-particle production amplitude

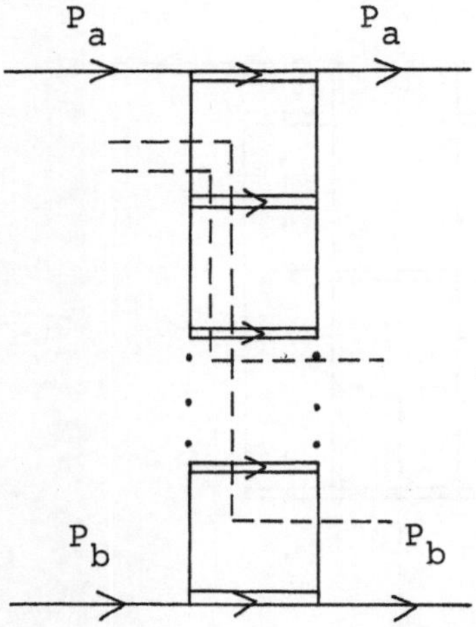

Fig. 3: Cutkosky singularities of the ladder diagram, in the crossed channel

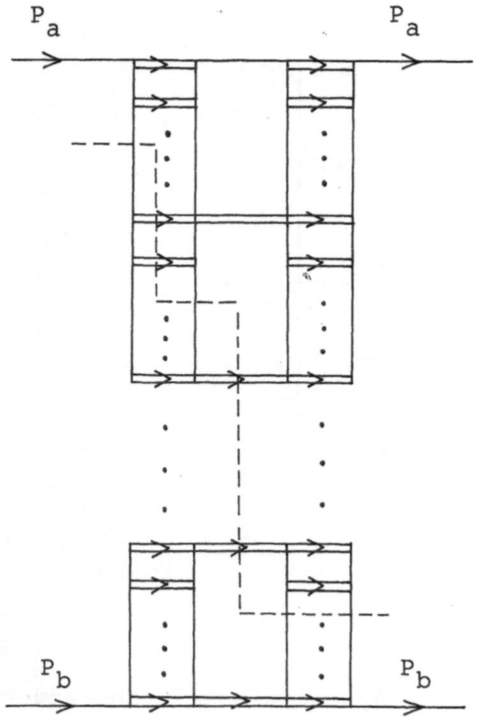

Fig. 4: Cutkosky singularities of the multi-ladder diagram

Acta Physica Austriaca, Suppl. XIII, 57–116 (1974)

PHENOMENOLOGY AND MODELS OF STRONG INTERACTIONS AT VERY HIGH ENERGIES*) +)

by

William R. FRAZER
Department of Physics
University of California, San Diego
La Jolla, California 92037

I. INTRODUCTION

In these lectures I will review some of the striking observations made in high-energy multiparticle reactions at the ISR and at NAL in the past two years - observations which have an immediate impact on our understanding of the strong interactions. First, and most striking, is the confirmation of the hypothesis of short-range correlations in rapidity. Direct and convincing confirmation came from measurement of two-particle correlations at the ISR. The second observation is that of high-energy diffraction dissociation. That this process should exist was no surprise, but some features of diffractive production

---

*) Work supported in part by the United States Atomic Energy Commission.

+) Lecture given at XIII. Internationale Universitätswochen für Kernphysik, Schladming, Austria, February 4-15, 1974.

had to be observed before the general picture became clear. One feature is the magnitude. The question, before we saw the data, was whether diffractive production is a small or large part of multiparticle production. The first estimates of the magnitude came from fits to multiplicity distributions from NAL, where it was found from two-component models that the diffractive component is only about 20% of the inelastic cross section, whereas the short-range correlation (SRC) component is about 80%. A second feature of diffractive dissociation, recently observed, is the existence of diffractive dissociation into large missing mass. As the beam energy increases, higher and higher masses can be produced, and the total diffractive section can rise with energy. There has been much discussion whether this rising diffractive cross section can account for the observed rising total cross section, and we shall discuss this question.

All these observations have stimulated the efforts of theorists. Multiperipheral-type models, which incorporate all the features, have been pursued with renewed vigor. Recent developments include the formulation in terms of an expansion in the number of large rapidity gaps, the recognition of the importance of the concept of a "bare" pomeron, and some attempts to sum the series to develop models of the pomeron. I shall review these developments, and present one model invented by Ball and Zachariasen which I find especially interesting.

II. SHORT-RANGE CORRELATIONS IN RAPIDITY

A. Review of evidence from inclusive spectra

Within the past year we have seen a very impressive accumulation of data from the ISR on inclusive spectra which strikingly confirm the hypothesis that multiparticle production is dominated by a mechanism involving short-range correlations (SRC) in rapidity. Let us review the evidence very briefly.

First, the usual kinematics: A particle's momentum q is resolved into components transverse to the incident beam, q_t, and the component along the incident beam direction, $q_\parallel$. The rapidity is then defined in terms of $q_\parallel$ and the energy E of the particle as

$$y = 1/2 \ \ell n \frac{E_q + q_\parallel}{E_q - q_\parallel} \qquad (2.1)$$

Very briefly stated, the SRC hypothesis says that when any two particles are separated in rapidity by a distance large compared to a correlation length L (to be specified), these particles are uncorrelated. In Fig. 2.1. I have drawn

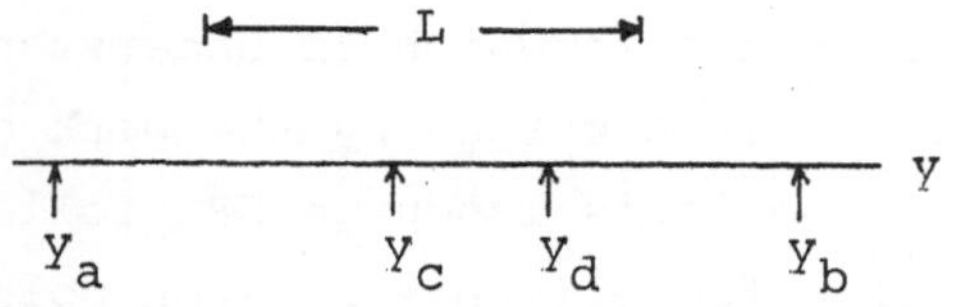

Fig. 2.1.: Possible rapidity configuration in two-particle inclusive reaction, a + b → c + d + ... Correlation length is indicated by L.

a possible rapidity configuration in a two-particle inclusive reaction, $a + b \to c + d + \ldots$. In the configuration shown, $\Delta y_{ac} \gg L$ and $\Delta y_{db} \gg L$, Δy_{cd} is not large. According to the SRC hypothesis, one can expect correlations between c and d, but not between any other pair.

Just a few words about the history and logical status of the SRC hypothesis: The SRC property was first noted in the multiperipheral model of Amati, Fubini, and Stanghellini [1]. That SRC was the fundamental feature of the model was first noted in 1963 by K. G. Wilson at this school [2], and discussed in detail by DeTar [3]. Multiperipheral models have a structure like that shown in Fig. 2.2. Clusters of

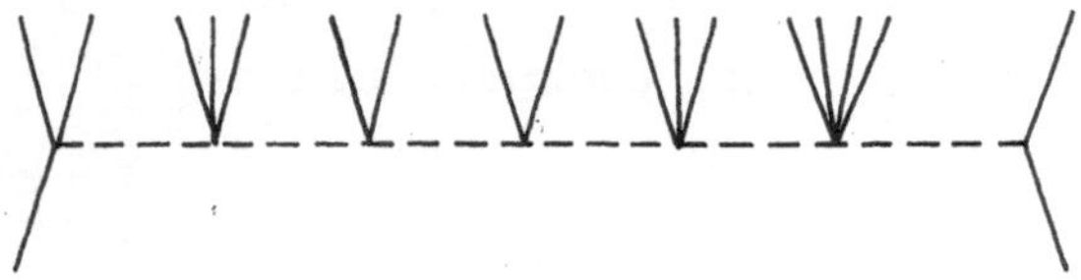

Fig. 2.2: A multiperipheral graph. Dotted lines indicate exchanged particle (a pion in original AFS model), which may differ from model to model.

particles are produced almost independently. Particles within a cluster are strongly correlated and are found closely spaced in rapidity. Correlations between particles in different clusters, which are more distant in rapidity, are weak.

Another language in which to understand short-range correlations is provided by the work of A. Mueller [4]. In Mueller's language SRC follows from the existence of a factorizable Pomeron pole. This connection between SRC and a simple Pomeron pole will be central to subsequent lectures.

To show the utility of the SRC concept in the simplest possible context, let's look at the simplest inclusive reaction, a total cross section.

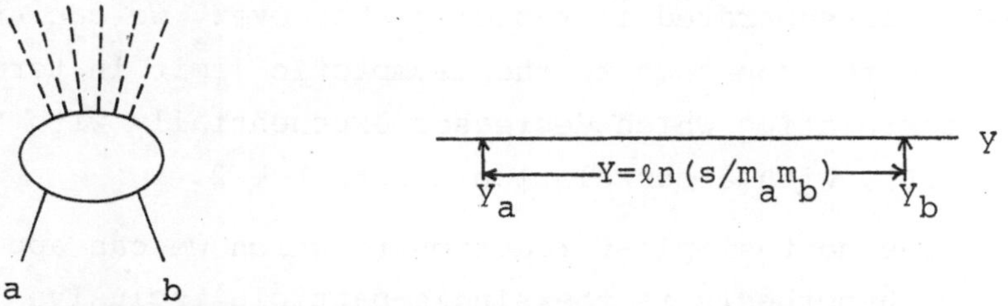

Fig. 2.3: Total cross section: spacing of observed (incident) particles in rapidity.

The total cross section depends only on the rapidity separation of the two incident particles, $Y \equiv \ell n(s/m_a m_b)$. The SRC hypothesis says that for Y very large, the cross section should become asymptotically independent of Y; that is, independent of energy. We can go beyond this simple conclusion to consider the approach to the asymptotic limit. Either empirically, or on the basis of Regge pole theory, one finds that all cross sections behave approximately as follows:

$$\sigma_{ab}(s) \sim \sigma_{ab}(\infty) + B_{ab}/\sqrt{s} \; . \qquad (2.2)$$

Let's rewrite this as a sort of "correlation function",

$$\begin{aligned} C(\Delta y) &\equiv \sigma_{ab}(s) - \sigma_{ab}(\infty) = B_{ab}/\sqrt{s} \\ &= B_{ab}\, e^{-1/2 \ell n\, s} \\ &= B_{ab}\, e^{-Y/L} \;, \; L \approx 2 \; . \end{aligned} \qquad (2.3)$$

The point I am trying to make is that some familiar facts about total cross sections can be re-expressed in SRC language. We can understand the constant total cross section as a lack of correlation when particles a and b are well-separated in rapidity. Moreover, we can understand the approach to the asymptotic limit in terms of a correlation which decreases exponentially with Y, $\exp(-Y/L)$, with a correlation length $L \approx 2$.

The next-simplest reaction to which we can apply the SRC hypothesis is the single-particle inclusive reaction, $a + b \to c + \ldots$.

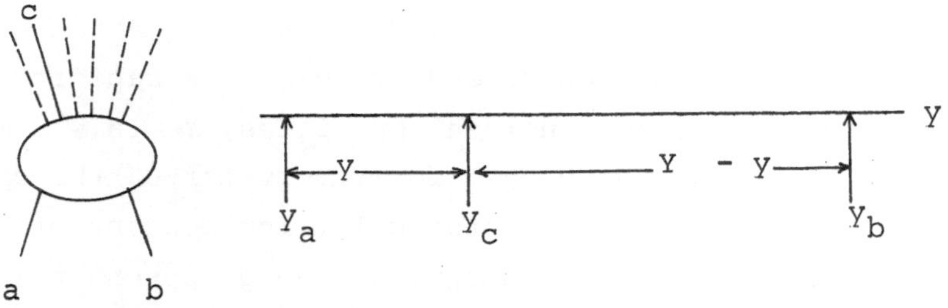

Fig. 2.4: Single-particle inclusive reaction in target fragmentation region.

Consider the configuration shown in Fig. 2.4, where the produced particle is in the target fragmentation region. As energy increases with y held fixed, the separation $\Delta y_{bc} = Y - y$ increases, and for $\Delta y_{bc} \gg L$, there should be no dependence of the cross section on this variable. Then

$$\frac{d\sigma}{dy}(y, Y-y) \sim \frac{d\sigma}{dy}(y), \quad \text{no s-dependence!} \qquad (2.4)$$

This is just what we call <u>scaling</u> in the fragmentation region.

Now consider the case shown in Fig. 2.5,

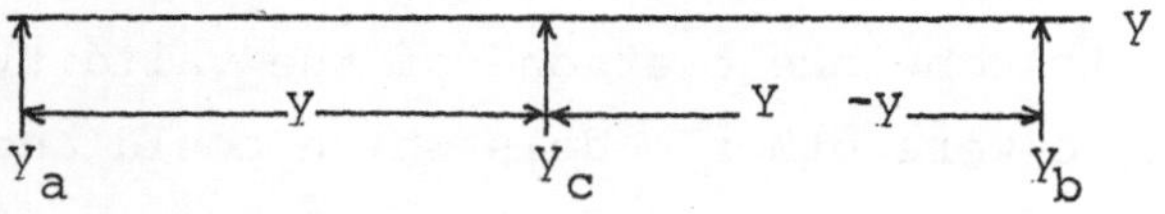

Fig. 2.5: Single-particle inclusive reaction in central region.

where both Δy_{ab} >> L and Δy_{bc} >> L. Then $d\sigma/dy$ is supposed to depend on neither of these variables - it is a universal constant (depending only on the nature of particle c)! That is, as s increases $d\sigma/dy$ should behave as follows

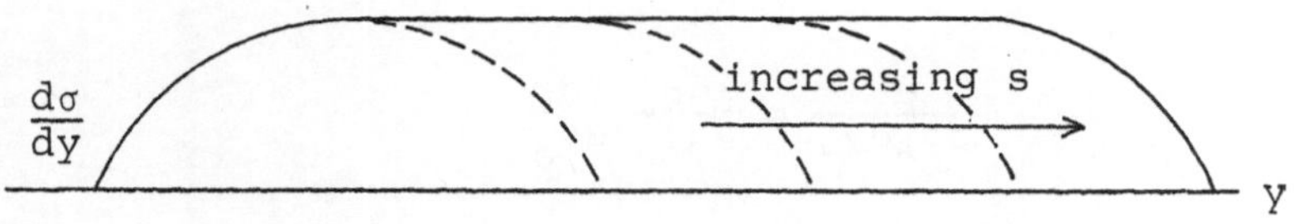

Fig. 2.6: Development of central plateau, using lab rapidity as variable.

Such behavior has been seen very clearly at the ISR (see Fig. 2.7). Note especially the reactions $pp \rightarrow \pi^{\pm} + \ldots$. Not only is the plateau seen very clearly, but also the expected equality of π^+ and π^- plateau heights is observed. Some energy dependence is observed in the plateau even at ISR energies, but this is not unexpected.* Looking back at Figs. 2.4 and 2.6, we can see that scaling

---

*) Curvature of the plateau is also expected, and is supposed to be correlated with the energy dependence. See Ref.[6], p.300.

in the central region requires $\ell n\ s >> 2L$, whereas scaling in the fragmentation regions requires only $\ell n\ s >> L$.

Impressive as these successes are, they were not sufficient to convince everyone of the validity of SRC, because there were other models which could accommodate all these phenomena. The definitive test came from measuring two-particle correlations. Consider now a two-particle inclusive reaction (see Fig. 2.1): $a + b \to c + d + \ldots$. Consider the kinematic configuration that one obtains when the energy is sufficiently high to permit $\Delta y_{ac} >> L$ and $\Delta y_{db} >> L$; i.e., both of the observed produced particles are in the central plateau region, as in Fig. 2.1. Then SRC says that if one considers $d\sigma/dy_c dy_d$ as a function of the rapidity separation Δy_{cd}, then for $\Delta y_{cd} >> L$, the particles should be uncorrelated: If SRC, then for $\Delta_{cd} >> L$,

$$\frac{d\sigma}{dy_c dy_d} \sim \frac{1 d\sigma}{\sigma dy_c} \frac{d\sigma}{dy_d} \quad . \tag{2.5}$$

It is then convenient to define a correlation function

$$C(y_c, y_d, s) = \frac{d\sigma}{dy_c dy_d} - \frac{1}{\sigma} \frac{d\sigma}{dy_c} \frac{d\sigma}{dy_d} \quad . \tag{2.6}$$

According to SRC, if both y_c and y_d are in the central region, $C(y_c, y_d, s)$ will depend <u>only</u> on Δy_{cd}, and for Δy_{cd} large,

$$C(\Delta y_{cd}) \sim \text{const.} \times e^{-|\Delta y_{cd}|/L} \quad . \tag{2.7}$$

Experimenters often find it easier to give reliable values of the normalized correlation function $R(y_c, y_d)$, defined as

$$R(y_c, y_d) = C(y_c, y_d)\,\sigma \Big/ \frac{d\sigma}{dy_c}\frac{d\sigma}{dy_d} . \qquad (2.8)$$

This function should have the same qualitative behavior in the central region.

The expected SRC behavior is beautifully confirmed by the ISR data of the Pisa-Stony Brook group, displayed by the contour maps in Fig. 2.8 [5]. They show in detail that 1) $R(y_c, y_d)$ depends strongly on Δy_{cd}, but very weakly on $y_c + y_d$ in the central region. 2) There is almost no energy dependence, except for the expected elongation of the ridge along $y_c = y_d$ as $\ell n\, s$ increases. 3) As a function of Δy_{cd}, R falls exponentially for separations $\Delta y_{cd} \gtrsim L$, and is consistent with a correlation length $L \approx 2$. This is, I believe, compelling evidence that SRC dominance is correct.

B. SRC and the Bare Pomeron

If we are convinced by all the evidence previously discussed that multiparticle production is dominated by a mechanism involving short-range correlations, what can we conclude about the J-plane singularity structure of the S-matrix? In fact, that was not the right question to ask - we cannot, with certainty, conclude anything, just as we can never prove on the basis of experimental data alone that resonances correspond to poles of the S-matrix. But just as the notion of an isolated simple

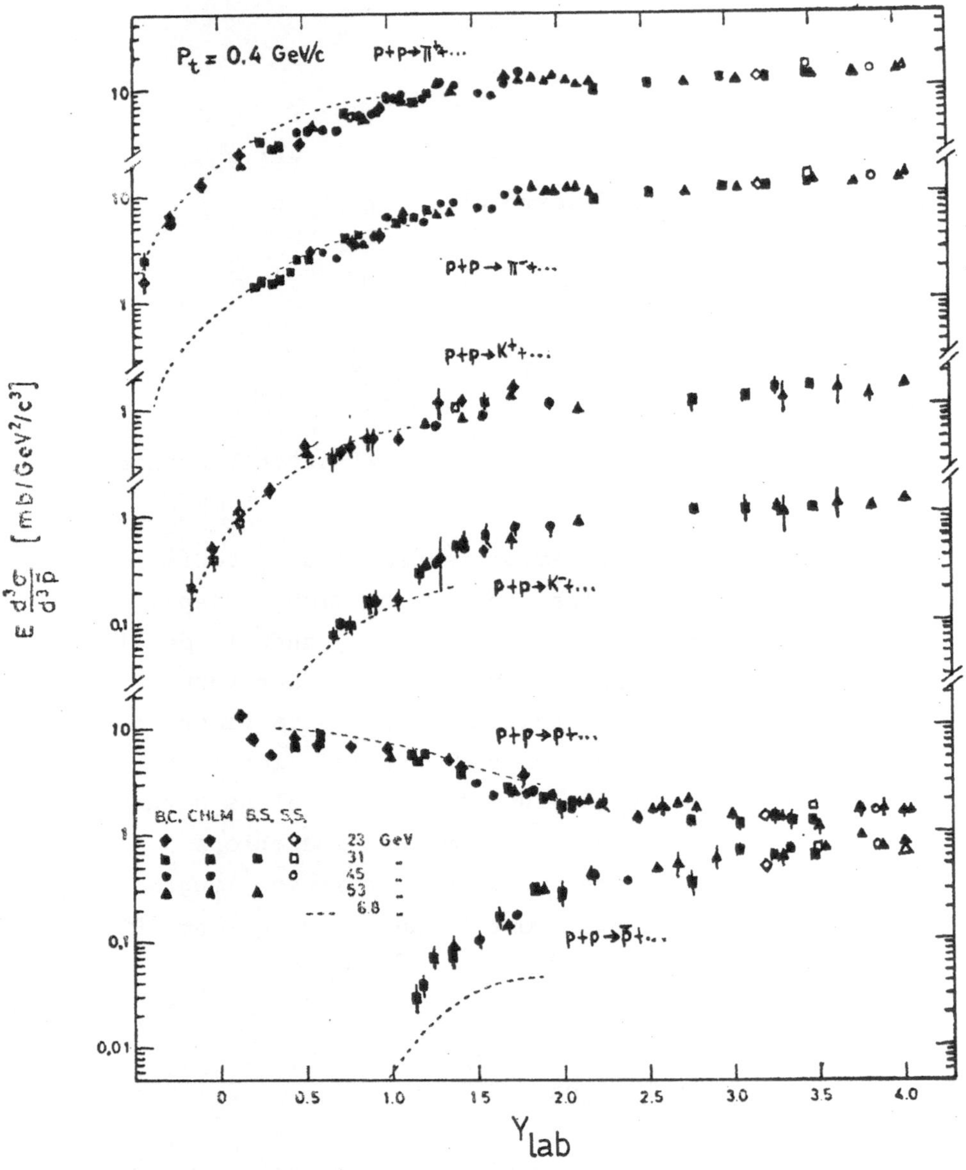

Fig. 2.7: The invariant single-particle inclusive cross-sections plotted versus laboratory rapidity Y_{lab} at P_t = 0.4 GeV/c for $\pi^{\pm}$, $K^{\pm}$, p, and $\bar{p}$ at various ISR energies. The dashed lines present data of Allaby et al. [19] at lower energy.

pole in the complex energy plane has proven to be the simplest, most powerful language in which to discuss resonances, so we seek the simplest way to understand the short-range correlation phenomenon.

Multiperipheral theorists have long realized (again, Wilson's paper [2] based on lectures at this school in 1963 is the earliest explicit reference) that factorizability, which results from simple S-matrix poles, gives rise to short-range correlations. The clear, elegant statement of this principle in inclusive reactions was first given by Mueller [4]. His work has been reviewed so often that I will not repeat it in detail here. Mueller's method consists of two steps: 1) the identification of the n-particle inclusive spectrum with a discontinuity of the forward $n + 2 \rightarrow n + 2$ amplitude, and 2) the asymptotic expansion of that forward amplitude in terms of J-plane singularities. The J-plane singularity which implies factorizability is, of course, a simple pole. Therefore, the natural explanation of SRC is that inclusive cross sections are dominated by a Regge pole. This pole must lie at or near $J = 1$ to account for the observed scaling, or near-scaling, of inclusive reactions. This Regge pole is called the "bare" pomeron, and we shall designate its trajectory $\alpha_0(t)$, with $\alpha_0(0) \approx 1$.

Let's go through in detail the way in which a simple pole in the J-plane implies short-range correlations. First consider the effect of a Regge pole with intercept at $J = 1$ on the total cross section for $a + b \rightarrow$ anything:*

*) The trajectory is labeled "P", without regard at this point for the distinction between pomeron and bare pomeron, to which we shall return later.

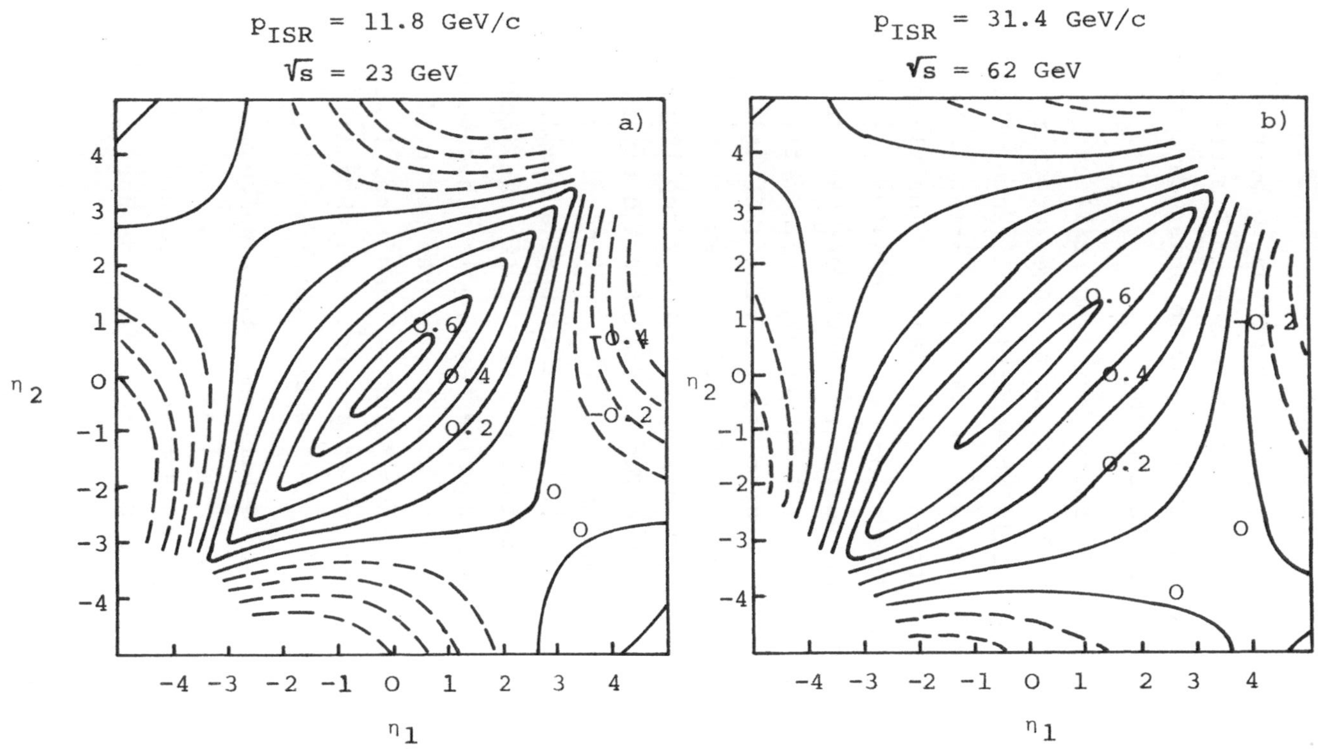

Fig. 2.8: Contour map of two-particle correlation function $R(y_1,y_2)$ measured by Pisa/Stony Brook group [5] at the ISR. The variables η_1, η_2 are approximately equal to y_1,y_2.

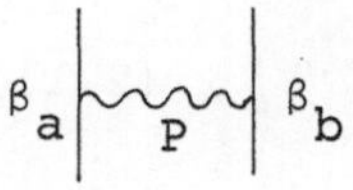

$$\Longrightarrow \sigma_{ab} \sim \beta_a \beta_b \, s^{\alpha(0)-1} = \beta_a \beta_b \; . \qquad (2.9)$$

Factorization of total cross sections is a well-known, but not easily tested, consequence of SRC. Going on to single-particle spectra for $a + b \to$ anything, in the central region, the appropriate Mueller-Regge diagram is

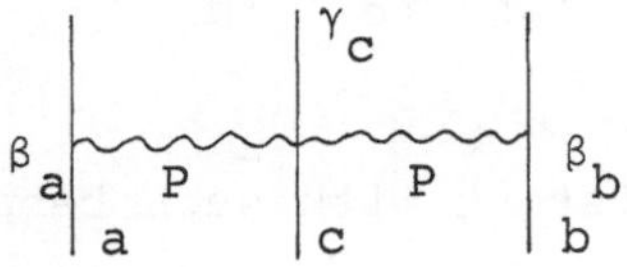

which implies a single-particle spectrum

$$\frac{d\sigma_{ab}}{dy_c} \sim \beta_a \beta_b \gamma_c \quad . \qquad (2.10)$$

This is the well-known prediction of the central plateau.*) Proceeding onward to two-particle spectra for $a + b \to c + d +$ anything, with both c and d in the central region but also separated from each other in rapidity by a distance large compared to a correlation length L (which is marginally possible at the ISR) the appropriate Mueller-Regge diagram is

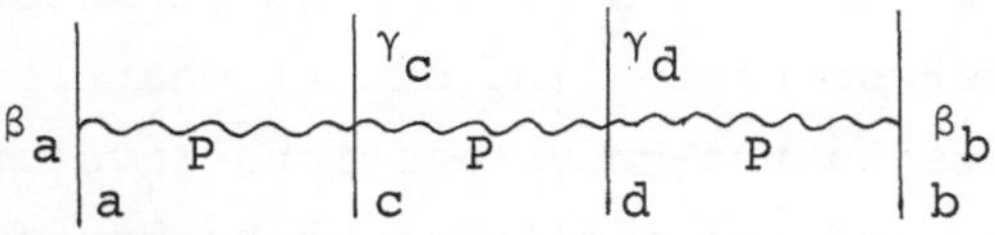

which implies a two-particle spectrum

*) We are focussing here on the contribution of the leading term. Secondary trajectories of course lend an energy dependence and curvature to the "plateau". See, for example, Ref. [1o], p. 3oo.

$$\frac{d\sigma_{ab}}{dy_c dy_d} \sim \beta_a \gamma_c \gamma_d \beta_b \quad . \tag{2.11}$$

This indeed satisfies the criterion for lack of correlation,

$$\frac{d\sigma_{ab}}{dy_c dy_d} \sim \frac{1}{\sigma_{ab}} \frac{d\sigma_{ab}}{dy_c} \frac{d\sigma_{ab}}{dy_d} \quad . \tag{2.12}$$

Experiments are consistent with such behavior, but do not yet provide a very stringent test [5,7]. Note that factorization is crucial in the derivation of the equality in Eq. (2.12).

If SRC and simple Regge poles could completely and consistently explain the data, these lectures would end here. The facts are otherwise, of course; in the next section we shall turn our attention to the long-range, diffractive component. Before leaving the topic of factorizable poles, however, I want to emphasize the importance of further experimental tests of factorization. No one expects that the pomeron is a pure, simple Regge pole; therefore, we do not expect factorization to be exact. It is rather surprising that the Mueller-Regge analysis based on factorization works as well as it does, and it will be very interesting to find out at what level it breaks down. Tests in two-body reactions involve resonance production, and are afflicted by background ambiguities. Tests in inclusive reactions are much richer, and NAL results on this subject will be most interesting.

The simplest tests involve single-particle spectra $a + b \to c$ + anything in the target fragmentation region;

that is, the region in which the rapidity separation $y_c - y_a$ is not large, but where the energy is sufficiently high that $y_b - y_c >> L$. This is the configuration shown in Fig. 4. Then SRC says that there should be no correlation between the beam and the produced particle c. The appropriate Mueller-Regge diagram is

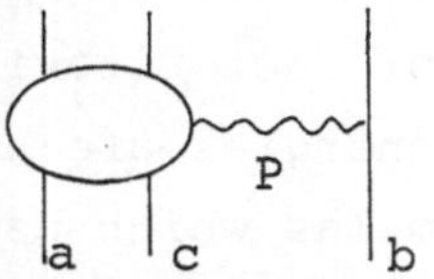

and the spectrum implied by this diagram is

$$\frac{d\sigma_{ab}}{dy_c d^2 q_\perp{}^c} = \gamma_{ac} (q_\perp{}^c, y_c - y_a) \beta_b \ . \qquad (2.13)$$

The factorizable dependence on the beam can be eliminated by dividing by the total cross section, forming the quantity

$$\frac{1}{\sigma_{ab}} \frac{d\sigma_{ab}}{dy_c d^2 q_\perp{}^c} = \gamma_{ac} (q_\perp{}^c, y_c - y_a) \ . \qquad (2.14)$$

Thus the prediction is that this quantity is independent of the nature of the beam [8]. For example, in the target fragmentation region at NAL energies all the following reactions should give identical differential spectra $\gamma_{p\pi^-}$, if factorization were exact:

$$\pi^+ p \to \pi^- + \ldots$$
$$\pi^- p \to \pi^- + \ldots$$
$$K^+ p \to \pi^- + \ldots$$

$$K^- p \to \pi^- + \ldots$$
$$p\ p \to \pi^- + \ldots$$
$$n\ p \to \pi^- + \ldots \quad .$$

Some such results are available, but at such low energies that the apparent violation of factorization is probably just an indication that the energies are not yet high enough to use asymptotic formulas which assume pomeron dominance [9].

III. MULTIPLICITY DISTRIBUTIONS

The study of multiplicity distributions had already enabled us to become convinced of the relevance of the SRC hypothesis before the beautiful data on two-particle inclusive spectra became available. Moreover, multiplicity data are most easily understood in terms of a two-component model, which has strongly influenced contemporary theoretical efforts toward understanding the nature of the pomeron [10,11]. For these reasons I will review some aspects of multiplicity distributions, despite the fact that most of the conclusions arrived at from these distributions can now be seen more clearly from analyses of inclusive spectra from the ISR.

Historically, the first encouragement that the SRC hypothesis and multiperipheral models were on the right track came from cosmic ray data - especially, from the Echo Lake data on charged-particle multiplicities in proton-proton scattering [12]. Although it has turned out that the systematic errors were underestimated, so that the Echo Lake results do not agree quantitatively with the newer accelerator data; nevertheless, they were the first to give strong support to the behavior $\langle n \rangle \propto \ell ns$ which is a consequence of SRC, and which, as we shall see, is compatible with recent accelerator data.

The fact that SRC implies that $\langle n \rangle \propto \ell ns$ follows from the prediction of a central plateau, and from the fact that the integral of the one-particle inclusive spectrum is just $\langle n \rangle$.

$$\int dy \, \frac{1}{\sigma} \frac{d\sigma}{dy} = \langle n \rangle \quad . \tag{3.1}$$

A central plateau which lengthens with ℓns will then give a contribution proportional to ℓns in this integral. There are, of course, correction terms associated with lower-lying Regge trajectories in the Mueller-Regge formalism. These corrections are approximately of the form [13]

$$\langle n \rangle = A \, \ell ns + B + C/\sqrt{s} + D \, \ell ns/\sqrt{s} + \ldots \tag{3.2}$$

A fit of this form is shown in Fig. 3.1. Although the data are compatible with this form, they are not sufficiently precise to rule out competing forms, such as $s^{1/4}$. Actually, this is not a very sensitive way to analyze the data. Since the logarithmic term is associated with the development of the central plateau, it is better to look directly at the one-particle spectrum to see if the plateau exists.

Now we turn to detailed phenomenology of multiplicity distributions. Look first at Fig. 3.2, which shows the cross sections as a function of energy for the production of a definite number of charged prongs. Compare it with Fig. 3.3, a Poisson distribution, which results from those over-simplified multiperipheral models is which SRC is carried to its logical extreme of no correlations at all. The qualitative resemblance between Fig. 3.2 and Fig. 3.3 is obvious. The most striking characteristic is that the Poisson distribution,

$$\sigma_n(s)/\sigma = \langle n\rangle^n e^{-\langle n\rangle}/n! \,, \quad \langle n\rangle \sim A \ln s \qquad (3.3)$$

gives each of the prong cross sections a decreasing power-law dependence, s^{-A}. In multiperipheral models a constant total cross section arises from partial cross sections which successively rise to a maximum, then fall off as a power (times a power of $\ln s$).

The general implications of SRC for multiplicity distributions are summarized in an elegant formalism due to Mueller [14]. Mueller forms a generating function $I(z,Y)$, which is defined as

$$I(z,Y) = \frac{1}{\sigma} \sum_n \sigma_n(Y)\, z^n \,, \qquad (3.4)$$

where $Y = \ln(s/m_a m_b)$. If the generating function were given, the prong cross sections could be found by differentiation. Now Mueller showed that the generating function is simply related to its moments f_n,

$$I(z,Y) = \exp\{\sum_{n=1}^{\infty} f_n (z-1)^n/n!\} \,, \qquad (3.5)$$

where the f_n's are integrals over correlation functions,

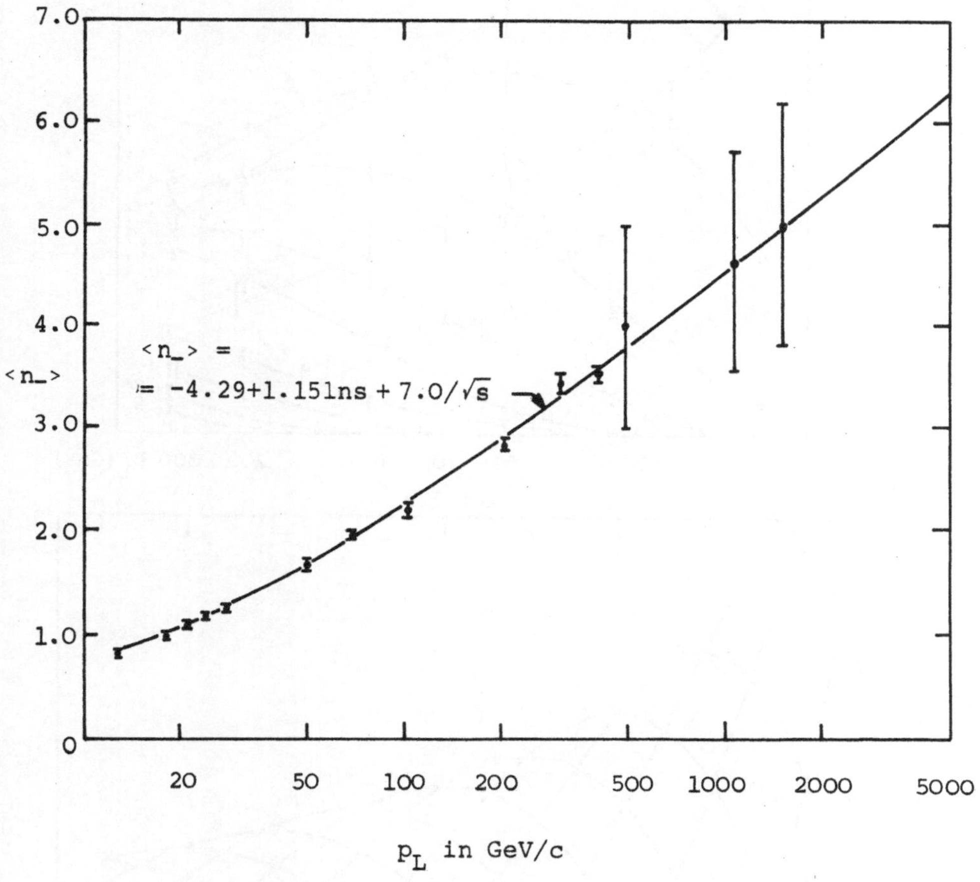

Fig. 3.1: A fit to the average multiplicity of negative particles produced in pp collisions. The fit was done before the 400 GeV point was measured. A fit including the 400 GeV point finds the parameters $\langle n_- \rangle = -(4.09 \pm .58) + (1.11 \pm .09)\,\ell n s + (6.6 \pm 1.7)/\sqrt{s}$.

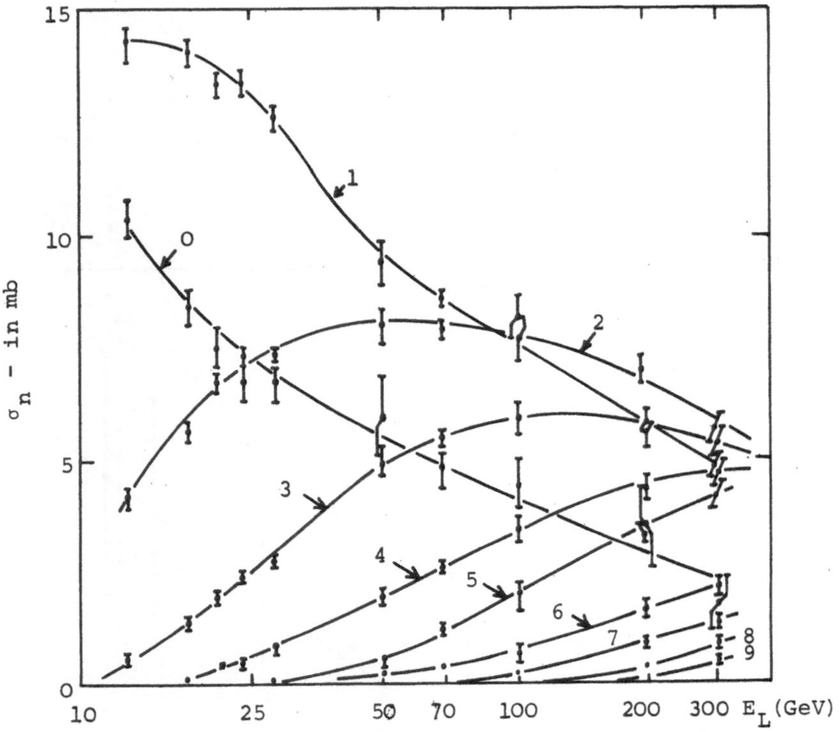

15
10
5
0
σ_n - in mb
10
25
50
70
100
200
300
E_L(GeV)
0
1
2
3
4
5
6
7
8
9

Fig. 3.2

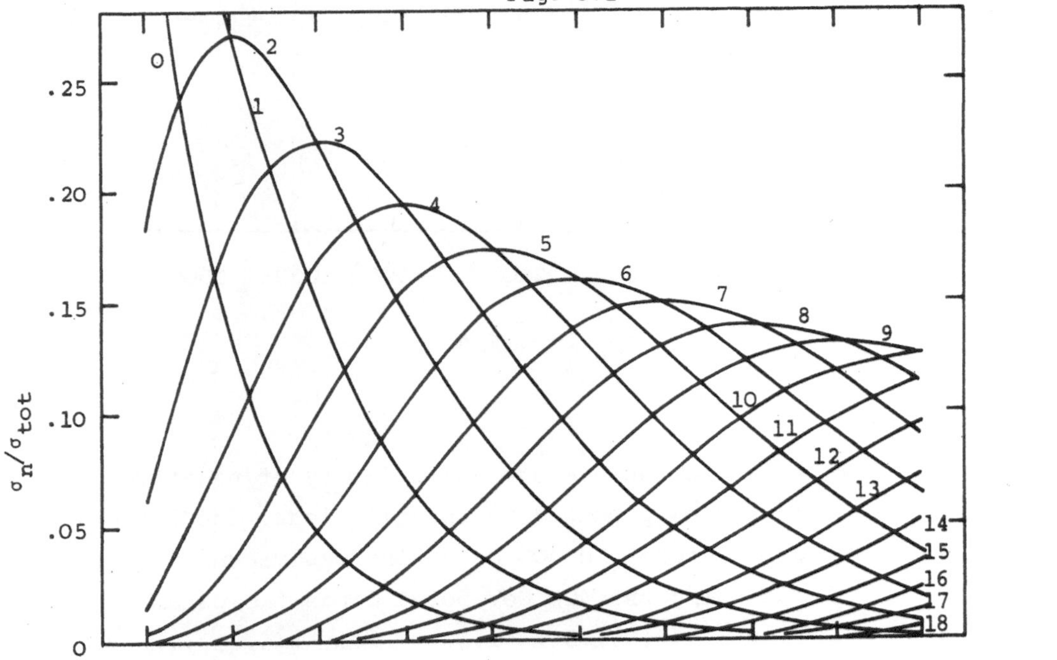

.25
.20
.15
.10
.05
0
σ_n/σ_{tot}
1
2
3
4
5
6
7
8
9
10
⟨n⟩~a ln s
0
1
2
3
4
5
6
7
8
9
10
11
12
13
14
15
16
17
18

Fig. 3.3

$$f_1 = \int dy_1\ C_1(y_1) = \langle n \rangle$$

$$f_2 = \iint dy_1 dy_2\ C_2(y_1,y_2) = \langle n(n-1) \rangle - \langle n \rangle^2, \quad (3.6)$$

etc.,

where $C_1 \equiv d\sigma/dy$, C_2 is defined in (2.6), etc. Since these correlation functions vanish when any pair of variables is separated by a distance large compared to the correlation length L, it follows that asymptotically

$$f_n(Y) \leq O(Y). \quad (3.7)$$

Since the next-order term is a constant, the form

$$f_n(Y) = a_n Y + b_n \quad (3.8)$$

is used for fits to the high-energy data. Recently published NAL data on f_2 show definite curvature; the linear form implied by pure SRC will not fit the data [15].

The curvature of f_2 forces us to introduce some production mechanism involving long-range correlations. The simplest such model is the two-component model, where a diffractive component is introduced in addition to the SRC component. Later we shall discuss the meaning of this diffractive component in more detail; here we shall merely adopt a definition which is convenient for phenomenological purposes: We here define "diffractive" to mean nothing more than a component of the partial cross sections which is constant as a function of energy

$$\sigma_n^{D}(s) \sim \sigma_n^{D} \quad . \tag{3.9}$$

Moreover, we assume that the diffractive component is confined to the low-multiplicity cross sections (later, in Sec. IV, we shall discuss another component: high-mass, high-multiplicity diffractive production). We then assume that

$$\sigma_n(s) = \sigma_n^{D} + \sigma_n^{M}(s) \quad . \tag{3.10}$$

For compactness we use the superscript "M" to designate the SRC component. This component is defined as being generated by a function I(z,Y) defined in Eq. (3.5) in terms of moments which obey Eq.(3.8).

A model of this form is not consistent with purely short-range correlations. It is not difficult to see that the bound in Eq. (3.7) is violated; the f_n's of the two-component model go asymptotically like Y^n, as is to be expected when the SRC hypothesis is violated. The curvature observed in f_2 in Fig. 3.4 is a direct measure of the relative strength of the two components.

A short calculation gives the result

$$\frac{f_2}{\langle n\rangle^2} \sim \frac{\sigma^{D}}{\sigma^{M}} \quad . \tag{3.11}$$

As is conventional, we normalize to the total inelastic cross section; σ^D is the total inelastic diffractive cross section, $\sigma^D \equiv \sum_n \sigma_n^{D}$; and similarly $\sigma^M \equiv \sum_n \sigma_n^{M}$. From the coefficient of the $\ell n^2 s$ term in f_2 in the fit

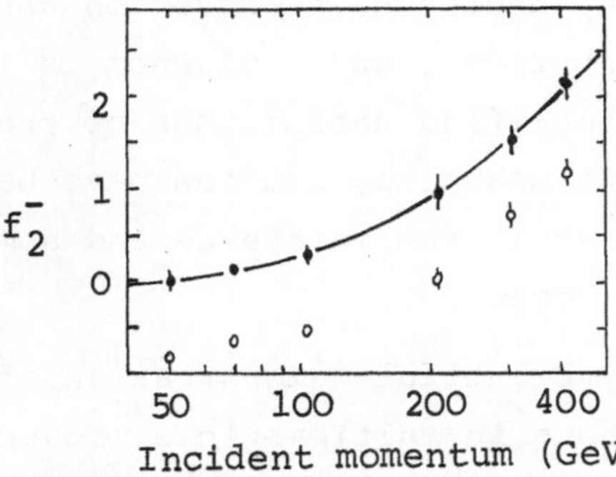

Fig. 3.4: The multiplicity moment f_2^- as a function of energy (see Ref.[15]). The fit is $f_2^- = (8.5\pm 3.5) + (-3.8\pm 1.3)\ell ns + (0.43\pm 0.12)(\ell ns)^2$.

in Fig. 3.4 and from the coefficient of the ℓns term in the fit to <n> in Fig. 3.1, I obtain the result

$$\frac{\sigma^D}{\sigma^M} = 0.35 \pm 0.11 \; . \qquad (3.12)$$

This is somewhat larger, but not inconsistent with, the value $\sigma^D/\sigma^M \approx 1/4$ obtained in various two-component model fits [10,11].

It is impressive to note that cosmic-ray physicists drew the same numbers out of relatively meager data many years ago [16]. Our two-component model may be more precisely defined, but it is not so different in spirit form the isobar + pionization model of Pal and Peters [17].

IV. DIFFRACTIVE DISSOCIATION

A. Some Experimental Evidence

In the previous section we discussed some evidence

in favor of two-component models, based on analysis of multiplicity distributions. Such evidence is indirect, and dependent on simplified definitions of the diffractive component. In the meantime, studies have been made which establish directly the existence and some properties of the diffractive component.

An especially beautiful demonstration of the existence of two components in multiparticle production has been given by the Pisa/Stony Brook group at the ISR [18]. Before I show these pictures I must explain what you are going to see. The two components have the following characteristics: A typical SRC event populates the rapidity space evenly (with some fluctuations and clustering which I shall not discuss in detail), as shown in Fig. 4.1. A typical event from the diffraction dissociation ("isobar") component, also shown in Fig. 4.1, has a single leading particle on

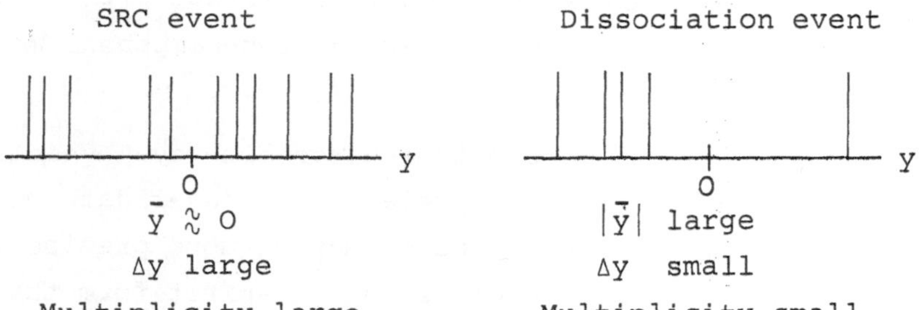

Fig. 4.1: Characteristics of SRC and diffractive dissociation components.

one end, and a cluster of particles on the other. Now suppose we remove the leading particle from each end of the y-distribution, and consider the average value $\bar{y}$ and width Δy of the remainder. The two components favor different values of $\bar{y}$ and Δy. A plot in which each event is represented as a dot in y - Δy space should then show three distinct clusters, as sketched below:

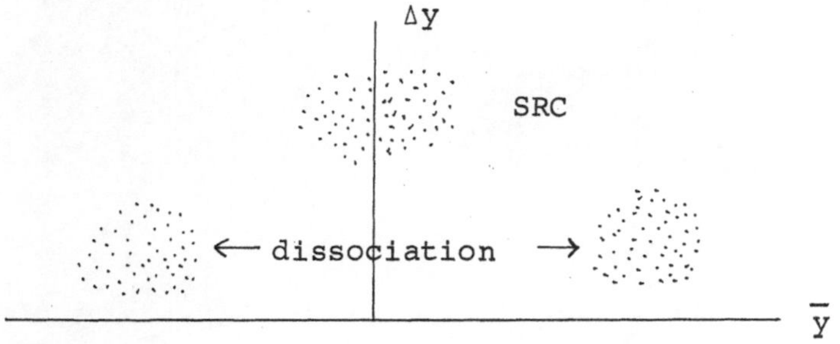

Fig. 4.2: Clustering of events in $\bar{y}$ - Δy plot, assuming two-component model.

The PSB group then makes a three-dimensional plot, with the density of events as the vertical coordinate [18]. Some of these plots are shown in Fig. 4.3. At low multiplicities the diffractive peaks dominate. As the multiplicity increases, the SRC peak takes over. One does not need any sophisticated theoretical interpretation to see the presence of two components in these data!

A second type of experiment which clearly shows the presence of the diffractive component, and which is capable of more quantiative analysis, is the measurement of inclusive spectra near $x = \pm 1$. Most of the data are for $p + p \to p + X$. An example is shown in Fig. 4.4 [19]. After remaining essentially flat at small x, the proton inclusive spectrum rises sharply near $x = 1$ (the elastic peak is subtracted out). This

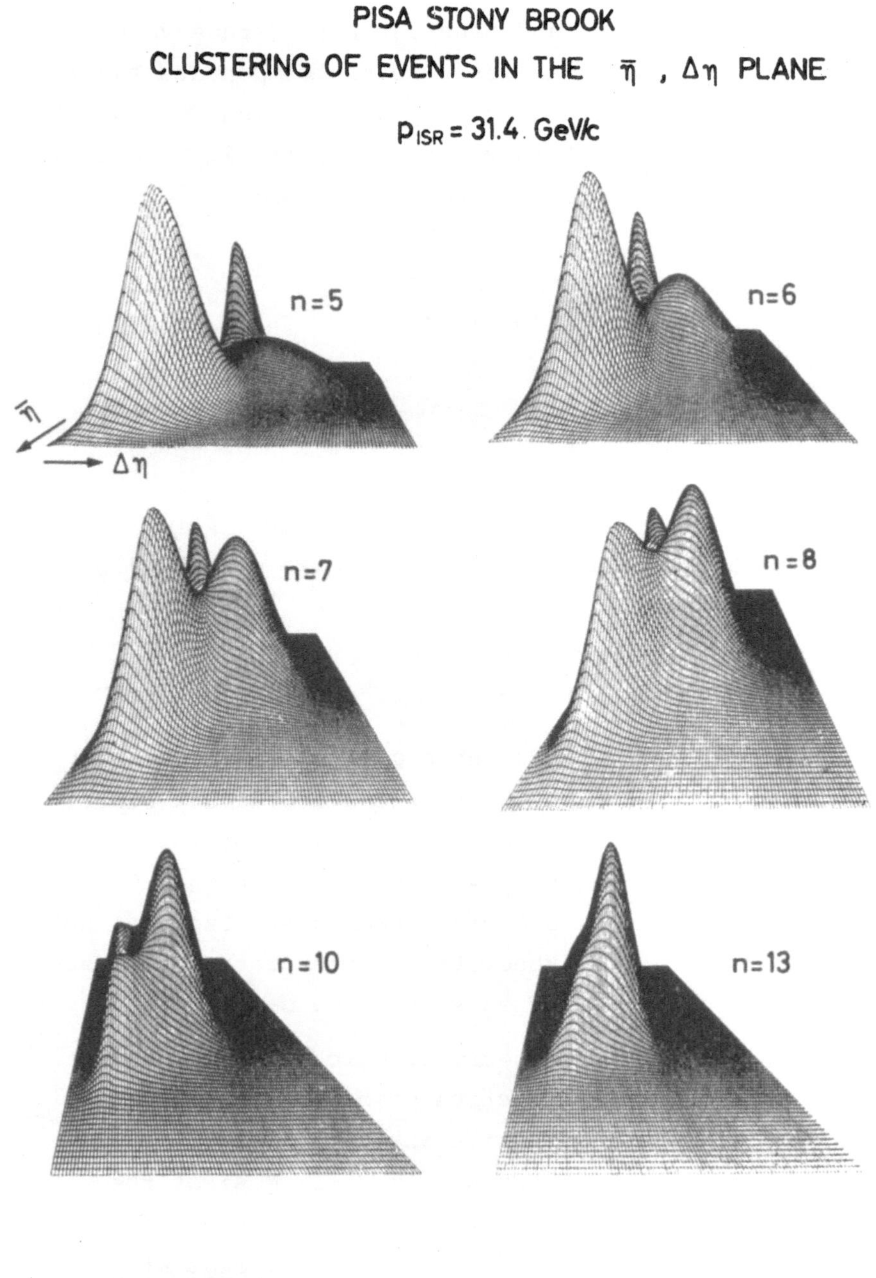

Fig. 4.3: Pisa/Stony Brook data plotted to show presence of SRC and diffractive components [18].

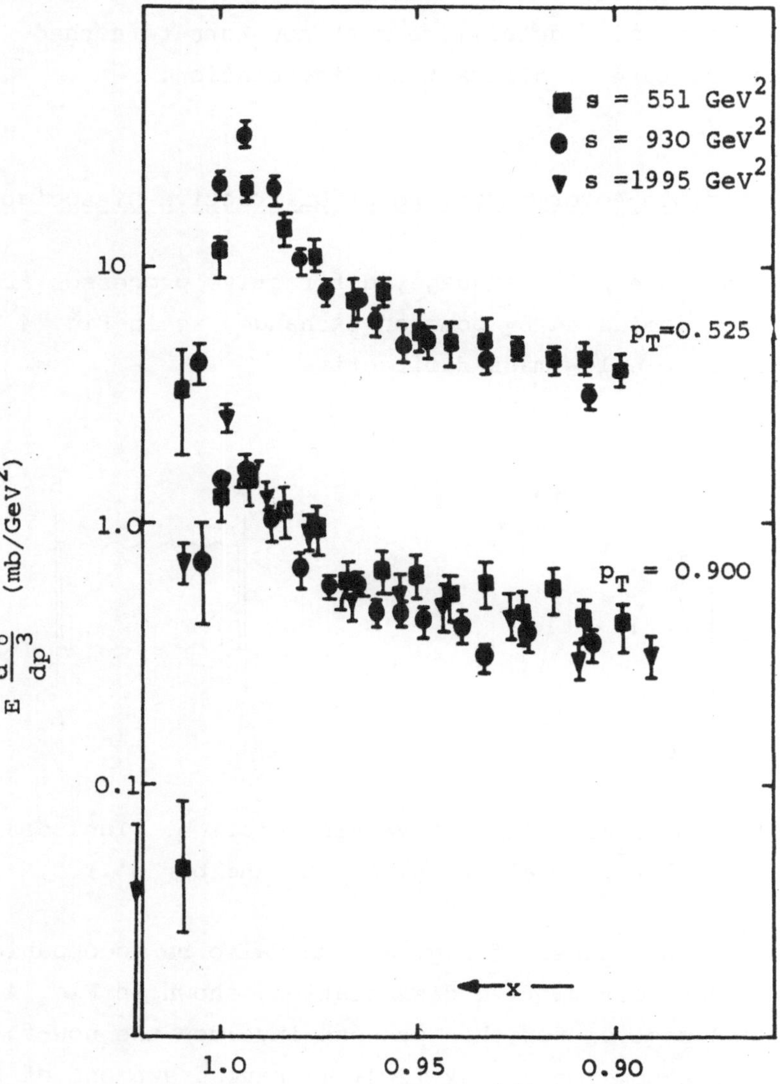

Fig. 4.4: Inelastic proton spectra measured by the CHLM group at the ISR [19].

rise is attributed to diffractive production. To discuss this in more detail, we must now turn to a theoretical picture of diffraction dissociation.

B. The Triple-Pomeron Picture of Diffractive Dissociation

In Regge pole language, diffractive processes are processes dominated by pomeron exchange, as in Fig. 4.5. But the simple low-mass diffractive

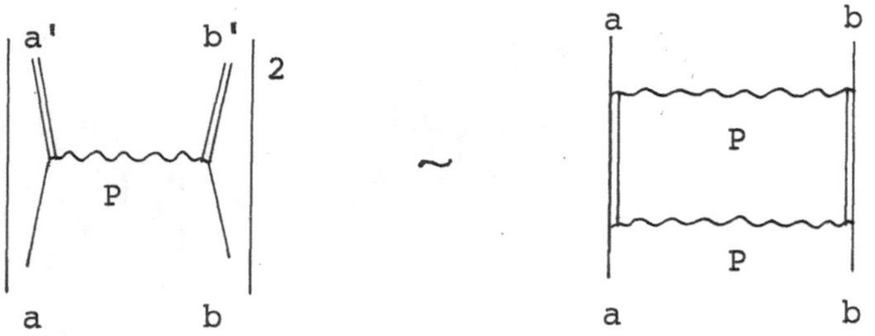

Fig. 4.5: Low-mass diffractive dissociation. (Includes elastic scattering, a = a' and b = b'.)

dissociation process of Fig. 4.5 may also be accompanies by high-mass diffractive dissociation, shown in Fig. 4.6. The cross section for this process involves the now-famous triple-pomeron vertex [20]. (I have drawn one of the pomerons as a bare pomeron, to indicate that the process I have in mind is one with only one large rapidity gap - this distinction is not important here, but I will discuss it in detail in the next section).

For simplicity I shall make the usual assumption

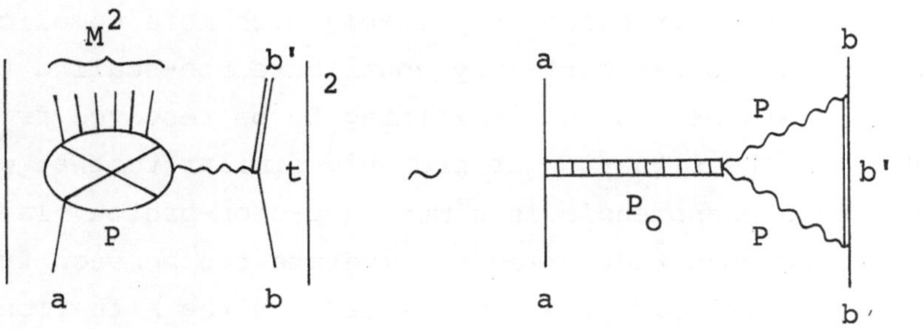

Fig. 4.6: High-mass diffractive dissociation.

that the pomeron can be adequately approximated by an effective pole. In this approximation one obtains the usual triple-Regge formula,

$$\frac{d\sigma}{dM^2dt} = \frac{s_0}{s^2}\frac{G_P^0(t)}{16\pi}\left(\frac{s}{M^2}\right)^{2\alpha_P(t)}\left(\frac{M^2}{s_0}\right)^{\alpha_0(0)}, \quad (4.1)$$

where

$$G_P^0(t) \equiv \beta_{bbP}^2(t)\ \beta_{aa0}(0)\,|\zeta_P(t)|^2\ g_{PP0}(t), \quad (4.2)$$

where $g_{PP0}(t)$ is the function describing the vertex of two pomerons and one bare pomeron. Using the fact that $x \approx 1 - (M^2/s)$, we can rewrite this as

$$s\frac{d\sigma}{dM^2dt} = \pi\frac{d\sigma}{d^3p/E} = \frac{G_P^0(t)}{16\pi}\left(\frac{s}{s_0}\right)^{\alpha_0(0)-1}(1-x)^{\alpha_0(0)-2\alpha_P(t)}. \quad (4.3)$$

Thus a peak approximately of the form $(1-x)^{-1}$ is implied for small t, with an invariant cross section which scales. Although this is roughly in accord with what is seen in Fig. 4.4, the situation is probably much more complicated. Lower trajectories certainly contribute non-scaling terms at NAL energies, which are waiting to be resolved from the ISR data. The situation is probably similar to that which Regge pole phenomenologists met in proton-proton elastic scattering when they tried to separate the pomeron from lowerlying trajectories. Two reviews of the situation can be found in the proceedings of the last Stony Brook conference, and I will not go into great detail here [21,22]. The qualitative conclusion seems sound - that we are seing the effect of the triple-pomeron coupling. Its magnitude is uncertain to at least a factor 2, and detailed questions (such as whether $g_{PPP}(0) = 0$) have not been convincingly answered.

Before leaving these data, let's look at one simple analysis which clearly shows the presence of the triple-pomeron effect. Looking back at Eq. (4.1), one sees that the M^2 dependence of the invariant cross section is

$$\sigma_{inv} \propto (M^2)^{\alpha_O(0)-2\alpha_P(t)} . \qquad (4.4)$$

From the M^2 dependence one can obtain a measure of the pomeron trajectory, $\alpha_P(t)$. The CHLM group has done this, [19] and find that a pomeron trajectory like that found in p-p elastic scattering, $\alpha_P(t) \approx 1 + .2t$, fits the data.

Now let's look at the contribution of high-mass diffractive dissociation to the total cross section. We need to integrate Eq. (4.1) over the kinematically allowed range of t, and over a range $M_0^2 < M^2 < rs$. The lower limit should be chosen high enough to make bare pomeron dominance an adequate approximation to the pomeron-proton "cross section", as shown in Fig. 4.6. The $M < M_0$ contribution is then "low mass diffractive dissociation", included in Fig. 4.5. The upper limit must be chosen such that s/M^2 is large enough to justify the assumption of pomeron dominance at the two pomeron legs in Fig. 4.6. For example, optimistic choices are $M_0^2 \approx 10$ GeV^2, and $r \approx 1/10$. Then if we carry out the t and M^2 integrals we obtain

$$\sigma_{D,M>M_0}(s) = \frac{1}{16\pi}\left(\frac{s}{s_0}\right)^{\alpha_0(0)-1} \int_{-\infty}^{0} dt\, \frac{G_P^0(t)}{\varepsilon(t)}\left[r^{\varepsilon(t)} - (s/M_0^2)^{-\varepsilon(t)}\right] \tag{4.5}$$

where

$$\varepsilon(t) \equiv 1 + \alpha_0(0) - 2\alpha_P(t) \ . \tag{4.6}$$

If we now assume a linear trajectory for small t, with $\alpha_0(0) \approx \alpha_P(0) \approx 1$,

$$\alpha_P(t) \approx 1 + \alpha_P' t \ , \tag{4.7}$$

we obtain

$$\sigma_{D,M>M_0}(s) = \frac{1}{32\pi\alpha_P'} \int_{-\infty}^{0} dt\, \frac{G_P^0(t)}{t}\left[(s/M_0^2)^{2\alpha_P' t} - r^{-2\alpha_P' t}\right] . \tag{4.8}$$

For the physically interesting case that the t-dependence of $G_0(t)$ is sufficiently steep to permit expansion about $t = 0$ of the term in brackets in Eq. (4.8), one obtains the result

$$\sigma_{D,M>M_0}(s) \approx \frac{\bar{G}_P^0}{16\pi} \ell n(\frac{rs}{M_0^2}) \quad , \tag{4.9}$$

where

$$\bar{G}_P^0 \equiv \int_{-\infty}^{0} dt \; G_P^0(t) \quad . \tag{4.10}$$

We see from Eq. (4.9) that the triple-pomeron formula predicts that the cross section for single diffractive dissociation into high masses should, to good approximation, show a logarithmic rise with energy over a finite energy range, regardless of the functional form of $G_P^0(t)$. The detailed functional form will, however, affect the asymptotic behavior for <u>very</u> large s. If, for example, we choose

$$G_P^0(t) = -G_0' t \; e^{bt} \quad , \tag{4.11}$$

then we find that

$$\sigma_{D,M>M_0}(s) = \frac{G_0' \ell n(rs/M_0^2)}{16\pi[b+2\alpha_P' \ell n(s/M_0^2)][b - 2\alpha_P' \ell n \; r]} \quad . \tag{4.12}$$

If, on the other hand, we choose

$$G_P^0(t) = G_0 \; e^{bt} \quad , \tag{4.13}$$

then we find that

$$\sigma_{D,M>M_0}(s) = \frac{G_0}{32\pi\alpha'_P} \ell n \left[\frac{b + 2\alpha'_P \ell n(s/M_0^2)}{b - 2\alpha'_P \ell n\, r}\right] . \qquad (4.14)$$

Although the choice in Eq. (4.11) has been a popular one because it leads asymptotically to a constant cross section and thereby avoids difficulty with the Froissart bound, the choice among functional forms of $G_0(t)$ may not be of great importance at energies available today. Both Eq. (4.12) and (4.14) reduce to Eq. (4.9) provided that

$$2\alpha'_P \ell n\, (s/M_0^2) << b , \qquad (4.15)$$

which seems likely to be satisfied at NAL and perhaps ISR energies, because of the small values reported for the slope of the Pomeranchuk trajectory.

To summarize, we have calculated the cross section for single diffractive dissociation into high-mass states in the approximation of pole dominance of the Pomeranchuk singularity, and found that it rises like ℓns as long as Eq. (4.15) is satisfied. Many multiperipheralists have excitedly made this observation, and rushed into print with varying degrees of temerity [23-26]. What causes hesitation, of course, is a) our uncertainty about the size of the triple-pomeron term, and b) our uncertainty about the behavior of the rest of the cross section. We are confident that we have identified a rising piece of the cross section; whether or not this piece accounts for the observed rise is another matter. Let me review the situation briefly.

As for the magnitude of the effective triple-pomeron coupling $\bar{G}_P$ defined in Eq. (4.10), one finds that $\bar{G}_P/16\pi \gtrsim$ 1mb would be needed to account for the rise observed at the ISR (provided the other terms which make up the cross section remained constant). It is impossible to extract this number from the data we have discussed earlier without making some unwarranted assumptions. For example, even if one were to assume that the peak near $x = 1$ in ISR data shown in Fig. 4.4 were entirely due to the triple-pomeron effect, one could not immediately calculate $\bar{G}$ because the ISR measurements at the highest energy do not extend to small t values. If one makes an exponential extrapolation (unjustified, of course) one finds a value in the same ballpark as the 1mb needed to account for the rising cross section. Almost all corrections would tend to lower this estimate, however: inclusion of lower trajectories, possibility that $G_P(0) = 0$, etc. In short, it's a real strain to get a large enough coupling. Nevertheless, it's interesting that it comes out so close.

As for the "background" - the remainder of the cross section in addition to the diffractive dissociation component - the theoretical situation is not completely clear. I will return to this question in the framework of multiperipheral models in a later section. Suffice it to say that in a treatment which aims at a self-consistent asymptotic solution it is difficult to escape the conclusion that the background will fall at least as fast as the diffractive cross section rises. One exception to this behavior is found in the case of complex Regge poles, which lead to oscillations. Another is found in a self-consistent model of Ball and Zachariasen, which I shall discuss later [27]. A third philosophy, to which I shall now turn, is that the terms we have analyzed are the low-order terms of a perturbation expansion in powers of the triple-pomeron coupling. At present energies, the series can be truncated and higher-order terms dropped. If we plead ignorance (correctly, I believe) about how to sum this series at asymptotic energies, there is no reason not to try to understand NAL-ISR data in terms of a three-component picture - SRC, low-mass diffractive dissociation, and high-mass diffractive dissociation. Paradoxes come about when we try to guess how nature will deal with repeated pomeron exchange, a phenomenon which has not yet been observed. At each stage of these lectures I am trying to tie the theory of the various components as closely as possible to the data, in the hope that a thorough exploration of the properties of the production mechanisms operative at NAL-ISR energies will give us a more solid base from which to extrapolate to higher energies.

C. Multiplicity Distribution in Diffractive Dissociation

Looking at Fig. 4.6 one sees that when one describes diffractive dissociation in terms of pomeron exchange, one is tempted to talk about a "cross-section for pomeron-hadron scattering", associated with the blob on the left-hand side of Fig. 4.6. It is very interesting to try to see from the data whether this picture has any validity. We would, of course, expect such terminology to make sense if the pomeron were an isolated, factorizable singularity of the S-matrix. None of us believes this to be exactly true, but as we have seen from the success of the SRC hypothesis, the pomeron does seem to behave effectively as a pole - at least to a fair approximation. Another test of this picture is provided in diffraction dissociation: we can study the pomeron-hadron "cross section", to see if it behaves like a normal hadronic cross section. One way is to compare the multiplicity distribution in diffractive dissociation with the total multiplicity distribution observed in hadron-hadron collisions [23].

If we study the production of n particles in diffractive dissociation (Fig. 4.7), assuming factorization of the pomeron, we can write the cross

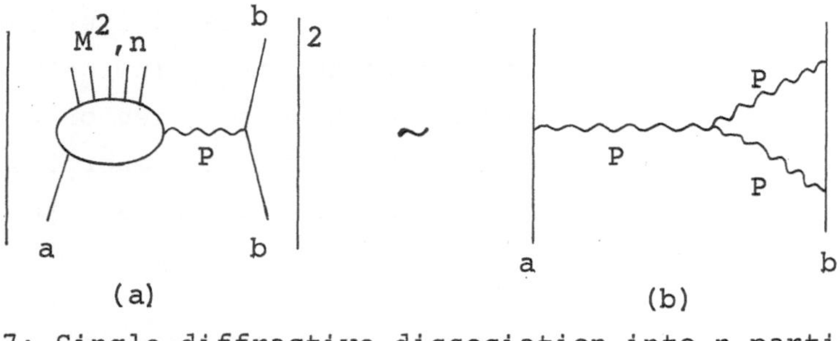

Fig. 4.7: Single diffractive dissociation into n particles.

section for single dissociation into a definite number of particles n, as

$$\frac{d^2\sigma_{ab}^{\ n}}{dM^2 dt} = \frac{|\beta_{bP}(t)|^2}{16\pi\ s^2} \left(\frac{s}{M^2}\right)^{2\alpha_P(t)} M^2\ \sigma_{aP}^{\ n}(M^2,t) \qquad (4.16)$$

where $\sigma_{aP}^{\ n}(M^2,t)$ is the cross section for production of n particles from hadron "a" and a pomeron. We further assume that hadron-pomeron cross sections are similar to hadron-hadron cross sections.

Consider now the average multiplicity of hadrons of type i produced in diffractive dissociation of hadron "a" into a state of mass M at a momentum transfer t

$$<n_a^i(M^2,t)>_D \equiv \sum_{n_i} n_i \frac{d^2\sigma_{ab}^{n_i}}{dM^2dt} \Big/ \frac{d^2\sigma_{ab}}{dM^2dt} , \qquad (4.17)$$

$$= \sum_{n_i} n_i \, \sigma_{aP}^{n_i}(M^2,t) \big/ \sigma_{aP}(M^2,t) .$$

Note that $<n_a^i>_D$ is defined with respect to the corresponding diffractive cross section, not the total cross section. The factorizable-pomeron hypothesis permits one to apply the standard multiperipheral or Mueller-Regge arguments to infer the asymptotic behavior for large M^2,

$$\sigma_{aP}(M^2,t) \; <n_a^i(M^2,t)> \; \sim$$

$$\beta_{aP}^{(0)} g_P(t) \; A^i \; \ln M^2 + B_a^i(t) + O\left[(M^2)^{\alpha_M(0)-1}\right] \qquad (4.18)$$

where $g_P(t)$ is the triple-pomeron coupling, and to infer that the pomeron-hadron cross section should behave as

$$\sigma_{aP}(M^2,t) \sim \beta_{aP}(0) g_P(t) + O\left[(M^2)^{\alpha_M(0)-1}\right] , \qquad (4.19)$$

where $\alpha_M(t)$ represents the highest-ranking secondary trajectory. For any desired level of accuracy there exists a value M_0 such that for $M > M_0$ the second term in Eqs. (4.18) and (4.19) representing the contribution of secondary trajectories, can be neglected. We then obtain the simple result:

The average multiplicity of hadrons of type i produced in diffraction dissociation into a state of mass M, where $M > M_0$, rises linearly with $\ln M^2$,

$$\langle n_a^i (M^2, s, t)\rangle_D = A^i \ln M^2 + B_a^i(t) \tag{4.20}$$

with the same coefficient A^i (independent of s,t, and incident particle type) found in the average multiplicity measured in hadronic reactions.

The average multiplicity for diffraction dissociation into states in a range of $M_0^2 \leq M^2 \leq rs$ at a given value of t can also be calculated, with the result

$$\langle n_a^i (M_0^2 \leq M^2 \leq rs, t)\rangle_D \sim A_i \frac{\int d(\ln M^2)(M^2)^{\varepsilon(t)} \ln M^2}{\int d(\ln M^2)(M^2)^{\varepsilon(t)}} + \ldots \tag{4.21a}$$

$$\approx \frac{1}{2} A^i \ln \frac{rs}{M_0^2} + C_a^i(t) + \ldots . \tag{4.21b}$$

Note the coefficient $A^i/2$, half the coefficient found in the average multiplicity measured in hadronic reactions. The simplification from Eq.(4.21a) to Eq.(4.21b) has been accomplished by taking the approximation $\varepsilon(t) \approx 0$, where $\varepsilon(t)$ is defined as

$$\varepsilon(t) \equiv 1 + \alpha_P(0) - 2\alpha_P(t) . \tag{4.22}$$

Provided that this approximation is adequate over the region in which $G_P(t)$ is significantly different from zero, one can integrate over t to obtain finally

$$<n_a^i \ (M_0^2 \leq M^2 \leq rs)>_D \sim \frac{1}{2} A^i \ \ell n \frac{rs}{M_0^2} + \bar{C}_a^i + \ldots \quad . \qquad (4.23)$$

The results above follow directly from the factorizable-pomeron hypothesis, and therefore provide definitive model-independent tests of that hypothesis. In order to clarify the origin of these results it is helpful to examine a simple model, which is offered in the spirit of exhibiting the simplest possible multiperipheral-type model of diffractive production. Like the Chew-Pignotti model, of which it is a simple extension, it should have illustrative value, but it must be recognized as an oversimplification rather than as a definitive prediction of the multiperipheral picture.

We construct the model by taking $\sigma_{aP}{}^n(M^2)$ to be the simplestmultiperipheral type of distribution, a Poisson distribution,

$$\sigma_{aP}{}^n(M^2,t) = \beta_{aP}(0) \ g_P(t) \ \frac{[A \ \ell n(M^2/s_a)]^n}{n!} \ e^{-A \ \ell n(M^2/s_a)} \ . \qquad (4.24)$$

Such a distribution has been found in Ref. [10] to be a good fit to $\sim$ 80% of the inelastic production in the range 100-300 GeV. The remainder of the production was taken, in that fit, to be diffractive. For simplicity we ignore the diffractive component in Eq. (3.24); this corresponds to calculation of the cross section for events with only one large rapidity gap, but such differences are not large enough to worry about in the context of the present crude model. The scale factor s_a is a parameter of the model, and need not necessarily have the same value as that needed to fit the multiplicity distribution in p-p collisions. This reflects the fact that the parameter B_a^i in Eq. (3.19) depends on the nature of the incident particles.

Substituting Eq. (4.24) in Eq. (4.16) one finds

$$\frac{d\sigma_{ab}^{\ n}}{dM^2dt} = \frac{G_P^{ab}(t)}{16\pi s^2} \left(\frac{s}{M^2}\right)^{2\alpha_P(t)} \left(\frac{M^2}{s_a}\right)^{1-A} \frac{[A\ \ell n\ (M^2/s_a)]^n}{n!} \ . \tag{4.25}$$

That is, the multiplicity distribution of hadrons produced diffractively into a state of high mass M is Poisson-distributed in this model. Again, we can integrate over a range of M, $M_0^2 \leq M^2 \leq rs$, to obtain, for $\varepsilon(t) \approx 0$ and for $n > 0$ (assuming $M_0^2 \leq s_a$),

$$\int_{M_0^2}^{rs} \frac{d^2\sigma_{ab}^{\ n}}{dM^2dt}\, dM^2 \approx \frac{G_P^{ab}(t)}{16\pi A}\, P\ (n+1,\ A\ \ell n\ \frac{rs}{s_a}) \tag{4.26}$$

where $P(n,x)$ is an incomplete gamma function. Integrating over t, we then obtain the following model for the cross section for the production of n particles by single diffraction dissociation,

$$\sigma_D^{\ n}(s) \approx \frac{\bar{G}_P^{ab}}{16\pi A}\, P\ (n+1,\ A\ \ell n\ \frac{rs}{s_a}) + (a \to b)\ . \tag{4.27}$$

The two terms represent the dissociation of particle a and particle b, respectively. The contribution of low-mass resonances, which are not included in Eq. (4.24), should be added to Eq. (4.27) to form the complete diffraction dissociation cross section. Although this contribution is not well known, it will affect only the low-multiplicity cross sections. For this reason, and because of the approximations which led to Eq.(4.27), we propose it as a reasonable model only for higher values of n.

The distribution given by Eq. (4.27) is well known to physicists; it is just the χ^2 distribution with the "number of degrees of freedom" $\nu/2 = n + 1$ and with $\chi^2/2 = A\,\ln(rs/s_a)$. It is shown in Fig. 4.8 for the case appropriate to E_L = 1500 GeV. As is cha-

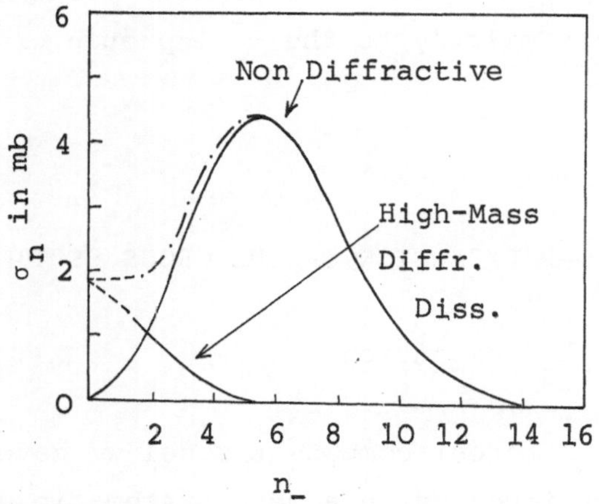

Fig. 4.8: High-mass diffractive contribution to the multiplicity distribution in p-p scattering at E_L = 1500 GeV, according to the model of Eq. (4.27). The nondiffractive term is a simple Poisson distribution from the fits of Ref. [10].

racteristic of χ^2 distribution, it falls to half its maximum at $n \approx A \ln(rs/s_a)$, which is just the average multiplicity of the non-diffractive component, shifted downward by a constant amount. Its mean is approximately

$$\langle n^i \rangle_{D,M > M_0} \lesssim \frac{1}{2} \langle n^i \rangle_{SRC} \quad . \tag{4.28}$$

Data are now becoming available on diffractive multiplicities. Fig. 4.9 shows data from $\pi^- p$ at 2o5 GeV from an NAL-Berkeley collaboration [28]. The multiplicity distribution in Fig. 4.9a is in qualitative agreement with the model we have discussed. In Fig. 4.9b one sees a striking similarity between the average multiplicity and dispersion observed in $\pi^- p \to$ anything and in the diffractive dissociation of the pion. Insofar as this experiment is isolating the diffractive component, we are seeing evidence that the π^- + pomeron cross section is behaving very similarly to the π^- + proton cross section!

V. RAPIDITY-GAP EXPANSION OF THE CROSS SECTION

A. Definition of the expansion

The two- (or three) component model we have been discussing can be imbedded in a more systematic expansion of the total cross section in terms of the number of large rapidity gaps [29]. That is, a given event is examined to see whether there are any intervals larger than some arbitrary value Δ in which no particles occur

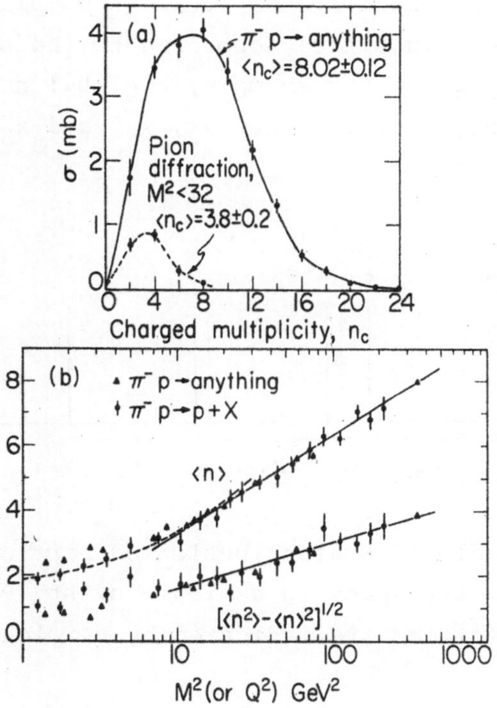

Fig. 4.9 (from Ref. [28]): (a) Charged multiplicity distribution at 205 GeV/c for $\pi^- + p \rightarrow$ anything and for pion diffractive dissociation. (b) Comparison of the energy dependence of the average charged multiplicity and dispersion for pion dissociation and for $\pi^- + p \rightarrow$ anything. The horizontal axis for pion dissociation is M^2; for $\pi^- + p \rightarrow$ anything it is Q^2, the square of the available center-of-mass energy.

(in practice, neutrals are of course a complication). Events are then classified according to the number of large rapidity gaps. For example, Fig. 5.1 shows a three-gap event, provided that x_2, x_4, and x_5 are greater than Δ.

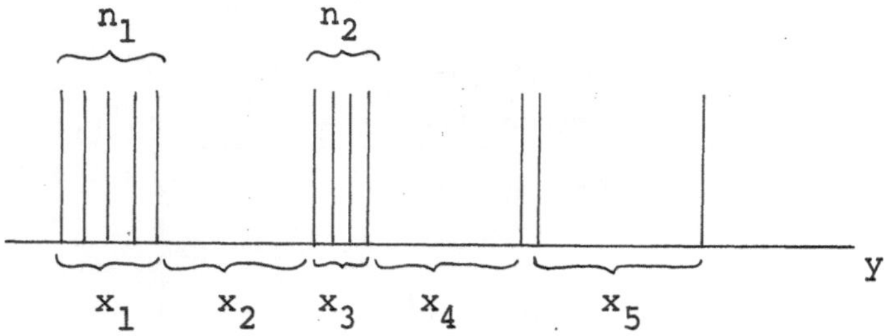

Fig. 5.1: Example of a distribution of particles in rapidity space in a single event. Particle rapidities are indicated by vertical lines.

The total cross section can then be decomposed unambiguously according to the number of large gaps,

$$\sigma_{ab}(s) = \sigma_{ab}^{0,\Delta}(s) + \sigma_{ab}^{1,\Delta}(s) + \ldots . \qquad (5.1)$$

Pictorially, we can represent this decomposition as follows:

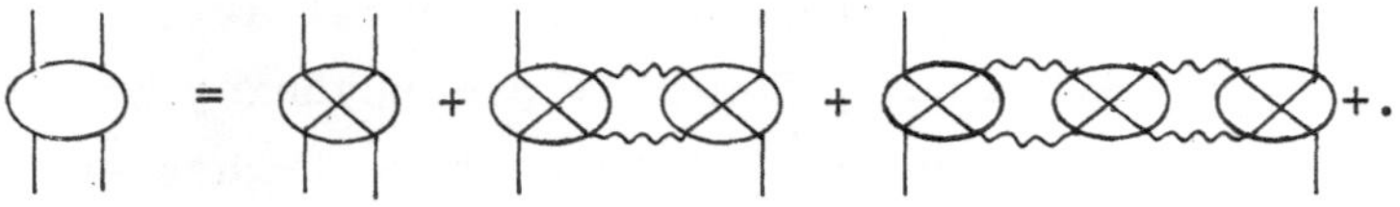

(5.2)

where the symbol 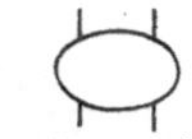 represents the forward absorptive elastic amplitude $A(s) = s\sigma(s)$, and where the symbol represents a cluster of particles produced with no large rapidity gaps. The wiggly line  represents a Pomeron, the assumption being that Δ is chosen sufficiently large that Pomeron exchange dominates the production of such large rapidity gaps.

It is useful to make a further decomposition of the gapless clusters into those of high invariant mass, denoted by 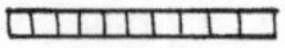 and those of low invariant mass, denoted by . Pictorially,

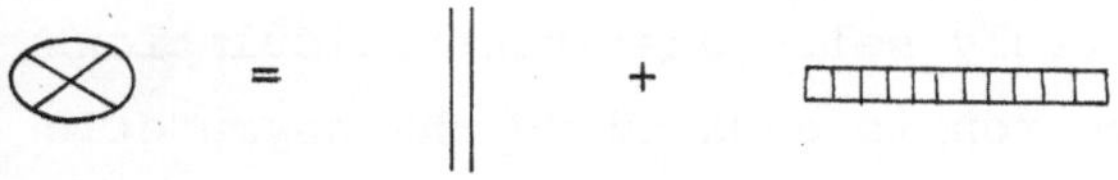

(5.3)

Equivalently, one can separate the clusters according to their spread in rapidity. For example, looking back at Fig. 5.1, one sees two high-mass and two low-mass clusters, provided that z_1 and z_3 are greater than some specified Δ_c, and that $z_5 < \Delta_c$.

Inserting Eq. (5.3) into Eq. (5.2) we then have the following decomposition, at high energies,

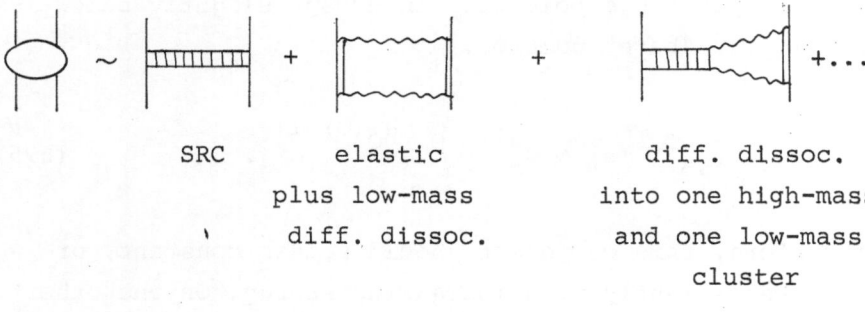

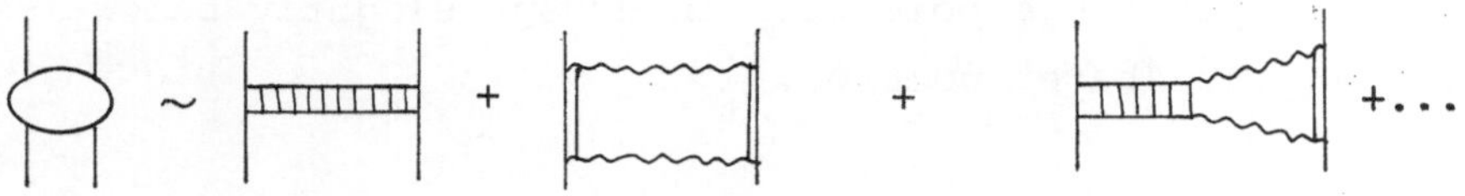

SRC — elastic plus low-mass diff. dissoc. — diff. dissoc. into one high-mass and one low-mass cluster

(5.4)

The three written explicitly on the r.h.s. of Eq. (5.4) correspond to those processes which have already been observed: the SRC component, elastic scattering, diffractive dissociation into low masses, and finally diffractive dissociation into high masses. The remaining terms are considerably smaller; some of them may nevertheless be detectable. I will return to this point later.

What is the utility of the decomposition in Eq. (5.4)? Three classes of applications have been made or attempted: a) Discussion of individual terms, their order of magnitude, their properties; b) Attempts to sum the series, usually in simplified models; and c) Attempts at truly self-consistent calculations, in which the input Pomeron is equated to the asymptotic limit of the sum of the series. This, too, has been accomplished only in simplified models, such as the Ball-Zachariasen model.

Let us turn now to consideration of the terms in Eq. (5.4):

SRC component: In model calculations, and according to plausible general arguments (see Sec. IIB) the asymptotic energy dependence of the no-gap component is governed by a simple Regge pole with intercept slightly below $J = 1$, the "bare" pomeron:

$$\sigma_{ab}^{O,\Delta}(s) \sim \beta_a^O \ \beta_b^O \ s^{\alpha_O(O)-1} \quad . \qquad (5.5)$$

Therefore, this component should remain constant, or decrease slightly with increasing energy. On the other hand, Gaisser and Tan have suggested that $N\bar{N}$ production through a multiperipheral mechanism could account for the rising pp cross section at the ISR energies. In this case, the SRC component would rise in this range. At any rate, the energy dependence of the SRC component is an interesting question which can be settled experimentally in the near future.

I have been using the terms "SRC component" and "no-gap component" interchangeably. Their approximate equivalence is indeed implied by models, but again this should be checked experimentally. A reasonable choice of Δ is $\Delta \approx 2$-3. For numerical estimates I will choose Δ such that $e^{\Delta} = 10$.

Elastic scattering and low-mass diffraction dissociation: There is very little new that I can say concerning elastic scattering or low-mass dissociation. Perturbative calculations of elastic scattering await an adequate zeroth approximation about which to perturb.

High-mass diffraction dissociation: See previous section.

B. Higher-order terms; small parameters

Consider those events characterized by one large rapidity gap, leading to large missing masses on both sides of the gap, $s_1 > M_O^2$ and $s_2 > M_O^2$, shown in Fig. 5.2. In the multi-Regge limit of large gap size and large s_1 and s_2, this process contributes

$$\frac{d\sigma}{dt\,ds_1\,ds_2} \quad \frac{g_{PPO}^2(t)\ \beta_{aaO}(0)\ \beta_{bbO}(0)}{16\pi s^2}\ \left(\frac{ss_0}{s_1 s_2}\right)^{2\alpha_P(t)} \ (s_1 s_2)^{\alpha_O(0)} \tag{5.6}$$

As in the previous section, $\alpha_O(0)$ occurs as the exponent of s_1 and s_2 because we are limiting our attention to

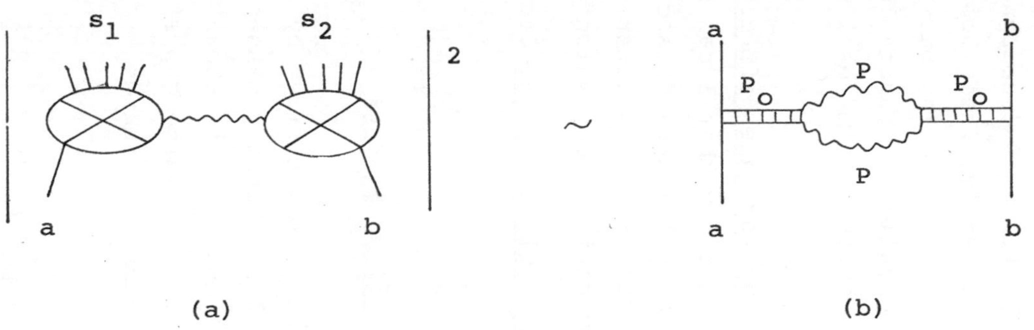

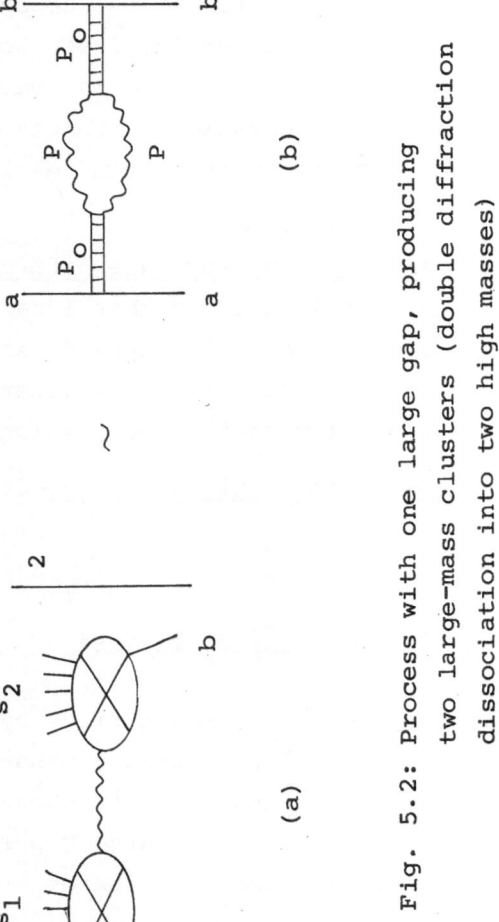

Fig. 5.2: Process with one large gap, producing two large-mass clusters (double diffraction dissociation into two high masses)

events with only one large rapidity gap. Hence the two blobs in Fig. 5.2 are governed by the same singularities as the SRC part of the cross section.

Performing the integrals over s_1, s_2 and t with the same approximation $\varepsilon(t) \approx 0$ as in Sec. IV, one finds for the contribution to the total cross section

$$\sigma_{DD,M>M_0} \approx \bar{\eta}_P \; \beta_{aa0}(0) \; \beta_{bb0}(0) \frac{1}{2} \left(\ln \frac{rss_0}{M_0^4}\right)^2 , \qquad (5.7)$$

where

$$\bar{\eta}_P \equiv \frac{1}{16\pi} \int_{-\infty}^{0} dt \; g^2_{PP0}(t) \; . \qquad (5.8)$$

The parameter $\bar{\eta}_P$ is very similar to the parameter $\eta_P \equiv g^2_{PPP}(0)/(32\pi\alpha'_P(0))$ introduced by Abarbanel et al.[20], but is more appropriate to the energy range and approximation scheme we are considering; that is, we are concentrating on an energy range (appropriate to today's accelerators) in which one can approximately neglect shrinkage whereas Abarbanel et al. tacitly assumed the energy to be high enough that the slope of the pomeron trajectory controlled the t-dependence.

The small dimensionless parameter $\bar{\eta}_P$ arises in higher-order terms whenever a pomeron bubble, as in Fig. 5.2b, is linked to a bare pomeron on both ends. Other types of diagrams in which two pomerons connect to an external line involve $\sqrt{\bar{\eta}_P}$. It is convenient, then to define as our basic expansion parameter

$$\lambda_P \equiv \sqrt{\bar{\eta}_P} \; . \qquad (5.9)$$

We could now proceed to calculate other higher-order contributions to the total cross section and catalog them according to their order in λ_P. The task is complicated, however, by our ignorance of the t-dependences involved. We cannot compare, for example, elastic scattering, single d.d., and double d.d., without knowledge of the t-dependence of $\beta(t)$ and $g_{PPO}(t)$.

Since differences in t-dependence seem unlikely to lead to order-of-magnitude differences among cross sections, it is useful to catalog the various terms under the assumption that all of them have roughly the same t-dependence which we characterize by slope parameters b_i. In this approximation one finds

$$\sigma_T \approx \beta^2 \tag{5.10a}$$

$$\sigma_{el} \approx \frac{\beta^4}{16\pi b_1} \tag{5.10b}$$

$$\sigma_{D,M > M_0} \approx \frac{\beta^3 g_{PPO}}{16\pi b_2} \ln \frac{rs}{M_0^2} \tag{5.10c}$$

$$\sigma_{DD,M > M_0} \approx \frac{\beta^2 g_{PPO}}{16\pi b_3} \frac{1}{2}\left(\ln \frac{rss_0}{M_0^4}\right)^2 . \tag{5.10d}$$

If, for the purpose of order-of-magnitude estimate, we take all the b_i's to be roughly equal, then

$$\lambda_P^2 \approx \frac{g_{PPO}}{16\pi b} , \tag{5.11}$$

and one finds the probabilities

	Process σ_i	Mueller-Regge Diagram	Order of σ_i/σ_T
1)		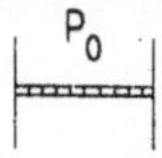	λ_0^2
2)	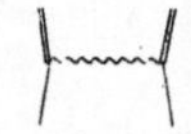	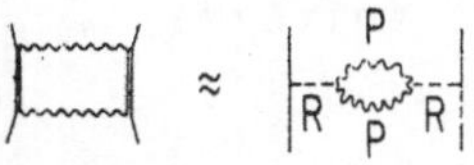	λ_D^2
3)		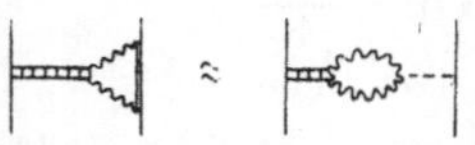	$\lambda_0 \lambda_D \lambda_P (Y-\Delta-\Delta_f)\ \theta(\ )$
4)	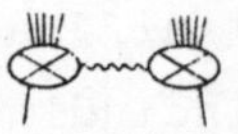	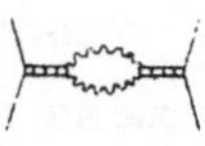	$\frac{1}{2} \lambda_0^2 \lambda_P^2 (Y-\Delta-2\Delta_f)^2\ \theta(\ )$
5)	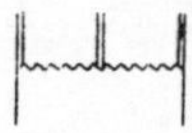	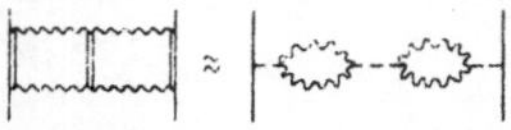	$\lambda_D^2 \lambda_I^2\ (Y-2\Delta)\ \theta(\ )$
6)	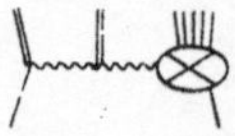	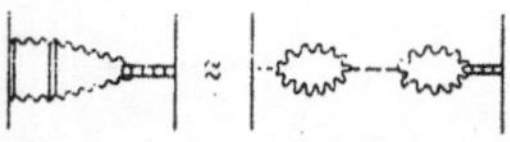	$\frac{1}{2} \lambda_0 \lambda_D \lambda_I^2 \lambda_P (Y-2\Delta-\Delta_f)^2\ \theta(\ )$
7)	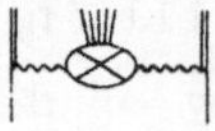	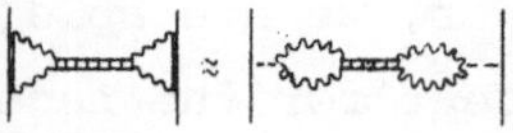	$\lambda_D^2 \frac{1}{2} \lambda_P^2 (Y-2\Delta-\Delta_f)^2\ \theta(\ )$
8)	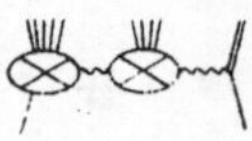	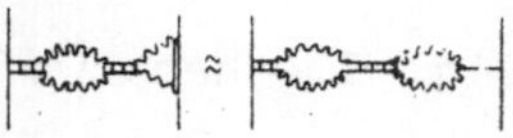	$\lambda_0 \lambda_D \frac{1}{3!} \lambda_P^3 (Y-2\Delta-2\Delta_f)^3\ \theta(\ )$
9)	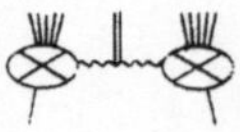	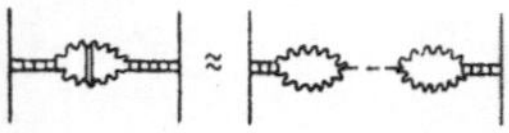	$\frac{1}{3!} \lambda_0^2 \lambda_I^2 \lambda_P^2 (Y-2\Delta-2\Delta_f)^3\ \theta(\ )$
10)	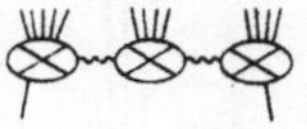	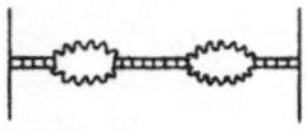	$\frac{1}{4!} \lambda_0^2 \lambda_P^4\ (Y-2\Delta-3\Delta_f)^4\ \theta(\ )$

Fig. 5.3: Relative order of all contributions to the total cross section with zero, one, or two large rapidity gaps. The parameters are defined in Sec.

$$\sigma_{el}/\sigma_T \approx \beta^2/16\pi b \equiv \lambda_{el}^2 \ , \tag{5.12a}$$

$$\sigma_{D,M>M_0} / \sigma_T \approx \lambda_P \lambda_{el} \ \ell n(rs/M_0^2) \ , \tag{5.12b}$$

$$\sigma_{DD,M>M_0}/\sigma_T \approx \cdot \frac{1}{2} \lambda_P (\ell n \frac{rss_0}{M_0^4})^2 \ . \tag{5.12c}$$

And, letting the symbol $\sigma_{DD,M<M_0}$ stand for all diffraction dissociation into clusters of mass $M < M_0$, <u>including</u> elastic scattering, we define

$$\sigma^{ab}_{DD,M<M_0}/\sigma_T \equiv \lambda_D^{ab\,2} \ . \tag{5.12d}$$

Continuing in a similar manner, we are able to fill in the catalog in Fig. 5.3, except for the necessity of defining a third parameter, to which we now turn.

The process shown in Fig. 5.4 has two large rapidity gaps, but does not

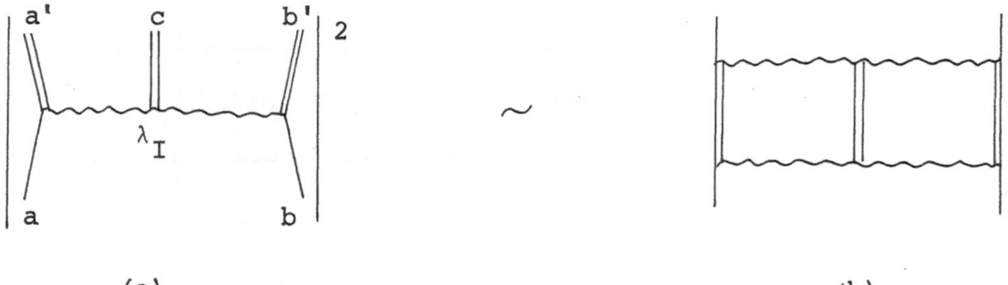

(a) (b)

Fig. 5.4: Double-pomeron process with two large rapidity gaps, leading to low-mass clusters only.

involve our parameter λ_P because there is no production of high-mass states. It is, however, believed to be small, since experimental searches have failed to find evidence for such a process. It seems unlikely that the smallness of this process is unrelated to the small parameter λ_P; nevertheless, we know of no way to make the connection with our present level of S-matrix technology and are forced to introduce another parameter λ_I to characterize the strength of the internal vertex in Fig. 5.4. (See Ref. [23] for further details.)

The magnitudes of the parameters we have introduced are not well known experimentally, but I will attempt some estimates. The best known of the parameters is λ_{el}, the ratio of elastic to total cross section. In p-p scattering over the NAL-ISR range the measured value is $\lambda_{el}^2 = 0.17 - 0.18$. The ratio of SRC to total cross section found in two-component fits implies $\lambda_O \approx 0.8$. The ratio of elastic plus low-mass dissociation to the total cross section is not well known. Adding in a reasonable guess for low-mass dissociation, I would guess $\lambda_D^2 \approx 1/4$. The parameter λ_P is related to $\bar{G}_P$ discussed in Sec. IV. Using the guess made in Sec. IV that $\bar{G}_P \approx 1/2$ mb, I find $\lambda_P \approx 1/40$, or $\bar{\eta}_P \approx .0006$.

The parameter λ_I^2, and hence the magnitude of the double-pomeron process, is very hard to estimate. My own guess for the totally inclusive process shown in Fig. 5.4 is a fraction of a millibarn. Searches in exclusive channels have set upper limits in these channels on the order of 50μb.

I think it would be very interesting to observe this reaction, since it is an example of double pomeron exchange, a process not contained in the optical picture of high-energy scattering. As evidence mounts that the pomeron is at least in part a factorizable singularity of the S-matrix, it becomes increasingly difficult to avoid inferring that double (and multiple) pomeron exchange processes exist. Nevertheless, direct observation of such a process would be a valuable stimulus toward further efforts to understand the pomeron.

C. Summing the series:

The perturbation series we are discussing is a finite series at finite energies; one can only produce a finite number of large rapidity gaps. Nevertheless, many efforts have been made to sum the series at infinite energy. Only by doing this can one learn the nature of the true J-plane singularities. The hope is, of course, that these true singularities might be simpler than the ones we infer from data fits in finite ranges of energy. In general, the series is too complicated to sum; therefore, realistic kernels in the resulting integral equations are approximated by degenerate kernels to reduce the problem to algebraic equations. Several such efforts can be found in the literature, and I do not want to repeat them here. Let me attempt, however, to tell you some of the results.

The bare pomeron, a simple J-plane pole, turns out not to be a singularity of the full absorptive amplitude. It is replaced by a physical pomeron, which is not a simple pole. The physical pomeron singularity always lies higher in the J-plane, and one finds

$$\alpha_P(0) - \alpha_O(0) \approx \lambda_P \qquad (5.13)$$

It follows that the SRC cross section must fall as a small negative power of the energy, $\sigma^{0,\Delta} \sim s^{-\lambda_P}$. In simple models this falling SRC term exactly compensates for the rising diffractive term discussed in Sec. IV. In more sophisticated models, with thresholds and complex poles, one can get a rising cross section over some range of energy, but in such models the cross section eventually proceeds via damped oscillations, to an asymptotically constant (or slightly decreasing) behavior.

Another interesting observation is that although both of the processes shown in Fig. 4.5 and 4.6 give rise to a branch point in the j-plane, their discontinuities have opposite sign [30]. The resulting cut is much weaker than one might have guessed from the ratios of the two components in two-component models. As a result, factorization of the pomeron is a better approximation in the total cross section than in the inelastic cross section. This supports normalizing to the total cross section, rather than the inelastic cross section, when using the Mueller-Regge formalism.

D. Thresholds and oscillations

The expressions given in Fig. 5.3 for some of the terms in the rapidity-gap expansion are all proportional to the theta-functions, which represent threshold factors. These are quite trivial: a process involving n large rapidity intervals of width Δ cannot occur until $Y \gg n\Delta$. Such threshold factors can, as shown by Chew and Snider, give rise to complex Regge poles. In the simplest multiperipheral model without thresholds, the Chew-Pignotti model, the partial cross sections add up to a total cross section which is constant (see Fig. 3.3). When thresholds are added to the model Chew and Snider found that the resulting oscillating cross section is very well represented by a leading real Regge pole plus a complex conjugate pair, as shown in Fig. 5.4.

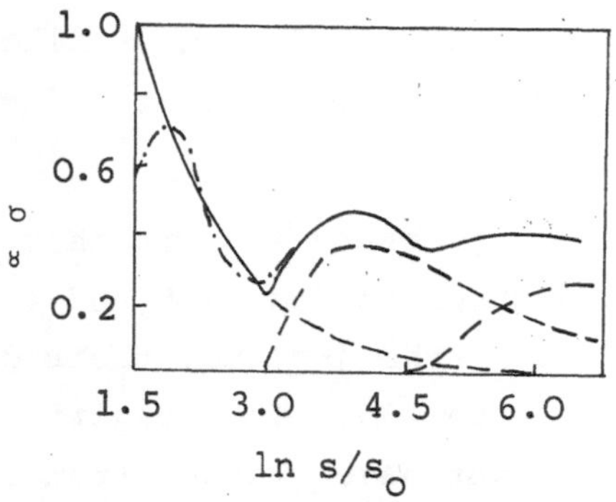

Fig. 5.5. Total cross section in Chew-Snider model with thresholds. Note how well the solid line (the exact cross section) is represented by one real pole plus one complex conjugate pair (dashdot line).

The question immediately arises as to whether the thresholds in the rapidity-gap expansion are real, since the rapidity-gap parameter Δ is an arbitrary separation of a continuous distribution into so-called low-mass and high-mass portions. It is, of course, hard to regard these apparent thresholds as anything but convenient devices which we have put in by hand. In that case, there should be no complex poles and no oscillations, and there is no chance of explaining the rising cross section in terms of diffraction dissociation - provided that we trust our understanding of multiperipheral models enough to reach such a definitive conclusion. I would still be very interested in experimental measurements of the behavior of the cross section with no large gaps, and the cross section with one large gap.

Gaisser and Tan called our attention to another threshold which might be related to the increasing total cross section: the $N\bar{N}$ production threshold. It is a fact that the cross section for $N\bar{N}$ production is increasing spectacularly in the region where the total cross section is rising. Moreover, such a mechanism might be able to produce oscillations of the long period observed. In multiperipheral models the period in Y is of the order

$$\Delta Y \approx 2\ln(M/\mu), \qquad (5.14)$$

where M is the mass of the cluster of particles produced (here $M = 2m_p$), whereas μ is the average momentum transfer in the multiperipheral chain. Pion exchange with a form factor could give a plausible result.

Chew and Koplik have recently pointed out that if complex poles are present in the total cross section, they should be present in all other inclusive amplitudes. This, of course, is the mechanism by which threshold effects (which are very important in some inclusive spectra, like $\bar{K}$ and $\bar{p}$) are incorporated into the Mueller-Regge formalism. If these complex poles are present, all inclusive spectra will oscillate. It may be, however, that the ISR energy range is not sufficiently great that one would expect to be able to tell an oscillation from a rise. The most favorable case seems to be the real part of the scattering amplitude, where the phase of the oscillation is expected to be advanced by $\pi/2$ ahead of the oscillation in the total cross section. This could be a crucial experiment in distinguishing among the plethora of models with rising cross sections.

REFERENCES

1. D. Amati, S. Fubini, A. Stanghellini, Nuovo Cim. 26, 896 (1962).
2. K. G. Wilson, Acta Physica Austr. 17, 37 (1963).
3. C. E. DeTar, Phys. Rev. D3, 128 (1971).
4. A. H.Mueller, Phys. Rev. D2, 2963 (1970).
5. Pisa/Stony Brook collaboration, Phys. Lett. 44B, 119 (1973).
6. W. Frazer, L. Ingber, C. Mehta, C. Poon, D. Silverman, K. Stowe, P. Ting, H. Yesian, Rev. Mod. Phys. 44, 284 (1972).
7. M. G. Albrow et al., Phys. Lett. B44, 518 (1973) and B44, 2o7 (1973).
8. H. M. Chan, C. S. Hsue, C. Quigg, J. M. Wang, Phys. Rev. Lett. 26, 672 (1971).
9. M. S. Chen et al., Phys. Rev. Lett. 26, 1585 (1971).
10. W. Frazer, R. Peccei, S. Pinsky, C.-Tan, Phys. Rev. D7, 2647 (1973).
11. K. Fialkowsky, H. Miettinen, Phys. Lett. 43B, 61 (1973) H. Harari, E. Rabinovici, Phys. Lett. 43B, 49 (1973).
12. L. W. Jones et al., Phys. Rev. Lett. 25, 1679 (197o).
13. D. M. Tow, Phys. Rev. D7, 3535 (1973); R. Cahn, SLAC-Pub-1121.

14. A. H. Mueller, Phys. Rev. D4, 150 (1971).

15. C. Bromberg et al., Phys. Rev. Lett. 31, 1563 (1973).

16. E. L. Feinberg, Physics Reports 5C, 240 (1972).

17. Y. Pal, B. Peters, Mat. Fys. Medd. Dan. Vid. Selskab 33, 15 (1964).

18. G. Bellettini, in High-Energy Collisions - 1972, AIP Conference Proceedings No. 15, ed. C. Quigg.

19. M. G. Albrow et al., Nucl. Phys. B54, 6 (1973) and B51, 388 (1973).

20. H. Abarbanel, G. Chew, M. L. Goldberger, L. M. Saunders, Phys. Rev. Lett. 26, 937 (1971).

21. J. Lee-Franzini, in High-Energy Collisions - 1972, AIP Conference Proceedings No. 15, ed. C. Quigg.

22. G. C. Gox, in High-Energy Collisions - 1972, AIP Conference Proceedings No. 15, ed. C. Quigg.

23. W. R. Frazer, D. R. Snider, Phys. Lett. 45B, 136 (1973); W. R. Frazer, D. R. Snider, C.-Tan, Phys. Rev. D8, 3180 (1973).

24. G. F. Chew, Phys. Rev. D7, 3525 (1973) and Phys. Lett. 44B, 169 (1973).

25. A. Capella, M.-S. Chen, Phys. Rev. D8, 2097 (1973); A. Capella, M.-S. Chen, M. Kugler, R. Peccei, Phys. Rev. Lett. 31, 497 (1973).

26. D. Amati, L. Caneschi, M. Ciafaloni, Nucl. Phys. B62, 173 (1973).

27. J. S. Ball, F. Zachariasen, Phys. Lett. 40B, 411 (1972); University of Utah preprint, 1973.

28. F. C. Winkelmann et al., Phys. Rev. Lett. 32, 121 (1974).

29. G. F. Chew, Phys. Rev. D7, 934 (1973).

30. M. Bishari, J. Koplik, Phys. Lett. 44B, 175 (1973).

31. G. F. Chew, D. R. Snider, Phys. Lett. 31B, 75 (1970).

32. T. Gaisser, C.-I. Tan, preprint BNL 18070 (1973).

33. G. F. Chew, J. Koplik, preprint LBL-2463 (1973).

Acta Physica Austriaca, Suppl. XIII, 117–289 (1974)

RECENT RESULTS OF THE EXPERIMENTAL ANALYSIS OF HADRONIC HIGH ENERGY INTERACTIONS*)

by

M. MARKYTAN
Institut für Hochenergiephysik der Österr.
Akademie der Wissenschaften
Wien

CONTENTS

---

*) Lectures given at XIII. Internationale Universitätswochen für Kernphysik, Schladming, Austria, February 4-15, 1974.

2. Polarisation Measurements in Strangeness Exchange Reactions

IV. PION EXCHANGE REACTIONS

1. Phase Shift Analysis of the $(\pi\pi)$ System
2. $K^- \pi^-$-Scattering
3. SU_3-Relation between Pion Exchange Dominated Quasi-Two-Body Reactions

V. ANALYSIS OF THE THREE MESON SYSTEM IN THE REACTIONS $\pi p \to (3\pi)p$ AND $Kp \to (K2\pi)p$ AND OF THE $(\pi\omega)$ AND $(K\omega)$ SYSTEMS IN $\pi(K)p \to \pi(K)\omega p$

1. Partial Wave Analysis of the (3π)-System
2. Decomposition of the $(K\pi\pi)$-Structure into Spin-Parity Amplitudes
3. Structures in the $(\pi\omega)$- and $(K\omega)$-Systems

VI. DIFFRACTION DISSOCIATION

1. Analysis of the Reaction $KN \to QN$ in Terms of Reggeized Helicity Amplitudes and the Cross-Over Effect
2. Energy Dependence of $\pi^- p$ Elastic Scattering and of the Reaction $\pi^- p \to \pi^- \pi^+ \pi^- p$
3. Forward-Backward Ordering of Particle Momenta in Multi-Particle Final States and Energy Dependence
4. Separation of Diffraction Dissociation from Non-Diffractive Processes in Multi-Particle Final States
5. Impact Parameter Profile of Elastic Scattering and Diffraction Dissociation
6. Experimental Results of the Inclusive Diffractive Processes $pp \to p$+anything and $\pi p \to p$+anything

VII. MULTIPLICITY DISTRIBUTIONS

1. Multiplicity Distributions of the Diffractively Produced Multi-Particle System
2. Dispersion of the Multiplicity Distribution
3. Moments of the Multiplicity Distribution
4. Koba-Nielsen-Olesen Scaling
5. Dependence of the Average Charged Multiplicity on the Transverse Momentum and its Energy Dependence at Large Transverse Momenta
6. The True Multiplicity Distribution
7. Charge Transfer

I. INTRODUCTION

In the last year, an enormous amount of experimental information on hadronic high energy interaction has become available. The most spectacular findings have been achieved in experiments at the CERN Intersecting Storage Ring, and the appearance of a predicted diffraction minimum in pp elastic scattering and the observation of rising total pp cross-sections were the results. But, also, the experimental analysis of bubble chamber data, though of limited statistics, but containing all observable reaction channels, furnished most valuable contributions to the understanding of the reaction mechanisms of hadronic interactions. Energy dependences, splitting up of reaction channels into partial contributions of special reaction mechanism, partial wave analyses, and resonance spectroscopy results could mainly be achieved only by this detector because of its 4π geometry detection capability that does not necessitate angular corrections due to detection efficiency. A still increasing emphasy has been laid on multiparticle and inclusive production with the essential aim to clarify the nature of diffractive production and in general the mechanisms of such reactions. Because of the large number of reaction variables in multiparticle production and the limited statistics of (essentially bubble chamber) data, one is unable to perform the data analyses in a 3N-4 dimensional space, but aims at finding characteristic features of such reaction channels that allow a simpler analysis and give conclusions for the reaction mechanisms. In the course of these lectures, it is tried to present a survey of the experimental situation of the analysis of hadronic interactions, but also to discuss the physics' ideas and models that are applied to interpret the experimental results. But it is always understood in this course that the experimental findings are the premisses for the understanding of physics in this topic, though most of the experiments have been stimulated by theoretical and phenomenological ideas.

II. ELASTIC SCATTERING

1. Experimental Results on pp Elastic Scattering and on Rising Total pp Cross-Sections

Elastic scattering of elementary particles is, as far as the simplicity of its final state is concerned, the most plain of particle interactions, on the other hand, through its connectedness to the sum of all possible inelastic reaction channels by unitarity, a very complicated reaction that continuously revealed surprising findings. As centre of mass energies of 10 to 50 GeV became accessible through the Intersecting Storage Ring of CERN, a change of the slope of the momentum transfer distribution of pp elastic scattering at $-t = 0.15\ (GeV/c)^2$ was detected [1]. At the ISR energies 10.8 + 10.8, 22.6 + 22.6, and 26.7 + 26.7 GeV, the fitted slopes before and after the kink in the t-distributions of pp elastic scattering were found to be (11.6, 10.4), (12.9, 10.8), and (12.4, 10.8) $(GeV/c)^{-2}$ respectively. For the energy and momentum transfer dependence of the diffraction slope the following parametrisation was assumed:

$$B(s,t) = B_o(t) + 2\,\gamma\,\ln\,(s/s_o)\,,\quad s_o = 1\ GeV^2 \qquad (II.1.1)$$

which yielded, for this range of ISR energies, the following values for the free parameters: $B_o = 7.0 \pm 1.14\ (GeV/c)^{-2}$, $\gamma = 0.37 \pm 0.077\ (GeV/c)^{-2}$ for $-t > 0.15\ (GeV/c)^2$, and $B_o = 9.21\ (GeV/c)^{-2}$, $\gamma = 0.1 \pm 0.062\ (GeV/c)^{-2}$ for $-t < 0.15\ (GeV/c)^2$. These measurements only extended to momentum transfers smaller than 1 $(GeV/c)^2$. No significant indication of such a change in the diffraction slope has been found in elastic scattering processes at lower energies. As the measurement of the angular distribution of pp elastic scattering was extended to momentum transfers of more than 2 $(GeV/c)^2$, a significant minimum at $t = -1.4\ (GeV/c)^2$ was detected as result of the mentioned experiment of the Aachen-CERN-Harvard-Genova-Torino Collaboration (Fig. 1). A compilation of pp elastic differential cross-sections between 3 and 24 GeV/c [2] and the angular distribution of the p-n cross-section at 24 GeV/c [2] show a break or kink in this momentum transfer region. In comparison, K^+p elastic scattering is characterized by a break at $t = -2.5\ (\text{GeV}/c)^2$ and $\pi^{\pm}p$ elastic scattering by breaks at $t = -0.6\ (\text{GeV}/c)^2$ and $-2.5\ (\text{GeV}/c)^2$ at medium energies of the order of 10 GeV [3,4]. The physical situation now is that pp elastic scattering has structure in its angular distribution at $t = -0.15\ (\text{GeV}/c)^2$ and $-1.4\ (\text{GeV}/c)^2$, the first of which showed up only at the extreme ISR energies, whereas the second one was already indicated at medium energies; that K^+p elastic scattering has only structure in the (-t)-region of 2 to 3 $(\text{GeV}/c)^2$ with the possibility that new structures may show up at much larger energies than 14 GeV; and that $\pi^{\pm}p$ elastic scattering shows breaks at $t = -0.6$ and $-2.5\ (\text{GeV}/c)^2$. The Glauber Formalism yields, in the impact parameter space, for the scattering amplitude with respect to the total helicity flip $n = |\lambda_1 - \lambda_2 - \lambda_3 + \lambda_4|$:

$$f_n^{(s)}(s,t) = 2k^2 \int b\, db\, A_n(b) J_n(b\sqrt{-t}) \qquad \text{(II.1.2)}$$

(k... c.m. momentum, b... impact parameter). The zeroes for $n = 0$ are for $b = 1$ fm: $-t = 0.2,\ 1.2\ (\text{GeV}/c)^2$ etc, and for $n = 1$: $-t = 0, 0.6,\ 1.9\ (\text{GeV}/c)^2$ etc. In the diffraction model for high energy pp-scattering of Chou and Yang [5] and of Durand and Lipes [6], it is assumed that the scattering particles, imagined to be extended objects, penetrate each other during the scattering process that is therefore dependent on the matter distribution of the hadron. In an approximate way, this matter distribution can be regarded as the charge distribution in the scattering hadrons for parametrisation. (The neutron matter distribution has to be identified with that of the proton). The Fourier-Transform of the product of the two proton electric form factors to impact parameter space, $\rho(b)$, is used to parametrize the local absorption of the elastic scattering amplitude:

$$A(b) = i(1 - S(b)), \; S(b) = e^{-\alpha\rho(b)} \; . \qquad (II.1.3)$$

α is the (complex) absorption coefficient. With the following parametrisation of the electric form factor of the proton:

$$G_e(t) = \left(\frac{\mu^2}{\mu^2 - t}\right)^2 , \qquad (II.1.4)$$

where μ is a free parameter, Durand and Lipes [6] obtained for the absorption:

$$S(b) = e^{-\text{const}\cdot(\mu b)^3 \, K_3(\mu b)} \; . \qquad (II.1.5)$$

This expression predicts two diffraction minima in the t-regions 1-2 $(GeV/c)^2$ and (5-8) $(GeV/c)^2$. If also spin-dependence is included in this penetration formalism, by assuming the weak spin-orbit interaction amplitude:

$$f_{LS}^{(s)}(s,t) = i(\vec{\sigma}_1 + \vec{\sigma}_2)\cdot\vec{n}\cdot 2k^2 \int S(b)\,\alpha_1 \rho_1(b)\, J_1(b\sqrt{-t})\, b\,db \; , \qquad (II.1.6)$$

the ansatz for the distribution in b which fits best the polarization data for protons predicts two zeroes in the t-distribution of the polarization in the region of the first diffraction minimum of the angular distribution [6] and none at $-t = 0.2$ $(GeV/c)^2$: Polarization data of pp scattering at 10 - 17.5 GeV/c [7] seem to indicate such a structure since they show a dip at about $-t = 0.9\ (\mathrm{GeV}/c)^2$. The ansatz (II.1.6) is only a tentative for explaining the structure of polarization data though it is not very likely that the rather small p polarization at high energies (less than 0.1) reveals this structure unambiguously. The occurrence of these diffractive zeros from out the structure of the electric form factor of the proton is equivalent to Glauber's multi-scattering approach within the hadrons. The back of the extended hadronic object is hidden by the front side. The small change in the diffraction slope at $-t = 0.15\ (\mathrm{GeV}/c)^2$ could possibly be explained by recognizing concentrations of special inelastic reactions (diffraction dissociation and non-diffractive processes) in different "rings" of impact parameter space, that will be discussed with diffraction dissociation. The most analogue features with respect to pp elastic scattering, are exhibited by K^+p elastic scattering which is also, by duality and the exoticity of the s-channel, supposed to be operated by what is called Pomeranchukon exchange. $\pi^{\pm}p$ clearly also show the dips of vector-tensor-meson exchange in addition to Pomeranchukon terms whose zero at $-t = 1.4\ (\mathrm{GeV}/c)^2$ may only become apparent at much higher energies. The second half of Fig. 1 shows the logarithmic encrease, with respect to s, of the diffraction slope of pp elastic scattering, as parametrized by equ. (II.1.1). γ corresponds to the slope of the Pomeranchukon Regge trajectory. In the impact parameter picture, this shrinking of the momentum transfer distribution would correspond to a radial extension of the scatterer as seen from the incoming particle, and more peripheral scattering ensues.

The second surprising result from ISR experiments

was the finding of rising total pp cross sections. In Fig. 2, a compilation of pp total cross-sections [9] at Serpuchov and ISR energies are shown together with the rising cross-sections of proton air interactions from cosmic ray data [10]. The latter have been measured with respect to their inelastic component because elastic scattering processes are extremely hard to obtain. But it is possible to extract, by means of theoretical models, using Glauber's multi-scattering approach, the pp elastic cross-sections at cosmic ray energies (i.e. up to 3×10^4 GeV) from inelastic p-air interactions with the result:

$$\sigma_{pp} = 38.8 + 0.4 \ln (s/s_o) \ , \qquad \text{(II.1.7)}$$

where s_o normalizes σ_{pp} to the experimental value of 70 GeV. This coincident result gives confidence into the results of the ISR experiments which also showed a remarkable coincidence with respect to this finding. The best fit of the parametrisation of the pp total cross-section:

$$\sigma = \sigma_o + \sigma_1 (\ln s/s_o)^{\nu} \qquad \text{(II.1.8)}$$

is $\sigma_o = 38.5$ mb, $\sigma_1 = 0.9 \pm 0.3$ mb, $\nu = 1.8 \pm 0.4$ and $s_o = 200 \text{ GeV}^2$ [9], which gives the rising energy dependence that is maximally allowed by the Froissart bound. But, this bound is not yet saturated at these energies since in Froissart's formula

$$\sigma_{tot}(s) \leq \frac{\pi}{m_\pi^2} (\ln s)^2,$$

for $\pi\pi$-scattering, π/m_π^2 corresponds to 60 mb. To the contrary, analysing the unitarity relation in impact parameter space, the data suggest that cross-sections only rise in a limited impact parameter interval, as will be discussed later. In Fig. 3 [11], the characteristics of the energy dependence of the momentum transfer distributions of pp elastic scattering and of that of the pp cross section are compared, so that the increase of the forward scattering cross-section is clearly visible.

2. Determination of the Real Part of the πN Diffraction Amplitude

The imaginary part of the forward elastic scattering amplitude is related to the total cross-section by the optical theorem, and the real part of the scattering amplitude at fixed momentum transfer to the imaginary part by fixed t-dispersion relations. Usually, the real part of the elastic scattering amplitude is experimentally determined by analysing the interference of Coulomb scattering with the strong interaction amplitude for ab scattering:

$$f_{Coulomb} = -2\alpha \frac{G^2(t)}{|t|} e^{i\alpha[\ln \frac{t_o}{|t|} - .577]} \qquad \text{(II.2.1)}$$

with

$$t_o = .08 \text{ GeV}^2 , \qquad \alpha = \frac{1}{137} .$$

G(t) is the electromagnetic form factor of the proton. The interference of positive particles scattering on protons is constructive. This Coulomb scattering amplitude gives the absolute cross-section and can serve as a tool to determine the absolute cross-section of the strong interaction term from a fit to the data. The highest energy at which the real part of the pp scattering amplitude was determined is 4o5 GeV within the US-USSR experiment at NAL [8]. The complication of the ratio $\rho = \mathrm{Re}\, f_{strong}(s,0)/\mathrm{Im}\, f_{strong}(s,0)$, depending on the incident momentum in the laboratory, is given in Fig. 4 [12]. It can be seen that the ratio ρ passes zero from above at $p_{lab} \sim 1.5$ GeV/c and seems to cross the zero line again at the lowest ISR energy. It is interesting to analyze also the second relation, namely the dispersion relations, by means of which, it is possible to determine the real part of the elastic scattering amplitude at $t \neq 0$. The full line corresponds to a dispersion relation calculation by Söding [13] for pp scattering in the forward direction, assuming an $(\ln s)^2$, behaviour for the total pp cross section (the dashed curve corresponds to a constant asymptotic cross-section). For the πp diffraction amplitude, a dispersion relation calculation has been done by Höhler and Jakob [14]: the isospin even invariant amplitude that dominates diffraction scattering,

$$A'^{+} = A^{+} + \frac{m(s-u)}{4m^2-t} B^{+} \qquad \text{(II.2.2)}$$

fulfills the fixed t dispersion relation

$$\mathrm{Re}\, A'^{+}(\nu,t) = A'^{+}(0,t) - \frac{g^2}{m} \frac{\nu^2}{(1-t/4m^2)[\nu^2-(\frac{t}{4m} \frac{\mu^2}{2m})^2]} + $$

$$+ \frac{2\nu^2}{\pi} \int\limits_{\mu + \frac{t}{4m}}^{\infty} \frac{d\nu'}{\nu'} \frac{\mathrm{Im}A'^+(\nu';t)}{\nu'^2 - \nu^2} \qquad\qquad (\mathrm{II}.2.3)$$

(m... nucleon mass, μ ... pion mass, $\nu = (s-u)/4m$, $A^+(0,t)$... subtraction term, $g^2/4\pi = 14.6$). The integral is split into a low energy part for which mainly phase shift analysis results are used for determining Im A'^+, and into a large energy part in which Im A'^+ is determined from differential cross-sections:

$$\left(\frac{d\sigma}{dt}\right)^+ \equiv \frac{1}{2}\left(\frac{d\sigma_{\pi+p}}{dt} + \frac{d\sigma_{\pi-p}}{dt} - \frac{d\sigma_{\pi-p\to\pi^0 n}}{dt}\right) ,$$

that is the isospin even part of the differential cross section. In order to obtain the imaginary part of A'^+, the assumption of s-channel helicity conservation can be made that corresponds to $A^+ = 0$ in equ. (II.2.2) [15]. With these two kinds of input for the imaginary part of A'^+ the dispersion relation is solved for each t-value, and the separate t-entities of the subtraction function $A'^+(0,t)$ are treated as independent parameters. The results of the real part of A'^+ are shown in the Argand diagrams of Figs. 5 and 6 [14]. ($C^+ \equiv A'^+$). One can see the resonance structure above a large diffractive background, and the resonance structure becomes smaller with increasing t. One obtains as result a rather large ratio $\rho = |\mathrm{Re}\,A'^+/\mathrm{Im}\,A'^+|$ of the order of 0.5 at 6 GeV/c and $t = -0.3$ (GeV/c)^2. The predictions of the slope of Re A'^+ are sensitive with respect to the input of (slope of differential cross section) x (cross section) of the isospin even part of πN scattering. For pp-elastic scattering, the modulus of ρ is of the order of 0.3, and from a bubble chamber analysis of K^-p scattering [16] at 10 GeV/c, in which small angle scattering was measured by an automatic film scanning machine and the real part of the forward scattering amplitude determined by Coulomb interference, the value of $\rho = (+ 25 \pm 8)\%$ was obtained. In the laboratory momentum region between 500 and 1000 MeV, the ratio ρ varies between +0.2 and +0.8, $\rho = 0.2 \pm 0.08$ at 4.2 GeV for K^-p scattering. In a dispersion relation analysis by Chao and Pietarinen [17] in which the low energy K^-p scattering data and the K^+p high energy scattering data were fitted, a positive ratio ρ is obtained up to 10 GeV K^- momenta. For K^+p elastic scattering, ρ is negative.

III. AMPLITUDE ANALYSIS OF SOME QUASI-TWO BODY REACTIONS AT MEDIUM ENERGIES

1. K^*_{890}-Production

The Athens-Demokritos-Liverpool-Vienna Collaboration has performed an experimental separation of natural and unnatural parity exchange contributions to the quasi two body reaction $K^-p \rightarrow K^{*-}_{890}p$ [19]. This is achieved by obtaining the following products of K^*_{890} spin density matrix elements in the Gottfried-Jackson reference system (the spin quantization axis is the direction of the incoming K^- in the K^* rest frame) and $d\sigma/dt$ values:

$\rho^J_{00} \frac{d\sigma}{dt}$ .... differential cross-section for K^*-production with K^*-helicity 0 and unnatural parity exchange only.

$(\rho^{J}_{1,1} \pm \rho^{J}_{1,-1})\frac{d\sigma}{dt}$ .... differential cross-section for the K^*-production with K^*-helicity 1 and natural (+), unnatural (-) parity exchange.

These contributions are displayed in Fig. 7. All three distributions show the indication of a dip in the forward direction. This is an astonishing result for the unnatural parity exchange contribution which is composed of the $\lambda_{K^*} = 0$ and $\pm$ 1 amplitudes. The $\lambda_{K^*} = 0$ part is dominated near the forward direction by pseudoscalar (π) exchange that cannot generate a K^* helicity one with zero spin exchange. The exchange of a pion pole or Regge trajectory is given by an amplitude which has a dip in the forward direction, because of angular momentum conservation in the forward direction, but the maximum of the $d\sigma/dt$ distribution is situated at $t = - m_\pi^2$ so that this dip is usually not detected by bubble chamber experiments. Absorption in the initial or/and final state makes, however, Regge cut contributions that interfere destructively with the π-pole or Regge trajectory exchange so that a sharp spike of width of the order of m_π^2 is usually seen in the forward direction of momentum transfer distributions of pion exchange reactions. The unnatural parity exchange contributions to K^* production with K^* helicity one can either be effectuated by a pion trajectory at $t \neq 0$ (the pion trajectory turned out, however, in most experimental analyses as having a rather small slope of increase with t, i.e. rather a pion pole) or by cuts whose $d\sigma/dt$ contribution is finite and smooth in the forward direction. The result of experiments suggests the conclusion that cut contributions are dominating unnatural parity exchange.

The determination of the K^*_{890} spin density matrix elements in the helicity system (spin quantization axis is the negative direction of the final state proton in the rest frame of the K^*_{890}) served for applying Harari's parametrisation of s-channel helicity amplitudes [20,21] to the data of $d\sigma/dt$ and $\rho^H_{mn} \frac{d\sigma}{dt}$:

$$f_{\lambda_{\bar{K}^*}\lambda_p;O\lambda_p} = A_{\{\lambda\}} e^{B_{\{\lambda\}}t} J_n(R\sqrt{-t})\{i+\tan \frac{\pi}{2}\alpha_\omega + $$

$$+ i - \text{cotan} \frac{\pi}{2}\alpha_f\} \ . \qquad \text{(III.1.1)}$$

α_ρ and α_{A_2} are exchange degenerate and implicitly contained in (III.1.1) the free parameters of which are the couplings $A_{\{\lambda\}}$ and slopes $B_{\{\lambda\}}$. R can be chosen so that the first zero ($t \sim -1.2\ (\text{GeV/c})^2$ for R = 1 fm) cancels the cotan $\frac{\pi}{2}\alpha_f$-pole at $t \sim -1.25\ (\text{GeV/c})^2$ for $\alpha_f(t) = 0$. In Fig. 7, the result of this parametrisation is shown for $\rho^H_{1,\pm1} \frac{d\sigma}{dt}$, for which only the two amplitudes with s-channel K^* helicity one and no spin flip at the nucleon vertex have been used. The result is satisfactory, and crossing gives the appropriate t-channel helicity mixing for $t \neq 0$, natural parity exchange being dominant.

2. Polarization Measurements in Strangeness Exchange Reactions

A. The Reactions $\bar{p}p \to \bar{\Lambda}\Lambda,\ \bar{\Sigma}^+\Sigma^+,\ \bar{\Lambda}\Sigma^o + \bar{\Sigma}^o\Lambda$

In these strangeness exchange reactions, all the

interacting particles are spin $\frac{1}{2}$ fermions, and therefore these reactions are described by five independent helicity amplitudes: (++,++); (++,--); (+-,+-); (+-;-+); (++,+-) in the s-channel, for instance. The product of the polarization component in the y-direction (normal to the production plane) and the differential cross-section is:

$$P\frac{d\sigma}{dt}=2\ \mathrm{Im}[(f_{++;++}+f_{++;--}+f_{+-;+-}-f_{+-;-+})f^{*}_{++;+-}], \qquad \text{(III.2.1)}$$

and is a measure whether Regge exchanges of different phase and/or cuts of different phase contribute to the reaction. The polarization components of the hyperons are determined by the method of moments with respect to the coordinate system: z = direction motion of the hyperon, y = normal to the production plane, x = y ⊗ z.

$$P_i = \frac{3}{\alpha}\cdot\frac{1}{N}\cdot\sum_{\text{events}}\cos\theta_{(i)},\ i = x,y,z. \qquad \text{(III.2.2)}$$

The spin correlations between the two hyperon spins are:

$$A_{ij} = \frac{9}{\alpha_1\cdot\alpha_2}\cdot\frac{1}{N}\cdot\sum_{\text{events}}(\cos\theta_{(i)}\cdot\cos\theta_{(j)}) . \qquad \text{(III.2.3)}$$

$\alpha, \alpha_1, \alpha_2$ ...... α-parameters of the considered hyperon decays, N.... number of events, $\theta_{(i)}$ ..... angle between the direction of the decay products and the i-th coordinate axis.

$$\alpha_{\Lambda} = 0.645, \qquad \alpha_{\Sigma^* \rightarrow p\pi^0} = -0.991 .$$

For the Σ^0 polarization measurement, one has to consider the sequential decay into $\Lambda\gamma$ and $\Lambda \rightarrow p\pi^-$, where only the second is parity violating.

$$P_{\Sigma^o(i)} = \frac{9}{\alpha_\Lambda} \cdot \frac{1}{N} \cdot \sum_{events} (\cos\theta_\Lambda \cdot \cos\theta_P) \quad . \qquad (III.2.4)$$

In Fig. 8 and 9, the results of the $d\sigma/dt$ distributions of the reactions $\bar{p}p \rightarrow \bar{\Lambda}\Lambda$ and $\bar{\Lambda}\Sigma^o + \Lambda\overline{\Sigma^o}$ at 3.6 GeV/c show a significant kink at $t = -0.25$ $(GeV/c)^2$, whereas the $\overline{\Sigma^{\mp}}\,\Sigma^+$ - reaction even exhibits a dip in this t-region [22]. The polarization measurements for the Λ-reactions give polarizations, oscillating with increasing $|t|$, from negative to positive and again to negative values, while the Σ^+-polarization is positive and grows from 0.3 to 0.8 at -0.3 $(GeV/c)^2$. The full lines interpolating the $d\sigma/dt$ distributions correspond to a two-slope parametrisation, the dotted lines are predictions from a speculated similarity of the reactions $\bar{p}p \rightarrow \bar{\Lambda}\Lambda$ and $\bar{\Lambda}\Sigma^o + \bar{\Sigma}^o\Lambda$ to the pion induced reactions $\pi^-p \rightarrow K^o\Lambda$, $K^o\Sigma^o$, and $\pi^+p \rightarrow K^+\Sigma^+$ [23]. The quantity $S = \frac{1}{4}(1 - A_{yy} + A_{xx} + A_{zz})$ has been determined by ref. [22] to -0.03 ± 0.11 for the $\bar{\Lambda}\Lambda$ final state which proves the $\bar{\Lambda}\Lambda$ system being dominated by a triplet spin state. The spin content of the $\overline{\Sigma^{\mp}}\,\Sigma^+$ final state has not been determined because of lack of statistics in two-prongs with two visible Σ-decays (736 events with 165 $\Sigma^+p\pi^o$ and $\overline{\Sigma^{\mp}}\,\bar{p}\pi^o$ decays). It may be speculated that the quantity S exhibits a larger singlet spin contribution. The exchange possibilities for the $\bar{\Lambda}\Lambda$ and $\overline{\Sigma^{\mp}}\,\Sigma^+$ are K, K^*_{890}, and K^*_{1400} the latter two of which are apprximately exchange degenerate and give helicity amplitudes in (Regge) phase. In the case of the pion induced reactions, the kaon exchange possibility is absent, and polarizations should be explained by cut contributions if the K^*_{890} and K^*_{1400} are exchange degenerate. The sharp kink in these $\bar{p}p$ reactions is also indicative for cut contributions together with the difference of K, K^*_{890}, and K^*_{1400} couplings to the $p\Lambda$ and $p\Sigma^+$ systems.

B. $\pi^+p \to K^+\Sigma^+$ in the Backward Direction at 3.5 GeV/c

In the context to the reactions the results of which have been given in the previous chapter, this strangeness exchange reaction is interesting for comparing polarization measurements of the Σ^+ emitted in the backward direction to that in the forward direction. The Σ^+ polarization in its dependence on u (black circles) is given in Fig.10 [24] in juxtaposition to the backward Λ-polarization in the reaction $\pi^-p \to \Lambda K^0$ (open circles) at 4.0 GeV/c. The Σ^+ polarization now is compatible with zero over the whole measured u-range, whereas the backward lambdas show a polarization increasing from 0.5 at u = 0.07 $(GeV/c)^2$ to 1 at u = -0.1 $(GeV/c)^2$. The strong polarization in the latter case is caused by the strong interference of the contributions of the $\Sigma_{\alpha\gamma}$ and $\Sigma_{\beta\delta}$ trajectories [26] exchanged, whereas the additional exchange possibility of Λ exchange is supposed to be at the origin of interference terms with the Σ-trajectories, which approximately cancel the $\Sigma_{\alpha\gamma}$ and $\Sigma_{\beta\delta}$ interference terms.

C. $\pi^-p \to K^{*0}_{890}\Lambda$, $K^{*0}_{890}\Sigma^0$, and K^{*0}_{890} $Y^*_1(1385)$ at 3.93 GeV/c

Following the qualitative explanatory argument of Λ and Σ polarizations by the interlude of K and K^*_{890} plus K^*_{1400} exchange, the hyperon polarizations should be equal to those in the above mentioned $\bar{p}p$ reactions. Indeed, as can be inspected in Fig. 11 in which the differential cross-sections and hyperon polarizations of the reactions $\pi^- p \to K^{*o}_{89o}\Lambda$ and $K^{*o}_{89o}\,\Sigma^o$ of ref. [27] are given, the t'-dependence of the Λ polarization shows the characteristic oscillatory behaviour as already known, while the Σ^o polarization is characterized as rapidly increasing towards 1 within large errors.

For this type of quasi-two body reactions, one has, in the case the hyperon polarization is measured, enough information for determining the amplitudes which contribute. By parity conservation, one has to determine six complex helicity amplitudes, but the overall phase and, because of summation over the helicities of the initial state proton, the phase between the different spin projections of the ingoing proton have to remain undetermined. The measurement of the ten real numbers to determine the amplitudes goes via the determination of the three real spin density matrix elements of the K^*_{89o}, the Λ polarization and seven distributions that correspond to spin correlations. In terms of <u>transversity amplitudes</u> for which the spin quantization axis per interacting particle is the normal to the production plane and the y-axis the particle's direction of flight, the reaction is described by the six following complex entities:

$$T^o_{++},\ T^o_{--},\ T^1_{+-},\ T^1_{-+},\ T^{-1}_{+-},\ T^{-1}_{-+}\ .$$

The notation is $T^{\lambda_{K^*}}_{\lambda_\Lambda,\lambda_p}$. Parity conservation is expressed by the following relation:

$$T^{\lambda_{K^*}}_{\lambda_\Lambda\lambda_p} = (-1)^{\lambda_\Lambda-\lambda_p+\lambda_{K^*}}\ T^{\lambda_{K^*}}_{\lambda_\Lambda\lambda_p} \quad . \qquad \text{(III.2.5)}$$

Natural parity exchange which demands the K^*_{890} helicity to be ± 1 is expressed by $T^o_{\pm\pm}$, and the natural parity exchange contribution to the hyperon polarization simply is

$$P_\Lambda = |T^o_{++}|^2 - |T^o_{--}|^2 \quad . \qquad \text{(III.2.6)}$$

The result of ref. [27] of a transversity amplitude analysis can be inspected in Fig. 12. In the first instance, the transversity amplitudes squared coincide perfectly between the two reactions $\pi^- p \to K^*_{890}\Lambda$ and $K^- p \to \Phi\Lambda$ as has been suggested by ref. [28]. In terms of Harari-Rosner quark diagrams, both reactions should exhibit the same SU(3) structure in the exchange channel with approximately the same couplings. The main amplitude features are: $|T^o_{--}|^2$ large and $|T^o_{++}|^2$ essentially vanishing, relatively large contributions of $|T^1_{+-}|^2$ and $|T^{-1}_{-+}|^2$. All differences of amplitudes squared with opposite fermion transversities for given K^* transversity act together to make the characteristic oscillation of the Λ polarization. The transversity amplitudes squared of the corresponding Σ^o reaction are much simpler in the sense that only $|T^o_{++}|^2$ contributes which causes maximal Σ^o polarization. Fig. 13 elucidates this analysis by explicitly representing the natural and unnatural parity exchange contributions to the $d\sigma/dt$ distributions of the $K^{*o}_{890}\Lambda$ and $K^{*o}_{890}\Sigma^o$ reaction. Both parities are participating to an equal extent in the Λ-reaction, but the unnatural parity exchanges are depressed by a factor of the order of ten in the Σ^o-reaction. Therefore, in the absence of other strange meson trajectories than K^*_{890} and K^*_{1400}, natural parity cuts should be assumed to cause the emission of highly polarized Σ^o's. Similar results were obtained from the analysis of the K^{*o}_{890} $Y^{+o}_1(1385)$ final state [27].

IV. PION EXCHANGE REACTIONS

1. Phase Shift Analysis of the $\pi\pi$ System

Up to now only quasi-two body reactions with final state particles or resonances of definite isospin, mass, spin and parity have been discussed; but, in general, when analysing resonant particle systems, one always has to take into consideration a superposition of isospin- and spin-parity-states at a fixed effective mass that account for the resonances and the background in the effective mass distribution. A very challenging object of such a type of analysis in terms of isospin- and partial wave contributions are the reactions

$$\pi^- p \to \pi^- \pi^+ n, \quad \pi^+ p \to \pi^+ \pi^+ n, \qquad \text{and} \qquad \pi^- p \to \pi^- \pi^- \Delta^{++}(1236).$$

Since all three reactions are dominated by pion exchange at high energies and small momentum transfer between the ingoing and the outgoing fermions as can be seen, for instance, by determining the energy dependence in this kinematical region, the ultimate aim is to analyse meson resonance spectroscopy in the partial waves of definite isospin states of the $\pi\pi$ interaction.

A reaction of two particles leading to the production of N particles can be described by 3N-4 independent variables that can be chosen in the case of N = 3:

s, $t_{NN} \equiv t$, $M_{\pi\pi}$, and the decay angles θ and ϕ of the $\pi\pi$ system, in its centre-of-mass, with respect to:

a) z = p (incident pion), if the reaction is analysed in the t-channel <u>(Gottfried-Jackson System)</u>

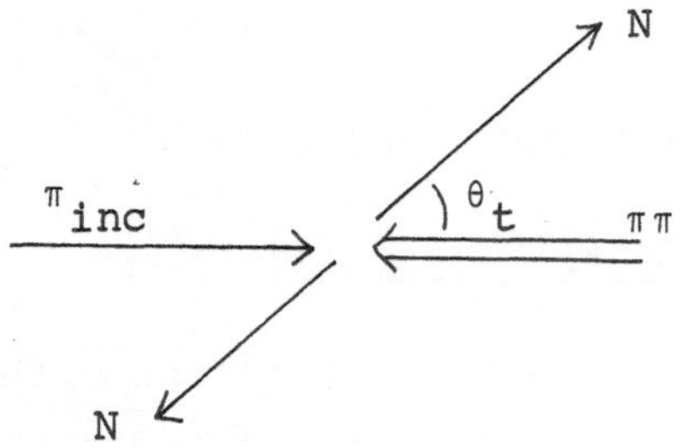

, or

b) z = -p (outgoing fermion), if the reaction is analyzed in the s-channel <u>(Helicity System)</u>

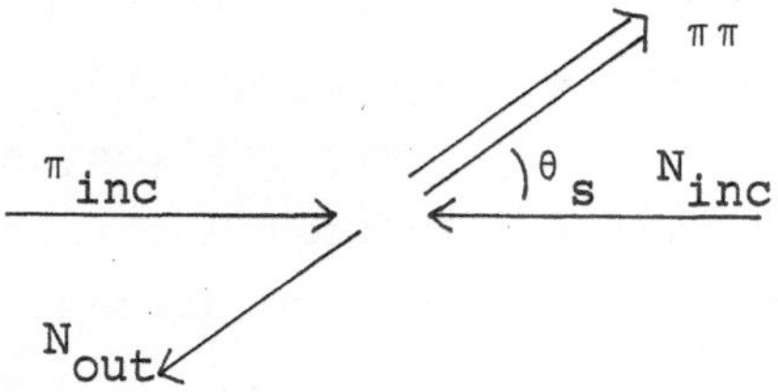

, and

y being normal to the reaction plane. The experimentally measured quantity

$$\frac{d^4\sigma}{dtdM_{\pi\pi}\,d\cos\theta d\phi} = \sum_{JM} \langle Y_J^M \rangle \, (M_{\pi\pi},t) \; \mathrm{Re}\; Y_J^M(\theta,\phi) \qquad \text{(IV.1.1)}$$

can be decomposed into its partial waves where, the pions being spinless particles: $0 \leq J \leq 2\ell$, $M = m_1 - m_2$. (ℓ.... angular momentum of the $\pi\pi$ system; m_1, m_2 ... z-axis projections or helicities of the angular momentum; $\langle Y_J^M \rangle$ ... moments of the angular distribution of the $\pi\pi$ system at fixed $M_{\pi\pi}$ and t). The moments are unnormalized, so that resonance structures become apparent in them. In Fig.14, the angular distribution of the reaction $\pi^- p \to \pi^- \pi^+ n$ at 17.2 GeV/c [29] in their $M_{\pi\pi}$ dependence shows strong resonance structures in the ρ^o, f^o and g^o regions in the appropriate moments of 2 * Resonance spin. But there are also pronounced interference terms of resonant states and background states which by showing resonance structure, reveal the presence of other partial wave contributions than the resonant ones. Thus, there is a strong signal of S-wave background beneath the ρ^o by $<Y_1^o>$, and beneath the f^o by $<Y_2^o>$, etc. In addition, there are present M components unequal zero, predominantly M = 1, which suggest other exchange mechanisms than one-pion exchange that would not generate helicities unequal zero of the $\pi\pi$ system. Furthermore, the t-dependence of the $\pi\pi$ moments is given in Fig. 14 both for the t-channel and the s-channel definition of the decay angles θ and ϕ. Only the moments with M = 1 show considerable variation by crossing since the $|\ell 1>$ states transform under the crossing transformation dominantly proportional to sin χ (χ... crossing angle of the $\pi\pi$ state) for small rotations by the y axis. The $<Y_1^o>$ moment shows a very sharp variation at the $\pi\pi$ mass passing the $K\bar{K}$-threshold at which this system forms the S^* resonance structure (unitarity limit).

The partial waves of the $\pi\pi$ elastic scattering angular distribution

$$\frac{d\sigma}{d\Omega} = \frac{1}{k^2} \left|\sum_\ell (2\ell+1)\, f_\ell\, P_\ell(\cos\theta)\right|^2 \qquad \text{(IV.1.2)}$$

(k.... pion momentum in the $\pi\pi$ centre of mass) are decomposed into the isospin contributions I = 0,1, and 2 under the condition that I + ℓ must be even because of G-parity of the two pion system:

$$f_\ell = \frac{2}{3} f_\ell^{I=0} + \frac{1}{3} f_\ell^{I=2} \qquad \text{for even } \ell$$

$$f_\ell = f_\ell^{I=1} \qquad \text{for odd } \ell . \tag{IV.1.3}$$

In the elastic region $M_{\pi\pi} < M_{4\pi} = 0.56$ GeV

$$f_\ell^I = e^{i\delta_\ell^I} \sin \delta_\ell^I \quad . \tag{IV.1.4}$$

The $\pi\pi$ system originating from $\pi\pi$ interactions has helicity strictly zero, but other exchanges or reaction mechanisms causing unequal zero helicities of this system can best be separated out by doing the experimental analysis in terms of natural parity exchange (+) and unnatural parity exchange (-) amplitudes:

$$A_\lambda^{\ell\pm} = [A_\lambda^\ell \pm (-1)^{\lambda+1} A_{-\lambda}^\ell]/\sqrt{2} \ , \tag{IV.1.5}$$

where λ is the helicity of the $\pi\pi$ system. $\lambda = 0$ must give a non-vanishing contribution for unnatural parity exchange (π). Equ. (IV.1.5) is valid for each fermion antifermion helicity combination.

Under the assumption that only S- and P-waves (with helicities of the $\pi\pi$ system 0, $\pm$ 1) are contributing, the first two moments having been expressed in terms of the partial wave contributions have the following shape:

$$\sqrt{4\pi} \ \langle Y_0^0 \rangle = |S|^2 + |P_0|^2 + |P_{+1}|^2 + |P_{-1}|^2$$

$$\sqrt{4\pi} \ \langle Y_1^0 \rangle = 2 \ \mathrm{Re} \ (SP_0^*) , \text{ etc.} \tag{IV.1.6}$$

A direct measure for unnatural parity exchange (π) is the combination (which can directly be obtained from the moments)

$|P_o|^2 + \frac{1}{3}|S|^2$ which dominates over the combination

$|P_{\pm 1}|^2 + \frac{1}{3}|S|^2$ in the t-region $|t| \leq 0.2\ (\mathrm{GeV/c})^2$

above which the second combination representing helicity $\pm$ 1 admixtures becomes the larger.

Before extrapolating the helicity zero contributions of the reaction amplitude to the pion pole in t, one first has to determine the helicity zero contributions from other reaction mechanisms such as absorption the inclusion of which in its dependence on s-channel helicities of the fermions and the $\pi\pi$ system according to the Williams-model [30,31] will be sketched. The amplitude for one pion exchange is

$$[\sum_{\ell} (2\ell+1) f_{\ell}^{\pi,\pi\mathrm{off}} P_{\ell}(\cos\theta)][\bar{u}(p_n)\gamma_5 u(p_p)] \cdot \frac{G_{\pi NN}/\sqrt{4\pi}}{m_{\pi}^2 - t}$$

$$= \sum_{\ell,\lambda} F_{\ell} A'_{\pi p\to(\ell,\lambda)+n} Y_{\ell}^{\lambda}(\omega,\psi) \ , \qquad \text{(IV.1.7)}$$

where ω is the angle between the incident π and the negative direction of the outgoing neutron in the reaction centre-of-mass. Formally in the helicity formalism, the one pion exchange amplitude is, by factorizing out half-angle singularity factors due to the boundary of the physical region of the reaction:

$$[\cos\frac{\omega}{2}]^{\frac{|\bar{\lambda}+\mu|}{2}} [\sin\frac{\omega}{2}]^{\frac{|\bar{\lambda}-\mu|}{2}} \frac{\text{polynomial in } (s,t,\bar{\lambda},\mu)}{m_{\pi}^2 - t} \qquad \text{(IV.1.8)}$$

where ω is the s-channel scattering angle in the reaction centre-of-mass, $\bar{\lambda} = \lambda - \lambda_n$, $\mu = \lambda_\pi - \lambda_p$, and the polynomial is of the Jacobi type. Due to these polynomials when inserting the expression of equ. (IV.1.8) into (IV.1.7), one obtains $\delta_{\ell\ell'}$ terms (see ref. [30]) which violate the unitarity limit of the low partial wave contributions, but are actually damped by absorption of these waves. Evaluating the Jacobi polynomial at $t = m_\pi^2$ does not leave a cos ω dependence, and absorbs the lowest partial wave completely, so that the absorbed one pion exchange amplitude is for small momentum transfers:

$$A_{0\lambda_p;\lambda\lambda_n} \simeq \left(\frac{-t}{s}\right)^{\frac{|\bar{\lambda}-\mu|}{2}} \cdot \frac{P(s,m_\pi^2,\bar{\lambda},\mu)}{m_\pi^2 - t} \cdot F(t) , \qquad \text{(IV.1.9)}$$

the approximation $[\cos \frac{\omega}{2}]^{\frac{|\bar{\lambda}+\mu|}{2}} \sim 1$, and $[\sin \frac{\omega}{2}]^{\frac{|\bar{\lambda}-\mu|}{2}} \sim (-\frac{t}{s})^{\frac{|\bar{\lambda}-\mu|}{2}}$ having been used. F(t) is an extrapolation function parametrised to exp $[B(t - m_\pi^2)]$, so that $F(m_\pi^2) = 1$. This amplitude is compatible with the amplitude, obtained by Kane [32], who assumed that nucleon spin-flip amplitudes dominate the s-channel high energy reaction [31]. For the reaction $\pi^- p \to \rho^0 n$, the s-channel angular distribution of the $\pi\pi$ system is in terms of the spin density matrix elements of the S- and P-wave contributions to the ρ-region:

$$W(\theta,\phi) = \frac{1}{4\pi} [(\rho_S^H + 3\rho_{00}^H) + 2\sqrt{3}\, \rho_{S0}^H \cos\theta - 3(\rho_{00}^H - \rho_{11}^H) \cdot$$

$$\cdot \sin^2\theta + \phi \text{dep.terms}] . \qquad \text{(IV.1.10)}$$

In Fig. 15, it can be seen that the s-channel spin density matrix elements as calculated with the helicity dependent absorbed one pion exchange amplitudes of the Williams model describe the experimental data of ρ^o-production at 17.2 GeV/c [33] very well. In the t-channel helicity zero part of the absorbed one pion exchange amplitude the phase shifts can be extracted from the Williams extrapolation to the pion pole of $f_\ell^{\pi,\pi\ off}$ in (IV.1.7), and in Fig. 15, the results of the $\pi\pi$ I = O S-wave phase shifts, of the slope B and the cut contribution C are given as a function of $M_{\pi\pi}$, use having been made of a simple t-channel pole - cut parametrisation [34] according to the Williams model. The $\pi\pi$ I = O S-wave phase shifts obtained by this method, agree quite well with Estabrooks and Martin's solution 1 [35] of these phase shifts (Fig. 16) which is characterized by P_o and P_{-1} being approximately in phase, whereas solution 2 would be characteristic by P_o and S being approximately in phase. Fig. 16 also shows the P-wave $\pi^-\pi^+$ phase shifts [35], together with Protopopescu et al.'s best solution [36], which rise steeply through 90 degrees at the ρ mass. In Fig. 17, the results of the parametrisation $P_o = \pi + C \sin\chi$, $P_{+1} = A_2 + C$, and $P_{-1} = C \cos\chi$ of ref. [37] in which the cut contribution C has been obtained from crossing the s-channel not helicity non-flip amplitude $M_{01;+-}$, expected to be the most important cut contribution at high energies [38], into the t-channel are shown. π and A_2 are the π- and A_2-exchange amplitudes in the t-channel of the reaction $\pi^- p \to \rho^o n$, and parametrized in Regge formalism with Regge phases. The analysis of the energy dependence of ρ^o production [37] yields effective trajectories for the three $\pi^-\pi^+$ P-wave contributions: $\alpha_{P_o}^{eff} \sim \alpha_\pi(t)$, $\alpha_{P_{-1}}^{eff} \sim \alpha_\pi(t)$, so that the cut-trajectory ressembles the pion trajectory, whereas $\alpha_{P_{+1}}^{eff}$ first rises with -t to about the values of the A_2

trajectory (because the cut is dominating at small -t) and then falls approximately along the A_2 trajectory, the exchange amplitude of which is proportional to -t.

The I = 2 S- and D-wave $\pi\pi$ phase shifts can be determined by analysing the reactions $\pi^+ p \to \pi^+\pi^+ n$ and $\pi^- p \to \pi^-\pi^-\Delta^{++}$. By Fig. 18, it is shown that the same procedure of analysis of the experimental angular distributions of the $\pi^+\pi^+$ system at 12.5 GeV/c [39] leads, by a parametrisation of the reaction $\pi^+ p \to \pi^+\pi^+ n$ in terms of π, A_2 and cut contributions to the S-, D_0 and $D_{\pm 1}$ waves of the $\pi^+\pi^+$ system and by extrapolation to the pion pole, to the exhibited experimental phase shifts δ_S^2 and δ_D^2. Both are, in the mass region $0.3 \leq M_{\pi^+\pi^+} \leq 1.5$ GeV, negative (in agreement with current algebra predictions [40], and $-5^\circ \leq \delta_D^2 \leq 0^\circ$ and $-30^\circ \leq \delta_S^2 \leq 0^\circ$. No resonance structure is apparent. For comparison, the results of the phase shift analysis of the $\pi^-\pi^-$ system in the reaction $\pi^- p \to \pi^-\pi^-\Delta^{++}(1236)$ at 3.93 GeV/c [41] are shown in Fig. 19. δ_0^2 has the same variation as in the $\pi^+\pi^+$ system, δ_2^2 and δ_4^2 are only slightly deviating from zero to negative values, and the inelasticity η_0^2 is therefore one nearly up to $M_{\pi^-\pi^-} = 1.0$ GeV. As can be seen by the solid curve in Fig. 20, the current algebra prediction [40] describes the mass dependence of δ_0^2 very well. The cross section of $\pi^-\pi^-$ interaction is, according to ref. [41], smoothly increasing from threshold to approximately a constant (Fig. 21), and the solid curve of the current algebra prediction of ref. [40] interpolates smoothly the experimental data of most of the experiments.

2. $K^-\pi^-$ Scattering

The Athens-Demokritos-Liverpool-Vienna Collaboration analysed in their K^-p experiment at 8.25 GeV/c, the reaction $K^-p \to K^-\pi^-\Delta^{++}$ [45] in order to determine the angular distribution and the mass dependence of the $K^-\pi^-$ cross-section. The steps of the analysis and the experimental results are given in Fig. 25. It was concluded from the spin density matrix elements of Δ^{++} that the $K^-\pi^-$ system is dominantly produced by pion exchange. The $K^-\pi^-$ interaction was parametrized by a real Veneziano term, taking into account ρ-exchange in the $K\bar{K}$-t-channel and K^* exchange in the $K\pi$-u-channel:

$$T_{K^-\pi^-} = -g \frac{\Gamma(1-\alpha_\rho(t_{KK}))\ \Gamma(1-\alpha_{K*}(u_{K\pi}))}{\Gamma(1-\alpha_\rho(t_{KK}) - \alpha_{K*}(u_{K\pi}))} , \qquad \text{(IV.2.1)}$$

and, in addition, by a diffractive amplitude having a phase γ with respect to the Veneziano amplitude:

$$T^{diffr}_{K^-\pi^-} = c\, e^{i\gamma} \cdot 2k \sqrt{s_{K\pi}}\ e^{at_{KK}} . \qquad \text{(IV.2.2)}$$

$a = 2.5\ (\text{GeV/c})^{-2}$, and $\gamma = 0.6$ rad, k...kaon momentum in the $(K^-\pi^-)$ centre-of-mass. In Fig. 25, the mass dependence of the elastic cross-section of $K^-\pi^-$ interaction is given with the least squares fit to the data (dotted line), the description with the Veneziano amplitude alone (dashed curve), and with the Veneziano plus the diffraction component (solid curve) the last of which is supposed to give the right description of the mass dependence. The total $K^-\pi^-$ cross-section is accessible through the inclusive reaction $K^-p \to \Delta^{++}$ + anything, and, again from the determination of the spin density matrix elements of the Δ^{++}, one obtains, by means of the Chew-Low extrapolation from the off-shell to the on-shell pion, the total cross section. The falling mass dependence of the elasticity σ_{el}/σ_{tot} of the $K^-\pi^-$ interaction is the expression of a continuously rising $K^-\pi^-$ total cross-section in the mass region $0.6 < M(K^-\pi^-) < 1.8$ GeV. Also shown in Fig. 25 are the moments of the $K^-\pi^-$ angular distribution $\langle Y_J^0 \rangle$, determined in the Gottfried-Jackson frame. The parametrisation of the $K^-\pi^-$ interaction in terms of diffraction scattering and a Veneziano amplitude reproduces reasonably the increase of all the moments with increasing $(K^-\pi^-)$ mass. Only, there seems to be apparent a threshold enhancement in $\langle Y_1^0 \rangle$, i.e. in the S- and P-wave interference term, but no such structure is observed in the P-wave contribution $\langle Y_2^0 \rangle$. Therefore, higher partial wave contributions show a weaker rise with mass, and all moments are structureless.

This determination of the elastic $K^-\pi^-$ cross-section coincides, within errors, with the one obtained in an analysis of the reaction $K^-p \to K^-\pi^-\Delta^{++}(1236)$ at 14.3 GeV/c [46].

3. SU_3-Relation between Pion Exchange Dominated Quasi-Two Body Reactions

The joint production of a vector meson and $\Delta(1236)$ in the reactions

$$K^+p \to K^{*0}(890)\ \Delta^{++}$$
$$K^-n \to K^{*0}(890)\ \Delta^-$$

$$\pi^+ p \rightarrow \rho^o \Delta^{++}$$
$$\pi^+ p \rightarrow \omega\Delta^{++}$$

can be mediated, at high energies, by the exchange of G = + 1 and G = - 1 octets. In terms of the two exchange amplitudes A_+ and A_-, the four reaction amplitudes squared can be written, use being made of SU(3) Clebsch-Gordan coefficients (ideal ω-ϕ mixing):

$$\begin{aligned} |A_{K^{*o}\Delta^{++}}|^2 &= |A_+ + A_-|^2 \\ |A_{\overline{K^{*o}}\Delta^{-}}|^2 &= |A_+ - A_-|^2 \\ |A_{\rho^o\Delta^{++}}|^2 &= 2|A_-|^2 \\ |A_{\omega\Delta^{++}}|^2 &= 2|A_+|^2 \;. \end{aligned} \qquad \text{(IV.3.1)}$$

This leads to the SU(3) relation:

$$\rho^J_{ik} \frac{d\sigma}{dt}(K^{*o}\Delta^{++}) + \rho^J_{ik} \frac{d\sigma}{dt}(\overline{K^{*o}}\Delta^-) =$$

$$= \rho^J_{ik} \frac{d\sigma}{dt}(\rho^o\Delta^{++}) + \rho^J_{ik} \frac{d\sigma}{dt}(\omega\Delta^{++}) \;. \qquad \text{(IV.3.2)}$$

Since $\rho^J_{oo} \frac{d\sigma}{dt}$ is negligible for G = + 1 amplitudes in the forward direction (natural parity exchange only contributes to helicity one states of the vector meson), one has in this kinematical region:

$$\frac{1}{2}\rho^J_{oo} \frac{d\sigma}{dt}(\rho^o\Delta^{++}) = \rho^J_{oo} \frac{d\sigma}{dt}(K^{*o}\Delta^{++}) = \rho^J_{oo} \frac{d\sigma}{dt}(\overline{K^{*o}}\Delta^-) \,. \qquad \text{(IV.3.3)}$$

According to Fig. 22, the approximate relation (IV.3.3) is indeed obeyed in the region $0 < -t < 0.2$ $(GeV/c)^2$ for the kaon and pion reactions at 5.5 GeV/c [44]. The differential cross section of the reaction $\pi^+ p \rightarrow \omega\Delta^{++}$ (1236) that only can occur via G = + 1 exchange, gradually rises from values which are two orders of magnitude smaller than those of the two other reactions in the very forward direction, to the same order of magnitude at $t \sim -0.18$ $(GeV/c)^2$. This proofs the validity of the approximate relation (IV.3.3) in the forward direction. In Fig. 23, the comparison of the two reactions $K^+ p \rightarrow K^{*o}\Delta^{++}$ and $K^- n \rightarrow K^{*o}\Delta^-$ fulfills the forward direction relation (IV.3.3), and the experimental values of the differential cross-sections of the $K^- n$ reaction at 5.5 GeV/c are successfully predicted at larger momentum transfers by the SU(3) relation (IV.3.2) [44].

An analogous SU(3) relation can be written for the reactions

$$K^+ n \rightarrow K^{*o} p$$

$$K^- p \rightarrow \overline{K^{*o}} n$$

$$\pi^- p \rightarrow \rho^o n, \pi^+ n \rightarrow \rho^o p$$

$$\pi^- p \rightarrow \omega n, \ \pi^+ n \rightarrow \omega p \ .$$

It is also demonstrated in Fig. 23 that the differential cross-sections of the reaction $\pi^+ n \rightarrow \omega p$ [44] are satisfactorily predicted from those of the three other reactions. In addition to the good results one obtains from the SU(3) relation by comparing differential cross-sections, Fig. 24 proves that also the prediction of the ω and $\overline{K^{*o}}$ spin density matrix elements in their t-dependence are fairly well described by the SU(3) relation for the reactions $\pi^+ n \rightarrow \omega p$ at 4.2 and 5.1 GeV/c and $K^- n \rightarrow \overline{K^{*o}} \Delta^-$ at 5.5 GeV/c. Therefore, SU(3) relations of the type of equ. (IV.3.2) are a good means for separating natural and unnatural parity contributions in "pion exchange" dominated reactions in which the $\pi\pi$ or the $K\pi$ systems are formed.

V. ANALYSIS OF THE THREE MESON SYSTEM IN THE REACTIONS $\pi p \rightarrow (3\pi)p$ AND $Kp \rightarrow (K2\pi)p$ AND OF THE $(\pi\omega)$ AND $(K\omega)$ SYSTEMS IN $\pi(K)p \rightarrow \pi(K)\omega p$

1. Partial Wave Analysis of the (3π)-System

The experimental analysis of four-body final states is especially interesting because the production cross-section of reactions such as $\pi^\pm p \rightarrow \pi^\pm p\pi^+\pi^-$, $K^\pm p \rightarrow K^\pm p\pi^+\pi^-$, $pp \rightarrow pp\pi^+\pi^-$, and particle + nucleus $\rightarrow$ particle + nucleus + + $(\pi^+\pi^-)$ is only rather weakly decreasing (suggesting Pomeron exchange, i.e. diffractive production and dissociation and, in addition, the exchange of meson Regge trajectories), and because the involved three particle systems show resonant or resonance like structures. For instance, in the reaction $\pi^- d \rightarrow \pi^-\pi^+\pi^- d$ at 15 GeV/c [47], the effective mass distribution of the three pion system shows, as is displayed in Fig. 26, resonance like enhancements A_1, A_2, A_3 and A_4. In the $K^\pm\pi^+\pi^-$ system, a threshold enhancement Q at $M(K\pi\pi) = 1.3$ GeV, and a second enhancement, the L-meson, at $M(K\pi\pi) = 1.78$ GeV are significant, whereas the $p\pi^+\pi^-$ system in many of the four-body final state reactions is characterised only by weak enhancements N^* (1470) and N^* (1700), the production cross-sections of which are fairly energy independent. It is the aim of the partial wave analysis of these three particle systems to investigate the resonance or non-resonance structure of these enhancements, to determine all the partial wave contributions to the production of a fixed information on the quantum numbers of the exchange mechanism. And the determination of the energy dependence of these partial wave contributions should be the ultimate goal in order to clarify the nature of the exchange mechanism, but this has not yet been achieved in the scope of bubble chamber experiments, though it will certainly become feasible very soon.

The low mass diffractive production of a three particle system can be treated, in the helicity formalism, as a quasi-two body process $ab \rightarrow cd$ where the particle system c has, in general, many spin-parity contributions. It is exactly the same situation as that which was encountered in the treatment of the two pion production in the process $\pi^- p \rightarrow \pi^- \pi^+ n$. The spin density matrix of the particle system c is, if ρ_{ab} is the density matrix of the initial state:

$$\rho^{(c)}_{mm'} = \sum_{\lambda_a, \lambda_b, \lambda_d} f_{m\lambda_d;\lambda_a\lambda_b}\ \rho_{ab}\ f^*_{m'\lambda_d;\lambda_a\lambda_b} \quad . \qquad \text{(V. 1.1)}$$

Applying the reflection operator $Y = e^{-i\pi J_y}\cdot P$ on the x-z-plane (reaction plane) to the expression (V.1.1), then it is not altered if ρ_{ab} corresponds to an unpolarised initial state, so that $[\rho^{(c)}, Y] = 0$. The application of Y to the basis vectors $|J^P M\rangle$ gives: $Y|J^P M\rangle = P(-1)^{J-M}|J^P -M\rangle$. Linear superpositions give eigenstates of Y with the eigenvalues $-\eta$, with respect to which $\rho^{(c)}$ is diagonal:

$$|J^P M\eta> = c_M[\,|J^P M> + \eta P\cdot(-1)^{J+M+1}|J^P -M>] \quad . \qquad (V.1.2)$$

P is the intrinsic parity of the three meson system, J its spin and M the spin projection on the z-axis (Gottfried-Jackson or helicity), and $c_M = \frac{1}{2}$ for M = 0 and $\frac{1}{\sqrt{2}}$ for $M \neq 0$. For high energies and/or small momentum transfers, when the crossing matrix is close to diagonal, η is closely related to the naturality of the exchange. With respect to the basis $|J^P M>$, the density matrix of the three meson system c is

$$\rho^{J_1^{P_1}\ J_2^{P_2}}_{M_1\eta_1\ M_2\eta_2} = 2\ c_{M_1}\ c_{M_2}(\rho^{J_1^{P_1}\ J_2^{P_2}}_{M_1\ \ M_2} + \eta_1(-1)^{J_2+M_2+1}\ \cdot$$

$$\cdot\ P_1\cdot\rho^{J_1^{P_1}\ J_2^{P_2}}_{M_1\ \ -M_2})\ \delta_{\eta_1\ \eta_2}\ , \qquad (V.1.3)$$

that contains the whole information of the production process.

The production process is accessible by the analysis of the observed decay angular distributions and the Dalitz plot distribution. The decay of the $(\pi_1^-\pi^+\pi_2^-)$ system with spin J and projection M into π_1^- and the two particle system $(\pi^+\pi_2^-)$ of spin S and helicity λ in the three pion centre-of-mass is:

$$A^{JS}_{M\lambda}(\theta_1\phi_1) = \sqrt{\frac{2J+1}{4\pi}}\ D^{*J}_{M\lambda}(\phi_1\theta_1 0)\ T^J_{S\lambda}\quad , \qquad (V.1.4)$$

and the subsequent decay of the two pion system $(\pi_2^-\pi^+)$ of spin S and helicity λ into π_2^- and π^+ in the centre-of-mass of the $(\pi_2^-\pi^+)$ system:

$$B_\lambda^S(\theta_2\phi_2) = \sqrt{\frac{2S+1}{4\pi}}\, D_{\lambda o}^{*S}(\phi_2\theta_2 0)\, T_o^S \quad . \tag{V.1.5}$$

The partial wave decomposition of the three pion decay with respect to the orientation of the three pion plane in a defined coordinate system is:

$$M_M^{J^P}(s_{\pi_1^-\pi^+}, s_{\pi_2^-\pi^+}, \phi\theta\gamma) = \sum_\nu H_\nu^{J^P}(s_{\pi_1^-\pi^+}, s_{\pi_2^-\pi^+}) \cdot$$

$$\cdot \sqrt{\frac{2J+1}{8\pi}}\, D_M^{*J}(\phi\theta\gamma) \quad . \tag{V.1.6}$$

ϕ,θ,γ are the Euler angles of the orientation of the three pion plane, ν is the projection of J on the analyser direction, and $H_\nu^{J^P}$ are the Dalitz plot distributions depending on the spin-parity of the system. The partial wave amplitudes $T_{S\lambda}^{J^P}$ and $T_o^{S^{P_S}}$ are defined for individual spin-parities J^P and S^{P_S} in equs. (V.1.4) - (V.1.6). Comparison of the three particle decay amplitude (V.1.6.) with the two step decay of equs. (V.1.4) and (V.1.5) allows to express the Dalitz plot distribution $H_\nu^{J^P}$ by a sum (over λ for fixed S) of products of d-functions of appropriate angles and T-amplitudes. Those are parametrised by a Breit-Wigner expression in the (ℓ,S) basis:

$$T^{J^P}_{S\ell} T^{S}_{oo} = BW\,(s_{\pi_2^-\pi^+})\, p^{\ell} q^{S}\, C^{J^P}_{S\ell} \quad . \qquad (V.1.7)$$

p and q are the pion momenta in the appropriate centre-of-mass, and the $C^{J^P}_{S\ell}$ are free parameters. I = 2 interactions are neglected because of the smallness and negativity of the I = 2 phase-shifts, as was discussed above. Because of the indistinguishability of π_1^- and π_2^-, symmetrisation is introduced by adding the Dalitz plot expression corresponding to the exchange of π_1^- and π_2^-. All these steps combine to the decay amplitude $M^{J^P}_{M\eta}$ with respect to the basis $|J^P M\eta>$:

$$M^{J^P}_{M\eta} = c_M\{M^{J^P}_{M} + \eta P.(-1)^{J+M+1}\; M^{J^P}_{-M}\} \; , \qquad (V.1.8)$$

so that the cross-section expression of the overall process is:

$$W(s,t,M_{3\pi}, s_{\pi_1^-\pi^+}, s_{\pi_2^-\pi^+}) =$$

$$= \sum_{\substack{J^P{'}s \\ M's,\eta's}} M^{J_1^{P_1}}_{M_1\eta_1}\, \rho^{J_1^{P_1}\; J_2^{P_2}}_{M_1\eta_1\; M_2\eta_2}\, M^{J_2^{P_2}}_{M_2\eta_2} \quad . \qquad (V.1.9)$$

This is the famous Ascoli-analysis (see refs. [48, 49], by which the investigation of the experimental data is carried out in a maximum liklihood procedure in an extremely complicated program.

Partial Wave Analysis of the Three Pion System in the Reaction $\pi^-p \to \pi^-\pi^+\pi^-p$ at 25 and 40 GeV/c.

These experiments of the CERN-IHEP Collaboration (see refs.[49,50]) could only analyse the momentum transfer range $0.04 \leq -t \leq 0.3$ $(GeV/c)^2$. The following quasi-two particle decays were assumed: $\varepsilon\pi$, $\rho\pi$, $f\pi$, and $\varepsilon'\pi$ (where ε' is a daughter of the f).

Another alternative is to use $\pi\pi$ phase shifts instead of saturating the $(\pi^+\pi^-)$ system by resonances. In Fig. 2, the partial wave contributions in the notation $J^P\ell M^\eta$ are shown for the 40 GeV/c reaction. It should be noted that the results for 25 GeV/c are of the same physics contents. The 1^+SO^+ $(\rho\pi)$ wave dominates the cross section in the mass region $1.0 \leq M(3\pi) < 1.5$ GeV, but does not peak at the A_1 mass. The $2^+ D1^+$ $(\rho\pi)$ wave contribution shows a beautiful A_2 resonance. The other wave contributions that have been taken into account turned out small and flat in their mass distribution, so that the assumption is permitted that their phase is not varying significantly.

One of the most intriguing questions in meson spectroscopy has been whether A_1 is a resonance or not, and the answer could be attempted in the Ascoli analysis by determining the relative phase of the $1^+ SO^+$ wave contribution and those partial wave contributions which do not have appreciable variation in mass. As is displayed in Fig. 28, all these experimentally determined phases, as well at 40 GeV/c as at 25 GeV/c, have a smooth behaviour with respect to $M(3\pi)$, and do not go steeply through 90 degrees. One has, therefore, to conclude that the A_1 is not a resonance, but of the nature of a threshold enhancement of the ($\rho\pi$) system, probably according to the Deck production mechanism. The same analysis has been made for the ($f\pi$) system in the A_3 region [49,50], and the same conclusion has to be adopted for the A_3 as for the A_1.

All the partial wave analysis results displayed in Fig. 27 have been obtained with the choice of the Gottfried-Jackson system as reference coordinate system for spin quantization. Exact helicity conservation in the t-channel would require only occurrence of $M = 0$ partial wave contributions, but the small contributions of $M \neq 0$ states violate t-channel helicity conservation by a small amount, as is shown in Fig. 29 for two mass intervals of the three pion system. The average rotation angle for the 1^+ wave from the Gottfried-Jackson system towards the helicity system into a frame in which helicity conservation holds exactly is 5-10 degrees. Therefore, helicity conservation of the production of the 1^+ wave contributions is only approximate.

2. Decomposition of the ($K\pi\pi$) Structure into Spin-Parity Amplitudes

The Ascoli analysis of the $(K\pi\pi)^-$ system has been performed by the Aachen-Berlin-CERN-London-Vienna Collaboration [51] at 10 and 16 GeV/c. No symmetrisation procedure is required as to the indistinguishability of particles. The $(K^-\pi^+)$ system is assumed to be saturated by κ (or S-wave phase shifts), $K^*(890)$ (or P-wave phase shifts), $K^*(1400)$, and the $(\pi^+\pi^-)$ system is either expressed in terms of phase shifts or by resonance saturation (ε,ρ, f,ε') [52]. The $K^-\pi^-$ interaction is of the same structurelessness as the $\pi^-\pi^-$ interaction in the $(3\pi)^-$ system. The results are shown in Fig. 30 with the partial wave notation J^PM^η. The dominant contribution to the Q enhancement is furnished by $1^+\ 0^+$ whereas the peaking of the $0^-\ 0^+$ (mainly due to the "κ" π decay) is displaced to higher masses than the peaking of $1^+\ 0^+$. It is surprising that the L enhancement does not become apparent significantly in <u>one</u> partial wave contribution, but that many partial wave contributions act together to its formation in the $(K\pi\pi)^-$ mass distribution. That the L enhancement is not only a threshold effect of the J^P = 2^- S-wave $(K^*(1400)\pi)$ and/or the (Kf) systems, and not only a $J^P = 1^+$ S-wave $(K\rho)$ and $(K^*(890)\pi)$ state, can be seen in Fig. 31.

The $J^P = 1^+$ S-wave contribution can be decomposed into the $K^*(890)\pi$ and $K\rho$ components, and the experimental analysis yields the ratio of the $(K\rho)$ to the $(K^*(890)\pi)$ amplitudes to decrease from 1.0 at threshold to zero at $M(K\pi\pi)$ = 1.5 GeV, and also the relative phase of these two amplitudes to decrease from 60° to 0° in this mass region (Fig. 32).

In terms of the quantum number η, natural and unnatural parity exchange contributions can be separated out from the experimental data, and diffractive production of the $(K\pi\pi)^-$ system requires natural parity exchange and unnatural parity dominance in the $(K\pi\pi)^-$ system. The Pomeranchukon has naturality +L, and the unnatural parity dominance in the diffractively produced system is inferred by Morrison's selection rule which states that only spin-parity components in a particle system can be produced diffractively in a meson nucleon interaction, when they obey the relation

$$\Delta P = (-1)^{\Delta J} \qquad \text{(V.2.1)}$$

(ΔP is the relative parity of the outgoing partial wave component to that of the incident particle of same baryonic number, and ΔJ = J - Spin of incident particle). Fig. 33 proves that the $(K\pi\pi)^-$ system is predominantly in the unnatural parity state whereas the exchange mechanism is predominantly of natural parity. The next step should now be the determination of the energy dependence of the various separated partial wave contributions, produced by a definite naturality exchange, and decaying into definite decay channels. This would furnish a full description of the particles or cuts that are exchanged with respect to the formation of the $(K\pi\pi)^-$ system in a definite partial wave state. This would correspond to the analysis of the $(\pi\pi)$ system in terms of exchanges and partial wave contributions and their energy dependence, as it has been carried out by Estabrooks, Martin, and Michael [37].

But, due to the limited statistics in bubble chamber experiments, the data of the two analysed energies had to be lumped together in order to obtain reasonably small errors.

3. Structures in the $(\pi\omega)$ - and $(K\omega)$ - Systems

The $(\pi\omega)$ system cannot be produced diffractively because of the change of G-parity with respect to the incident pion. But the analysis of this system in the reaction $\pi^- p \to \pi^- \omega p$ at 4, 5, and 7.5 GeV/c, carried out by Chaloupka [53] with the program of Ascoli's analysis, unambiguously demonstrates that the B-meson is a 1^+ spin-parity state.

The $(\pi_1^-\pi^+\pi_2^-\pi^o)$ system is, in its first decay step, considered to decompose into π_1^- + $(\pi^+\pi_2^-\pi^o)$ in the four particle centre-of-mass. This is completely analogous to the three particle case. Only, the decay of the $(\pi^+\pi_2^-\pi^o)$ has to be treated with respect to the orientation of the three pion plane in a coordinate system whose z-axis is the direction of motion of the (3π) system in the (4π) rest frame, and whose y-axis has been obtained by rotating the production plane normal by the z-axis in the (4π) rest frame until the normal is perpendicular to the (3π) direction of motion, and can serve as y-axis in the (3π) rest frame.

The partial wave analysis has been performed in the regions $0 < t' < 0.2$ $(GeV/c)^2$ and $0.2 \leq t' < 0.6$ $(GeV/c)^2$ (Fig. 34), and the result is the occurrence of the B-meson enhancement at M (ωπ) = 1.24 GeV in the partial wave components $1^+ SO^+$, $1^+ S1^+$, and $1^+ S1^-$. The $1^+ SO^+$ resonance signal has considerably decreased in the higher t-region, whereas the $1^+ S1^+$ component shows an equally strong peaking in both t-intervals. The 1^+S1^- component is weak. B can be produced by absorbed π exchange, which is concluded by 1^+S1^- to be weak, and by ω and A_2 exchange which can produce the B-meson in both helicity states. The 1^+SO^+ component is more concentrated at small t than the 1^+S1^+.

The (Kω) system in the reaction $K^-p \rightarrow K^-\omega p$, however, can be produced diffractively, and was analysed, by the Athens-Demokritos-Liverpool-Vienna Collaboration, at 8.25 GeV/c [54]. The (Kω) mass distributions from the $K^-p \rightarrow K^-\omega p$ reactions at 8.25, 10.1, and 16 GeV/c are shown in Fig. 35. They are structureless, and only infer the existence of a broad threshold enhancement. The energy dependence of the (Kω) mass distribution can be parametrized according to Morrison [55] as

$$\frac{d\sigma}{dM} = c(M) \left[\frac{p_{Lab}}{1\ \mathrm{GeV}/c}\right]^{-n(M)} \quad . \qquad (V.3.1)$$

The exponent $n(M_{K\omega})$ and the slope $B(M_{K\omega})$ of the $\frac{d\sigma}{dt}$ distribution are shown in Fig. 36: the slope decreases with increasing $(K\omega)$ mass in the fashion that is already known from the decrease of the slope of $(K\pi\pi)^-$ or (πN) production, and $\frac{d\sigma}{dM}$ is only constant in energy near threshold. At higher masses, the diffractive component becomes less and less important. This again permits the conclusion that diffractive production of particle systems is only physically occurring at their lowest effective masses. A moments analysis of the $(K\omega)$ system in the low statistics $K^-p \rightarrow K^-\omega p$ reaction only permitted the inference of the existence of $J^P = 0^-$, 1^+, and 1^-.

VI. DIFFRACTION DISSOCIATION

1. Analysis of the Reaction $KN \rightarrow QN$ in Terms of Reggeized Helicity Amplitudes and the Cross-Over Effect

From the reaction $K_L^op \rightarrow K_S^o\pi^+\pi^-p$ which was analysed in the momentum range 4-14 GeV/c [56], the following two reactions can be extracted: $K^op \rightarrow Q^op$, $Q^o \rightarrow K^{*+}_{890}\pi^-$, and $\bar{K}^op \rightarrow \bar{Q}^op$, $\bar{Q}^o \rightarrow K^{*-}_{890}\pi^+$. The $\frac{d\sigma}{dt}$-distributions for both reactions showed a significant cross-over which could not be explained by a simple factorizable Deck model. The Deck graphs for both reactions are:

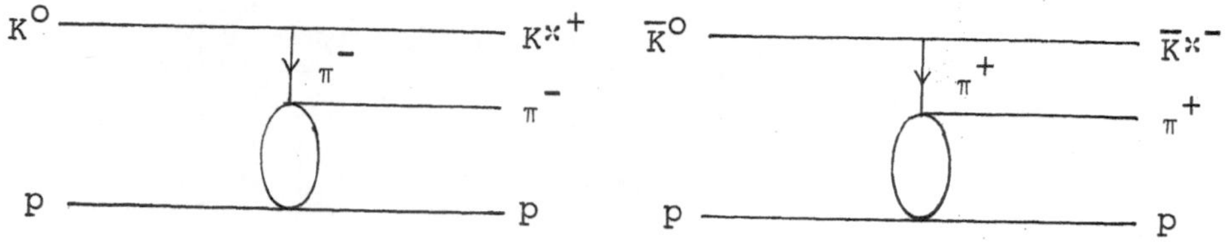

If pion exchange in the Deck graphs is factorizable, then the cross-over of $d\sigma/dt$ should be exactly the cross-over of $\pi^{\pm}p$ elastic scattering. Since π^-p elastic scattering dominates π^+p at $-t < 0.1$ (GeV/c)^2, Q^o production should dominate $\bar{Q}^o$ production at small momentum transfers. But, the effect is the other way round. K^*_{890} being emitted with smaller momentum than the pion is only a very small contribution to the reaction and could not account for the effect. On the other hand, if the Q^o and $\bar{Q}^o$ enhancements were excited states of the K^o and $\bar{K}^o$, respectively, and the $d\sigma/dt$ distributions the same as those for elastic scattering, then $\bar{Q}^o$ production being produced by a non-exotic s-channel reaction must be larger than Q^o production at small momentum transfers.

In the analysis of the cross-over effect of the $Q^{\pm}$ production $d\sigma/dt$ distributions, the result is very much dependent on how cleanly the non-diffractive reactions $K^+p \to K^{*o}_{890}\Delta^{++}$ (1236) and $K^-p \to K^-\pi^-\Delta^{++}$ (1236) can be separated off. This can be done by the following mass restrictions: $1.2 \leq M(K\pi\pi) \leq 1.5$ GeV, $0.84 \leq M(K^{\pm}\pi^{\mp}) \leq 0.94$ GeV, and $M(p\pi^{\pm}) \geq 1.4$ GeV; or it can be achieved by demanding that, in the $(K\pi\pi)$ rest frame, the π^+ be emitted into the forward hemisphere with respect to the incident kaon, by which procedure exactly half of the angular distribution is accepted [57]. It is displayed in Fig. 37 that the cross-over effect for $Q^{\pm}$-production is indeed very weak if it is present at all. Moreover, the data of the $d\sigma/dt$ distributions show a slight depression in the forward direction. The copious contribution of pion exchange reactions like $K^+p \rightarrow K^{*o}_{890}\Delta^{++}$ and $K^-p \rightarrow K^-\pi^-\Delta^{++}$ is reflected in the energy dependence of the cross-section of the reactions $K^{\pm}p \rightarrow K^{\pm}\pi^+\pi^-p$ (Fig. 38). Also, the vector meson exchange components of $Q^{\pm}$ production contribute to the falling of the channel cross-sections with energy.

It is therefore tempting to determine these meson exchange contributions to $Q^{\pm}$ production by analysing the $\frac{d\sigma}{dt}$ distributions, the spin density matrix elements of the Q, and the energy dependence of the differential cross-sections in terms of reggeized helicity amplitudes (in the s-channel) [58]. This is the other side of approach for analysing Q production than Ascoli's partial wave analysis: the dominance of $J^P = 1^+$ in the Q region is assumed, and the contribution of the various helicity amplitudes are fitted to the mentioned pieces of information. By the experimental reason that no significant Q enhancement is apparent in the reaction $K^-p \rightarrow \bar{K}^o\pi^+\pi^-n$, $I = 1$ exchanges are neglected, and only the exchange of the Pomeranchukon and the exchange degenerate ω and f trajectories is assumed. $M_{+,1+}(s,t)$ is parametrized in the usual Regge fashion in terms of ω-f and is real by exchange degeneracy. For Pomeranschukon exchange, <u>s-channel helicity conservation</u> is assumed so that it only contributes to $M_{+,o+}$:

$$M^P_{+,o+} = \beta\, e^{-i\frac{\pi}{2}\alpha_P(t)} \cdot e^{\gamma t'}\, s^{\alpha_P(t)} \quad . \qquad \text{(VI.1.1)}$$

The imaginary parts of the ω and f components of $M_{+,o+}$

are in terms of the dual absorptive model of Harari et al. [59]:

$$\mathrm{Im}\, M^{f}_{+,0+} = -\mathrm{Im}\, M^{\omega}_{+,0+} = \beta_1 J_0(R\sqrt{-t'})\, s^{\alpha_\omega(t)}\, e^{\gamma_1 t'} \quad . \qquad \text{(VI.1.2)}$$

The real parts of the ω and f components of $M_{+,0+}$ are parametrized in the form:

$$\mathrm{Re}\, M^{f,\omega}_{+,0+} = \pm\beta_1\, \mathrm{Re}\, F^{T,V}_{++}(s_0,t)\left(\frac{s}{s_0}\right)^{\alpha_\omega(t)} , \qquad \text{(VI.1.3)}$$

where the structure functions $F^{T,V}_{++}$ have been calculated from the Barger-Phillips model [60] of a dispersion relation analysis. A further assumption was that ω and f do not couple to the nucleon spin flip in the s-channel, and the helicity amplitude $M_{+,0+}$ was viewed as having the structure of the elastic s-channel non-flip amplitude. From the absence of unnatural parity exchange contributions, the simple equality $\rho_{11} = -\rho_{1,-1}$ must hold.

This parametrisation gives quite good a description of the experimental $\rho_{ik}\frac{d\sigma}{dt}$ - distributions for Q^+ production in the energy range 7 - 9 GeV/c (Fig.39) and for Q^- production at 8.25 GeV/c (Fig. 40).

$\rho_{11} = -\rho_{1,-1}$ is reasonably obeyed by the experimental data, which proves the absence of unnatural parity exchanges in Q-production, since, moreover, these exchanges would mainly have isospin one (π, A_1). The conclusion is that the ω and f exchange components effectuate the approximate t-channel helicity conservation of Q production, but the prediction can be made that s-channel helicity conservation of Q production may become important at much higher energies.

2. Energy Dependence of $\pi^- p$ Elastic Scattering and of the Reaction $\pi^- p \to \pi^- \pi^- \pi^+ p$

In Fig. 38, the energy dependence of the cross-section of the reactions $K^{\pm}p \to K^{\pm}\pi^+\pi^- p$, in which diffractive production plays an important role, was given for laboratory momenta up to 16 GeV/c. The energy dependence of the cross-section of the reaction $\pi^- p \to \pi^- \pi^- \pi^+ p$ has a falling tendency up to laboratory momenta of 2o5 GeV/c [61] (Fig. 41) which most probably has to be attributed to the diffractive component, because the reaction energy is already so high that exchange components of lower lying trajectories must be already negligible. The elastic $\pi^- p$ cross section, to the contrary has already reached constancy, and could be argued to rise at higher energies. These qualitative features permit the conclusion that the Pomeranschukon elastic scattering is of different nature from the Pomeranschukon in diffractive production of particle systems. This argument will lead to explicitely considering three terms on the right hand side of the unitarity relation: elastic scattering, diffractive production and non-diffractive production of multiparticle systems, i.e. one will have to consider a three-component Pomeranschukon ([62], [63], and many others).

It has been noted that the fall-off with energy of the differential cross-section becomes much more pronounced with increasing mass of the diffractively produced particle system. Therefore, at very high energies, the threshold enhancements only should survive, and Fig. 42 proves that, in the case of the reaction $\pi^- p \rightarrow \pi^- \pi^- \pi^+ p$ at 2o5 GeV/c, only the threshold enhancements $N^*(1470)$ of the $(p\pi^+\pi^-)$ and A_1 of the $(\pi^-\pi^-\pi^+)$ systems have become neatly separated out. Therefore, it is worth to study the diffractive low mass production of particle systems at as high energies as possible in order to deal with the pure diffraction components.

3. Forward-Backward Ordering of Particle Momenta in Multi-Particle Final States and Energy Dependence

The four-body final state reactions could reasonably well be analysed by Ascoli's partial wave analysis program or in terms of Reggeized helicity amplitudes, because the number of competing quasi-two and three body processes was rather low and a more or less clean separation of non-diffractive from diffractive reactions could thus be obtained. Turning to the analysis of reaction mechanisms in five-body final state interactions

$$K^- p \rightarrow p K^- \pi^+ \pi^- \pi^0 ,$$

$$K^- p \rightarrow n K^- \pi^+ \pi^+ \pi^- ,$$

$$K^- p \rightarrow p \bar{K}^0 \pi^+ \pi^- \pi^- ,$$

one best approaches this analysis from measuring the energy dependence of reaction cross-sections with respect to the quantum number transfer from the incident kaon to the particle system produced into the forward hemisphere with respect to the incident kaon in the total centre-of-mass. Charge, strangeness, and baryon number transfer are considered, and the charge transfer is defined as:

$$Q_t = (Q_{K^-_{inc}} - \sum_{f=1}^{n} Q_f) \;, \qquad \text{(VI.3.1)}$$

where n is the number of particles emitted per event into the forward hemisphere. This already allows a separation into the following reaction mechanisms:

$Q_t = S_t = B_t = 0$: diffractive production plus neutral non-strange meson exchange

$Q_t = \pm 1$, $S_t = B_t = 0$: charged non-strange meson exchange

$S_t = \pm 1$, $B_t = 0$: strange meson exchange

$S_t = \pm 1$, $B_t = -1$: hyperon exchange, etc.

Even cut-contributions of the type $Q_t = \pm 2$ etc. can be estimated by this method.

In order to determine the energy dependence of cross-sections corresponding to these particle momenta orderings, one has to be aware that only the amplitudes have the Regge-like power behaviour in p_{lab} (see Morrison's parametrisation: $c(p_{lab}/1\ \text{GeV/c})^{-n}$), and that cross-sections have to be corrected for phase space since not all particle orderings into forward and backward emission are equally probable.

In ref. [64] (Aachen-Berlin-CERN-London-Vienna

Collaboration) the following experimental analysis has been reported: by Monte Carlo generation of events according to cylindrical phase space ($\Pi e^{-4p_{Ti}^2}$ cut-off in transverse momenta), the phase space contribution ϕ to each ordering has been determined, and the average amplitude in each ordering sector is parametrized in its energy dependence:

$$\frac{\sigma \text{ ordering}}{\phi \text{ ordering}} \propto p_{Lab}^{-n_{\text{ordering}}} \qquad \text{(VI.3.2)}$$

(ϕ ... cylindrical phase space contribution). In Fig. 43, the ordering is expressed by a sector number which is defined by

$$\sum_{i=1}^{5} \alpha_i \ 2^{i-1}$$

where i numbers the outgoing particles in the series ($\pi_1 \ \pi_2 \ \pi_3$ KN) and $\alpha_i = 1$ if $p^*_{Li} > 0$ (otherwise 0). It can be seen that, for all three reactions considered, the possibly diffractively produced ordering of forward emission of the (K3π) system (Sector number 15) is more andmore favoured at higher energies. Double diffraction dissociation does not appear to be significantly realised. The energy dependence analysis in the various ordering sectors displays, in Fig. 44, the hierarchy of strength of the cross-section decreasing with energy depending on what quantum number transfer sector has been considered. Clearly, the quantum number transfer which allows diffractive production is characterized by the weakest fall-off of its cross-section. On the basis of this analysis which allows approximate insight into the reaction mechanism of five-body final state interactions, more subtle methods of analysis like partial wave analyses, cluster investigation in exclusive reaction channels, etc. can be envisaged to investigate the diffractive production component.

4. Separation of Diffraction Dissociation from Non-Diffractive Processes in Multi-Particle Final States

It has been shown up to now in the experimental analyses of the four-body and five-body final state reactions and of the reaction $K^-p \rightarrow K^-\omega p$ that diffractive production of multiparticle systems occurs predominantly in the threshold regions of these systems and falls off rather strongly with increasing particle system mass. But in the region of strong decrease lies the threshold of the particle system of the next higher multiplicity which can be produced diffractively. The cross-section rises to its maximum in the neighbourhood of the threshold of this diffractively produced particle system, and falls off in a mass region in which the diffraction channel of the next higher multiplicity opens. This observation led Pokorski and Van Hove [65] to the notion of competing diffractive exclusive channels which is illustrated in Fig. 45. The shaded band corresponds to the assumption that the mass distribution for diffractive ($\pi^+\pi^-p$) production (as analysed by partial wave decomposition, longitudinal phase space ordering, etc.) is also approximately the same for the overall ($N\pi\pi$) diffraction production (with inclusion of the $p\pi^o\pi^o$ and $\pi^+\pi^on$ systems). By this method, it is possible to decompose the inclusive reaction $\pi^- p \to \pi^- X^+$ into the various competing diffractive exclusive channels.

The process of diffractive production can be viewed as a two-step process which consists firstly of the formation of a meson or baryon excitation which can be a fireball, and secondly of the decay of this excited system into the competing exclusive channels, assumedly governed by (cylindrical) phase space.

$$\frac{d\sigma_{n_X}}{dM_X} = \rho(s,t,M_X)\ B_{n_X}(M_X) \ , \qquad \text{(VI.4.1)}$$

where ρ is the mass spectrum of the excited system (fireball) and B_{n_X} are the branching ratios into the competing exclusive decay channels [65]. Factorisation is expressed by $\rho(s,t,M_X) = \rho(t,M_X)$ only, and ρ is given by the sum

$$\frac{d\sigma}{dM_X} = \sum_{n_X} \frac{d\sigma_{n_X}}{dM_{n_X}} = \rho \qquad \text{(VI.4.2)}$$

for masses up to 1.75 GeV, above which the non-diffractive channels of higher multiplicity than 3 become important (from an energy dependence analysis). In Fig. 46, the experimental separation of exclusive diffractively produced particle systems together with the prediction of model (VI.4.1.) are given. Line (2) corresponds to the fall-off with increasing mass of the inclusive diffraction dissociation cross-section after 1.75 GeV with respect to the experimental mass distribution (1). The probability of decay into n_X particles is [65]:

$$P_{n_X} = c(M)h_{n_X}\int \delta^{(4)}(Q-\sum_1^{n_X} k^{(i)})\prod_1^{n_X} e^{-\beta_i k_T^{(i)2}} \frac{d^3k^{(i)}}{2E^{(i)}} . \quad (VI.4.3)$$

c(M) effectuates the normalization of $\sum P$ to one. The couplings h_{n_X} are either determined from the data, or, in an independent emission model, are [65]:

$$h_{n_X} = \frac{\lambda^{n_X}}{(n_X-1)!} . \quad (VI.4.4)$$

Fig. 47 shows the cylindrical phase-space decay probabilities as a function of mass for the lowest multiplicity decays. Indeed, when the probability of one decay of a fixed multiplicity has passed its maximum, the channel of the next multiplicity starts its steep rise to its maximum.

5. Impact Parameter Profile of Elastic Scattering and Diffraction Dissociation

In the unitarity relation which relates the imaginary part of the elastic scattering amplitude (pp, for instance) to the modulus squared of the elastic scattering amplitude plus the overlap function of all inelastic channels, this second contribution is split up into the diffractive and non-diffractive overlap-function in impact parameter space:

$$\mathrm{Im}\, A_{el}(b) = |A_{el}(b)|^2 + G_{Diffr}(b) + G_{Non\text{-}diffr}(b). \tag{VI.5.1}$$

The impact parameter profile of elastic scattering is known from the analyses of Yang, Chou [5], Durand, and Lipes [6] to be given by the opacity function $\chi(b) \propto \mu^2(\mu b)^3 K_3(\mu b)$. For the analysis of the diffractive overlap function, the $\frac{d^2\sigma}{dtdM^2}$ data of the reaction $pp \to pX^+$ at $s = 930\ \mathrm{GeV}^2$ [66] were used by Sakai and White [63] in the following parametrisation:

$$\frac{s}{\pi} \frac{d^2\sigma}{dtdM^2} = c(M) \left[\left(\frac{1}{1 - \frac{t}{\mu^2}} \right)^2 \cdot \left(1 - \frac{t}{Y(M)} \right)^{-2} \right]^2 . \tag{VI.5.2}$$

$c(M)$ and $Y(M)$ were determined by the fit to the experimental data, by which it was found that $Y(M)$ increases with mass like $\frac{\mu^2 M}{m_p}$.

Assuming s-channel helicity conservation for diffractive production ($\lambda_1 = \lambda_3$ and $\lambda_2 = \lambda_4$ in the reaction $1 + 2 \to 3 + 4$), and that the t-channel is dominated by natural parity exchange, one has to deal only with the one helicity amplitude $f^{(s)}_{1/2\ 1/2;1/2\ 1/2}$, and its profile function $A(b)$ is the same as that for elastic scattering, namely the Fourier transform of $f^{(s)}_{1/2\ 1/2;\ 1/2\ 1/2}$:

$$A(b) \propto \int_0^\infty d(-t)\ f(s,t)\ J_0(b\sqrt{-t}), \tag{VI.5.3}$$

where, by the approximate assumption that the helicity amplitudes are imaginary, $|f|^2$ is either taken from the $d\sigma/dt$ distribution of elastic scattering, or from $\frac{s}{\pi}$.

$\frac{d^2\sigma}{dtdM^2}$ of the inclusive small t diffractive production of the particle system X.

$$G_{Diffr}(b) = 2\int dM^2 |A_{Diffr}(b,M)|^2, \qquad (VI.5.4)$$

where the factor 2 takes into account dissociation at each p vertex. Inclusion of the diffractive phases is obtained by writing

$$f(t) = i\sqrt{\frac{d^2\sigma}{dtdM^2}}\; e^{-\frac{1}{2}i\pi\alpha' t},$$

where α' is the slope of the Pomeranschukon trajectory. The $\alpha'.\ln s$ term yields the shrinkage of $\frac{d\sigma}{dt}$ distributions. The results for the average impact parameters of the three components are: $\langle b^2\rangle_{el} = 0.67$ fm, $\langle b^2\rangle_{Diffr} = 0.5$ fm, and $\langle b^2\rangle_{Non\text{-}diffr} = 1$ fm, as obtained by subtraction of $|A_{el}|^2$ and $G_{Diffr}(b)$ from the left hand side of the unitarity relation (VI.5.1). This means that the non-diffractive overlap function is much more peripheral than the diffractive one for high energy proton-proton interactions.

Applying t-channel helicity conservation, it is assumed that only natural parity exchanges contribute, and that the proton which is not diffractively dissociated conserves the helicity so that only one t-channel helicity amplitude $f^{(t)}_{\bar\lambda_1\lambda_3;\lambda_2\bar\lambda_4}$ $(\bar\lambda_1 = \lambda_3,\ \lambda_2 = \bar\lambda_4)$ has to be considered by parity conservation. Crossing into the s-channel gives the helicity amplitude contributions $f^{(s)}_{\lambda 1/2;1/2\ 1/2}$, and with $n = |\lambda - \frac{1}{2}|$ one has

$$A_n(b) \propto \int_0^\infty d(-t) f^{(s)}_{\lambda 1/2;1/2\ 1/2}(s,t)\, J_n(b\sqrt{-t}). \qquad (VI.5.5)$$

Using the model dependent results of Ajduk for the helicity amplitudes $f^{(s)}_{\lambda 1/2;\ 1/2\ 1/2}$ from an analysis of the helicity dependent elastic scattering diffraction slope by means of the unitarity relation [67], Sakai and White [63] obtain a much better profile distribution of the elastic, diffractive, and non-diffractive components in impact parameter space: $\langle b^2\rangle_{\text{Diffr.}} = 0.97$ fm and $\langle b^2\rangle_{\text{non-diffr.}} = 0.98$ fm. These values also yield the peripherality of elastic scattering. The contributions of the various s-channel helicity amplitudes with $n \leq 3$ are given in Fig. 48 for the reaction $pp \to pX^+$ at $s = 930\ \text{GeV}^2$ in the case of t-channel helicity conservation. The zeros of the functions $J_n\ (b\sqrt{-t})$ give appropriate zeros in the respective helicity amplitude contribution. The impact parameter profiles for the three components of the overlap function are also given in Fig. 48 for t-channel helicity conservation.

6. Experimental Results of the Inclusive Diffractive Processes $pp \to p$ + anything and $\pi^- p \to p$ + anything

In the phenomenological analysis of impact parameter profiles which was discussed in the last chapter by following Sakai and White [63], use was made of the parametrisation (VI.5.2) of the diffractive component of the inclusive reaction $pp \to p$ + anything. The diffractive component of the inclusive reactions $a + p \to p$ + anything can be visualized by occurring via Pomeranchukon exchange according to the following diagram:

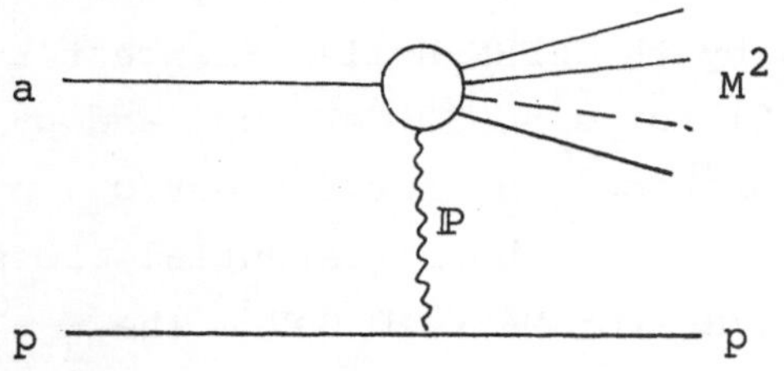

where M^2 is the effective mass squared of the undetected particle system in the fragmentation region of the projectile particle a. The Feynman variable $x = 2p_L(cm)\sqrt{s}$ and M^2 are related by $M^2 \simeq s(1-x_p)$ for $|p_L| >> |p_T|$ and $M >> m_p$, so that the differential cross-section $\frac{d^2\sigma}{dxdt}$ and $\frac{d^2\sigma^p}{dtd(M^2/s)}$ are equivalent.

According to the arguments of Pokorski and Van Hove [65] that, in exclusive channels, diffraction dissociation into low masses is dominant as has been discussed in chapter VI.4, the diffractive component has its kinematical region near $x_p \simeq 1$ for very high energies. And there is, indeed, observed a characteristic spike in the region $0.96 \leq |x_p| \leq 1.0$ for the inclusive processes $pp \to pX^+$ and $\pi^- p \to pX^-$ at $s = 551\ GeV^2$ and $380\ GeV^2$ [68,69], respectively, as is shown in Fig. 49. Both reactions exhibit the same shape of the x_p-distribution in the spike region, suggesting the same structure in the mass distributions of pion and proton fragmentation. Furthermore, it has been proved by the CERN-Holland-Lancaster-Manchester ISR Collaboration [70] that the x_p-distributions of the invariant cross-sections $Ed^3\sigma/dp^3$ are energy independent in the range 22 - 45 GeV for $pp \to pX^+$. The M^2 - distribution of the processes $pp \to pX^+$ and $\pi^- p \to pX^-$ show the same shape in the region $M^2 < 20\ GeV^2$ as has been found by S. J. Barish et al. [72] for 200 GeV/c pp, by the Columbia-Stony Brook Collaboration [73] for $s = 563\ \mathrm{GeV}^2$ pp, by the CERN-Holland-Lancaster-Manchester Collaboration [68] for $s = 930\ \mathrm{GeV}^2$ pp, and by the Berkeley-NAL Collaboration [69] for 205 GeV/c $\pi^- p$ interactions. In this M^2-region, the differential cross-section $d\sigma/dM^2$ falls as $1/M^2$ for $M^2 < 20\ \mathrm{GeV}^2$. The $\frac{d^2\sigma}{dt\,d(M^2/s)}$ distribution can be expressed, in its momentum transfer dependence, by $\exp(b(M^2)t)$, as it was studied by the NAL experiments at 102, 205, 303 and 405 GeV/c pp (see Refs. [18,73], and by the ISR experiments of the CERN-Holland-Lancaster-Manchester [70] and Aachen-CERN-Genova-Harward-Torino [74] Collaborations. The slope parameter b varies, as a function of M^2, from $9.1 \pm 0.7\ (\mathrm{GeV/c})^{-2}$ for $M^2 < 5\ \mathrm{GeV}^2$ to $5.8 \pm 0.6\ (\mathrm{GeV/c})^{-2}$ for the mass region $50 \leq M^2 < 100\ \mathrm{GeV}^2$ in the case of $pp \to pX^+$ at 205 GeV/c [72]. The experimental situation is therefore the following: in the region $0.96 < X_p < 1$, the invariant cross-section is fairly energy independent whereas, outside this region, there seems to be an energy dependent contribution; the mass dependence of the invariant cross-section is $1/M^2$ for the fall-off of the diffractively produced peak, and the momentum transfer dependence is exponential with decreasing slopes as M^2 is increased.

In the triple Regge limit (M^2 large and s/M^2 large) the invariant cross-section is:

$$s\,\frac{d^2\sigma}{dt\,dM^2} = \sum_{\alpha_1\alpha_2\alpha_3}$$

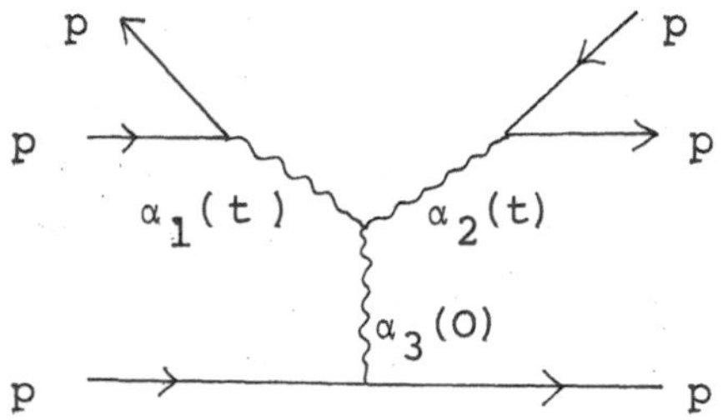

$=$

$$= \sum_{\alpha_1\alpha_2\alpha_3} \frac{\beta_{\alpha_1\alpha_2\alpha_3}(t)}{s} \left(\frac{s}{M^2}\right)^{\alpha_1(t)+\alpha_2(t)} (M^2)^{\alpha_3(0)} . \qquad \text{(VI.6.1)}$$

$\beta_{\alpha_1\alpha_2\alpha_3}$ includes the p-p-α and the three Reggeon α_1-α_2-α_3 couplings which depend on the momentum transfer t, and s/M^2 is the asymptotic expression of cos θ_t [75,76] for the "quasi-two body" reaction pp $\rightarrow$ p + anything (M^2). The contribution of $\alpha_1 = \alpha_2 = \alpha_3$ = Pomeranchukon for small momentum transfers leads to scaling because $\frac{s}{M^2} = (1-x_p)^{-1}$ so that the invariant cross-section is independent of the energy, and is only a function of x_p and t in the triple Regge limit. In the case of $\alpha_1 = \alpha_2 = 0.5 + 0.9t \equiv \alpha_R$ and α_3 = Pomeranschukon, the invariant cross-section is proportional to $(s/M^2)^{2\alpha_R-1} = (1-x_p)^{1-2\alpha_R} \rightarrow$ const for $t \rightarrow 0\,(\mathrm{GeV/c})^2$, so that one has also scaling in the triple Regge limit. The contribution of $\alpha_1 = \alpha_2$ = Pomeranschukon, but $\alpha_3 = 0.5 + 0.9t \equiv \alpha_R$, is, in its energy dependence, proportional to $(s/M^2)^{2\alpha_P - 1} \cdot M^{2\alpha_R - 1} = \frac{1}{\sqrt{s}}(1-x_p)^{\frac{1}{2}-2\alpha_P} \rightarrow$ $\rightarrow \frac{1}{\sqrt{s}}(1-x_p)^{-\frac{3}{2}}$ for $t \rightarrow 0\ (\mathrm{GeV/c})^2$. The most obvious parametrisation of the invariant cross-section is, for experimental purposes, in terms of a scaling and a non-scaling Regge exchange dominated contribution to particle production of mass M [77]:

$$E\frac{d^3\sigma}{dp^3} = \frac{s}{\pi}\frac{d^2\sigma}{dM^2dt} = A(x_p)e^{b(x_p)t}\left[1+\frac{B(x_p)}{\sqrt{s}}\right]. \qquad \text{(VI.6.2)}$$

The special energy dependence of $s^{-\frac{1}{2}}$ is based on the dominance of vector and tensor meson trajectory exchange in the partial process $\alpha_1(t) + p \rightarrow \alpha_2(t) + p$ of (VI.6.1), but could be different in other inclusive reactions, in which, for instance, ϕ exchange could contribute. $B(x_p)$ should be a rising function in x_p like $(1 - x_p)^{-3/2}$ for large M^2 and $t \rightarrow 0$. An analysis of the energy dependence of the data of the reaction $pp \rightarrow p$ + anything from 50 to 400 GeV by the Rutgers-Imperial College Collaboration [77] yielded $B(0.83) = 1.9 \pm 0.7$ GeV and $B(0.91) = 4.3 \pm 0.4$ GeV, the ratio of which is well explained, within errors, by the triple Regge contribution which is proportional to $s^{-1/2}$. In the parametrisation (VI.6.2), the slope $b(x_p)$ turned out [77] to be, independently of x_p, 6 $(GeV/c)^{-2}$ because the flattening off of the slope in the parametrisation $\exp(b(M^2)t)$ which has been reported by refs. [28,72,73] is expressed by the rising of $B(x_p)$. The parametrisation of Sakai and White [63] which is given in (VI.5.2) is equivalent to the parametrisation $\exp(b(M^2)t)$, taking into account the M-dependence of the slope by the function Y(M).

In order to test factorization of Pomeranschukon exchange with respect to the inclusive production of particle systems in the proton fragmentation region, there is need of measuring the inclusive process $\pi^- p \rightarrow \pi^-$ + anything which has not yet been reported for NAL energies. Factorization to be good is, however, reported by Leith [18] from the Pomeranschukon exchange dominated exclusive channels $pp \rightarrow pp\pi^+\pi^-$ and $\pi^- p \rightarrow \pi^-\pi^+\pi^- p$ at 2o5 GeV/c, analysed by Winkelmann of the Berkeley-NAL Collaboration. Thus, an older analysis by Ljung and Yamdagni [78] is confirmed for the NAL energy region.

VII. MULTIPLICITY DISTRIBUTIONS

1. Multiplicity Distributions of the Diffractive Produced Multi-Particle System

Multi-particle production has been discussed in the previous chapters either for exclusive reactions or with respect to the dynamics of diffraction dissociation. Here, the question for the multiplicity distribution in multi-particle systems which are predominantly diffractively produced, is posed to experiment, and the average multiplicity and the higher multiplicity moments in their dependence on the kinematical variables of the multi-particle production reaction allow global insights into the reaction mechanism. The maximal, kinematically possible multiplicity is required to rise proportional to M (where M is the mass of the unspecified, for instance, diffractively produced particle system). But, from Fig. 50, it is evident that, for the inclusive reaction $p + p \rightarrow p + X(M)$ at 102 [79], 205 [72], and 303 GeV/c [73], the average multiplicity of charged particles in the multiparticle system X increases logarithmically with its mass. Barish et al. [72] prove in addition, that the average multiplicity as a function of M^2 from Fig. 50 coincides with the average multiplicity of charged particles as obtained from the reaction $p + p \rightarrow$ n charged particles of reaction energy M. The average multiplicity is therefore mainly dependent on the available energy, and not on the type of incident particles, because in the first case the incident particles were a Pomeranchukon and a proton, and in the second case two protons. A comparison of the average multiplicity of charged particles from the reactions $p + p \rightarrow p + X(M)$ and $\pi^- p + X(M)$ at 205 GeV/c [72] again exhibits, as is shown in Fig. 51, an analogous behaviour with respect to its variation with M. The solid curves in Fig. 51 are fits to the experimental data of the expressions $\langle n_{charged} \rangle = a + b.\ln Q^2$ and $a + b.\ln M^2$ for the π^- and p reaction, respectively, and the broken line shows the impossibility of the average charged multiplicity being proportional to M. ($Q = \sqrt{s} - m_\pi - m_p$ for the reaction $\pi^- p \to n$ charged particles, and $Q = \sqrt{M^2} - m_\pi$ for the diffractively produced system in $\pi^- p \to pX^-$ at the same energy). Again, the average charged multiplicity as a function of Q^2 is independent of the incident objects. As was seen in the good description of the invariant cross-section of inclusive reactions by the generalized optical theorem in the triple Regge limit (VI.6.1), s/M^2 has to be large, and the average multiplicity of diffractively produced particle systems must be considerably lower than that of the overall inclusive reaction. This conclusion is proved by Fig. 52, that shows the multiplicity distributions for $\pi^- p \to pX^-$ (diffractively) and $\pi^- p \to n$ charged particles at 205 GeV/c [72]. For the first experimental situation $\langle n_{ch} \rangle = 3.8 \pm 0.2$, whereas $\langle n_{ch} \rangle = 8.02 \pm 0.12$ is obtained for the overall unspecified reaction. This result is also in qualitative agreement with Pokorski and Van Hove's conjecture [65] of only low mass diffractive excitation being of importance in the series of multiparticle exclusive reaction channels.

The average multiplicity of charged particles

$$\langle n_{ch} \rangle = \sum_{n=2}^{N(M)} n \frac{\sigma_n}{\sigma_{inel}} \qquad \text{(VII.1.1}$$

is the experimental first approximation to how the cross-sections σ_n of the various prong events of a reaction are related amongst one another. In (VII.1.1), N(M) is the maximal possible charged multiplicity. Hwa [80] and Wróblewski [81] suggested that, if $\sigma_n = \sigma_o/n^2$, one obtains, by virtue of

$$\lim_{N(M) \to \infty} \sum \frac{1}{n} \sim \ln M \quad ,$$

the logarithmic increase of the average charged multiplicity for asymptotic reaction energies. But it was displayed in Fig. 52, that the relation $\sigma_n = \sigma_o/n^2$ seems to be valid only for $n > \langle n \rangle$, where the average multiplicity $\langle n \rangle$ in turn tends to infinity logarithmically. Therefore, the logarithmic increase with M of $\langle n \rangle$ is, in this argumentation, especially guaranteed by the very small cross-sections of prongs of much larger mutliplicity than the average. On the other hand, the experimental result of the M^2-dependence of $d\sigma/dM^2$ was $1/M^2$ [69] for masses higher than the diffraction maximum: this behaviour is also conjectured, in the triple Regge limit, by the triple Pomeranchukon term $\frac{s}{M^2}$ in the generalized optical model expression (VI.6.1), and also by the Pomeranchukon-Pomeranchukon-Reggeon contribution, proportional to $(s/M^2)^{2\alpha_P - 1} \cdot M^{2\alpha_R - 1}$, which has a $1/M^2$ dependence for $t \to 0$. If this behaviour is valid for the entire mass range, then

$$\sigma_{inel} \propto \int_{M_o^2}^{M_{max}^2} \frac{dM^2}{M^2} \propto \ln s \qquad \text{(VII.1.2)}$$

for asymptotic energies, and the average charged multiplicity

$$\langle n_{ch} \rangle \sim \frac{1}{\sigma_{inel}} \int_{M_o^2}^{M_{max}^2} \frac{n(M)}{M^2} \, dM^2 \propto \ln s \ , \qquad \text{(VII.1.3)}$$

if $n(M) \propto \ln M^2$. This approximative argumentation takes into consideration asymptotically rising inelastic cross-sections and the same energy and mass variation of the average multiplicity. The $1/M^2$ behaviour of $d\sigma/dM^2$ and the $1/n^2$ behaviour are therefore related for high masses and high multiplicities.

The logarithmic increase of the mean particle multiplicity is, for instance, the result of the "bare-bone" multiperipheral model of Chew and Pignotti [82,83], in which the n-particle production amplitude is

$$A_n = \prod_{i=1}^{n} g^{n+1} \; s_{i,i+1}^{\alpha} \ , \qquad \text{(VII.1.4)}$$

where g is the same coupling constant for all vertex couplings, α the same average trajectory exchanged between all vertices, and $s_{i,i+1} = (p_i + p_{i+1})^2$. Phase space integration in the approximation of negligible transverse momenta leads to a Poisson multiplicity distribution, to the condition $2\alpha - 1 + g^2 = 1$ due to asymptotic constancy of the total cross-section (which has been disproved by experiment at ISR energies), and to $\langle n \rangle = g^2 . \ln s$ for the average multiplicity at very high energies.

In the statistical approach to inclusive distributions by Sivers and Thomas [84] who use for the reaction p + p → N + N + nπ the following n pions production amplitude squared:

$$|A_{n+2}|^2 = s^\delta \frac{g^n}{n!} W(n_+,n_-,n_o) \cdot$$

$$\cdot \prod_{i=1}^{2} e^{2\lambda \cdot ps^{-\frac{1}{2}} - R^2 p_{iT}^2} \prod_{j=1}^{n} e^{-R^2 q_{jT}^2} \qquad \text{(VII.1.5)}$$

(λ_μ... four-vector that has only a time component in the centre-of-mass frame, δ... phenomenological parameter for energy dependence, g... gives probability for pion emission, W...isospin weighting factor for emission of n_+ positive pions, etc.), that takes into account peripheral production of the two nucleons N, again the logarithmic increase of the average multiplicity with energy is obtained. Due to the peripheral production of "leading particles", the increase in energy available for particle production does not seem, up to the present energies, to be proportional to a power of s, but the logarithmic increase of the average charged multiplicity seems to favour models with a dominant component of independent emission of particles or particle clusters. These particle clusters [85] are viewed, from the analysis of two particle correlations, to be composed of the order of four particles, and the decay of these clusters can, as the simplest approximation and in good agremment with experiment [86], assumed to be isotropic for small cluster masses. The multiperipheral production mechanism is in this model in terms of clusters, whose average number rather than whose mass is predicted to increase logarithmically with increasing reaction energy.

2. Dispersion of the Muliplicity Distribution

In the last chapter, it has been argued in terms of the two particle correlation function with respect to the rapidities y_1 and y_2 $(y = \ln \frac{E + p_L}{\sqrt{m^2+p_T^2}})$:

$$f(y_1,y_2) = \frac{1}{\sigma_{inel}} \frac{d^2\sigma}{dy_1 dy_2} - (\frac{1}{\sigma_{inel}} \frac{d\sigma}{dy_1})(\frac{1}{\sigma_{inel}} \frac{d\sigma}{dy_2}), \quad \text{(VII.2.1)}$$

the integration of which yields

$$f_2 = \langle n(n-1)\rangle - \langle n\rangle^2, \quad \text{(VII.2.2)}$$

that models assuming completely independent particle emission have to be excluded because of experimentally verified positive correlations. Sivers and Thomas' [84] calculation of the multiplicity moments for negative pions production

$$f_k = g^k_{(-)} \frac{\partial^k}{\partial g^k_{(-)}} \ln \Omega, \quad \text{(VII.2.3)}$$

where

$$\Omega(s) = \sum_{n=0}^{\infty} \int \prod_{i=1}^{2} \frac{d^3p_i}{2E_i} \prod_{j=1}^{n} \frac{d^3g_j}{2\omega_j} \cdot$$

$$\delta^4(P - \sum_{i=1}^{2} p_i - \sum_{j=1}^{n} g_j)|A_{n+2}|^2, \quad \text{(VII.2.4)}$$

yield

$$f_k = A_k \ln s + B_k + O(\frac{1}{\ln s}) \qquad \text{(VII.2.5)}$$

with

$$A_k = (-1)^{k+1} \frac{\pi}{2R^2} \sqrt{g_{(+)} g_{(-)}} \frac{\Gamma(k-\frac{1}{2})}{\Gamma(\frac{1}{2})} . \qquad \text{(VII.2.6)}$$

$g_{(+)}$, $g_{(-)}$, and $g_{(o)}$ are used to approximate the isospin weight $W(n_+, n_-, n_o)$ by independent coupling parameters for each charge. The result of the ratio f_2/f_1 is:

$$\frac{f_2}{f_1} = \frac{\langle n_-(n_--1)\rangle - \langle n_-\rangle^2}{\langle n_-\rangle} \underset{s\to\infty}{\to} -\frac{1}{2} , \qquad \text{(VII.2.7)}$$

so that the constraints by energy momentum and charge conservation to independent pion production imply negative two-particle correlations with energy dependence $f_2 \sim -\frac{1}{2} \cdot \langle n_-\rangle \sim -0.7 \ln s$. This is the behaviour which has been observed for proton-antiproton annihilation and for virtual antiproton-proton annihilation in the reaction $K^- p \to \Lambda$ (forward) + pions [87].

Positive two particle correlations follow from the multiperipheral cluster production model [88], in which, assuming multiperipheral cluster production with a cluster multiplicity obeying approximately a Poisson distribution, and assuming the average number of particles coming from one cluster to be $\langle m\rangle$:

$$f_2 = \langle m(m-1)\rangle \cdot g^2 . \ln s . \qquad \text{(VII.2.8)}$$

In expression (VII.2.8), g^2 represents, in analogy to the multiperipheral "bare-bone" model, the coupling constant of clusters to vertices, and f_2 has the ln s - factor since the average cluster multiplicity rises logarithmically with energy in the multiperipheral cluster production model. f_2 has to be positive.

Using, as in the previous arguments on the energy dependence of the average multiplicity, the approximation $\sigma_n \propto \frac{1}{n^2}$ (for large n), one finds from

$$\langle n^2 \rangle = \frac{1}{\sigma_{inel}} \sum_{n=2}^{N(M)} n^2 \sigma_n : \; f_2 \propto \sqrt{s}$$

at asymptotic energies, at which the $(\ln s)^2$ - (from $\langle n \rangle^2$) and the ln s - behaviour (from $\langle n \rangle$) are negligeable. Such a behaviour is the result of a low mass excitation model. A thorough treatment of the various classes of models with respect to the energy dependence of the average multiplicity and two particle correlation function is given by N. Schmitz [89].

The <u>dispersion</u> of the average multiplicity is given by

$$D = \sqrt{\langle n^2 \rangle - \langle n \rangle^2} = \sqrt{f_2 + \langle n \rangle} \; . \qquad \text{(VII.2.9)}$$

In the multiperipheral cluster production model, $D^2 \propto \ln s$ by (VII.1.4) and (VII.2.8), whereas, in the low mass excitation model, $D^2 \propto \sqrt{s}$ asymptotically by the $\sqrt{s}$-behaviour of f_2.

One can therefore distinguish between the two production mechanisms by the energy dependence of the multiplicity moments. The multiperipheral cluster production model and the low mass excitation model can also be combined to a two component picture of particle production:

$$\sigma_{inel} = \sum_n \sigma_n = \sigma_{exc} + \sigma_{multiper} , \qquad \text{(VII.2.10)}$$

so that

$$\langle n\rangle = a\langle n\rangle_{exc} + b\langle n\rangle_{multiper} \qquad \text{(VII.2.11)}$$

$$(a = \sum_n \sigma_n^{exc}/\sigma_{inel},\ b = \sum_n \sigma_n^{multiper}/\sigma_{inel}) \quad \text{and}$$

$$f_2 = a\, f_2^{(exc)} + b\, f_2^{(multiper)} + ab[\langle n\rangle_{exc} - \langle n\rangle_{multiper}]^2$$

$$\text{(VII.2.12)}$$

$$D^2 = aD^2_{exc} + bD^2_{multiper} + ab[\langle n\rangle_{exc} - \langle n\rangle_{multiper}]^2 .$$

For asymptotic energies, the last term in (VII.2.12) would yield a $(\log s)^2$ - behaviour for D^2, if the low mass excitation component is conjectured to be the already discussed inclusive diffractive production mechanism. Together with the logarithmic increase with s of the average multiplicity, $\langle n\rangle/D$ should approach a constant value at asymptotic energies; for the experimentally obtained values of 20% diffractive production and 80% multiperipheral cluster production $\langle n\rangle/D \to 2$ from above, since $D_{multiper} \ll \langle n\rangle_{multiper}$, $D_{exc} \ll \langle n\rangle_{exc}$, and $\langle n\rangle_{exc} \ll \langle n\rangle_{multiper}$ at asymptotic energies. The picture of the two component model is supported by the experimental results, a part of which is represented in Fig. 53 for all particle-proton collisions except $\bar{p}p$ [90]. It can be seen that the ratio $\langle n \rangle / D$ indeed approaches the value 2 from above, and, in the case of pp collisions, is exactly this value for $p_{lab} \gtrsim 30$ GeV/c. In Fig. 54, a compilation of the laboratory momentum dependence of the ratio $\langle n \rangle / D$ is again shown for the various types of collision with the inclusion of $\bar{p}p$ data. Whereas it is proven that the $\langle n \rangle / D$ ratio drops to the value of two for pp- and $\pi^- p$-interactions, a rising tendancy of this ratio in the momentum region $p_{lab} < 10$ GeV/c is discovered in the case of $\bar{p}p$-collisions. It was argued for antiproton-proton annihilation whose contribution is relevant at low energies, and whose multiplicity moments behaviour is mainly governed by the energy-momentum and charge conservation laws [84], that f_2 is negative, as is also the case for the other types of collisions at medium energies (Fig. 55). But the $\bar{p}p$ annihilation component contributes a considerably larger average multiplicity at low energies than non-annihilation processes, so that also $D^2 = f_2 + \langle n \rangle$ becomes enhanced. $\bar{p}p$-collisions should require a three-component interpretation: diffractive excitation, multiperipheral cluster production, and annihilation. Using $f_2 = -\frac{1}{2}\langle n \rangle$ from (VII.2.7), $D_2 = \frac{1}{2}\langle n \rangle$, in which the leading particle behaviour of the final state nucleons is included. In the type of models leading to (VII.2.7), $\langle n \rangle / D$ is suppressed to the order of magnitude of two. For pp-collisions, it has been shown by Malhotra [92] and Wróblewski [93] that (see Fig. 57) the ratio $(\langle n_{ch} \rangle - 1)/D$ is a constant (~ 1.75) with respect to $\langle n_{ch} \rangle$, leading to $\langle n_{ch} \rangle / D \to 1.75$ for $\langle n_{ch} \rangle \gg 1$.

3. Moments of Multiplicity Distributions

The central moments of the multiplicity distribution are defined as

$$\mu_k = <(n - <n>)^k> \; , \qquad \text{(VII.3.1)}$$

and $D = \sqrt{\mu_2}$. All these moments can be determined experimentally from the multiplicity distribution, and a compilation by Buras et al. [94], and by Fialkowski and Miettinen [95] of the second to the fourth moment of the multiplicity distribution for pp-collisions in the energy range 15 - 300 GeV/c is shown in Fig. 56. The $(\mu_k)^{1/k}$ are linear in $<n_{ch}>$, and in analogy to the case k = 2 that has been treated by Wróblewski [97], they have been fitted by Buras et al. [94] with the linear ansatz:

$$(\mu_k)^{1/k} = a_k \, (<n_{ch}> - \alpha) \; , \qquad \text{(VII.3.2)}$$

where all a_k are positive. The correlation parameters f_k should therefore depend on the average charged multiplicity $<n_{ch}>$ like a polynomial of order k in $<n_{ch}>$ so that, in contradiction to a pure multiperipheral cluster production mechanism, but in favour of a two component picture with diffraction excitation, f_2 should be proportional to $(\ln s)^2$ thus allowing $<n>/D$ to become constant. The left figure of Fig. 57 expresses the result of the fit by means of the ansatz (VII.3.2) for D, yielding $a_2 = 0.57$ and $\alpha \sim 1$. There is, therefore, a significant difference in the $<n_{ch}>$ dependence of D for pure multiperipheral cluster production or short-range ordering for which $D \propto \sqrt{\langle n_{ch} \rangle}$, and the Wróblewski result $D \propto (\langle n_{ch} \rangle - 1)$ which corresponds to a two-component picture.

4. Koba-Nielsen-Olesen Scaling

Up to now, only the mutual relation of $\langle n_{ch} \rangle$, D, f_2, and higher moments and their energy dependence has been discussed, but not yet the energy dependence of the whole multiplicity distribution, that is of course contained implicitly in energy independent expressions like Wróblewski's empirical relation: $(\langle n_{ch} \rangle - 1)/D$ is constant, and Buras, Dias de Deus, and Møller's results. All these relations do not contain explicitly the cross-section σ_n of n particle production. Koba, Nielsen, and Olesen [98] investigated the implementation of Feynman scaling [99] onto multiplicity distributions by determining all multiplicity moments by partial integration of the Feynman function $f^{(q)}$ $(x_1, \vec{p}_{T1}, \dots, x_q, \vec{p}_{Tq})$ over all x's and p_T's. This yields for the q-th moment: $\langle n(n-1)\dots(n-q+1) \rangle = \tilde{f}^{(q)}(0,\dots,0).(\ln s)^q + O((\ln s)^{q-1})$, after integration over p_T has been performed. This is also the asymptotic expression for $\langle n^q \rangle$. In terms of the variable $z = n/\tilde{f}^{(1)}(0) \cdot \ln s$, Koba, Nielsen and Olesen obtain from $\langle n^q \rangle = \sum n\, \sigma_n/\sigma_{inel} = \tilde{f}^{(q)}(0,\dots,0)(\ln s)^q + O((\ln s)^{q-1})$ that corresponds asymptotically to the linear ansatz (VII.3.2), the identity:

$$\int_0^\infty dz\ z^q . \tilde{f}^{(1)}(0).\ln s \cdot \frac{\sigma_n(s)}{\sigma_{inel}(s)} = \frac{\tilde{f}^{(q)}(0,\dots,0)}{[\tilde{f}^{(1)}(0)]^q} + O\left(\frac{1}{\ln s}\right). \tag{VII.4.1}$$

Because of scaling, the moments of the right hand side of equ. (VII.4.1) demand the function $\psi(z) = \tilde{f}^{(1)}(0).$ $\ln s.\sigma_n(s)/\sigma_{inel}(s)$ to be not explicitly dependent on s, so that

$$\frac{\sigma_n(s)}{\sigma_{inel}(s)} = \frac{1}{\tilde{f}^{(1)}(0).\ln s} \Psi \left(\frac{n}{\tilde{f}^{(1)}(0).\ln s}\right) = \frac{1}{<n>} \Psi \left(\frac{n}{<n>}\right). \tag{VII.4.2}$$

Taking into consideration the phenomenological results of Buras et al. [94], and Fialkowski and Miettinen [95], one obtains the corrected Koba-Nielsen-Olesen scaling formula

$$\frac{\sigma_n(s)}{\sigma_{inel}(s)} = \frac{1}{<n_{ch} - \alpha>} \Psi \left(\frac{n_{ch} - \alpha}{<n_{ch} - \alpha>}\right) , \tag{VII.4.2}$$

that gives, in correspondance also to the linear ansatz for the moments (VII.3.2), $\alpha = 0.9$. The not explicitly energy dependent (scaling) multiplicity distribution for $pp \to n_{ch}$ is displayed in Fig. 57b, in which the solid curve has been applied to guide the eye. The data compilation in Fig. 57b is due to Buras et al. [94]. Along a similar line of derivation, Buras and Koba [100] obtained the scaling multiplicity distribution $\phi = (<n>-1)^2 \cdot \sigma_n(s)/\pi(n-1)\sigma_{inel}(s)$ vs. $u = \frac{\pi}{4}\left(\frac{n-1}{<n>-1}\right)^2$, which is <u>exponential.</u> The data of $pp \to n_{ch}$ in the energy region up to NAL energies (300 GeV/c) are well described by Buras and Koba's scaling multiplicity expression.

5. Dependence of the Average Charged Multiplicity on the Transverse Momentum and its Energy Dependence at Large Transverse Momenta

In the previous sections, emphasis was laid upon the peripherality of the diffractively produced particle system in the inclusive reactions $pp \to p$ + anything and $\pi^- p \to p$ + anything by analysing the $\frac{d^2\sigma}{dt d(M^2/s)}$ distributions in the region $0.96 < x_p < 1.0$. The result was the logarithmic increase of the average charged multiplicity with increasing mass, whereas the average multiplicity was fairly independent of the momentum transfer t. For the reaction $p + p \to p$ + anything at 28.5 GeV/c, A. Ramanauskas et al. [101] observed that the average charged multiplicity is rising with increasing missing mass, but also, within a fixed missing mass interval, with increasing transverse momentum in the projectile frame, as can be inspected in Fig. 58a. The energy dependence of the invariant cross-section $E\frac{d^3\sigma}{dp^3}$ for large transverse momenta in the reaction $pp \to \pi^\pm$ + anything at 90° at ISR energies has been measured by the British-Scandinavian Collaboration [103], and their results that are displayed in Fig. 58b show that the invariant cross-section is rising with energy at large transverse momenta. The energy dependence of large transverse momentum π^0 production at ISR energies has been verified by the CERN-Columbia-Rockefeller Collaboration [105] to be much more pronounced than the $\pi^\pm$ production at large p_T. It is not yet possible to resolve the functional behaviour of the increase with energy. But it must be attributed to central collisions, because small impact parameters due to high multiplicity production of particles from energy and angular momentum conservation, allow large momentum transfers.

6. The True Multiplicity Distribution

In order to determine the total multiplicity, neutrals included, the n particle production cross-section from K^-p collisions is decomposed into a non-diffractive contribution σ_n^M and the kaon and proton diffraction contributions σ_n^{KD} and σ_n^{pD}, respectively, where n is the true multiplicity. The single reaction channels, numbered by i, have, in general, contributions of all three components, so that

$$\sigma_n^i = \alpha_n^i \sigma_n^M + \beta_n^i \sigma_n^{KD} + \gamma_n^i \sigma_n^{pD} \quad , \qquad \text{(VII.6.1)}$$

with

$$\sum_i \alpha_n^i = \sum_i \beta_n^i = \sum_i \gamma_n^i = 1 \; .$$

The α-factors, that give the relative contributions of the channels i to the non-diffractive cross-section, are calculated from the statistical isospin model [106] in which, for fixed initial isospin (I,I_3), it is assumed that each isospin state of the n particle final state has equal weight. The probability that, for initial state isospin (I,I_3), the final state particles possess (t_i,θ_i) as isospins and third components, is

$$P^{(I)}_{\theta_1,\ldots,\theta_n} = (2I+1).2^{-2I_3-1} . \int_{-1}^{+1} dx \, (1+x)^{2I_3} \, .$$

$$\cdot \, P^{(0,2I_3)}_{I-I_3}(x) . [\prod_{i=1}^{n} P^{(0,2\theta_i)}_{t_i-\theta_i}(x)] \qquad \text{(VII.6.2)}$$

where $P_J^{(\mu,\nu)}$ are Jacobi polynomials and $x = \cos\theta$ in I-space. For the case of two nucleons and n pions (that is analogous to one nucleon and one kaon plus n pions), formula (VII.6.2) yields:

$$P^{(I)}_{\theta_1,\theta_2;n_+,n_o,n_-} = (2I+1).2^{-(n_++n_-+m)-2} \int_{-1}^{+1} dx \cdot \qquad \text{(VII.6.3)}$$

$$\cdot\ (1+x)^{n_++n_-+m+1} . x^{n_o} . P_{I-m}^{(0,2m)}(x) \ ,$$

where $n_++n_-+m = 2n_++m_1+m_2$.

For three body final states, the β and γ factors that are the relative contributions of the various reaction channels to the diffractive kaon and proton dissociation cross-sections, respectively, are calculated from Clebsch-Gordan coefficients. For higher multiplicities, these factors can be obtained either by the statistical isospin model of Cerulus [106] or by the cascade decay hypothesis in which an isospin 1/2-cluster, produced diffractively, is assumed to decay into an isospin 1/2-cluster plus a pion.

The following parametrisation has been adopted by ref. 107 according to Hansen, Kittel, and Morrison [108]:

$$\sigma_n^M(p_{Lab}) \propto p_{Lab}^{-a} . R_n(p_{Lab}) . p_{Lab}^{2-n} \qquad \text{(VII.6.4)}$$

(R_n is the n particle phase space volume and a the Regge-like energy dependence exponent in the amplitude squared).

$$\sigma_n^{KD}(p_{Lab}) = \sigma_n^{KD}(\infty) . e^{b.t_{max}(p_{Lab},M)}$$

$$\sigma_n^{pD}(p_{Lab}) = \sigma_n^{pD}(\infty).e^{b'.t_{max}(p_{Lab},M')}, \qquad (VII.6.5)$$

where M and M' are the missing masses with respect to the final state nucleon or final state kaon, respectively. In Fig. 59, the laboratory momentum dependence of the cross-section of the different charge final states of the reaction $K^-p \to N\bar{K}\pi\pi$ is displayed. The diffractive components are seen to dominate the cross-section already at 20 GeV/c, when proton and kaon diffraction are possible in a reaction channel; when only proton diffraction is possible as in the reaction $K^-p \to nK^-\pi^+\pi^o$, the diffractive and non-diffractive components are about equal at 20 GeV/c. The cross-section of channels in which no diffraction is possible as $K^-p \to n\bar{K}^o\pi^+\pi^-$ drops steaply and comparably to the non-diffractive components of the other channels.

The multiplicity distribution (the multiplicity of neutrals being included) can be inspected in Fig. 60 to be of the same qualitative characteristics as the overall and diffractive multiplicity distributions for the processes $\pi^- p \to$ anything and $\pi^-p \to pX^-$ at 205 GeV/c (Fig. 52). From the fit by means of the parametrisation (VII.6.4-5) of the channel cross-sections, the following average multiplicities were obtained for 10 GeV/c K^-p interactions:

$\langle n\rangle_{total} = 5.6 \pm 0.2$ $\qquad \langle n_{\pi^o}\rangle = 1.25 \pm 0.08$

$\langle n\rangle_{non\text{-}diffr.} = 5.9 \pm 0.3$ $\qquad \langle n_n\rangle = 0.45 \pm 0.03$

$\langle n\rangle_{diffr.} = 4.3 \pm 0.3$ $\qquad \langle n_{\bar{K}^o}\rangle = 0.43 \pm 0.03$

An interesting result is that the multiplicity of neutrals is decreasing with increasing charged multiplicity. Dao and Whitmore [109] show in their compilation of $\langle n_{\pi^0} \rangle$ vs. n_- distributions from pp interactions how, at 12 GeV/c, $\langle n_{\pi^0} \rangle$ is weakly decreasing with increasing n_-, but that, with increasing collision energy, $\langle n_{\pi^0} \rangle$ becomes strongly rising with n_- (see also Fig. 18 of ref. [96]). As an illustration, this rise of $\langle n_{\pi^0} \rangle$ with n_{ch} is shown in Fig. 61 for pp interactions at two ISR energies ($\sqrt{s}$ = 23 and 53 GeV/c) from an experiment of the CERN-Hamburg-Vienna Collaboration [110]. In Fig. 62 is displayed the laboratory momentum dependence of the total multiplicity distribution of the reaction $K^-p \to N\bar{K}$ + (n-2) pions: since the non-diffractive component is strongly falling off at high energies, it is predicted to observe a separate diffractive multiplicity maximum at $n \sim 4$ at very high energies which has not yet been observed in the region of three, four, and five body final states.

7. Charge Transfer

The charge transfer in a reference system (the centre-of-mass system, for example) can be defined by the following diagram for the reaction $a + b \to$ anything inelastic:

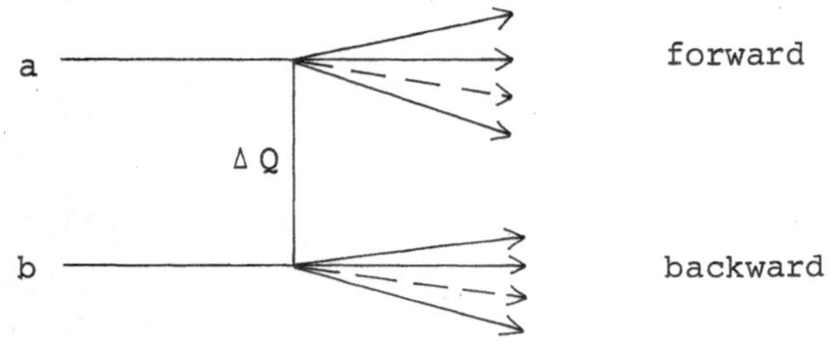

The dispersion of the charge transfer is: $D^2 = \langle \Delta Q^2 \rangle - \langle \Delta Q \rangle^2$, where $\langle \Delta Q \rangle$ is the average charge transfer, defined in the usual way:

$$\langle \Delta Q \rangle = \sum_{\Delta Q} \Delta Q \frac{\sigma_{\Delta Q}}{\sigma_{inel}} \quad .$$

The charge transfer is a special case of an arbitrary quantum number transfer, such as strangeness, baryon number transfer etc., which are related to the charge transfer by the Gell-Mann and Nishijima formula. The analysis of the interaction $K^- p \rightarrow$ anything inelastic at 10 and 16 GeV/c in terms of charge transfer, its average, and dispersion has been developped by Counihan in the Aachen-Berlin-CERN-London-Vienna Collaboration [111]. The charge transfer distributions in the c.m. system are given in Fig. 63, from which it can be inferred that 80% of the $K^- p$ inelastic cross-section are characterized by $\Delta Q = 0$ and +1. At lo GeV/c, the mean charge transfer is 0.64 while, at 16 GeV/c, the mean charge transfer is decreased to the value 0.55. This decrease is due to the increase with energy of the $\Delta Q < 0$ and the decrease with increasing energy of the $\Delta Q > +1$ contributions. The conclusion therefore is that ΔQ has the tendancy to approach zero at asymptotic energies.

The multiperipheral model is characterized by large subenergies in the average, i.e. by a short range ordering by which it is conjectured [85,99] that the dynamics of multiparticle production is independent of the incident particles. Therefore, at asymptotic energies at which short range ordering may by valid, $\Delta Q = 0$ ensues. Assuming the multiperipheral picture Quigg and Thomas [112] predict the squared dispersion of the charge transfer to be

$$D^2 = \frac{8\Delta}{3Y} n_{ch} , \qquad \text{(VII.7.1)}$$

where Δ is a mobility parameter, and Y is the difference of the beam and target rapidities. Fig. 64 shows the squared dispersion of the charge transfer as a function of n_{ch}, and the energy behaviour which is inferred by the logarithmic increase of $Y = y_{beam} - y_{target}$ in (VII.7.1) is reproduced by the experimental data since the increase of D^2 with n_{ch}, which is also guaranteed by formula (VII.7.1), becomes flatter at the higher energy. Formula (VII.7.1) can be "generalized" to the differential form

$$D^2(y) = \frac{8\Delta}{3} [\frac{dn_{ch}}{dy} - (f_p + f_K)] \equiv \frac{8\Delta}{3} \frac{dn_f}{dy} , \qquad \text{(VII.7.2)}$$

in which subtraction has been made of the leading proton and kaon diffraction contributions f_p and f_K, respectively. The result is the distribution of free charged particles per unit in y-space. Defining $D^2(y)$ from $\Delta Q(y) = \sum_i q_i \theta(y_i^{cm} - y)$, where q_i are the charges of the forward moving particles and y is the rapidity of the frame relative to the centre-of-mass, one obtains good correspondance of the left and right hand sides from the experimental data. Figs. 65 and 66 show the charge profiles per unit of rapidity in the incident kaon rest frame (yielding the charge distribution for kaon fragmentation), and in the target rest frame (yielding the charge distribution for proton fragmentation). It followed from the definition of $\Delta Q(y)$:

$$\frac{d}{dy} <\Delta Q(y)> = - \frac{dQ}{dy} , \qquad \text{(VII.7.3)}$$

and $f_p(y)$ and $f_K(y)$ were determined from the following equations:

$$f^{(10\ \mathrm{GeV}/c)}(y) = f_p(y-y_p^{(10\ \mathrm{GeV}/c)})-f_K(y_K^{(10\ \mathrm{GeV}/c)}-y)$$

$$\text{(VII.7.4)}$$

$$f^{(16\ \mathrm{GeV}/c)}(y) = f_p(y-y_p^{(16\ \mathrm{GeV}/c)})-f_K(y_K^{(16\ \mathrm{GeV}/c)}-y) .$$

In the fragmentation models, the overspills into the opposite hemispheres are analysed: the mean charge transfer also tends to zero because the fragmentation frames of the kaon and proton become sufficiently separated at asymptotic energies, and

$$D^2 \sim \frac{1}{2} n_{ch} <\Delta Q>, \qquad \text{(VII.7.5)}$$

because $<\Delta Q>$ is twice the spillover of fragments into the opposite hemisphere, and D^2 twice the spillover of proton <u>and</u> kaon fragments times the respective multiplicities. The linear increase of D^2 with n_{ch} is verified by Fig. 64 so that essentially analogous predictions are made by the fragmentation models and the multiperipheral class of models.

I thank very much Dr. Widder for the many enlightening discussions I had with him on the subjects presented in this review, and my warmest gratitude may be expressed to Dr. Cautis who helped a great deal preparing the manuscript.

REFERENCES

1. G. Barbiellini et al. (Aachen-CERN-Harvard-Genova-Torino Collab.), "Small-Angle Proton-Proton Elastic Scattering at Very High Energies (460 GeV^2 < s < 2900 GeV^2)", Phys. Lett. 39B, 663 (1972).

2. J. V. Allaby et al. (CERN-Rome Collaboration), "Energy Dependence of the Structure in High Energy Proton-Proton Elastic Scattering", Phys. Lett. 34B, 431 (1971).

3. C. Baglin et al. (CERN-Ecole Polytechnique-Orsay-Stockholm Collaboration), "Large Angle π^+p Elastic Scattering at 10 GeV/c" (Paper Nr. 162, submitted to the 2-nd Aix-en-Provence Elementary Particle Conference), Phys. Lett. 47B, 85 (1973).

4. P. Cronillon et al. (Cornell-Brookhaven-Northeastern Collaboration), "Large Angle π^-p Elastic Scattering at 14 and 23 GeV/c", Phys. Rev. Lett. 30, 403 (1973).
R. Rubinstein et al. (Brookhaven-Cornell-Northeastern Collaboration), "Large Angle π^+p and K^+p Elastic Scattering at 13.8 GeV/c", Phys. Rev. Lett. 30, 1010 (1973).

5. T. T. Chou, C. N. Yang, "Model of Elastic High Energy Scattering", Phys. Rev. 170, 1591 (1968).
T. T. Chou, C. N. Yang, "Possible Existence of Kinks in High Energy Elastic pp-Scattering", Phys. Rev. Lett. 20, 1213 (1968).

6. L. Durand III, R. Lipes, "Diffraction Model for High Energy pp Scattering", Phys. Rev. Lett. 20, 637 (1968).

7. see: G. Giacomelli, "High Energy Hadron Physics", Proceedings of the Amsterdam International Conference on Elementary Particles (June 30 - July 6, 1971), North-Holland Publishing Company, p. 1 - 38.

8. V. Bartenev et al. (USSR-USA Collaboration at NAL), "Measurement of the Slope of the Diffraction Peak for Elastic pp Scattering from 8 to 400 GeV", Phys. Rev. Lett. 31, 1088 (1974).

9. U. Amaldi et al. (CERN-Rome Collaboration), "The Energy Dependence of the Proton-Proton Total Cross-Section for Centre-of-Mass Energies between 23 and 53 GeV", Phys. Lett. 44B, 112 (1974).

10. G. B. Yodh, Yash Pal, J. S.Trefil, "Evicence for Rapidly Rising p-p Total Cross Section from Cosmic Ray Data", Phys. Rev. Lett. 28, 1005 (1972).

11. U. Amaldi, "Elastic Scattering and Low Multiplicities", Rapporteur's talk at the 2-nd Aix-en-Provence International Conference on Elementary Particles, 1973.

12. U. Amaldi et al. (CERN-Rome Collaboration), "Measurements of the Proton-Proton Total Cross Section by Means of Coulomb Scattering at the CERN Intersecting Storage Ring", Phys. Lett. 43B, 231 (1973).

13. P. Söding, "Real Part of the Proton-Proton and Proton Anti-Proton Forward Scattering Amplitude at High Energies", Phys. Lett. 8, 285 (1964).

14. G. Höhler, H. P. Jakob, "The Real Part of the πN Diffraction Amplitude", Preprint of the University of Karlsruhe, Germany, TKP 13/73.

15. G. Höhler, R. Strauss, Zeitschrift f. Physik 232, 2o5 (1970).

16. J. R. Campbell et al., "Measurement of the Real Part of the K^-p Elastic Scattering Amplitude at 10 GeV/c", Nucl. Phys. B64, 1 (1973).

17. Yung-An Chao, E. Pietarinen, "Extraction of ΛK^-p and ΣK^-p Coupling Constants from $K^{\pm}p$ Forward Amplitudes", Phys. Rev. Lett. 26, 1060 (1971).

18. D. W. G. S. Leith, "Diffractive Processes", invited paper presented at the 1973 Meeting of the Division of Particles and Fields of the APS, Berkeley, August 1973: SLAC-PUB-1330 (T/E).

19. E. Zevgolatakos et al. (Athens-Demokritos-Liverpool-Vienna Collaboration), "A Study of the Reactions $K^-p \to \bar{K}^0\pi^-p$ and $K^-p \to K^{*-}$ (890)p at 8.25 GeV/c", Nucl. Phys. B55, 15 (1973).

20. H. Harari, Annals of Physics 63, 432 (1971).

21. M. Davier, H. Harari, "Elastic $K^{\pm}p$ Scattering and a Dual Absorptive Model", Phys. Lett. 35B, 239 (1971).

22. H. W. Atherton et al., "Study of the Reactions $\bar{p}p \to \bar{\Lambda}\Lambda$, $\bar{\Lambda}\Sigma^0$+ c.c, $\Sigma^+ \Sigma^+$ at 3.6 GeV/c", CERN/D.Ph. II/Phys. 73-12; to be submitted to Nuclear Physics B.

23. G. Plaut, "Exchange Degeneracy and Absorption in $p\bar{p} \to \bar{Y}Y$", Nucl. Phys. B35, 221 (1971).

24. F. Bradamante et al. (CERN-Trieste High Energy Group), "Polarisation in $\pi^+p \to K^+\Sigma^+$ Backward Scattering at 3.5 GeV/c", Phys. Lett. 44B, 202 (1973).

25. W. Beusch et al. (CERN-ETH-IC Collaboration), "Angular Distribution and Polarisation in the Back-

ward Peak of $\pi^- p \to \Lambda K^o$ at 4 and 6.2 GeV/c", Nucl. Phys. B19, 546 (1970).

26. C. Schmid, J. K. Storrow, "Exchange Degeneracy Tests in the Resonance Region of the Reactions $\bar{K}N \to \pi\Sigma$ and $\bar{K}N \to \pi\Lambda$", Nucl. Phys. B29, 219 (1971).

27. D. Yaffe et al., "A Study of the Reactions $\pi^- p \to K^{*o}_{890}\Lambda$, $K^{*o}_{890}\Sigma^o$, and K^{*o}_{890} Y_1^{*o} (1385)", to be submitted to Nuclear Physics B. (CERN/D.Ph.II/Phys. 73-24).

28. R. D. Field, R. L. Eisner, M. Aguilar-Benitez, "Study of Vector Meson Production with Hypercharge Exchange", Phys. Rev. D6, 1863 (1972).

29. G. Grayer et al. (CERN-Munich Collaboration), " $\pi\pi$ Phase Shift Analysis from an Experiment $\pi^- p \to \pi^-\pi^+ n$ at 17.2 GeV/c", AIP-Conference Proceedings No. 8, Particles and Fields Subseries No. 3, Experimental Meson Spectroscopy - 1972 (Third Philadelphia Conference), Edited by A. H. Rosenfeld and K.-W. Lai, American Institute of Physics, pp. 5-16.

30. P.K. Williams, "Form Factors, Kronecker-δ Terms, and the Absorptive Peripheral Model", Phys. Rev. 181, 1963 (1969).
L. Chan, P. K. Williams, "Decays of High Spin Objects Produced by Pion Exchange", Phys. Rev. 188, 2455 (1969).

31. P. K. Williams, "Extrapolation Model for $\pi\pi$ Scattering", Phys. Rev. D1, 1312 (1970).

32. G. L. Kane, "Simplified Procedure for Performing Absorption Corrections", Phys. Rev. 163, 1544 (1967).

33. P. Estabrooks, A.D. Martin, "Analysis of $\pi^- p \to \pi^-\pi^+ n$ Data at 17.2 GeV/c", Phys. Lett. 41B, 350 (1972).

34. P. Estabrooks et al., CERN preprint TH-1661, to be

published in: Proceedings of the International Conference on $\pi\pi$ Scattering and Associated Topics, Tallahassee, 28-30 March, 1973.

35. A.D. Martin, P. Estabrooks, "$\pi\pi$ Phase Shifts from $\pi^- p \to \pi^- \pi^+ n$ Data", Proceedings of the Pion Exchange Meeting, Daresbury Nuclear Physics Laboratory, March 1973, pp. 77-122.

36. S. D. Protopopescu et al., "$\pi\pi$ Partial Wave Analysis from Reactions $\pi^+ p \to \pi^+ \pi^- \Delta^{++}$ and $\pi^+ p \to K^+ K^- \Delta^{++}$ at 7.1 GeV/c", Phys. Rev. D7, 1279 (1973).

37. P. Estabrooks, A. D. Martin, C. Michael, "Exchange Mechanisms for $\pi^- p \to \rho^o n$ and ρ-ω Interference", CERN Preprint Ref. TH 1732 (1973).

38. M. Ross, F. S. Henyey, G. L. Kane, "On the Structure of High Energy Two-Body Non-Diffractive Reactions", Nucl. Phys. B23, 269 (197o).

39. W. Hoogland et al. (Amsterdam-CERN-Munich Collaboration), "Isospin-2 $\pi\pi$ Phase Shifts from an Experiment at 12.5 GeV/c", CERN Preprint, 1973, submitted to Physics Letters.

40. R. Arnowitt, Proceedings of the Conference on $\pi\pi$ and $K\pi$ Interactions, Argonne National Laboratory (1969), p. 619.

41. M. J. Losty et al. (CERN-Saclay Collaboration), "A Study of $\pi^- \pi^-$ Scattering from $\pi^- p$ Interactions at 3.93 GeV/c", CERN Preprint D. Ph.II/PHYS 73-26.

42. J. P. Baton, G. Laurens, J. Reignier, "$\pi\pi$ Phase Shifts from Chew-Low Extrapolations of $\pi^- p \to \pi\pi N$ at 2.77 GeV/c", Phys. Lett. 33B, 528 (1970).

43. O. R. Sander et al., "S- and D-Wave Phase Shifts for

$\pi^+\pi^+ \rightarrow \pi^+\pi^+$ T = 2 Strong Elastic Scattering at Low Dipion Energies", Paper No.476, submitted to the 16th Int. Conference on High Energy Physics, Batavia (1972).

44. G. Ciapetti, R. L. Eisner, A. C. Irving, J. B. Kinson, D. C. Watkins, "Study of Reactions Involving a Vector Meson Produced in Association with the Δ(1236) or Nucleon", Nucl. Phys. B66, 350 (1973).

45. C. Brankin et al. (Athens-Demokritos-Liverpool-Vienna Collaboration), "A Study of the $K^-\pi^-$ Interaction from K^-p Data at 8.25 GeV/c", Nucl. Phys. B63, 211 (1973).

46. M. Bardadin-Otwinowska et al. (Saclay-Ecole Polytechnique-Rutherford Collaboration), "Determination of the $K^-\pi^-$ Inelastic Cross Sections for C. M. Energy up to 2.8 GeV", Saclay Preprint D.Ph.PE 73-11.

47. P. L. Bastien et al. (Seattle-Berkeley Collaboration), "Diffractive Dissociation of Pions at 15 GeV/c", Paper No. 376, submitted to the 2nd Aix-en-Provence Conference on Elementary Particles, September 1973.

48. D. V. Brockway, "Study of the Three Pion Final State Interactions in the Reaction $\pi^-p \rightarrow \pi^-\pi^+\pi^-p$ at 5 and 7.5 GeV/c", University of Illinois report COO-1195-197 (1970).

49. R. Klanner, "Analysis of the Reaction $\pi^-p \rightarrow \pi^-\pi^-\pi^+p$ at 25 and 40 GeV/c", NP Internal Report 73-9, CERN.

50. G. Ascoli et al., "Summary of the Experimental Situation Regarding A_1, A_2 and A_3 Production", papers 442 and 444 submitted to the 16th Int. Conference on High Energy Physics, Batavia, September 1972.
G. Ascoli et al. (Illinois-Toronto-Genova-DESY-Milan-Saclay-Harvard-Aachen-Berlin-Bonn-CERN-Heidelberg-

Notre Dame-Wisconsin Collaboration), "Spin Parity Analysis of the A_3", Phys. Rev. D7, 669 (1973).

51. M. Deutschmann et al. (Aachen-Berlin-CERN-London-Vienna Collaboration), "Spin Parity Structure of the Q and L Enhancements in $K^-p \rightarrow (K^-\pi^+\pi^-)p$ at 10 and 16 GeV/c", CERN Preprint D.Ph.II/PHYS 73, submitted to Phys. Letters.

52. J. D. Hansen, G. T. Jones, G. Otter, G. Rudolph, "Formalism and Assumptions Involved in Partial Wave Analysis of Three-Meson Systems", CERN Preprint D.Ph.II/PHYS. 73-34, to be published.

53. V. Chaloupka, "Analysis of the $\pi\omega$ System Produced in the Reaction $\pi^-p \rightarrow p\omega\pi^-$", CERN Preprint D.Ph.II/PHYS 73-33, submitted to the 2nd Aix-en-Provence Conference on Elementary Particles, September 1973.

54. A. Apostolakis et al. (Athens-Demokritos-Liverpool-Vienna Collaboration), "A Study of the Reaction $K^-p \rightarrow K^-\omega p$ at 8.25 GeV/c", Preprint of the University of Athens, submitted to Nuclear Physics B.

55. D. R. O. Morrison, Phys. Lett. 22, 528 (1966).

56. G. W. Brandenburg et al. "Observation of a Crossover in the Differential Cross Sections for Q-Meson Production", Phys. Rev. Lett. 28, 932 (1972).

57. A. Stergiou et al. (Athens-Bruxelles-CERN-Demokritos-Liverpool-Mons-Vienna Collaboration), "Comparison of the Reactions $K^{\pm}p \rightarrow Q^{\pm}p$ at Incident Momentum 8.25 GeV/c", CERN preprint D.Ph.II/PHYS 73-1, Paper No. 314, submitted to the 2nd Aix-en-Provence Conference on Elementary Particles, September 1973.

58. A. P. Contogouris et al. (Athens-Demokritos-Liverpool-Vienna Collaboration), "A Model for the Reaction KN → QN" Paper No. 281, submitted to the 2nd Aix-en-Provence Conference on Elementary Particles, September 1973.

59. H. Harari, "Dual Absorptive Model for Dips in Inelastic Hadron Processes", Phys. Rev. Lett. 26, 1400 (1971).
M. Davier, H. Harari, "Elastic $K^{\pm}p$ Scattering and a Dual Absorptive Model", Phys. Lett. 35B, 239 (1971).
H. Harari, A. Schwimmer, "Properties of Hadronic Amplitudes in an Absorptive Model", Phys. Rev. D5, 2780 (1971).

60. V. Barger, R. J. N. Phillips, "Meson Regge Exchanges from Simultaneous Analysis of πN Scattering and Dispersion Sum Rules", Phys. Rev. 187, 2210 (1969).

61. G. S. Abrams et al. (Berkeley-NAL Collaboration), "Diffraction Dissociation in 205 GeV/c $\pi^- p$ Two- and Four Prong Interactions", Paper No. 359, submitted to the 2nd Aix-en-Provence Conference on Elementary Particles, September 1973.

62. E. H. De Groot, H. I. Miettinen, "Shadow Approach to Diffraction Scattering", Proceedings of the VIIIth Recontre de Moriond, Méribel-les-Allues, March 1973, pp. 193-234.

63. N. Sakai, J. N. J. White, "An Investigation of Unitarity at ISR Energies", Nucl. Phys. B59, 511 (1973).

64. H. Grässler et al. (Aachen-Berlin-CERN-London-Vienna Collaboration), "Quantum Number Transfer in $K^- p$ Reactions at 6, 10 and 16 GeV/c", Nucl. Phys. B59, 333 (1973).

65. S. Pokorski, L. van Hove, "A Two-Step Dynamical Picture of Diffraction Dissociation and the High Energy Behaviour of Transverse Momenta", Nucl. Phys. B60, 379 (1973).

66. M. G. Albrow et al. (CERN-Holland-Lancaster-Manchester ISR Collaboration), "The Spectrum of Protons Produced in pp Collisions at 31 GeV/c Total Energy", Nucl. Phys. B54, 6 (1973).

67. Z. Ajduk, "Structure of the Inelastic Overlap Function for πp Collisions in a Diffraction Dissociation Model", Nuovo Cimento 16A, 111 (1973).

68. M. G. Albrow et al. (CERN-Holland-Lancaster-Manchester Collaboration), "Missing Mass Spectra in p-p Inelastic Scattering at Total Energies of 23 GeV and 31 GeV", paper Nr. 375, presented at the 2nd Aix-en-Provence International Conference on Elementary Particles, September 1973.

69. F. C. Winkelmann et al. (Berkeley-NAL Collaboration), "Pion Diffraction Dissociation in 205 GeV/c π^-p Interactions", Phys. Rev. Lett. 32, 121 (1974).

70. M. G. Albrow et al. (CERN-Daresbury-Holland-Lancaster-Manchester Collaboration), "The Spectrum of Protons Produced in pp Collisions at 31 GeV Total Energy", Nucl. Phys. B54, 6 (1973).

71. L. Foà, "Experimental Review of High Multiplicity Reactions", Rapporteur's talk at the 2nd Aix-en-Provence International Conference on Elementary Particles, September 1973, Proceedings: pp. 317-338.

72. S. J. Barish, D. C. Colley, P. F. Schultz, J. Whitmore (ANL-NAL Collaboration), "Characteristics of the Re-

action p + p → p + X at 205 GeV/c", Phys. Rev. Lett. 31, 1080 (1974).

73. S. Childress, P. Franzini, J. Lee-Franzini, R. McCarthy, R. D. Schamberger, Jr. (Columbia-Stony Brook Collaboration), "Small-Momentum-Transfer p-p Inelastic Scattering at 300 GeV/c", Phys. Rev. Lett. 32, 389 (1974).

74. Aachen-CERN-Genova-Harvard-Torino ISR Collaboration, communicated to ref. [18] by G. Goldhaber.

75. Ph. Salin, "Phénomenologie de Mueller-Regge", Cours de l'Ecole de Physique de Gif-sur-Yvette, September-October 1973.

76. K. Gottfried, "An Introduction to Multiple Production Processes", Lectures given in the Academic Training Program, November-December 1972; Ref. TH. 1615-CERN.

77. K. Abe, T. DeLillo, B. Robinson, F. Sannes, J. Carr, J. Kleyne, I. Siotis (Rutgers-Imperial College Collaboration), "Measurement of p + p → p + X Between 50 and 400 GeV/c", Phys. Rev. Lett. 31, 1527 (1974).

78. N. K. Yamdagni, S. Ljung, "Factorization of the Diffraction Dissociation Amplitude in Four-Body Final States", Phys. Lett. 37B, 117 (1971).

79. C. M. Bromberg et al. (Rochester-Michigan Collaboration), "Cross Sections and Charged-Particle Multiplicities at 102 and 405 GeV/c", Phys. Rev. Lett. 31, 1563 (1973); Phys. Rev. Lett. 32, 83 (1974).

80. R. C. Hwa, "Multiplicity Distribution and Single Particle Spectrum in the Diffractive Model", Phys. Rev. Lett. 26, 1143 (1971).

81. A. K. Wróblewski, "Multiple Production and Processes at Ultra High Energies", Rapporteur's talk at the

XV. International Conference on High Energy Physics, Kiev 1970, pp. 42-94.

82. G. F. Chew, A. Pignotti, "Multiperipheral Bootstrap Model", Phys. Rev. 176, 2112 (1968).
N. F. Bali, G. F. Chew, A. Pignotti, "Multiple-Production Theory via Toller Variables", Phys. Rev. Lett. 19, 614 (1967).

83. C. E. DeTar, "Momentum Spectrum of Hadronic Secondaries in the Multiperipheral Model", Phys. Rev. D3, 128 (1971).

84. D. Sivers, G. H. Thomas, "Analysis of Inclusive Distributions Using a Statistical Approach", Phys. Rev. D6, 1961 (1974).

85. E. L. Berger, "Multiparticle Production Processes at High Energy", Lectures presented at Erice (July 1973) Basko Polje (September 1973), and Gif-sur-Yvette (September 1973); Ref. TH.1737-CERN.

86. E. L. Berger, G. C. Fox, A. Krzywicki, "Quantitative Measure of Cluster Formation in Multiparticle Production", Phys. Lett. 43B, 132 (1973).
E. L. Berger, A. Krzywicki, "Kinematic Aspects of Inclusive Phenomenology", Phys. Lett. 36B, 380 (1971)

87. H. Muirhead, "Antiproton-Proton Interactions", Proceedings of the 2nd Aix-en-Provence International Conference on Elementary Particles, September 1973, pp. 365-368.

88. E. L. Berger, G. C. Fox: "Multiplicity Distributions and Inclusive Spectra in a Multiperipheral Cluster Emission Model", Phys. Lett. 47B, 162 (1973).

89. N. Schmitz, "An Introduction to Proton-Proton-Collisions at High Energies", Lectures given at the

XIII. Cracow School of Theoretical Physics, Zakopane, 1-12 June 1973, preprint of the Max-Planck Institute for Physics and Astrophysics, Munich: MPI/PAE/Exp. El. 32, May 1973, and Acta Physica Polon. B4, 689 (1973).

90. V. V. Ammosov et al. (Aachen-Berlin-CERN-London-Mons-Saclay-Serpuchov-Vienna Collaboration), "Charged Particle Multiplicity Distributions for 33.8 GeV/c K^-p and 50 GeV/c π^-p Interactions", Nucl. Phys. B58, 77 (1973).

91. F. T. Dao, J. Lach, J. Whitmore, "Study of Charged Multiplicity Distributions in High Energy Particle Collisions", Phys. Lett. 45B, 513 (1973).

92. P. K. Malhotra, "Dependence of Multiplicity on Primary Energy in Nucleon-Nucleon and Pion-Nucleon Collisions", Nucl. Phys. 46, 559 (1963).

93. A. Wróblewski, "Multiplicity Distributions in Proton-Proton Collisions", Lecture given at the XIII. Cracow School of Theoretical Physics, Zakopane, June 1-12, 1973, Acta Phys. Polon. B4, 857 (1973).

94. A. J. Buras, J. Dias de Deus, R. Møller, "Multiplicity Scaling at Low Energies, a Generalized Wroblewski Formula, and the Leading Particle Effect", Phys. Lett. 47B, 251 (1973).

95. K. Fialkowski, H. I. Miettinen, "High Energy Multiplicity Distributions and the Two-Component Picture of Particle Production", Phys. Lett. 43B, 61 (1973).

96. A. Białas, "Correlations in Particle Production at High Energies", invited talk given at the IVth International Symposium on Multiparticle Hadrodynamics, Collegio Ghislieri, Pavia, 31 August - 4 September

1973; Ref. TH. 1745-CERN.

97. A. Wróblewski, Proceedings of the III. Colloquium on Multiparticle Reactions, Zakopane 1972, p. 81.

98. Z. Koba, H. B. Nielsen, P. Olesen, "Scaling of Multiplicity Distributions in High Energy Hadron Collisions", Nucl. Phys. B40, 317 (1972).

99. R. P. Feynman, "Very High-Energy Collisions of Hadrons", Phys. Rev. Lett. 23, 1415 (1969).
J. Benecke, T. T. Chou, C. N. Yang, E. Yen, "Hypothesis of Limiting Fragmentation in High-Energy Collisions", Phys. Rev. 188, 2159 (1969).

100. A. J. Buras, Z. Koba, "Scaling Behaviour of Charged-Prong Multiplicity Distributions in High Energy Hadronic Collisions and Local Excitation Model", Nuovo Cimento Lett. 6, 629 (1973).

101. A. Ramanauskas et al. (Brookhaven-Virginia-Wisconsin-Purdue-Pennsylvania Collaboration), "Observation of Increasing Charged Multiplicity as a Function of Transverse Momentum in 28.5 GeV/c pp Interactions", Phys. Rev. Lett. 31, 1371 (1973).

102. P. Schübelin, "Observation of Increasing Charged Multiplicity as a Function of Transverse Momentum in 28.5 GeV/c pp Interactions", Proceedings of the 2nd Aix-en-Provence International Conference on Elementary Particles, September 1973, pp. 4o1-4o3.

103. B. Alper et al. (British-Scandinavian Collaboration), "Preliminary Results on Charged Particle Production at High Transverse Momenta in p-p Collisions at 90° at the CERN ISR", Paper No. 423, presented at the 2nd Aix-en-Provence International Conference on Elementary Particles, September 1973.

104. L. J. Caroll, "Preliminary Results on Charged Particle Production at High Transverse Momenta in p-p Collisions at 90° at the CERN ISR", Proceedings of the 2nd Aix-en-Provence International Conference on Elementary Particles, September 1973, pp. 4o7-4o8.

105. F. W. Büsser et al. (CERN-Columbia-Rockefeller Collaboration), "Observation of π^{o} Mesons with Large Transverse Momentum in High Energy Proton-Proton Collisions", Phys. Lett. 46B, 471 (1973).

106. F. Cerulus, "Closed Formulae for the Statistical Weights", Il Nuovo Cimento XIX, 528 (1961).
J. Bartke, "Statistical Model and Cross-Sections for High Energy Reactions", Inst.of Nucl. Physics Cracow Report No. 719/PH (1970).

107. M. Deutschmann et al. (Aachen-Berlin-CERN-London-Vienna Collaboration), "The "True" Multiplicity Distribution in $K^{-}p$ Interaction and the Two-Component Model", CERN preprint D.Ph.II/PHYS 73-27, to be submitted to Nucl. Phys. B.

108. J. D. Hansen, W. Kittel, D.R.O. Morrison, "Study of Many Body Cross Sections and Link with Two-Body Cross Sections and with the Reaction Mechanism", Nucl. Phys. B25, 605 (1971).

109. F. T. Dao, J. Whitmore, "Study of Neutral-Charged Particle Correlations in High Energy Collisions", Phys. Lett. 46B, 252 (1973).

110. H. Dibon et al. (CERN-Hamburg-Vienna Collaboration), "Correlations between Photons and Charged Particles Measured at the CERN ISR", Paper No. 338, presented at the 2nd Aix-en-Provence International Conference on Elementary Particles, September 1973.

111. P. Bosetti et al. (Aachen-Berlin-CERN-London-Vienna Collaboration), "Charge Exchange and Charge Distributions in K^-p Interactions at 10 and 16 GeV/c", Nucl. Phys. B62, 46 (1973).

112. C. Quigg, G. H. Thomas, "Charge Transfer in a Multiperipheral Picture", Phys. Rev. D7, 2752 (1973).

FIGURE CAPTIONS

Fig. 1. a) Differential cross-sections of pp scattering at the centre-of-mass energies: 21.5, 23.5, 30.7, 44.9, and 53 GeV. Taken from ref. [11].
b) Energy dependence of the diffraction slope of $d\sigma/dt$ of pp elastic scattering [18]. Taken from ref. [18].

Fig. 2. a) Energy dependence of the total pp cross-section. Taken from ref. [9].
b) Energy dependence of the total pp cross-section in the region of cosmic ray energies ($E = 10^3$ to 4.10^4 GeV) as calculated from inelastic p-air collisions. Taken from ref. [10].

Fig. 3. a) Characteristics of the energy dependence of the forward elastic pp differential cross-section and of the position of the diffraction minimum in the momentum transfer region $-t \sim 1.4\ (GeV/c)^2$. Taken from ref. [11].
b) Characteristics of the rising of the elastic, total inelastic, and total pp cross-sections

with increasing laboratory momentum. Taken from ref. [11].

Fig. 4. Laboratory momentum dependence of the ratio real part/imaginary part of the forward pp elastic scattering amplitude. Taken from ref. [12].

Fig. 5. Argand Diagram of the isospin even diffraction amplitude of pion nucleon scattering at t = 0 $(GeV/c)^2$. Taken from ref. [14].

Fig. 6. Argand Diagram of the isospin even diffraction amplitude of pion nucleon scattering at t = -0.1 $(GeV/c)^2$ as calculated from dispersion relation. Taken from ref. [14].

Fig. 7. Momentum transfer dependence of the $\frac{d\sigma}{dt}$ distribution of the reaction $K^- p \rightarrow K^{*-}_{890} p$ at 8.25 GeV/c together with its separation into natural and unnatural parity exchange, and of $\rho^H_{1,\pm 1} \cdot d\sigma/dt$ distributions. (H ... Helicity frame). Taken from ref. [19].

Fig. 8. t'-dependence of the differential cross-section of the reactions $\bar{p}p \rightarrow \bar{\Lambda}\Lambda$ and $\bar{p}p \rightarrow \Sigma^+\overline{\Sigma^+}$ at 3.6 GeV/c and of the polarization of $\Lambda + \bar{\Lambda}$ and $\Sigma^+ + \overline{\Sigma^+}$, respectively. Full line: Two-slope parametrisation of the differential cross-section. Dotted line: Prediction of ref. [23]. Taken from ref. [22].

Fig. 9. t'-dependence of the differential cross-section of the reaction $\bar{p}p \rightarrow \bar{\Lambda}\Sigma^0 + \overline{\Sigma^0}\Lambda$ at 3.6 GeV/c and of the $\Lambda + \bar{\Lambda}$ polarization. Full line, dotted line: see Fig. 8. Taken from ref. [22].

Fig. 10. u-dependence of the backward Σ^+ polarization in the reaction $\pi^+ p \to \Sigma^+ K^+$ (full circles) in comparison to the polarization of backward Λ's from the reaction $\pi^- p \to \Lambda K^o$ (open circles). Taken from ref. [24].

Fig. 11. t'-dependence of the differential cross-section and of the hyperon polarization of the reactions $\pi^- p \to K^{*o}_{890}\Lambda$ and $\pi^- p \to K^{*o}_{890} \Sigma^o$ at 3.93 GeV/c. The solid line is the prediction by ref. [28] from an amplitude analysis of reaction $K^- p \to \phi\Lambda$. Taken from ref. [27].

Fig. 12. t'-dependence of the transversity amplitudes squared of the reactions $\pi^- p \to K^{*o}_{890}\Lambda$, $K^- p \to \phi\Lambda$, and $\pi^- p \to K^{*o}_{890} \Sigma^o$ at 4 GeV/c. Taken from ref. [27].

Fig. 13. Natural parity and unnatural parity exchange contributions to the $d\sigma/dt$ distributions of the reactions $\pi^- p \to K^{*o}_{890}\Lambda$ and $\pi^- p \to K^{*o}_{890} \Sigma^o$. Taken from ref. [27].

Fig. 14. Moments of the $\pi^+\pi^-$ angular distribution in the reaction $\pi^- p \to \pi^-\pi^+ n$ at 17.2 GeV/c in its dependence on $M_{\pi^+\pi^-}$ and t. Taken from ref. [29]

Fig. 15. a) Spin density matrix elements of the S- and P-wave contributions to the ρ^o-region in the reaction $\pi^- p \to \pi^-\pi^+ n$ at 17.2 GeV/c. Taken from ref. [33].

b) Mass dependence of the $\pi^+\pi^-$ I = 0 S-wave phase-shift, the slope of the extrapolation function F(t), and the cut contribution according to a Williams model parametrisation. Taken from ref. [34].

Fig. 16. a) $I = O$ S-wave phase shifts of the $\pi^+\pi^-$ system in the reaction $\pi^- p \to \pi^-\pi^+ n$ at 17.2 GeV/c, determined in their $M_{\pi\pi}$ dependence by two solutions from the moments of the $\pi^-\pi^+$ angular distribution. The broken line represents the best solution obtained by ref. [36]. Taken from ref.[35].
b) P-wave phase shifts of the $\pi^+\pi^-$ system. Taken from ref. [35].

Fig. 17. t-dependence of the $\pi\pi$ partial wave amplitudes P_o, P_{+1} and P_{-1} for the reaction $\pi^- p \to \rho^o n$ at 17.2 GeV/c. Taken from ref. [37].

Fig. 18. a) t-dependence of the s-channel and t-channel moments of the $\pi^+\pi^+$ system in the reaction $\pi^+ p \to \pi^+\pi^+ n$ at 12.5 GeV/c. The full line curves originate from an S and D_o, $D_{\pm 1}$ Regge parametrisation. Taken from ref. [39].
b) $M_{\pi\pi}$ dependence of δ_S^2 and δ_D^2 of the $\pi^+\pi^+$ system in the reaction $\pi^+ p \to \pi^+\pi^+ n$ at 12.5 GeV/c. Taken from ref. [39].

Fig. 19. $M(\pi^-\pi^-)$ dependence of the phase shifts δ_0^2, δ_2^2, δ_4^2 and of the inelasticity η_0^2 from a phase shift analysis of the $\pi^-\pi^-$ system in the reaction $\pi^- p \to \pi^-\pi^- \Delta^{++}$ (1236) at 3.93 GeV/c. Taken from ref. [41].

Fig. 20. Mass dependence of δ_0^2 of $\pi^-\pi^-$ scattering in the reaction $\pi^- p \to \pi^-\pi^-\Delta^{++}$(1236) at 3.93 GeV. The solid curve is the current algebra prediction of ref.[40]. The dashed curves represent the error boundaries of the analysis of ref. [42]. Taken from ref. [41].

Fig. 21. Mass dependence of $\sigma(\pi^-\pi^-)$ from a phase shift analysis of the reaction $\pi^- p \rightarrow \pi^-\pi^-\Delta^{++}(1236)$ at 3.93 GeV/c. The full squares are from ref. [41], the open squares from ref. [43]. The solid curve originates from the algebra prediction of ref. [40].

Fig. 22. Comparison of the t-depending differential cross-sections and $\rho^J_{00}(d\sigma/dt)$ of the reactions $K^+p \rightarrow K^{*0}(890)\Delta^{++}$, $\pi^+p \rightarrow \rho^0\Delta^{++}$ and $\pi^+p \rightarrow \omega\Delta^{++}$ at 5.5 GeV/c with respect to an SU(3) relationship among them. Taken from ref. [44].

Fig. 23. Prediction of the t-dependence of the differential cross-section of the reaction $\pi^+n \rightarrow \omega p$ at 4.2 and 5.1 GeV/c, and of the reaction $K^-n \rightarrow \bar{K}^{*0}\,\Delta^-$ at 5.5 GeV/c from appropriate SU(3) relations. Taken from ref.[44].

Fig. 24. Prediction of the t-dependence of the ω and $\bar{K}^{*0}$ spin density matrix elements in the reactions $\pi^+n \rightarrow \omega p$ at 4.2 and 5.1 GeV/c, and $K^-n \rightarrow \bar{K}^{*0}\,\Delta^-$ at 5.5 GeV/c from appropriate SU(3) relations. Taken from ref. [44].

Fig. 25. Results of the analysis of the $K^-\pi^-$ system in the reaction $K^-p \rightarrow K^-\pi^-\Delta^{++}(1236)$ at 8.25 GeV/c:
a) Definition of the Gottfried-Jackson coordinate system for the determination of the Δ^{++} spin density matrix elements from the Δ^{++} decay angular distribution;
b) experimental results of the Δ^{++} spin density matrix elements;
c) moments of the $K^-\pi^-$ angular distribution as functions of $M(K^-\pi^-)$;
d) elasticity of the $K^-\pi^-$ interaction as a

function of $M(K^-\pi^-)$;

e) elastic cross-section of the $K^-\pi^-$ interaction as a function of the $K^-\pi^-$ mass.

Fig. 26. Effective mass distribution of the three pion system in the reaction $\pi^-d \to \pi^-\pi^+\pi^-d$ at 15 GeV/c. Taken from ref. [47].

Fig. 27. Results of the partial wave analysis of the (3π) system in the reaction $\pi^-p \to \pi^-\pi^+\pi^-p$ at 40 GeV/c. The partial wave notation is $J^P \ell M^\eta$. Taken from ref. [49].

Fig. 28. Dependence on the three pion mass of the relative phases of the 1^+SO^+ wave with respect to smooth partial wave contributions in the reaction $\pi^-p \to \pi^-\pi^+\pi^-p$ at 25 and 40 GeV/c. Taken from ref. [49].

Fig. 29. Momentum transfer dependence of the rotation angle by which the (3π) system is rotated from the Gottfried-Jackson system into a frame (Donohue-Högaasen frame) in which exact helicity conservation holds. The solid curve corresponds to helicity conservation in the s-channel. Taken from ref. [49].

Fig. 30. Decomposition of the $(K\pi\pi)$ mass distribution into the partial wave contributions J^PM^η in the reaction $K^-p \to K^-\pi^+\pi^-p$ at 10 and 16 GeV/c. Taken from ref. [51].

Fig. 31. Mass distributions of the $J^P = 1^+$ S-wave $(K\rho)$ and $(K^*(890)\pi)$ contributions and of the $J^P=2^-$ S-wave $(K^*(1400)\pi)$ and $(K\rho)$ contributions to the $(K\pi\pi)^-$ system in the reaction $K^-p \to K^-\pi^+\pi^-p$ at 10 and 16 GeV/c. Taken from ref. [51].

Fig. 32. Analysis of the 1^+S-wave contribution to the $(K\pi\pi)^-$ system in terms of the $(K^*(890)\pi)$ and $(K\rho)$ decay amplitudes, the ratio and the relative phase of which are given in d) and e). Taken from ref. [51].

Fig. 33. Mass distributions of the unnatural and natural parity contributions to the $(K\pi\pi)^-$ system, and of natural and unnatural parity contributions to the exchange mechanism in the reaction $K^-p \to K^-\pi^-\pi^+p$ at 10 and 16 GeV/c. Taken from ref. [51].

Fig. 34. Partial wave contributions to the $(\pi\omega)$ system in the reaction $\pi^-p \to p\omega\pi^-$ at 4, 5 and 7.5 GeV/c. Taken from ref. [53].

Fig. 35. $(K\omega)$ mass distributions of the reaction $K^-p \to K^-\omega p$ at 8.25, 10, and 16 GeV/c.

Fig. 36. Slope of the $d\sigma/dt'$ distribution of $(K\omega)$ production as a function of its mass in the reaction $K^-p \to K^-\omega p$ at 8.25, 10.1, and 16 GeV/c, and energy dependence exponent n as a function of the $(K\omega)$ mass.

Fig. 37. t'-dependence of the differential cross-section of the reactions $K^\pm p \to Q^\pm p$ at 8.25 GeV. Taken from ref. [57].

Fig. 38. Energy dependence of the cross-sections of the reactions $K^\pm p \to K^\pm p\pi^+\pi^-$. Taken from ref. [57].

Fig. 39. Momentum transfer distributions of the ρ_{ik} $d\sigma/dt$ of Q^+ production in the reaction $K^+p \to Q^+p$ in the energy range 7 - 9 GeV/c. Taken from ref. [58].

Fig. 40. Momentum transfer distributions of the ρ_{ik} $d\sigma/dt$ of Q^- production in the reaction $K^-p \rightarrow \bar{Q}^-p$ at 8.25 GeV/c. Taken from ref. [58].

Fig. 41. Laboratory momentum dependence of π^-p elastic scattering and of the cross-section of the reaction $\pi^-p \rightarrow \pi^-\pi^-\pi^+p$ up to 205 GeV/c [61]. Taken from ref. [18].

Fig. 42. Effective mass distributions of the $(p\pi^+\pi^-)$ and $(\pi^+\pi^-\pi^-)$ systems in the reaction $\pi^-p \rightarrow \pi^-\pi^-\pi^+p$ at 2o5 GeV/c. Taken from ref. [61].

Fig. 43. Distribution of weighted events with respect to the possible quantum number transfer orderings in the five-body final state reactions $K^-p \rightarrow pK^-\pi^+\pi^-\pi^o$, $K^-p \rightarrow p\bar{K}^o\pi^+\pi^-\pi^-$ and $K^-p \rightarrow nK^-\pi^+\pi^+\pi^-$ at 6, 10, and 16 GeV/c. Taken from ref. [64].

Fig. 44. Momentum dependence of the cross-section of the reactions $K^-p \rightarrow K^-p\pi^+\pi^-\pi^o$, $K^-p \rightarrow p\bar{K}^o\pi^+\pi^-\pi^-$, and $K^-p \rightarrow nK^-\pi^+\pi^+\pi^-$ depending on the quantum number transfer. Taken from ref.[64].

Fig. 45. Mass dependence of the inclusive cross-section of the reaction $\pi^-p \rightarrow \pi^-X$, and $(p\pi^o)$ plus $(n\pi^+)$ mass dependence of the two pion production cross-section in the π^-p interactions at 16 GeV/c. The difference corresponds to the more-than-two pion mass distribution of the cross-section. An analogous separation of the inclusive mass distribution into three and more-than-three pion production is made. The shaded band corresponds to the argued distribution of the diffractive three pion system. Taken from ref. [65].

Fig. 46. Decomposition of the inclusive system X in the

reaction $\pi^- p \to \pi^- X$ at 16 GeV/c into the lower multiplicity particle systems and determination of the mass distribution fall-off of the diffractively produced exclusive particle systems. Phase space predictions are also given. Taken from ref. [65].

Fig. 47. Cylindrical phase space calculations for the probabilities of decay into the lower multiplicities and their dependence on the mass. Taken from ref. [65].

Fig. 48. a) Contribution to $d^2\sigma/dtdM^2$, for $M^2 = 4$, of s-channel helicity amplitudes in terms of the net helicity flip n in the case of t-channel helicity conservation of the diffractively produced particle system in the reaction $pp \to pX$ (M^2) at $s = 930$ GeV^2. Taken from ref. [63].
b) Profiles in impact parameter of the components of the overlap function in the case of t-channel helicity conservation. Taken from ref. [63].

Fig. 49. x-distribution of $E\frac{d^3\sigma}{dp^3}$ in arbitrary relative scale for the two inclusive reactions $pp \to pX^+$ ($p_T = 0.15$ GeV/c, full circles, ref.[68]) and $\pi^- p \to pX^-$ (integrated over p_T, full squares, ref. [69]) at $s = 551$ GeV^2 and 380 GeV^2, respectively. Taken from ref. [71].

Fig. 50. Average charged multiplicity of the diffractively produced multi-particle system as a function of its mass squared, for $p + p \to p + X$ (M) at 102 [79], 205 [72], and 303 GeV/c [73]. Taken from ref. [18].

Fig. 51. a) Comparison of the average charged multiplicity as a function of Q^2 for the reaction $\pi^- p \rightarrow n$ charged particles at 205 GeV/c [69], for which $Q = \sqrt{s} - m_\pi - m_p$ (crosses), and of the average charged multiplicity $\langle n_f \rangle$ of the diffractively produced multi-particle system in the reaction $\pi^- + p \rightarrow p + X(M)$, for which $Q = \sqrt{M^2} - m_\pi$ (full circles). Taken from ref. [71].

b) Comparison of the logarithmic (solid line) and linear (broken line) behaviour of the average charged multiplicity of the diffractively produced multi-particle system in the reaction $p + p \rightarrow p + X(M)$ at 102 [79], 205 [72], and 303 GeV/c [73]. Taken from ref. [71].

Fig. 52. Comparison of the charged multiplicity distributions of the inclusive process $\pi^- p \rightarrow$ anything and of pion diffraction in $\pi^- p \rightarrow pX^-$ at 205 GeV/c [69]. Taken from ref.[71].

Fig. 53. Laboratory momentum dependence of the ratio $\langle n \rangle / D$. Taken from ref. [90].

Fig. 54. Laboratory momentum dependence of the ratio $\langle n \rangle / D$ for pp-(full circles), $\bar{p}$p-(open circles), π^-p-(triangles pointing downwards), π^+p- (squares), and K^+p-collisions (triangles pointing upwards) [91]. Taken from a compilation of data presented at the 2nd Aix-en-Provence International Conference on Elementary Particles, September 1973, by J. Lach (see ref. [71]). The broken lines are eye-guiding interpolations of the pp-, π^-p-, and $\bar{p}$p-collision data.

Fig. 55. Two-particle correlation function f_2 in its

dependence on the laboratory momentum for $\pi^{\pm}p$, $K^{\pm}p$, and pp-collisions. Taken from ref. [90].

Fig. 56. Dependence of the second to the fourth moment of multiplicity distribution for pp scattering (5.5 to 303 GeV/c) on the charged multiplicity. The data correspond to data compilations by Buras et al. [94], and Fialkowski and Miettinen [95], whose fits are given by the solid and broken line curves, respectively. Taken from ref. [96].

Fig. 57. a) $<n_{ch}>$ dependence of the expression $(<n_{ch}>-1)/D$ for $pp \rightarrow n_{ch}$ according to ref. [93]. Taken from ref. [71].

b) Dependence of the multiplicity distribution $(<n_{ch}> - 0.9)\sigma_n/\sigma_{inel}$ on the variable $z' = (n_{ch} - 0.9)/(<n_{ch}> - 0.9)$ for $pp \rightarrow n_{ch}$ in the energy regions $5.5 < p_{lab} < 28$ GeV/c (full circles), $28 < p_{lab} < 70$ GeV/c (triangles), and $150 < p_{lab} < 350$ GeV/c (crosses) according to ref. [94]. Taken from ref. [71].

Fig. 58. a) p_T dependence of the average charged multiplicity in the projectile frame of the reaction $p + p \rightarrow p$ + anything at 28.5 GeV/c [101] for various missing mass intervals. Taken from ref. [102].

b) Energy dependence of the invariant cross-section $E\frac{d^3\sigma}{dp^3}$ for high transverse momenta in $pp \rightarrow \pi^{\pm}$+ anything at 90° at ISR energies according to ref. [103]. Taken from ref. [104].

Fig. 59. Laboratory momentum dependence of the cross-sections of the different charge states of the reaction $K^-p \rightarrow N\bar{K}\pi\pi$ and their decomposition into the diffractive and non-diffractive components. Taken from ref. [107].

Fig. 60. Multiplicity distribution (with inclusion of the multiplicity of neutrals) of the reaction $K^-p \rightarrow N\bar{K} + (n-2)$ pions at 10 GeV/c, and its decomposition into the "true" diffractive and non-diffractive multiplicity distributions. Taken from ref. [107].

Fig. 61. Average π^0 multiplicity as a function of the charged multiplicity for pp interactions at two ISR energies from an experiment of the CERN-Hamburg-Vienna Collaboration [110]: $\sqrt{s}$ = 23 GeV/c (open circles), $\sqrt{s}$ = 53 GeV/c (crosses). Taken from ref. [71].

Fig. 62. Laboratory momentum dependence of the "true" multiplicity of the reaction $K^-p \rightarrow \bar{K}N + (n-2)$ pions as predicted from the parametrization (VII.6.4-5). Taken from ref. [107].

Fig. 63. Centre-of-mass charge transfer distributions of the reaction $K^-p \rightarrow$ anything inelastic at 10 GeV/c (broken lines) and at 16 GeV/c (full lines). Taken from ref. [111].

Fig. 64. Squared dispersion of the charge transfer depending on the charged multiplicity for the reaction $K^-p \rightarrow$ anything inelastic at 10 GeV/c (broken lines) and 16 GeV/c (full lines). Taken from ref. [111].

Fig. 65. Charge per unit of rapidity in the incident kaon rest frame. Taken from ref. [111].

Fig. 66. Charge per unit of rapidity in the target proton rest frame. Taken from ref. [111].

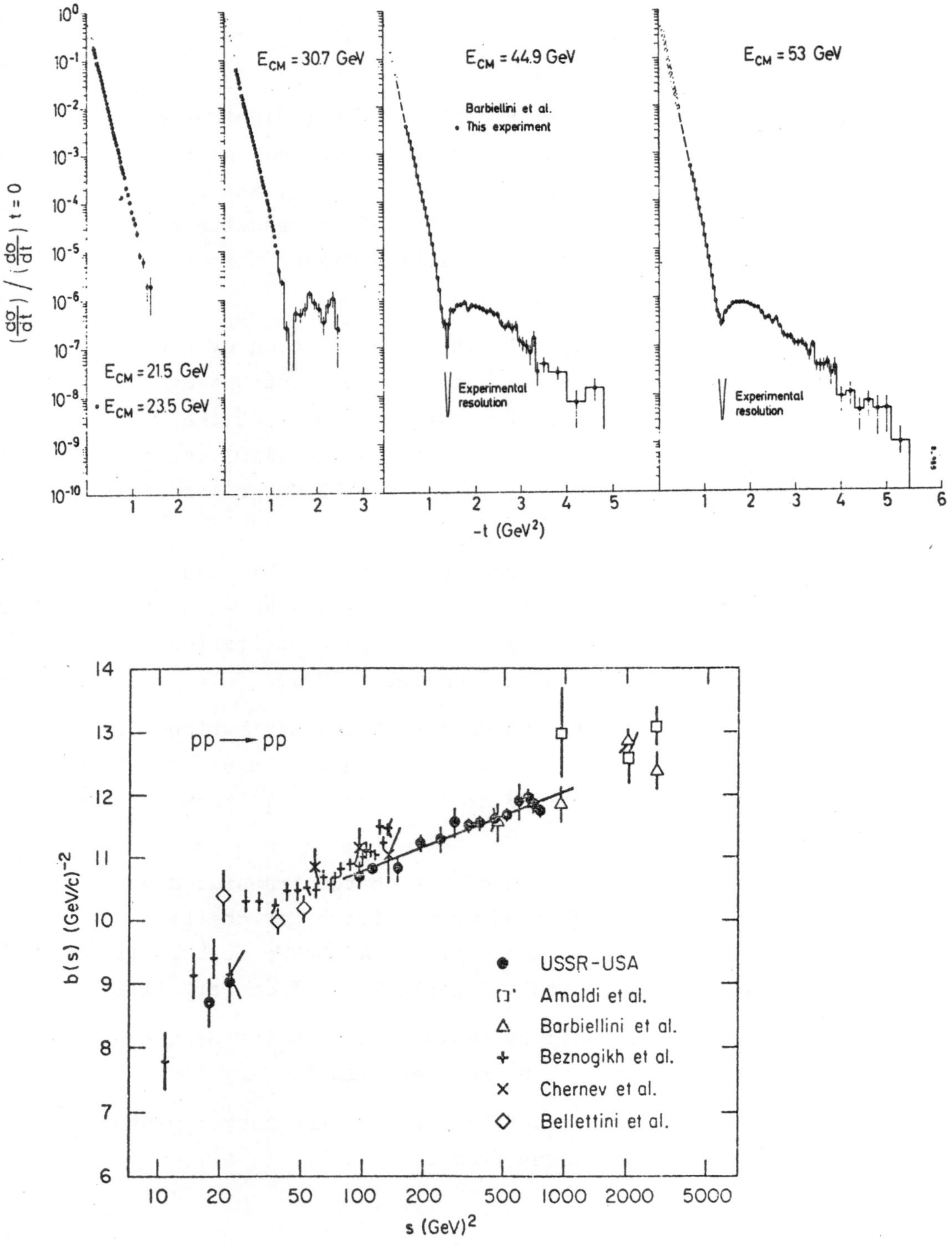

E_{CM} = 30.7 GeV
E_{CM} = 44.9 GeV
Barbiellini et al.
• This experiment
E_{CM} = 53 GeV
E_{CM} = 21.5 GeV
• E_{CM} = 23.5 GeV
Experimental resolution
Experimental resolution
-t (GeV²)
pp → pp
USSR-USA
Amaldi et al.
Barbiellini et al.
Beznogikh et al.
Chernev et al.
Bellettini et al.
b(s) (GeV/c)⁻²
s (GeV)²

Fig. 1

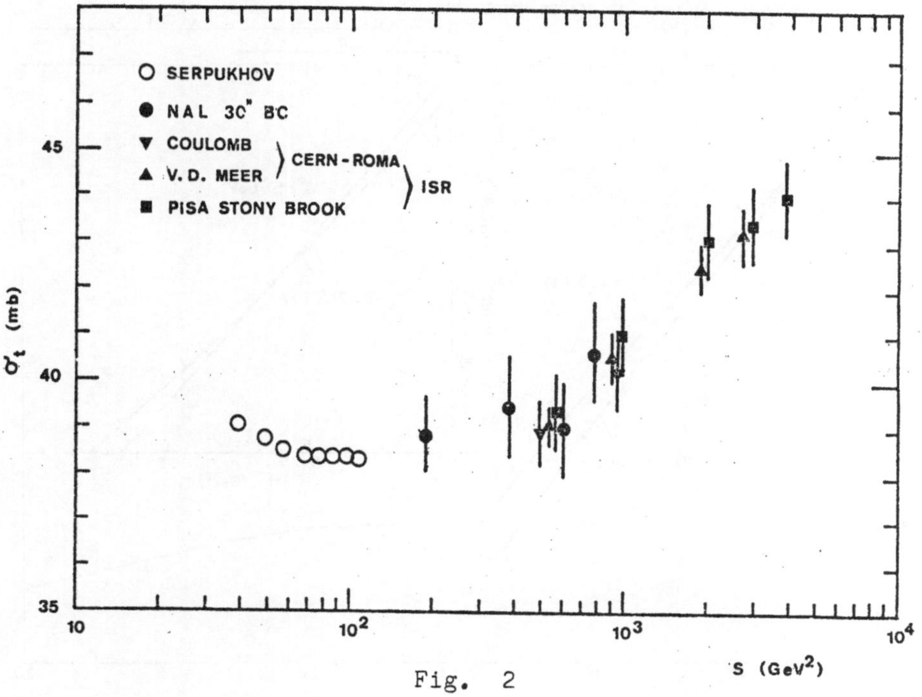

SERPUKHOV
NAL 30" BC
COULOMB
V. D. MEER
CERN-ROMA
PISA STONY BROOK
ISR
σt (mb)
45
40
35
10
10^2
10^3
10^4
S (GeV^2)

Fig. 2

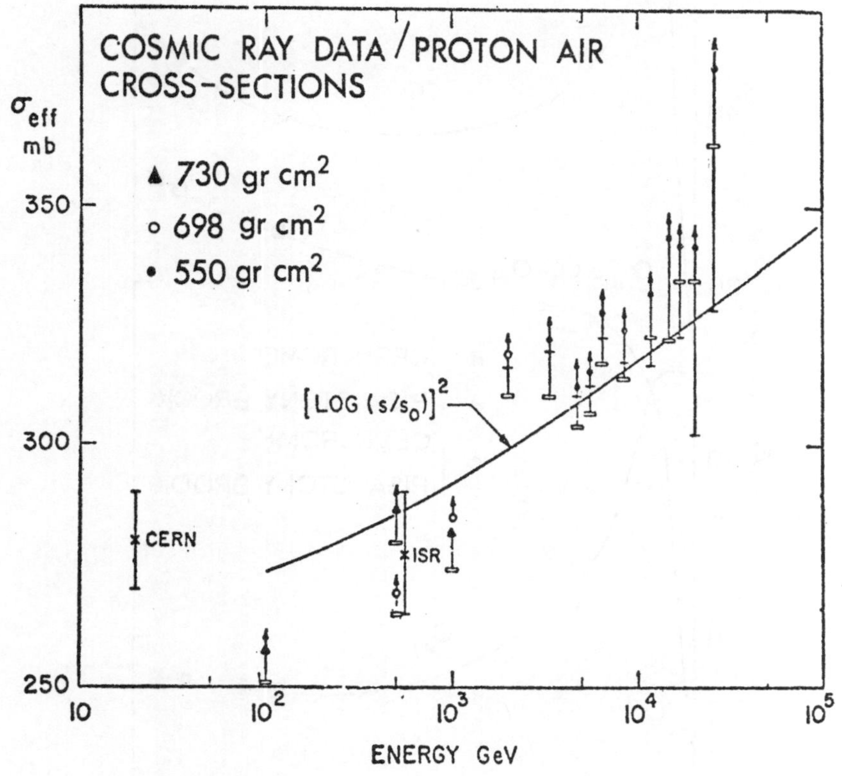

COSMIC RAY DATA / PROTON AIR CROSS-SECTIONS
σeff mb
730 gr cm^2
698 gr cm^2
550 gr cm^2
350
300
250
[LOG (s/s0)]^2
CERN
ISR
10
10^2
10^3
10^4
10^5
ENERGY GeV

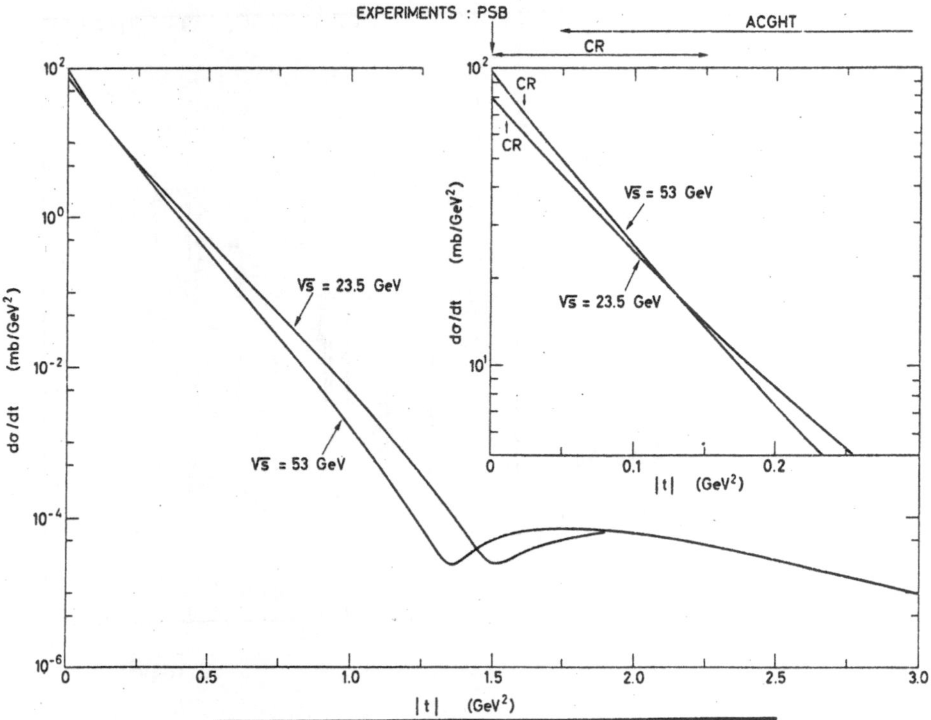
EXPERIMENTS : PSB
ACGHT
CR
CR
CR
√s = 53 GeV
√s = 23.5 GeV
dσ/dt (mb/GeV²)
|t| (GeV²)
√s = 23.5 GeV
√s = 53 GeV

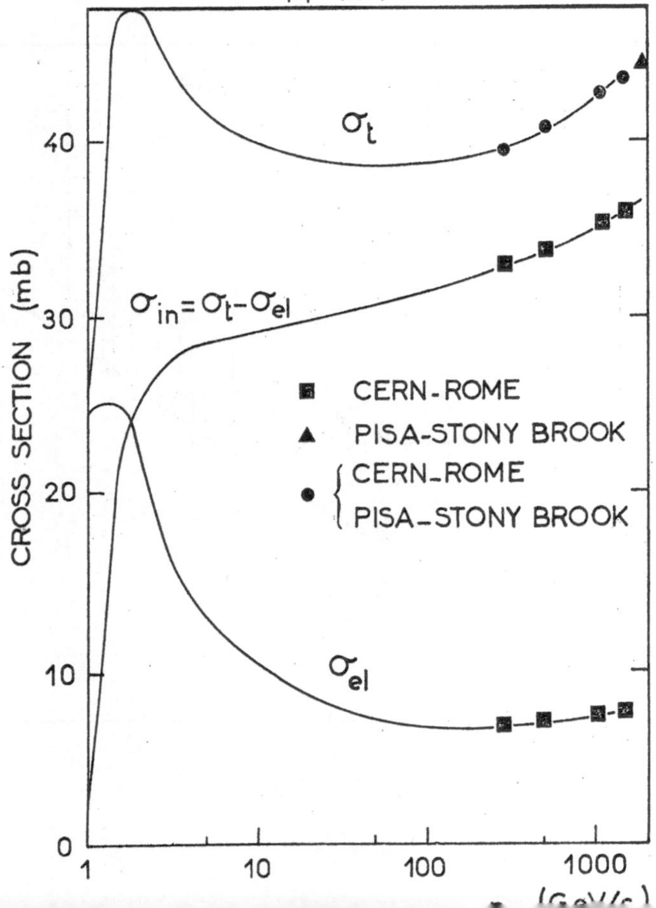
σt
σin=σt−σel
σel
CROSS SECTION (mb)
CERN-ROME
PISA-STONY BROOK
CERN-ROME
PISA-STONY BROOK

Fig. 3

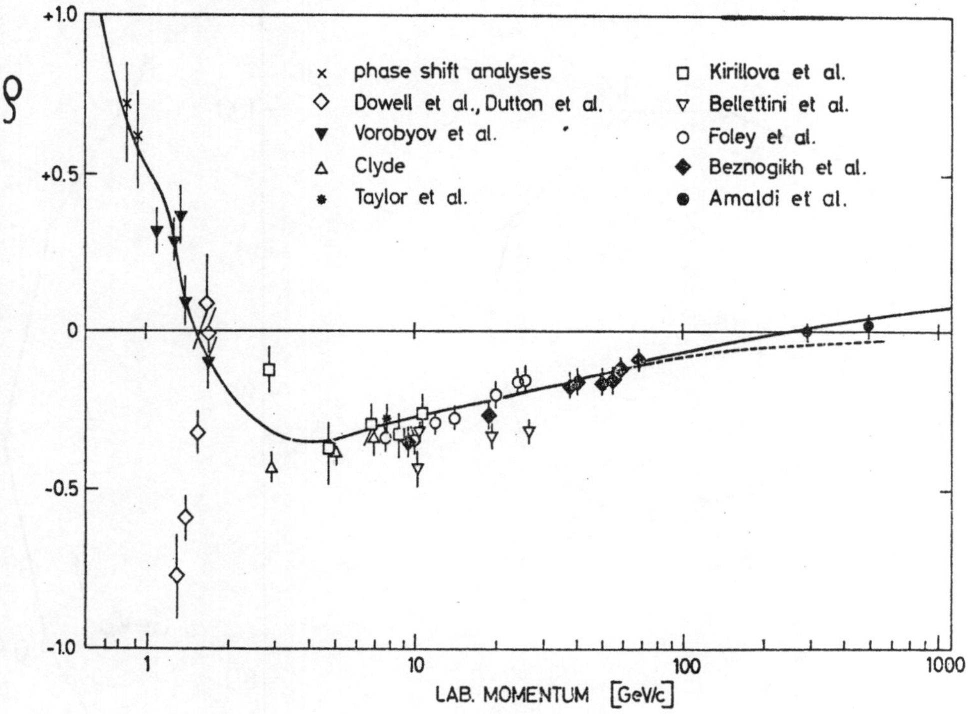
ρ
+1.0
+0.5
0
-0.5
-1.0
× phase shift analyses
◇ Dowell et al., Dutton et al.
▼ Vorobyov et al.
△ Clyde
* Taylor et al.
□ Kirillova et al.
▽ Bellettini et al.
○ Foley et al.
◆ Beznogikh et al.
● Amaldi et al.
1
10
100
1000
LAB. MOMENTUM [GeV/c]

Fig. 4

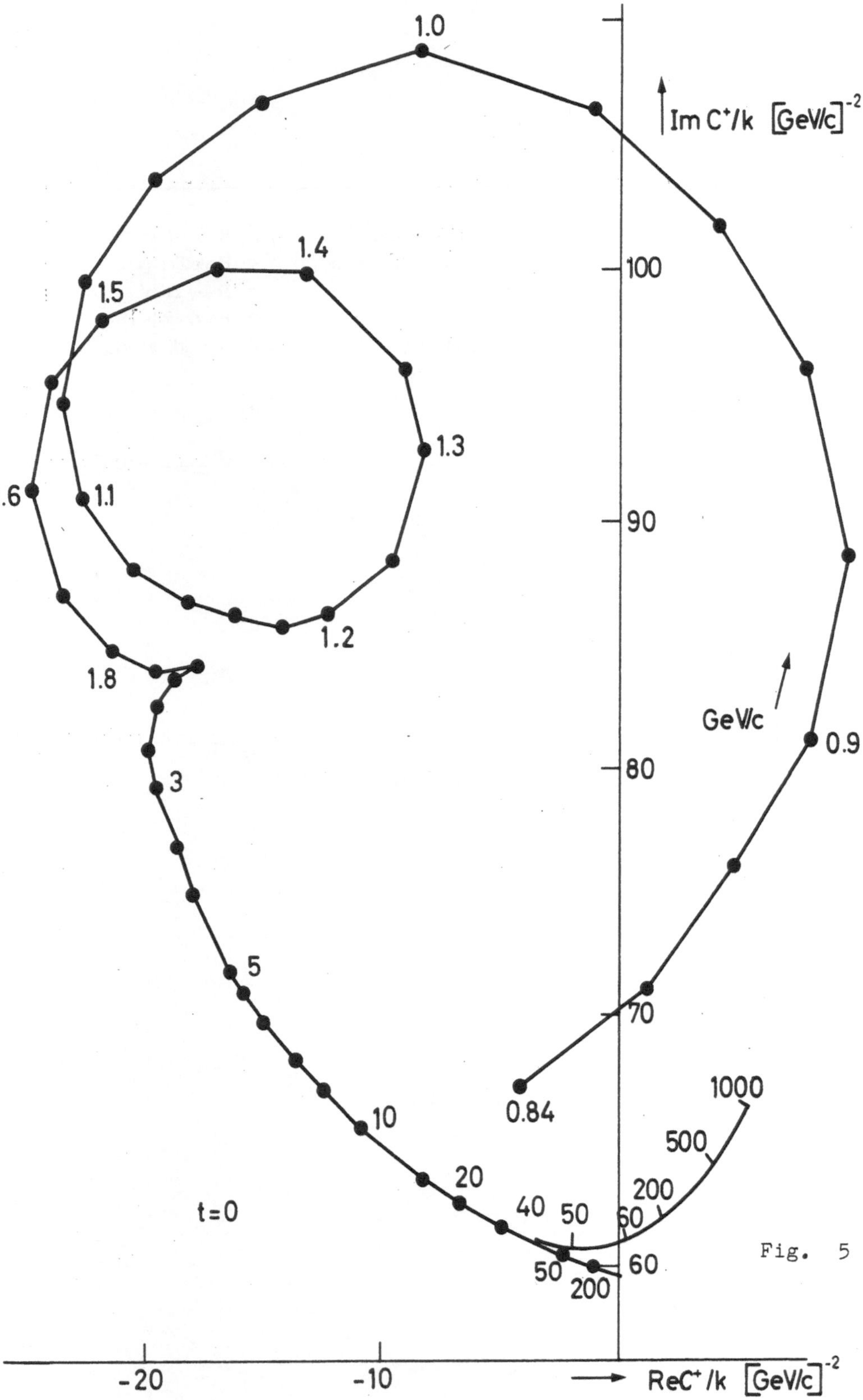
1.0
Im C⁺/k [GeV/c]^-2
1.4
100
1.5
1.3
1.6
1.1
90
1.2
1.8
GeV/c
0.9
3
80
5
70
0.84
1000
10
500
20
200
t=0
40
50
60
50
60
200
-20
-10
ReC⁺/k [GeV/c]^-2

Fig. 5

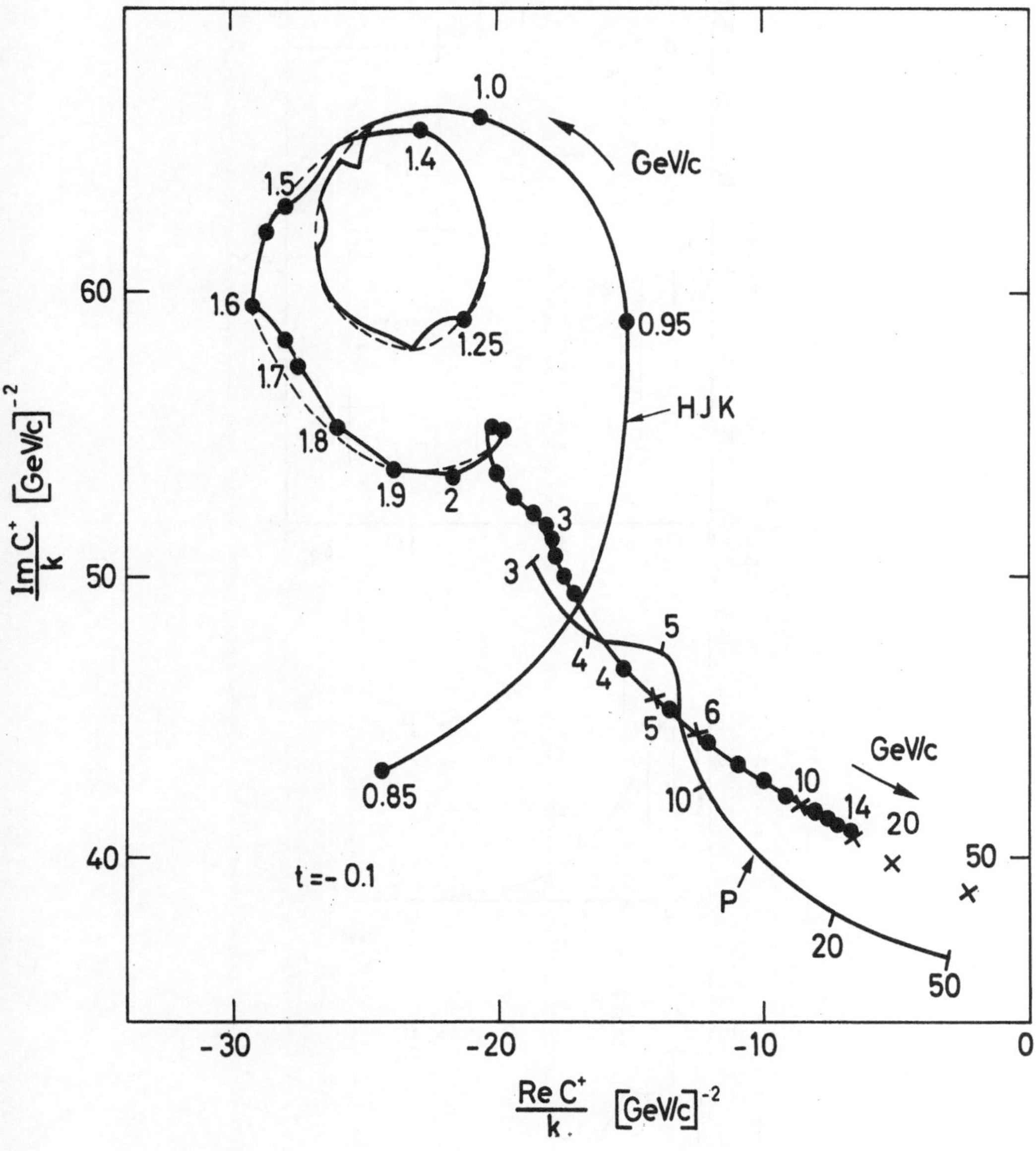
1.0
GeV/c
1.4
1.5
60
1.6
0.95
1.25
1.7
HJK
1.8
1.9
2
3
50
3
5
4
4
6
5
GeV/c
0.85
10
10
14
20
40
50
t = - 0.1
P
20
50
- 30
- 20
- 10
0
Re C+ / k [GeV/c]-2
Im C+ / k [GeV/c]-2

Fig. 6

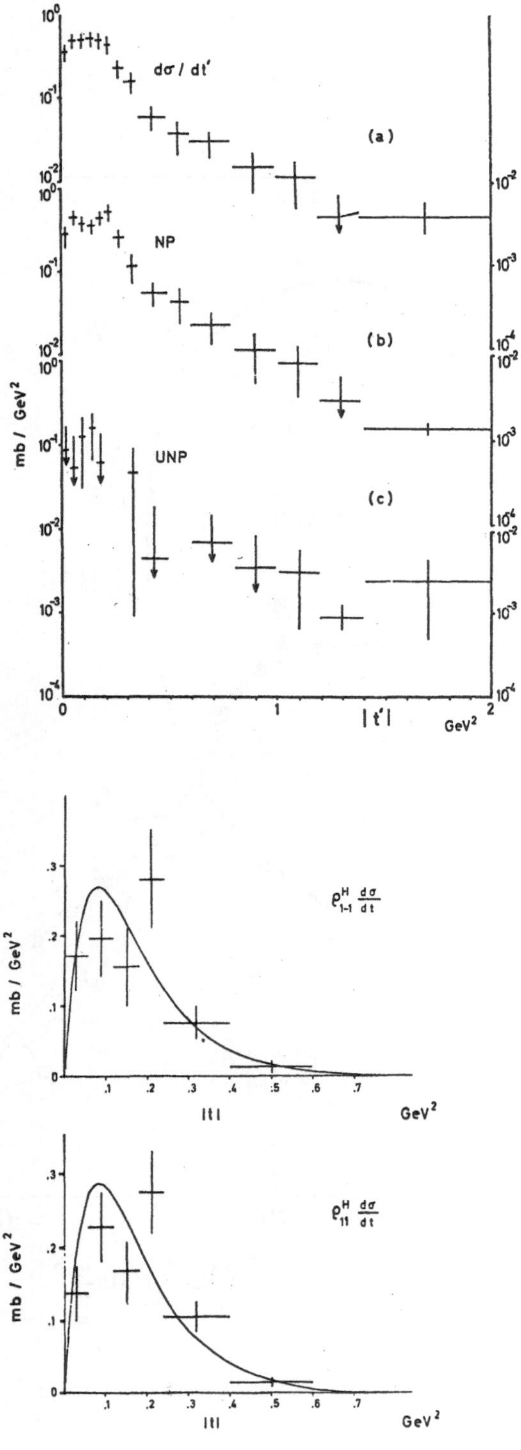
dσ / dt'
(a)
NP
(b)
UNP
(c)
mb / GeV²
|t'|
GeV²
$\rho^H_{1-1} \frac{d\sigma}{dt}$
$\rho^H_{11} \frac{d\sigma}{dt}$
|t|
GeV²

Fig. 7

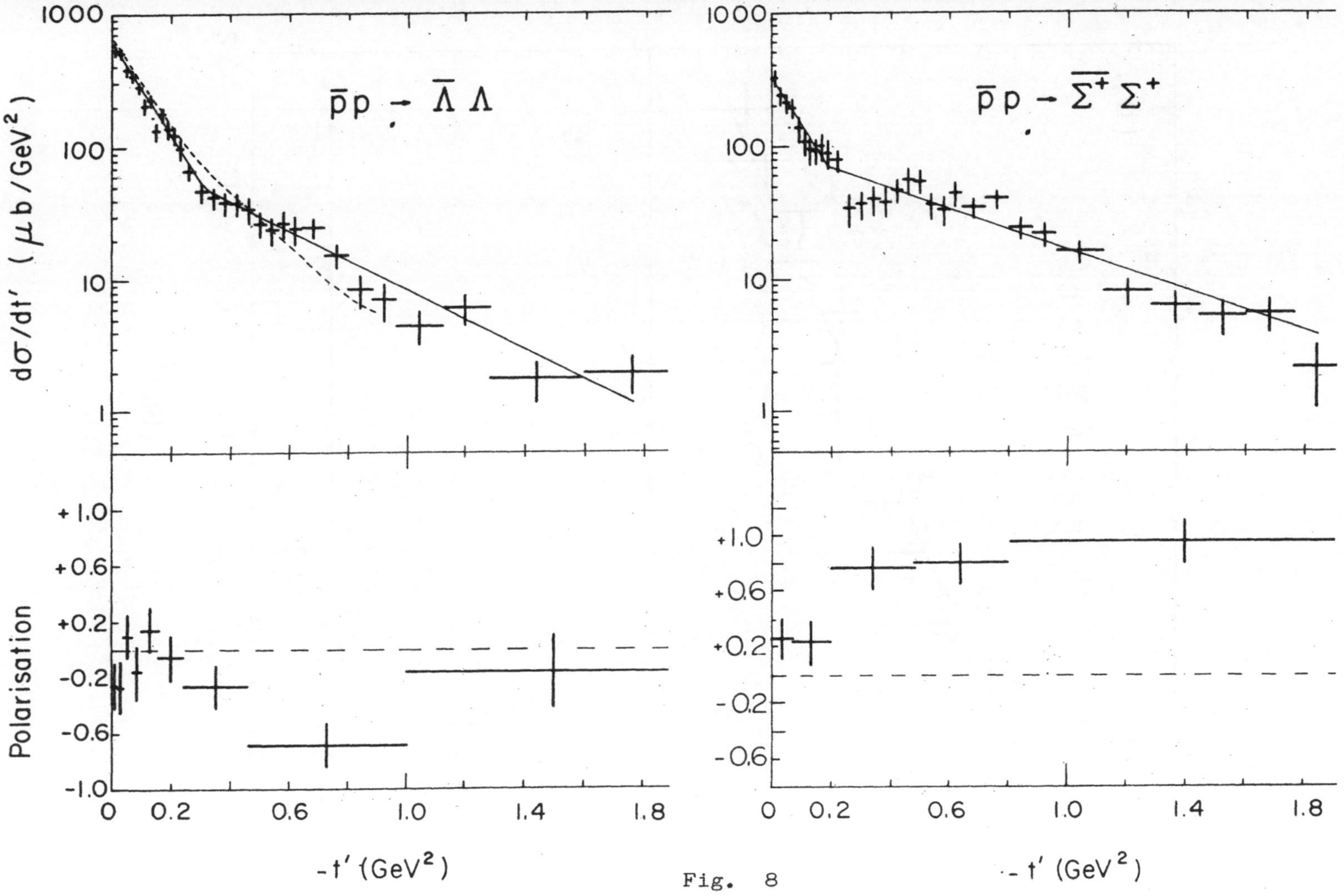

$\bar{p}p \to \bar{\Lambda}\Lambda$
$\bar{p}p \to \overline{\Sigma^+}\Sigma^+$
dσ/dt′ (μb/GeV²)
1000
100
10
1
Polarisation
+1.0
+0.6
+0.2
-0.2
-0.6
-1.0
0
0.2
0.6
1.0
1.4
1.8
-t′ (GeV²)

Fig. 8

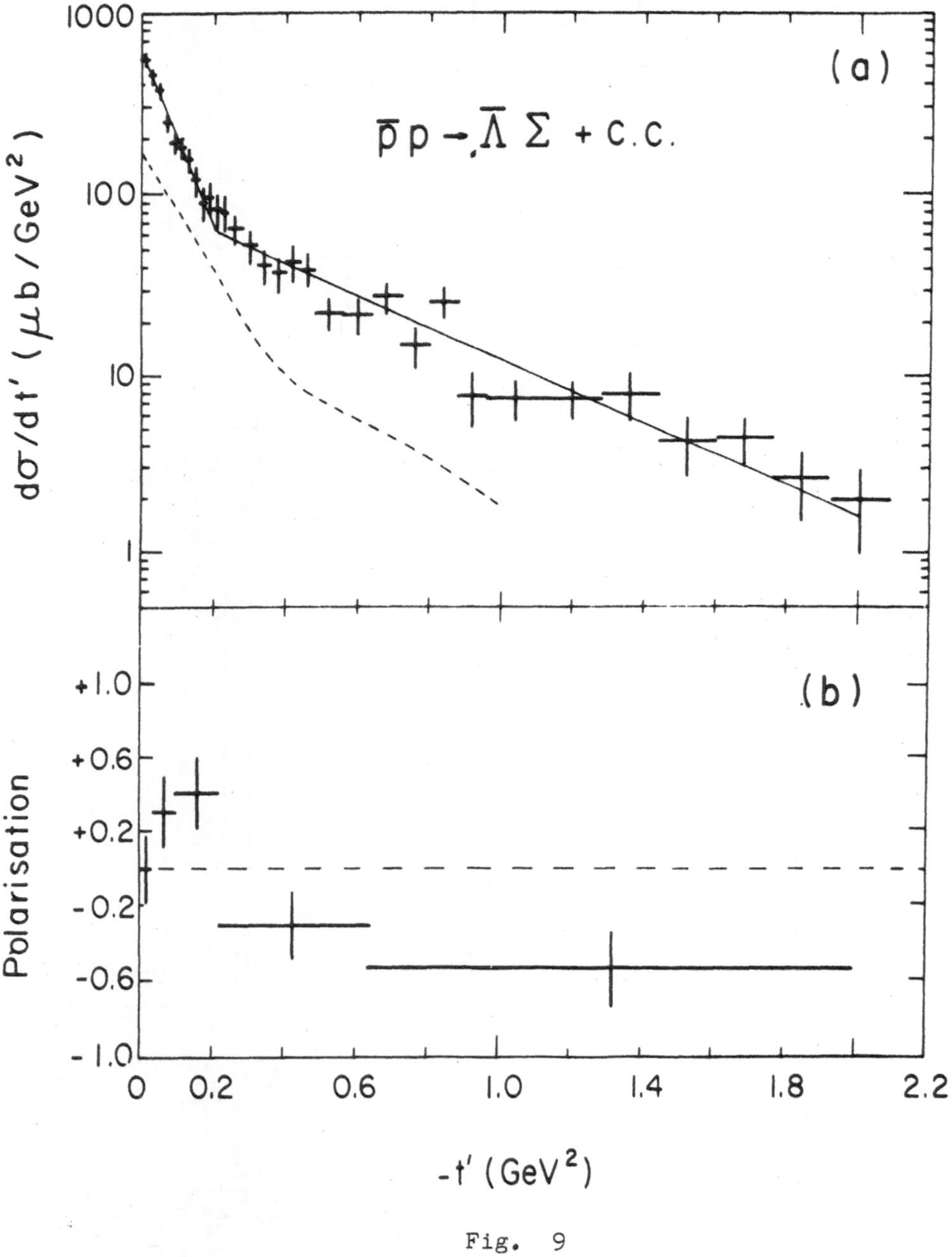

1000
100
10
1
(a)
p̄p → Λ̄ Σ + c.c.
dσ/dt′ (μb / GeV²)
+1.0
+0.6
+0.2
-0.2
-0.6
-1.0
(b)
Polarisation
0
0.2
0.6
1.0
1.4
1.8
2.2
-t′ (GeV²)

Fig. 9

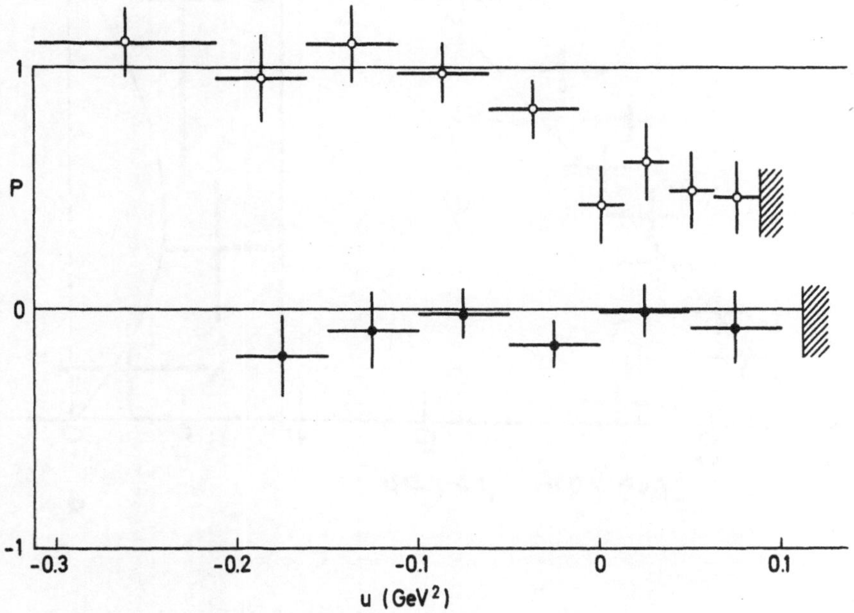

Fig. 10

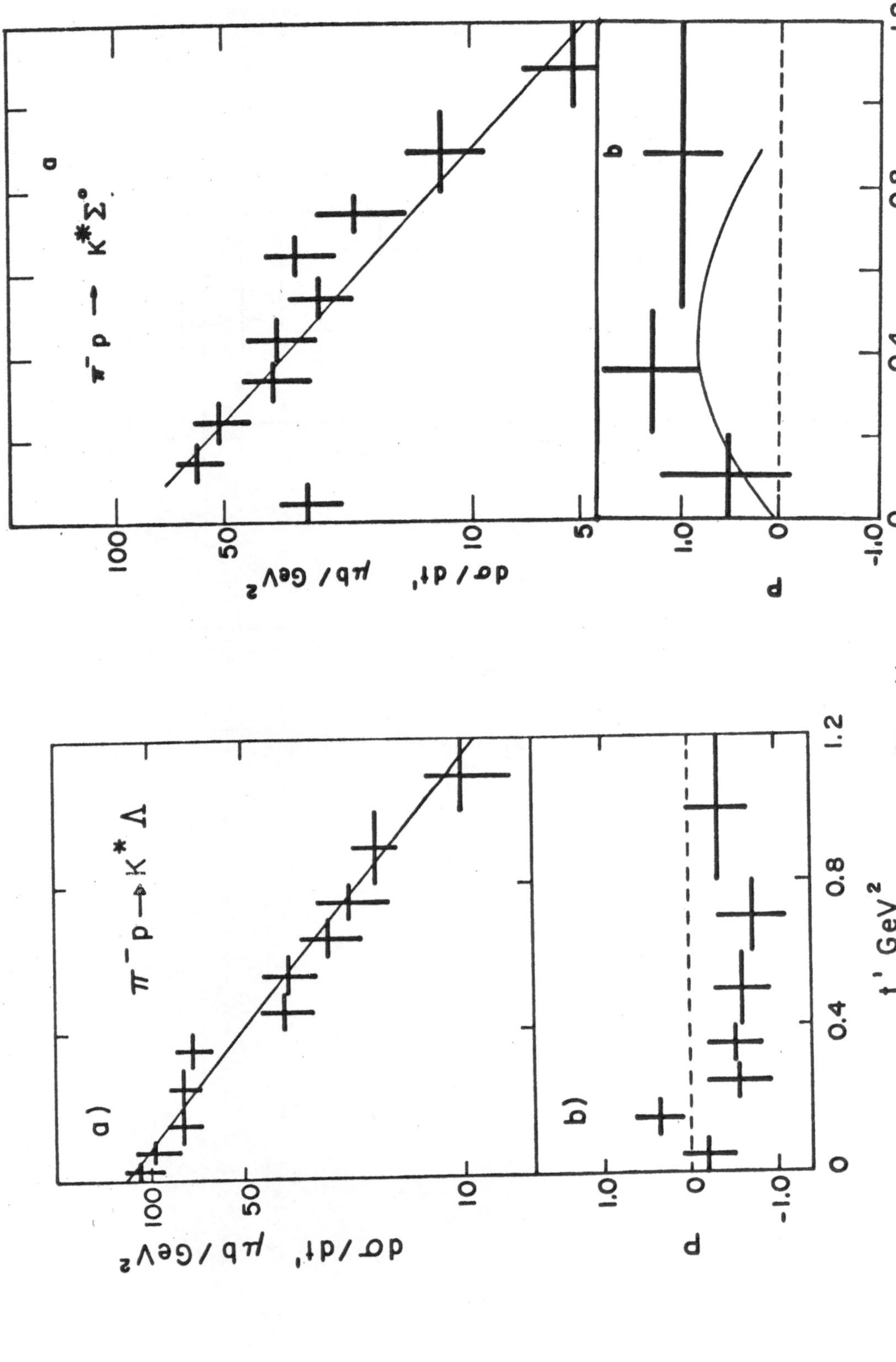

Fig. 11

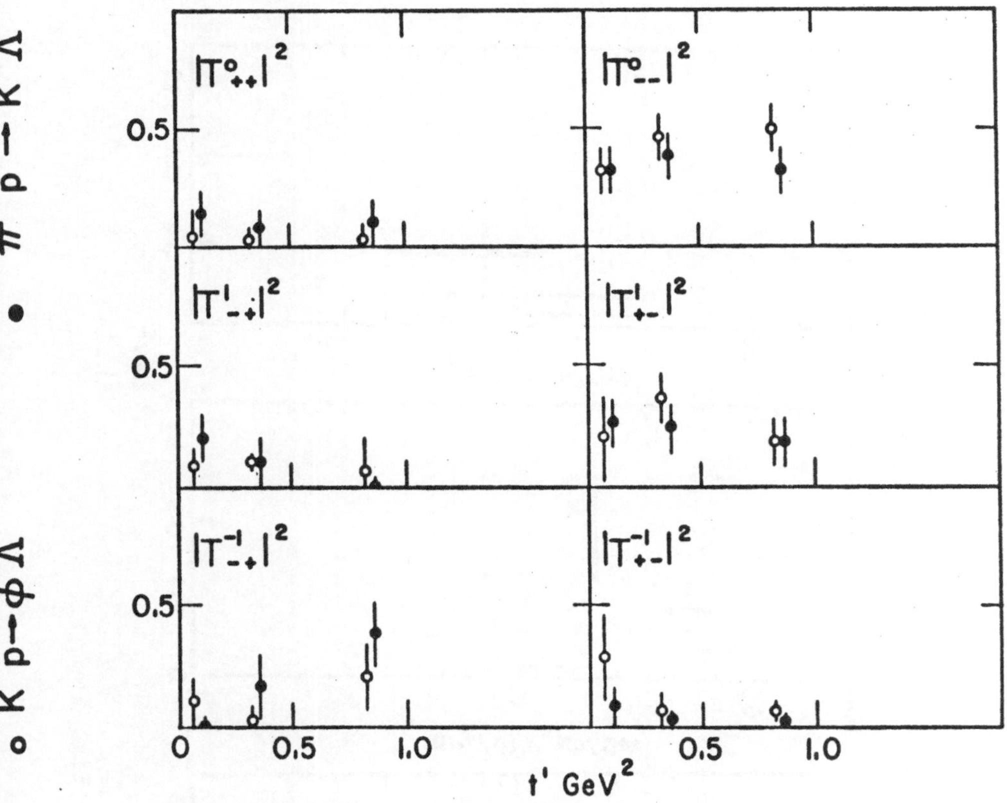
• π⁻p → K*Λ
∘ K⁻p → φΛ
|T⁰₊₊|²
|T⁰₋₋|²
|T¹₋₊|²
|T¹₊₋|²
|T⁻¹₋₊|²
|T⁻¹₊₋|²
0.5
0
1.0
t' GeV²

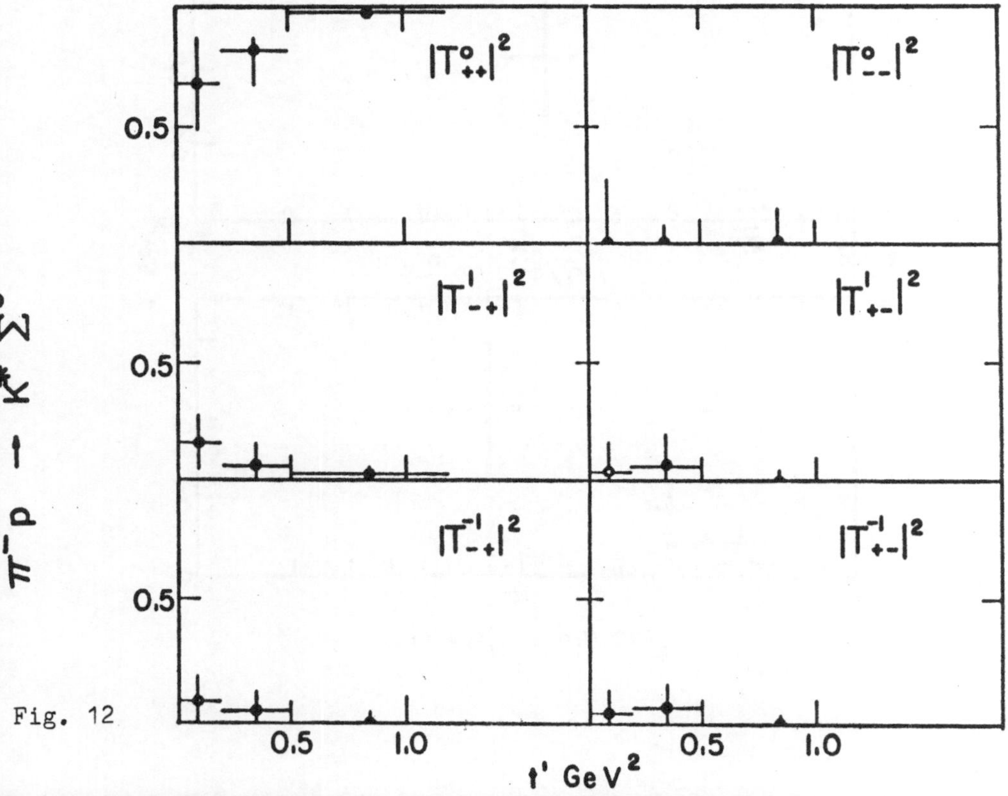
π⁻p → K*Σ°
|T⁰₊₊|²
|T⁰₋₋|²
|T¹₋₊|²
|T¹₊₋|²
|T⁻¹₋₊|²
|T⁻¹₊₋|²
0.5
1.0
t' GeV²

Fig. 12

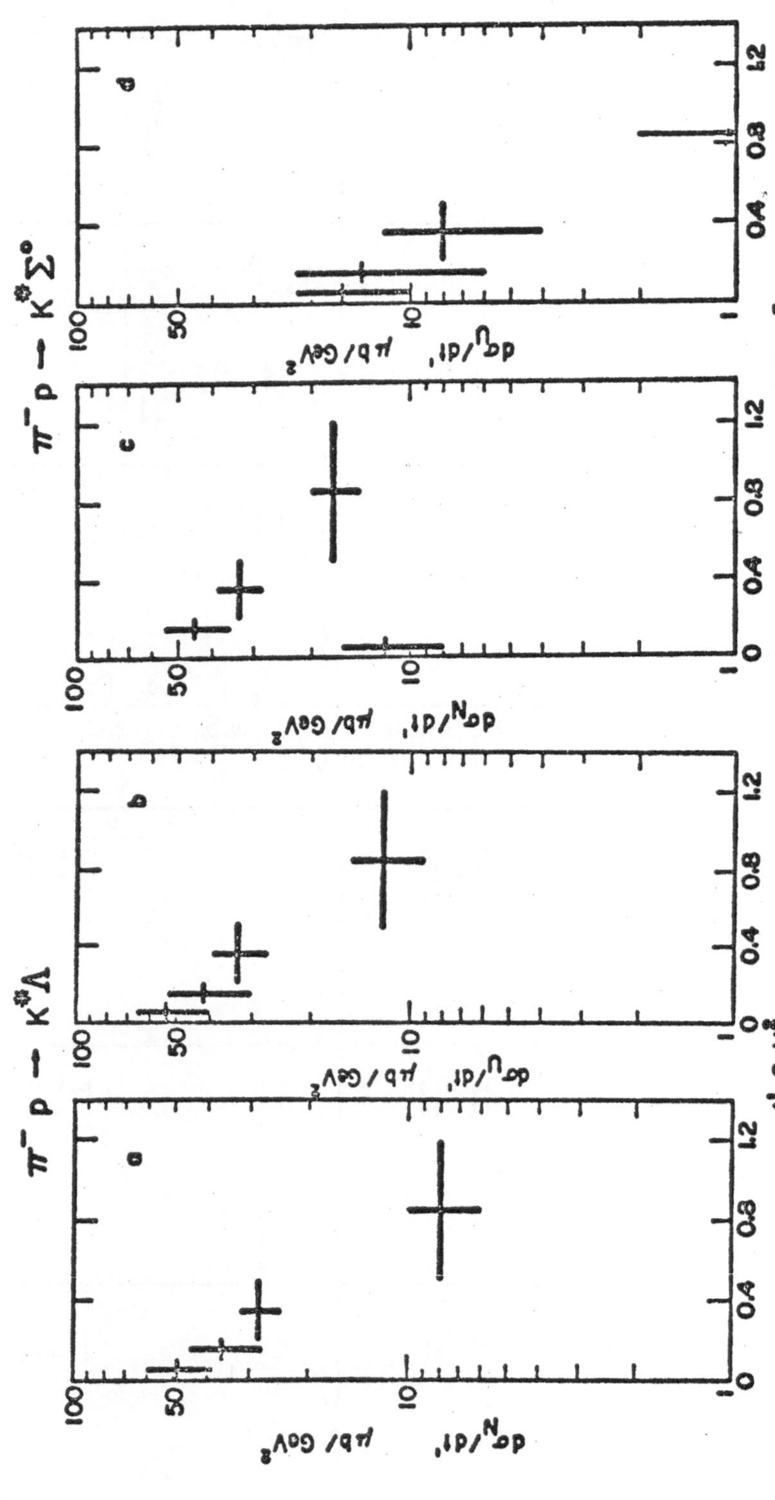
π⁻p → K*Λ
π⁻p → K*Σ°
a
b
c
d
dσ_N/dt', μb/GeV²
dσ_U/dt', μb/GeV²
t' GeV²
100
50
10
1
0
0.4
0.8
1.2

Fig. 13

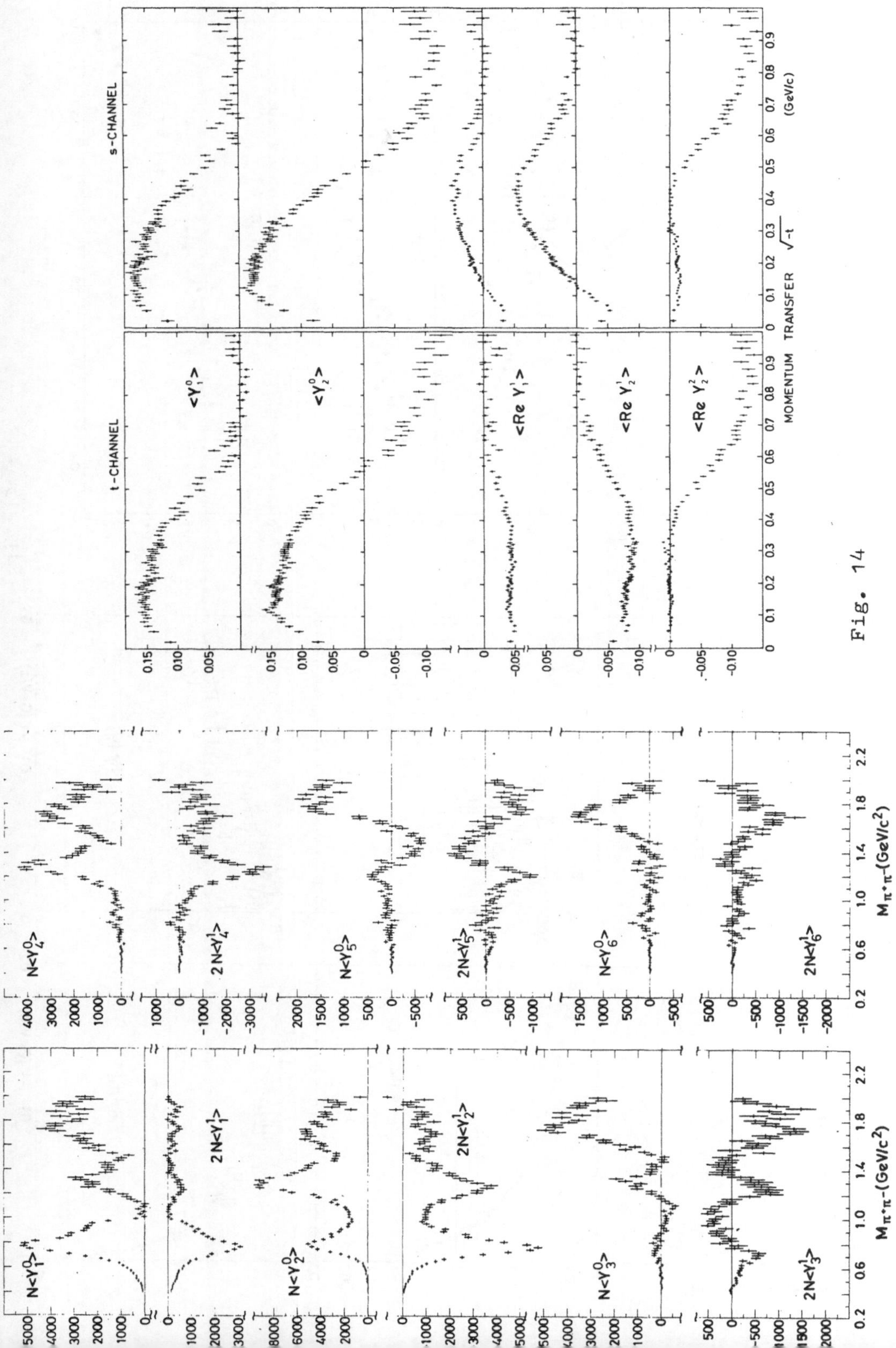

t-CHANNEL
s-CHANNEL
<Y$_1^0$>
<Y$_2^0$>
<Re Y$_1^1$>
<Re Y$_2^1$>
<Re Y$_2^2$>
MOMENTUM TRANSFER $\sqrt{-t}$ (GeV/c)
N<Y$_1^0$>
2N<Y$_1^1$>
N<Y$_2^0$>
2N<Y$_2^1$>
N<Y$_3^0$>
2N<Y$_3^1$>
N<Y$_4^0$>
2N<Y$_4^1$>
N<Y$_5^0$>
2N<Y$_5^1$>
N<Y$_6^0$>
2N<Y$_6^1$>
M$_{\pi^+\pi^-}$ (GeV/c^2)

Fig. 14

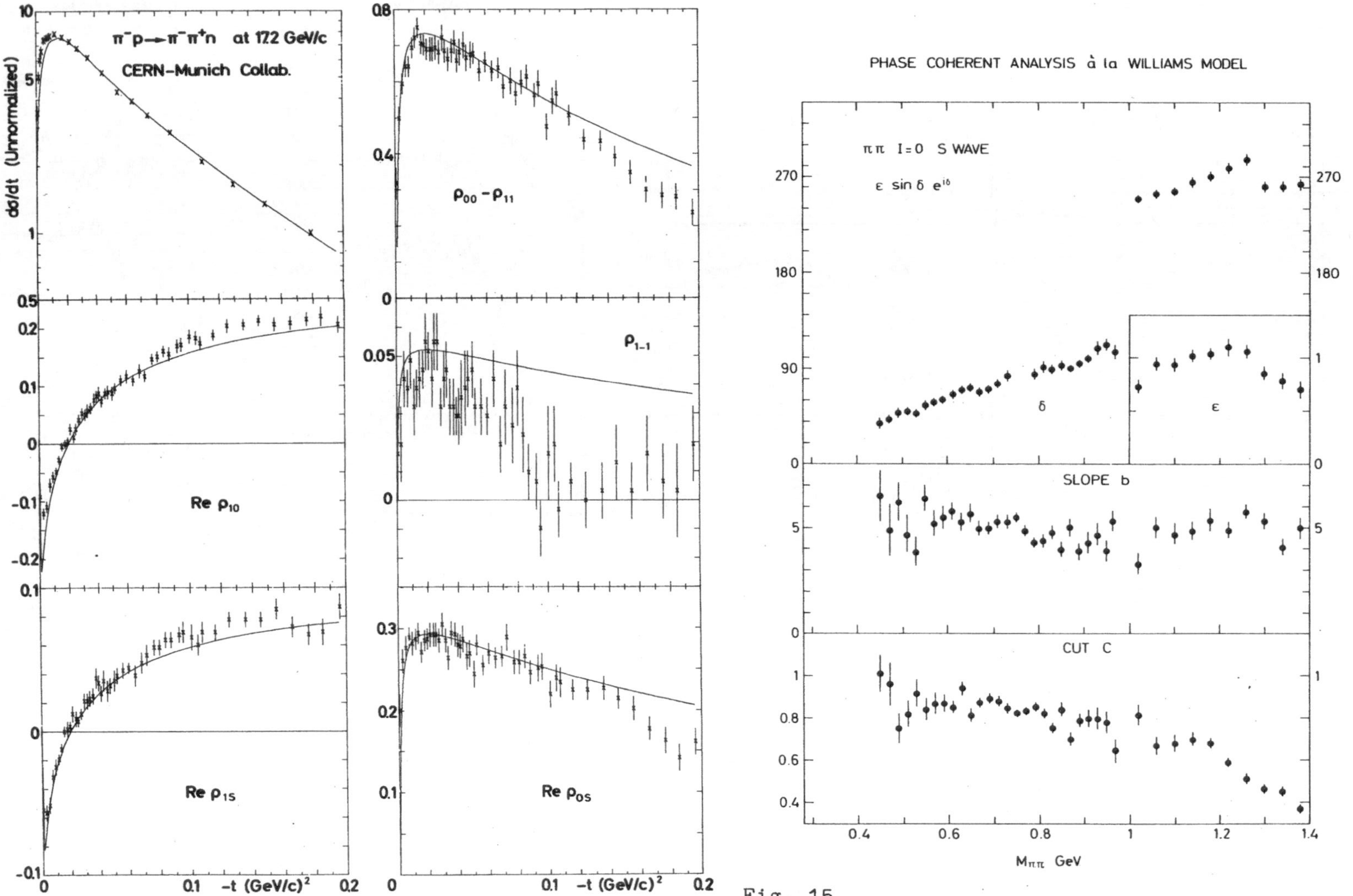

π⁻p → π⁻π⁺n at 17.2 GeV/c
CERN-Munich Collab.
dσ/dt (Unnormalized)
Re ρ10
Re ρ1S
-t (GeV/c)²
ρ00 - ρ11
ρ1-1
Re ρ0S
PHASE COHERENT ANALYSIS à la WILLIAMS MODEL
ππ I=0 S WAVE
ε sin δ e^iδ
δ
ε
SLOPE b
CUT C
Mππ GeV

Fig. 15

S- and P-Wave $\pi\pi$ Phases

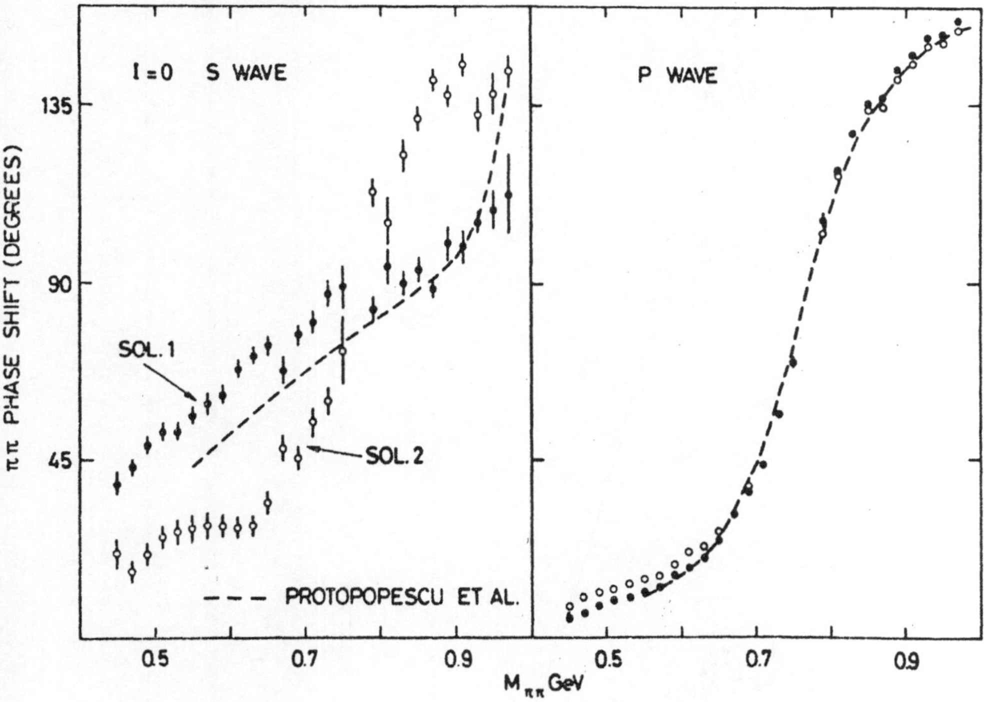

Fig. 16

ρ PRODUCTION AMPLITUDES AT 17 GeV/c

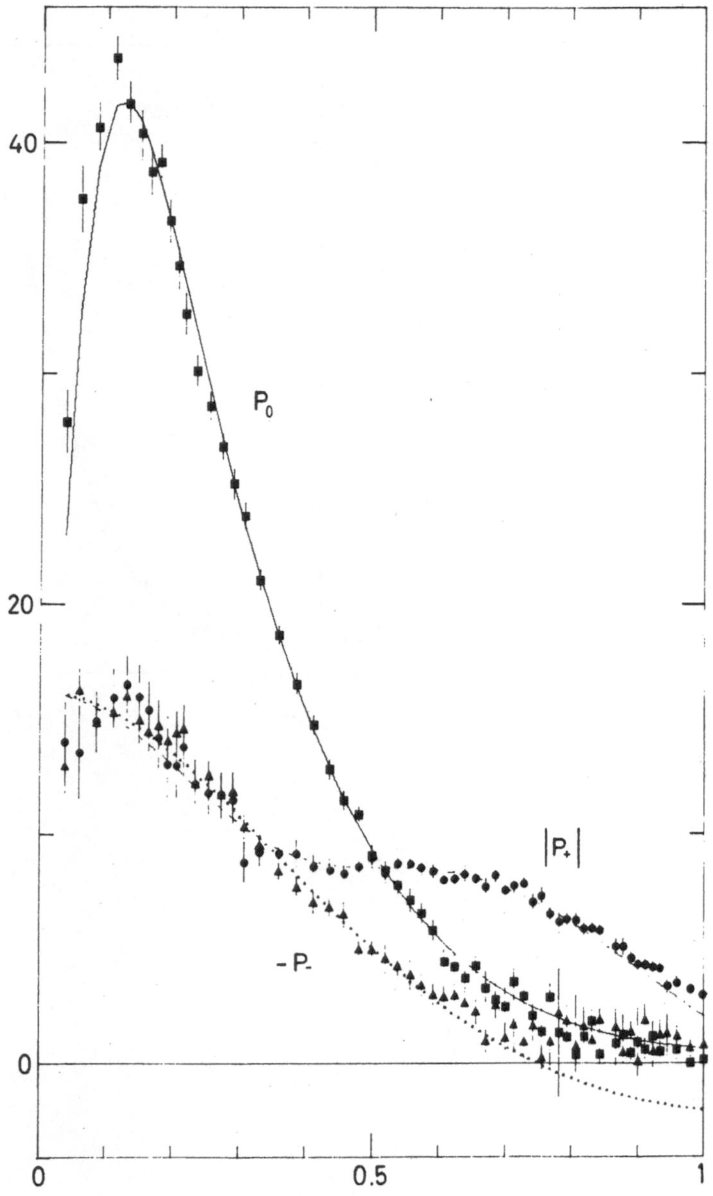

Fig. 17

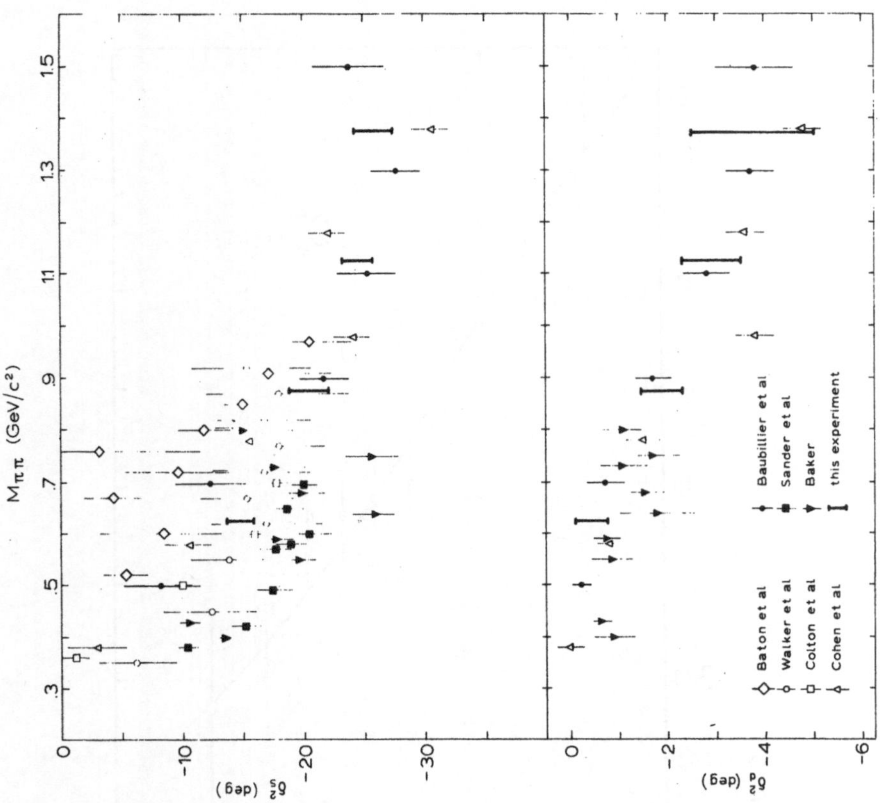

$M_{\pi\pi}$ (GeV/c²)
δ_2^S (deg)
δ_2^D (deg)
Baton et al
Walker et al
Colton et al
Cohen et al
Baubillier et al
Sander et al
Baker
this experiment

Fig. 18

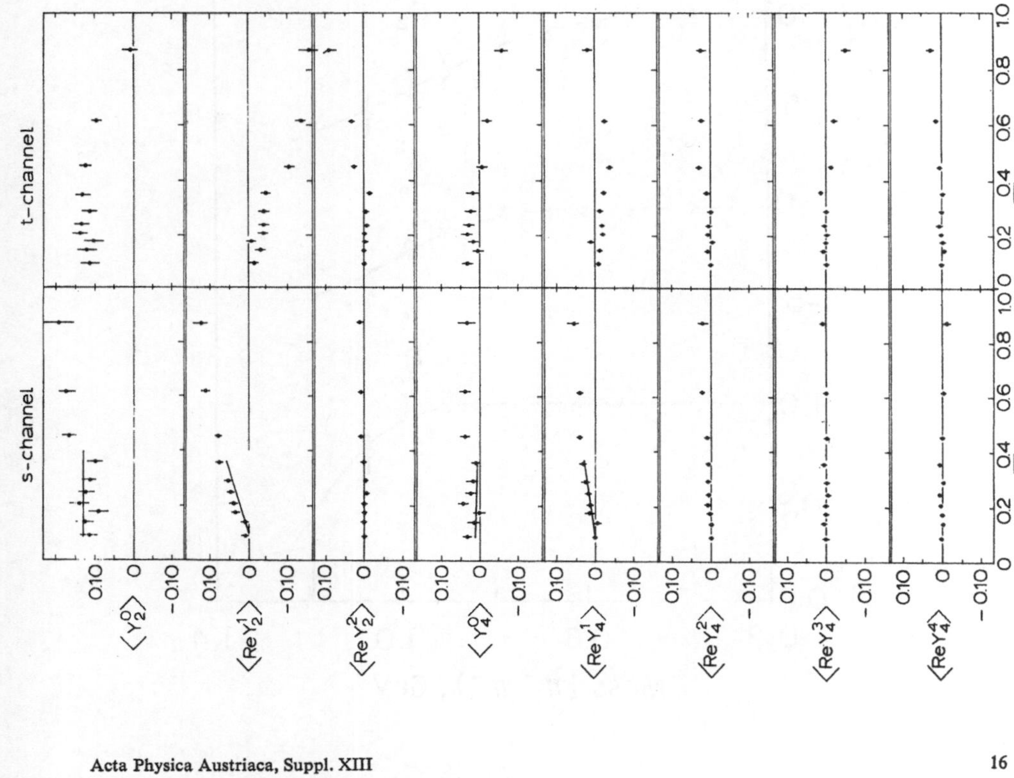

s-channel
t-channel
$\langle Y_2^0 \rangle$
$\langle ReY_2^1 \rangle$
$\langle ReY_2^2 \rangle$
$\langle Y_4^0 \rangle$
$\langle ReY_4^1 \rangle$
$\langle ReY_4^2 \rangle$
$\langle ReY_4^3 \rangle$
$\langle ReY_4^4 \rangle$
$\sqrt{-t}$ (GeV/c)

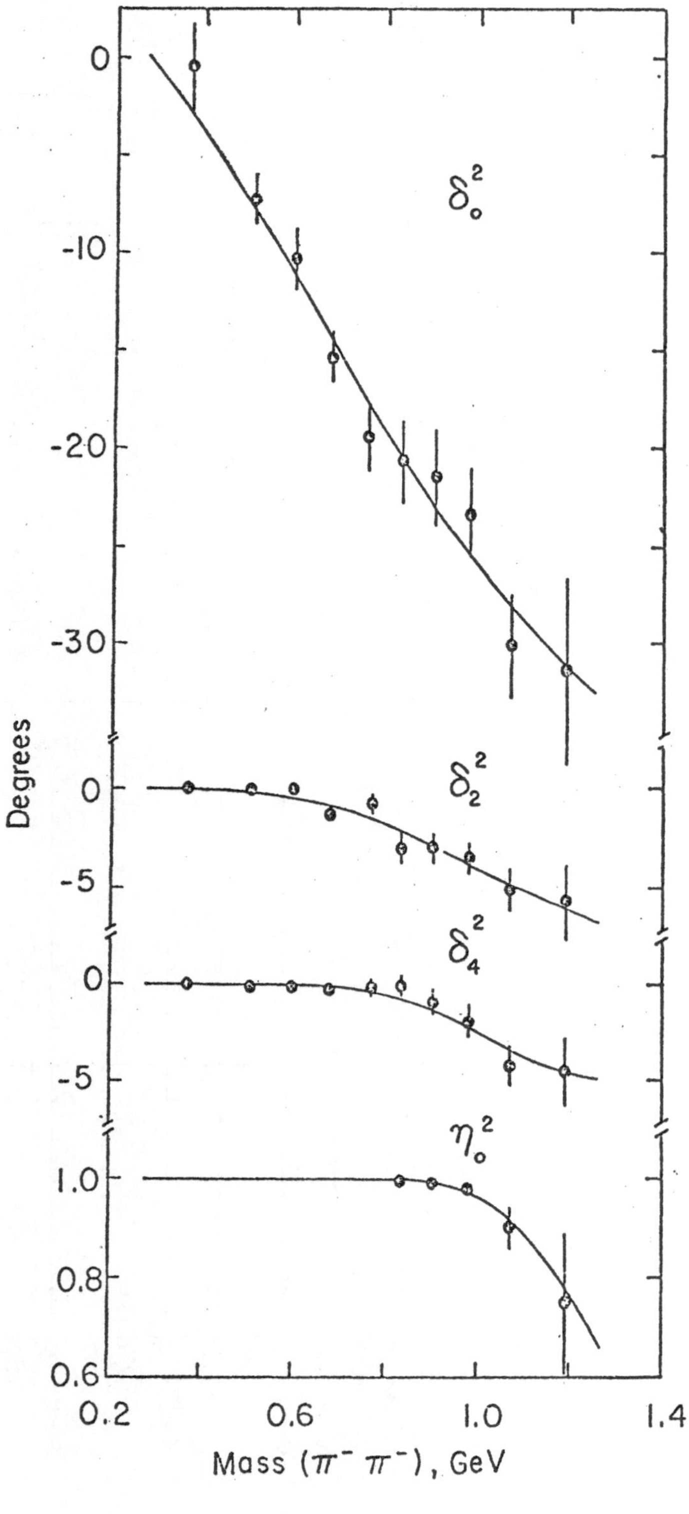

0
-10
-20
-30
0
-5
0
-5
1.0
0.8
0.6
0.2
0.6
1.0
1.4
Degrees
δ_0^2
δ_2^2
δ_4^2
η_0^2
Mass ($\pi^- \pi^-$), GeV

Fig. 19

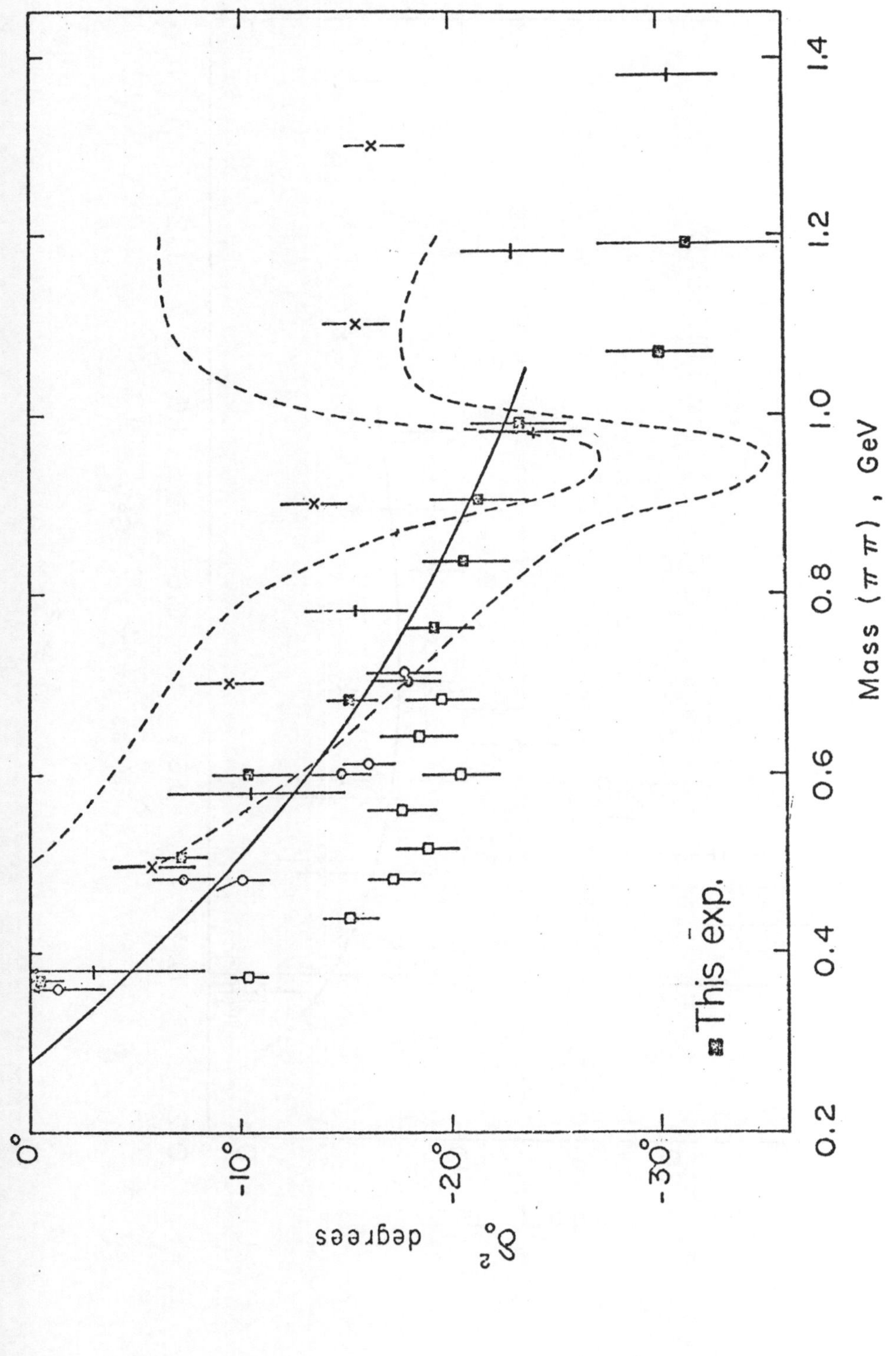

0°
-10°
-20°
-30°
δ_2^0 degrees
0.2
0.4
0.6
0.8
1.0
1.2
1.4
Mass (π π) , GeV
This exp.

Fig. 20

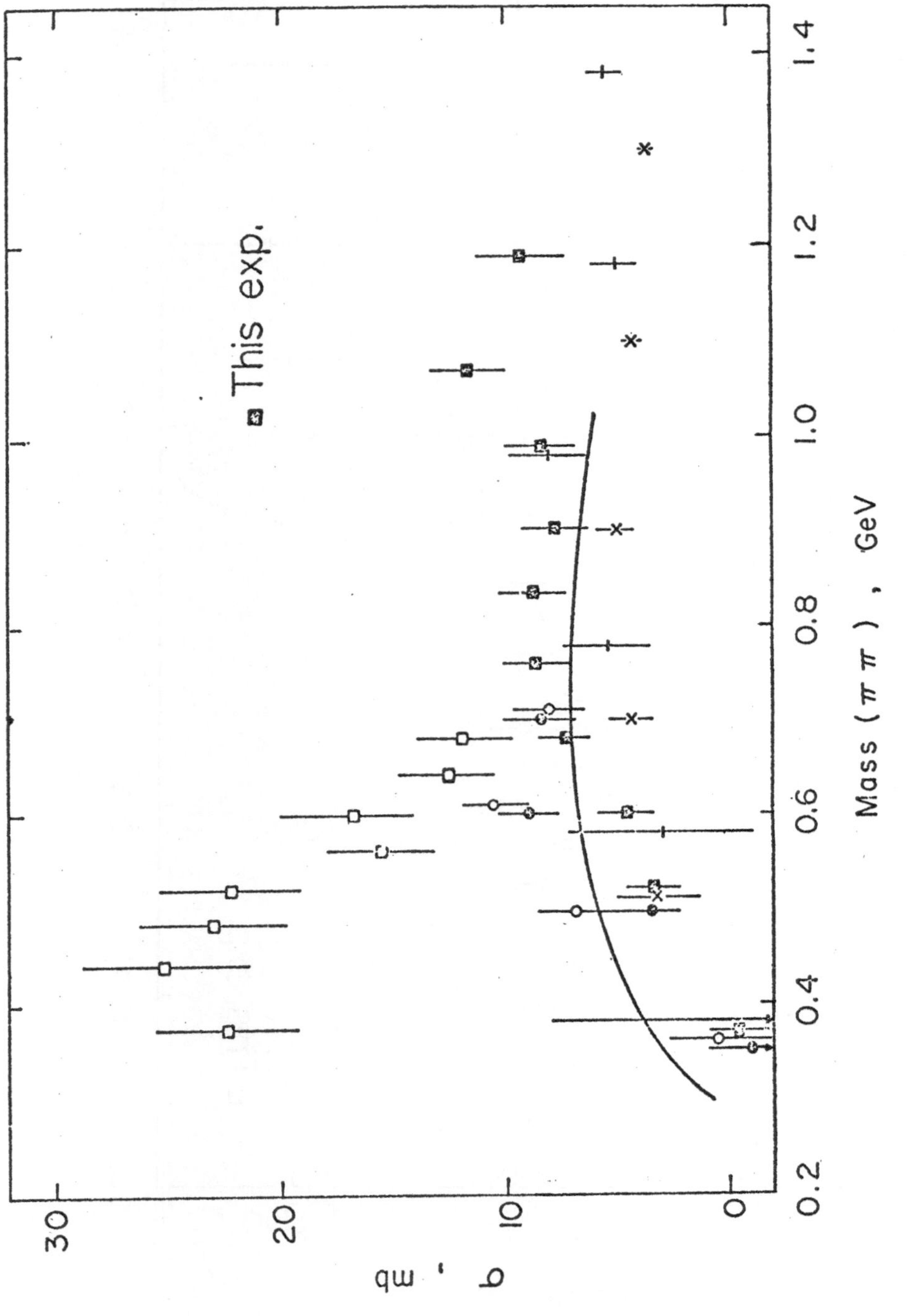

Fig. 21

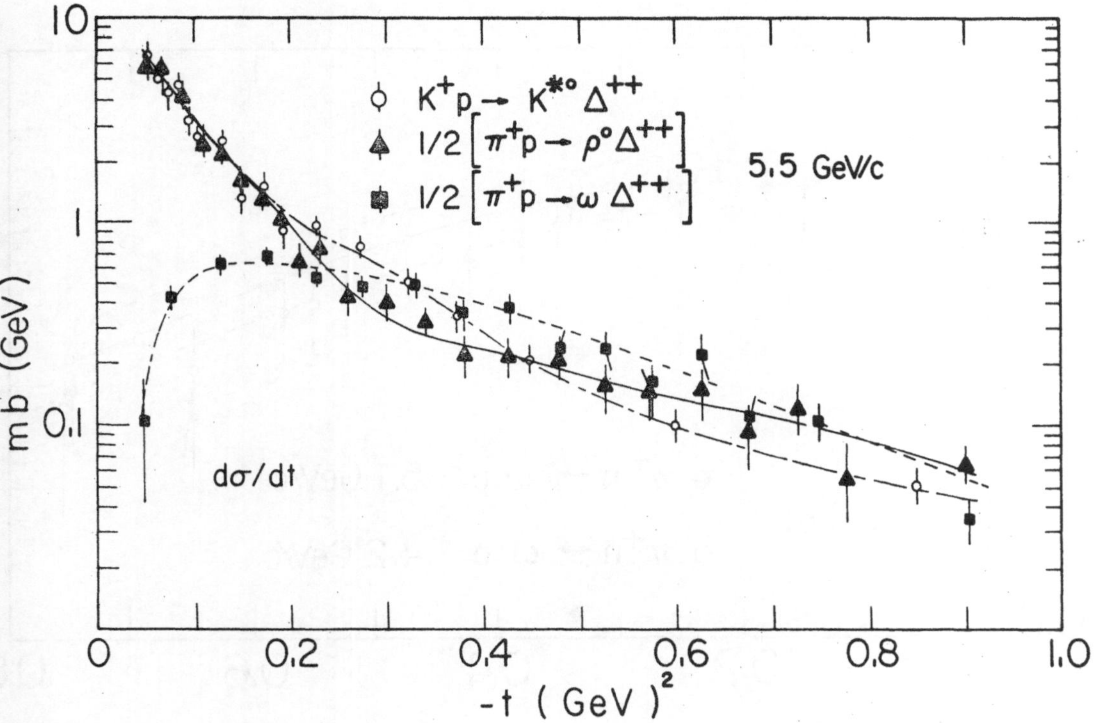
K⁺p → K*° Δ⁺⁺
1/2 [π⁺p → ρ° Δ⁺⁺]
1/2 [π⁺p → ω Δ⁺⁺]
5.5 GeV/c
dσ/dt
mb (GeV)⁻²
-t (GeV)²

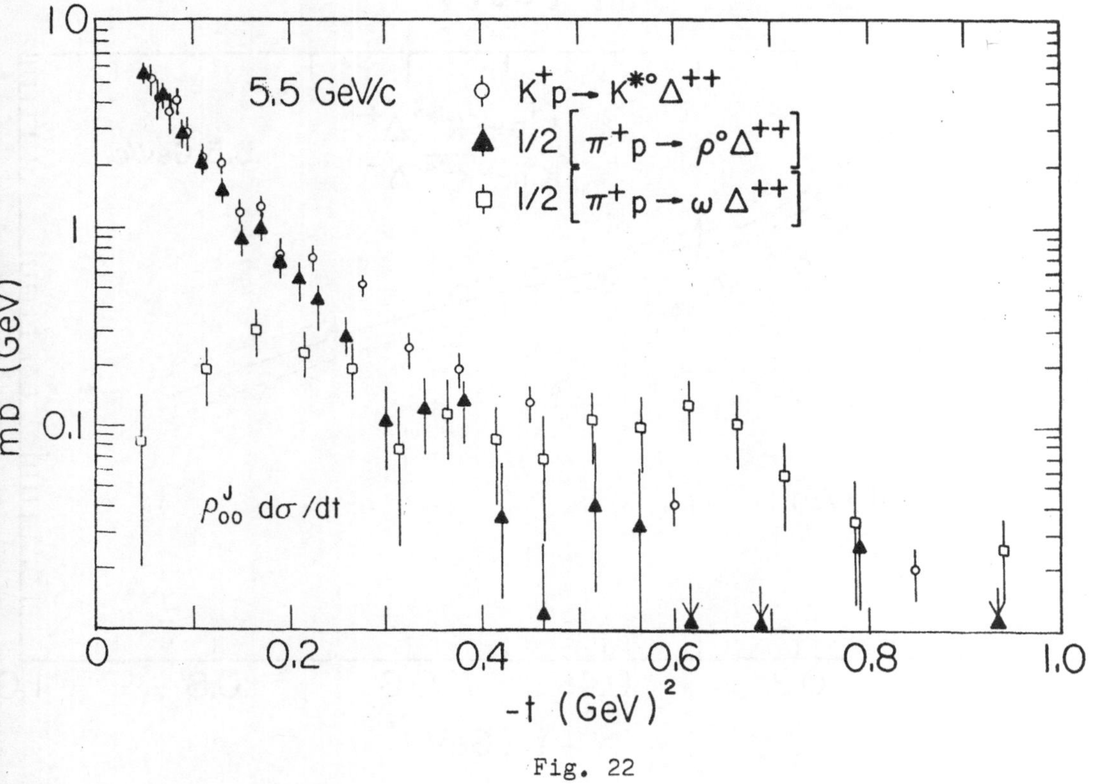
5.5 GeV/c
K⁺p → K*° Δ⁺⁺
1/2 [π⁺p → ρ° Δ⁺⁺]
1/2 [π⁺p → ω Δ⁺⁺]
ρ₀₀ʲ dσ/dt
mb (GeV)⁻²
-t (GeV)²

Fig. 22

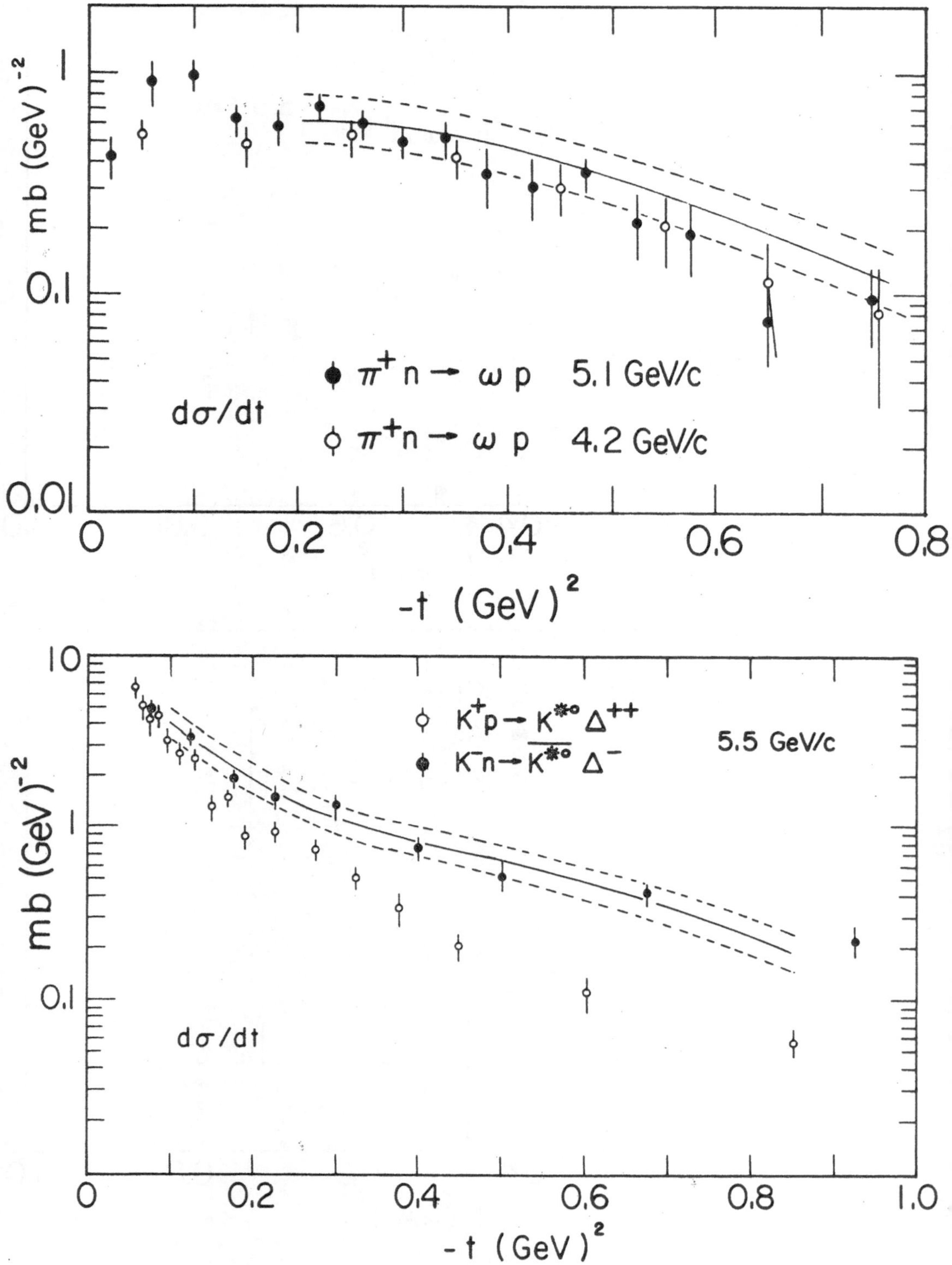
mb (GeV)$^{-2}$
1
0.1
0.01
0
0.2
0.4
0.6
0.8
-t (GeV)^2
dσ/dt
$\pi^+ n \to \omega p$ 5.1 GeV/c
$\pi^+ n \to \omega p$ 4.2 GeV/c
10
1
0.1
mb (GeV)$^{-2}$
0
0.2
0.4
0.6
0.8
1.0
-t (GeV)^2
$K^+ p \to K^{*0} \Delta^{++}$
$K^- n \to \overline{K^{*0}} \Delta^-$
5.5 GeV/c
dσ/dt

Fig. 23

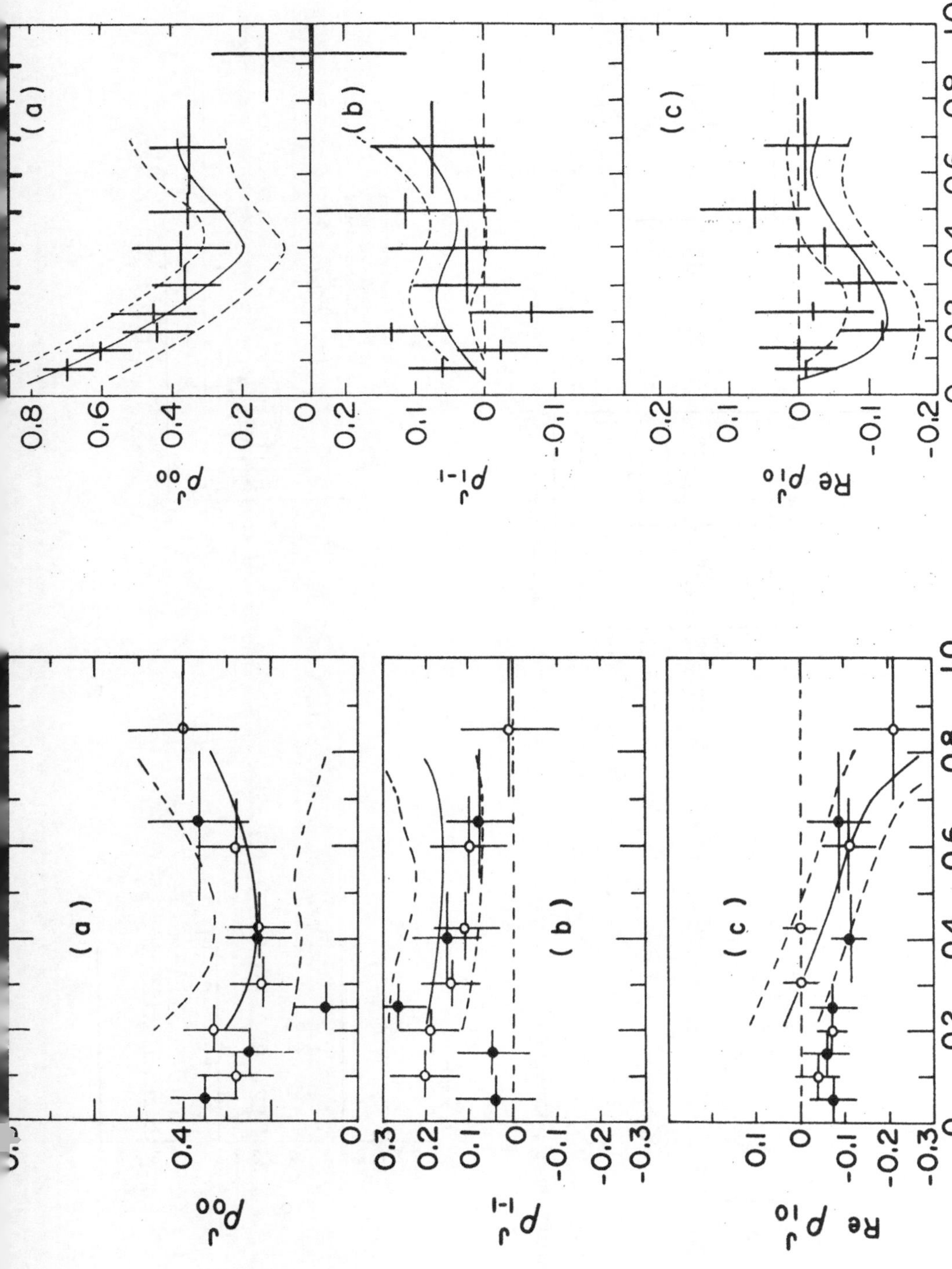

Fig. 24

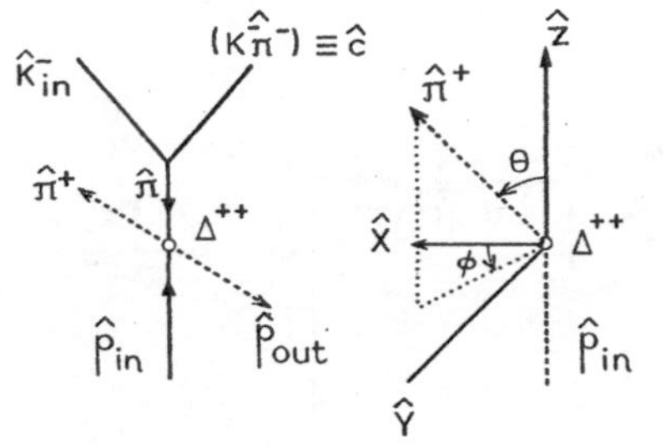

$$\cos\vartheta = (\hat P_{in}\cdot\hat\pi^+)/|\hat P_{in}|\,|\hat\pi^+|,$$

$$\cos\phi = ((\hat K^- \times \hat c)/|\hat K^- \times \hat c|)\cdot((\hat P_{in}\times\hat\pi^+)/|\hat P_{in}\times\hat\pi^+|).$$

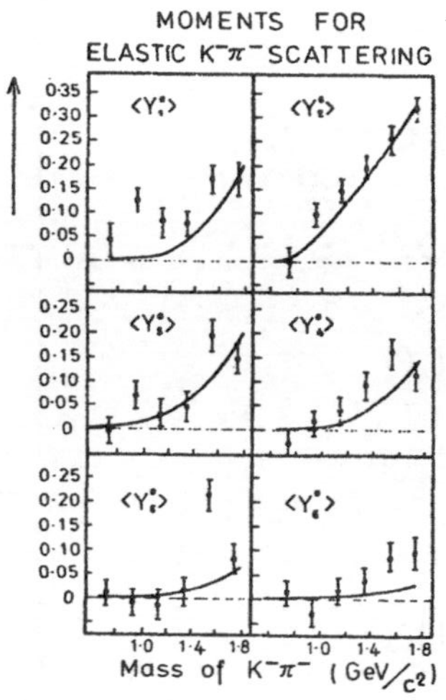

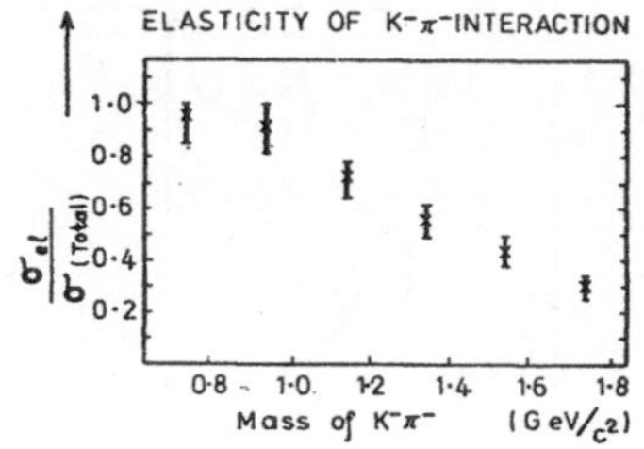

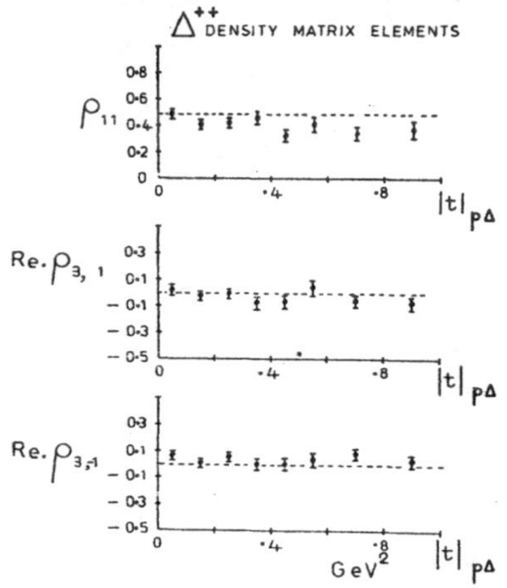

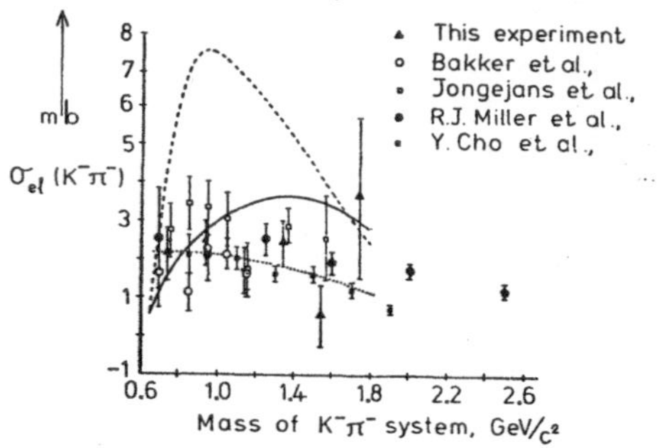

Fig. 25

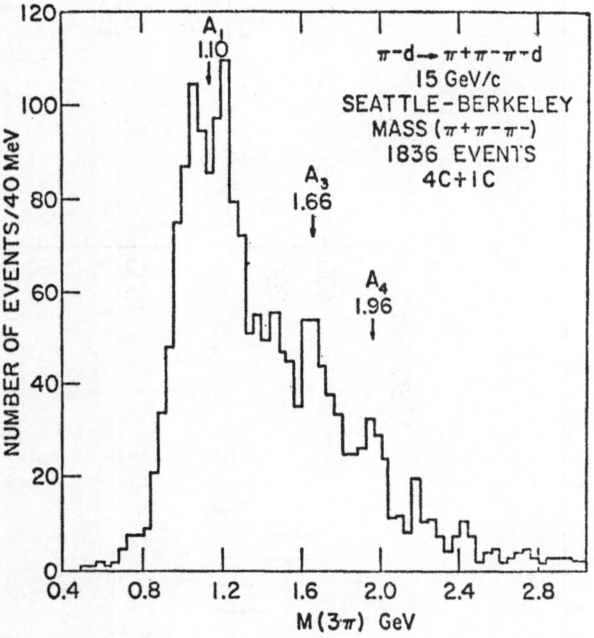
π⁻d → π⁺π⁻π⁻d
15 GeV/c
SEATTLE-BERKELEY
MASS (π⁺π⁻π⁻)
1836 EVENTS
4C+1C
A1
1.10
A3
1.66
A4
1.96
NUMBER OF EVENTS/40 MeV
M(3π) GeV
0
20
40
60
80
100
120
0.4
0.8
1.2
1.6
2.0
2.4
2.8

Fig. 26

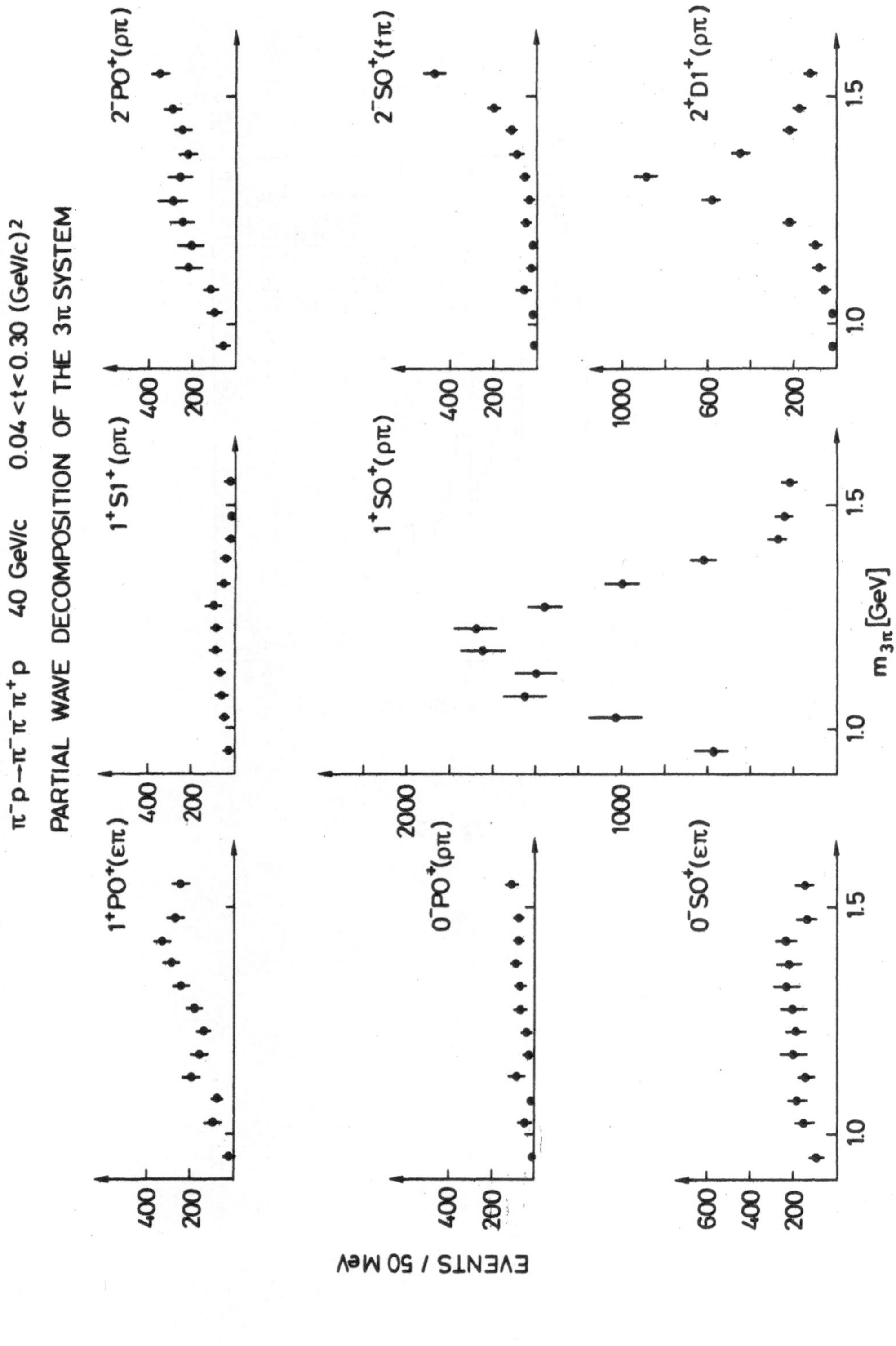
π⁻p → π⁻π⁻π⁺p   40 GeV/c   0.04 < t < 0.30 (GeV/c)²
PARTIAL WAVE DECOMPOSITION OF THE 3π SYSTEM
1⁺P0⁺(επ)
1⁺S1⁺(ρπ)
2⁻P0⁺(ρπ)
0⁻P0⁺(ρπ)
1⁺S0⁺(ρπ)
2⁻S0⁺(fπ)
0⁻S0⁺(επ)
2⁺D1⁺(ρπ)
EVENTS / 50 MeV
m3π [GeV]
1.0
1.5
200
400
600
1000
2000

Fig. 27

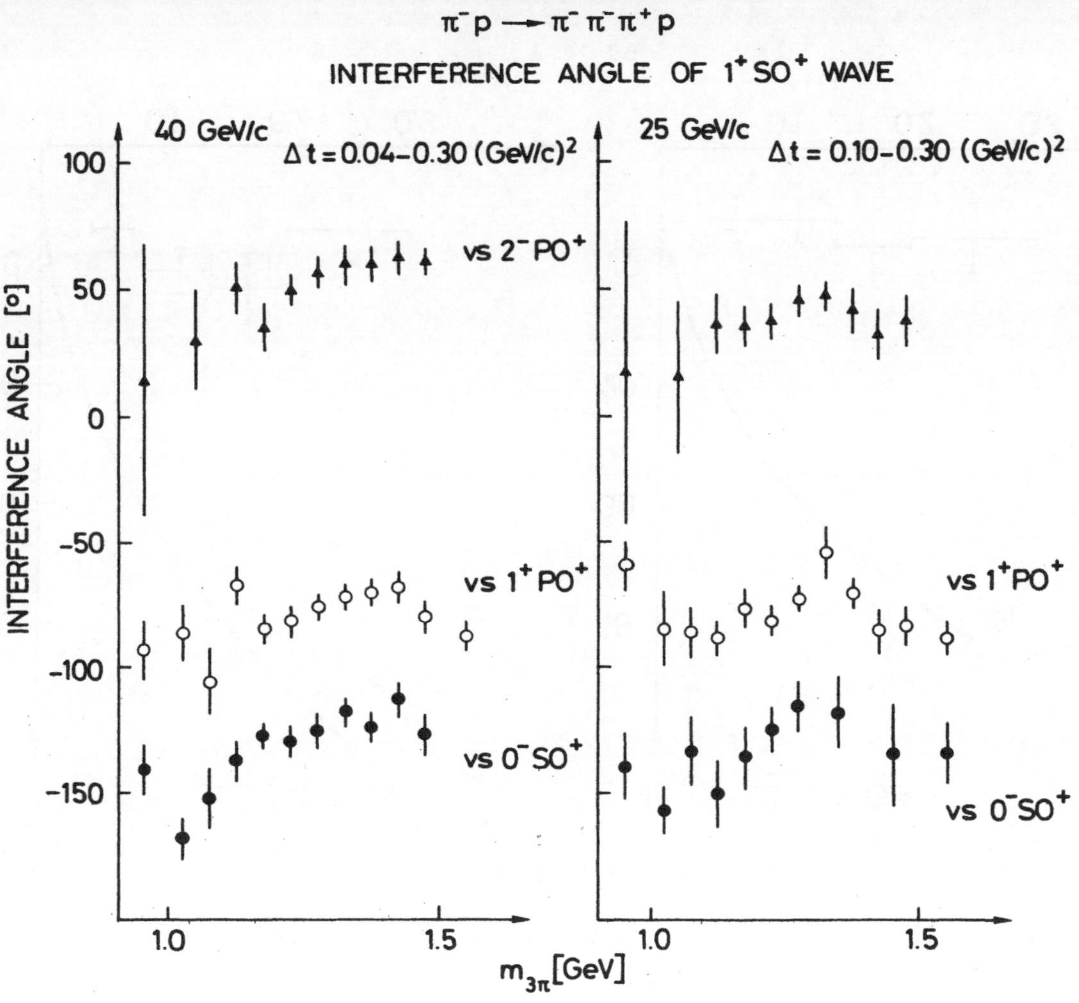
π⁻p → π⁻π⁻π⁺p
INTERFERENCE ANGLE OF 1⁺SO⁺ WAVE
40 GeV/c
Δt = 0.04–0.30 (GeV/c)²
25 GeV/c
Δt = 0.10–0.30 (GeV/c)²
vs 2⁻PO⁺
vs 1⁺PO⁺
vs 0⁻SO⁺
INTERFERENCE ANGLE [°]
100
50
0
-50
-100
-150
1.0
1.5
m3π [GeV]

Fig. 28

40 GeV $\pi^- p \longrightarrow \pi^- \pi^- \pi^+ p$

ROTATION ANGLE FROM G.J. SYSTEM TO D-H SYSTEM
FOR 1^+-WAVE

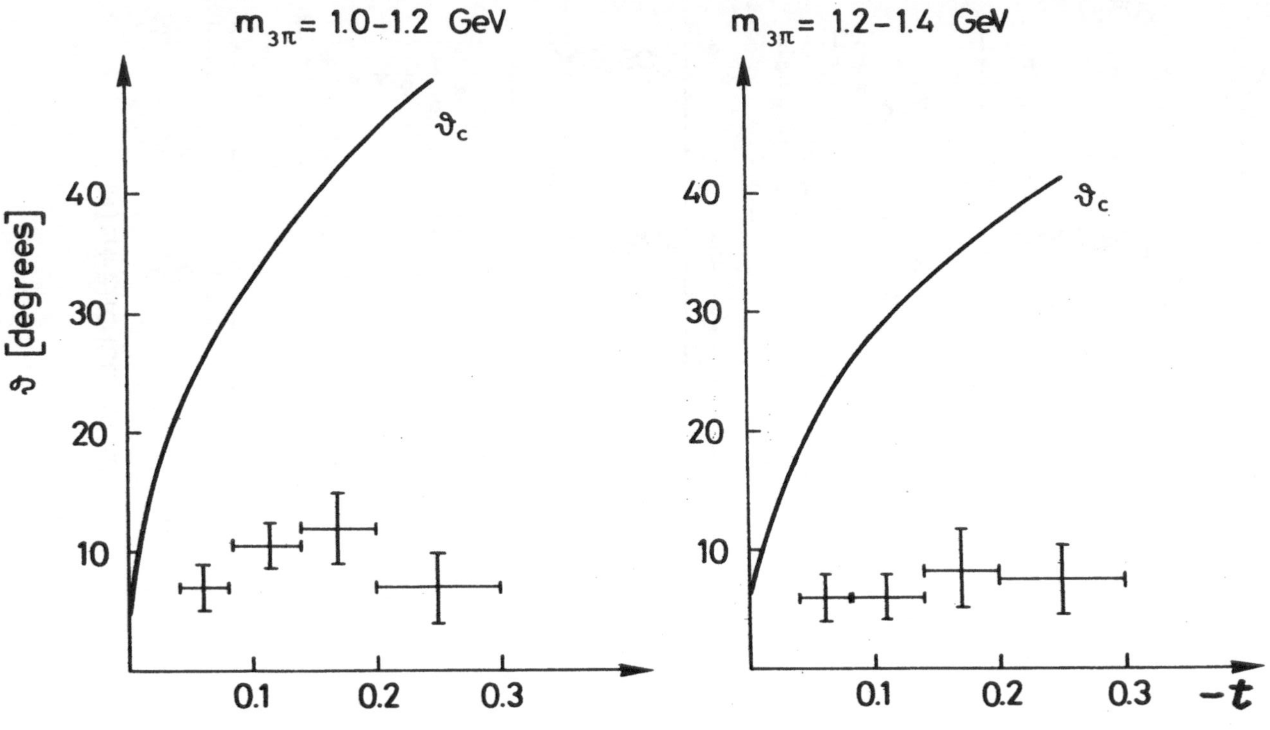

Fig. 29

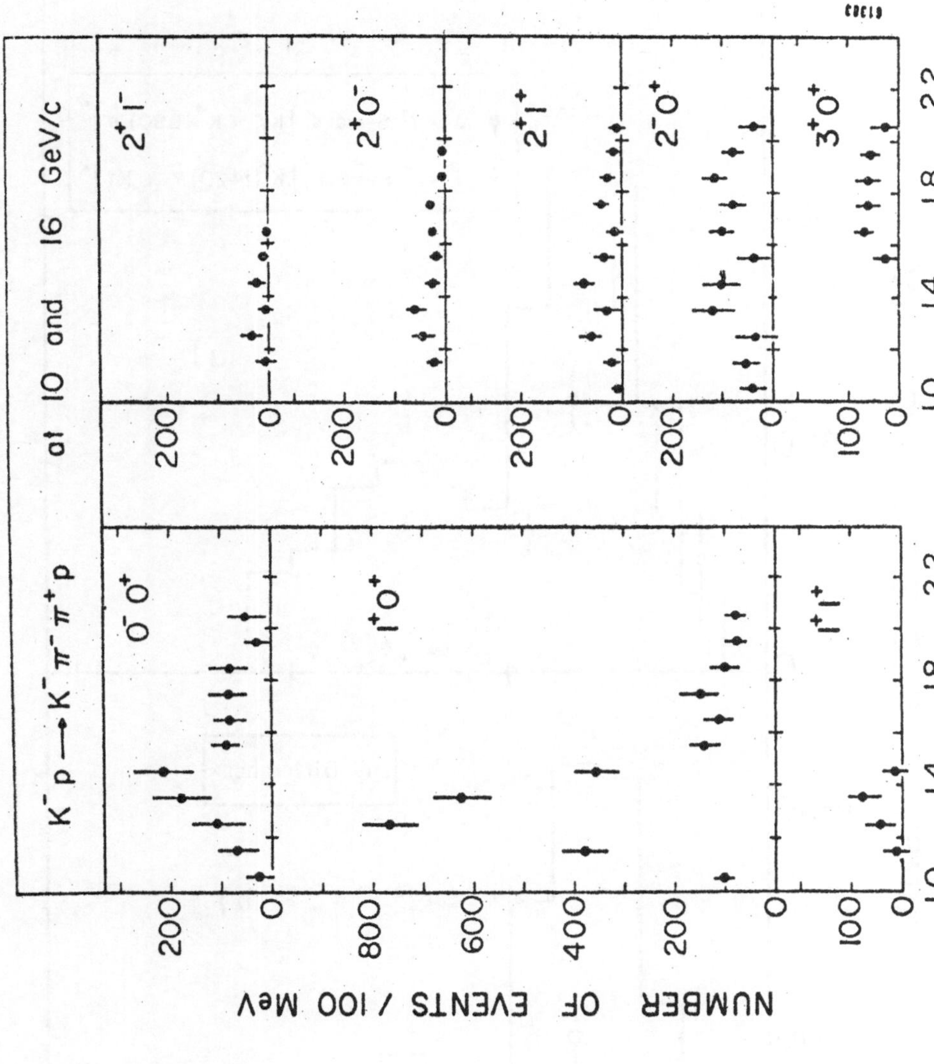

K⁻p → K⁻π⁻π⁺p at 10 and 16 GeV/c
NUMBER OF EVENTS / 100 MeV
M (Kππ) , GeV
0⁻0⁺
1⁺0⁺
1⁺1⁺
2⁺1⁻
2⁺0⁻
2⁺1⁺
2⁻0⁺
3⁺0⁺

Fig. 30

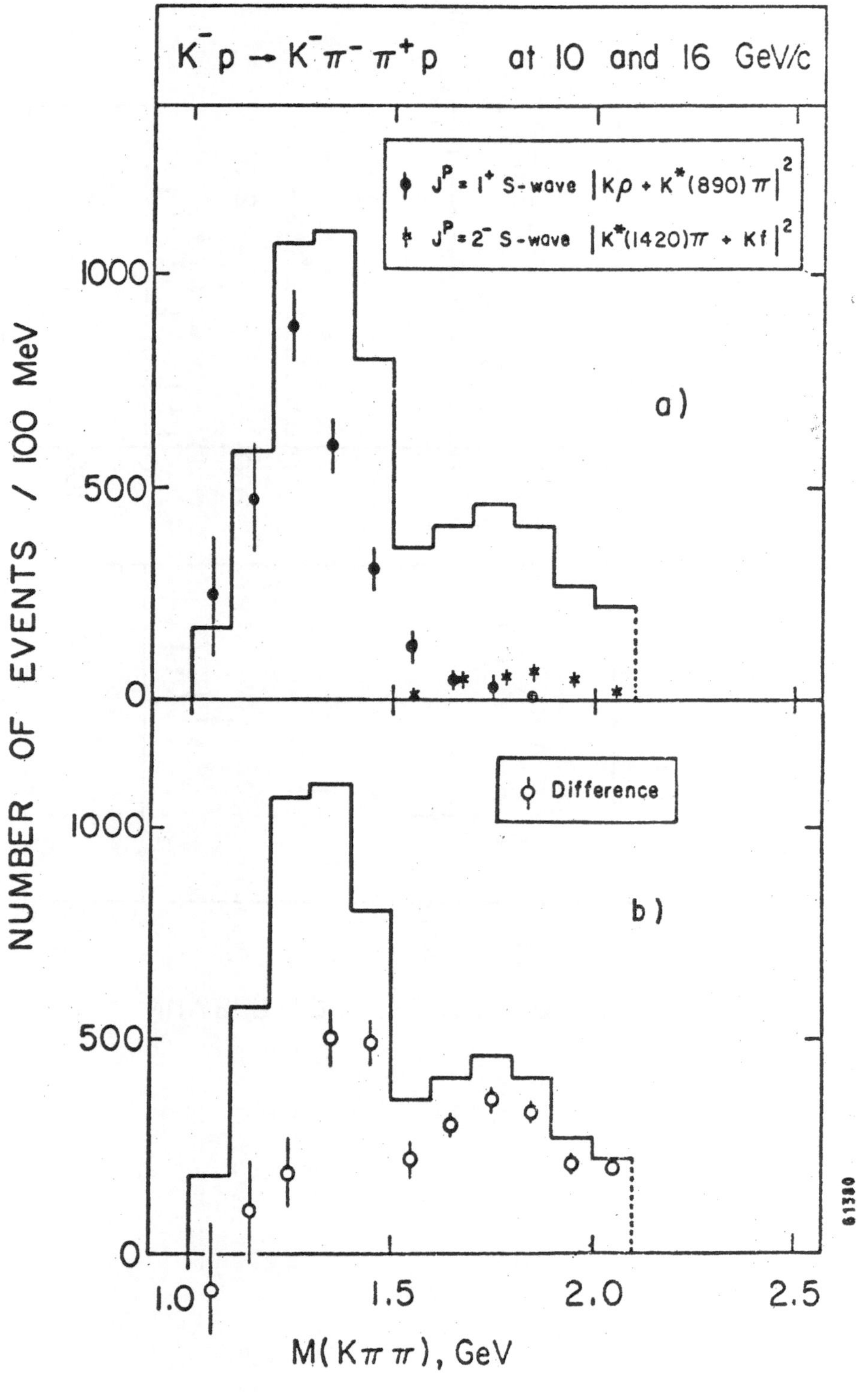

K⁻p → K⁻π⁻π⁺p at 10 and 16 GeV/c
$J^P = 1^+$ S-wave $|K\rho + K^*(890)\pi|^2$
$J^P = 2^-$ S-wave $|K^*(1420)\pi + Kf|^2$
a)
Difference
b)
NUMBER OF EVENTS / 100 MeV
1000
500
0
1.0
1.5
2.0
2.5
M(Kππ), GeV

Fig. 31

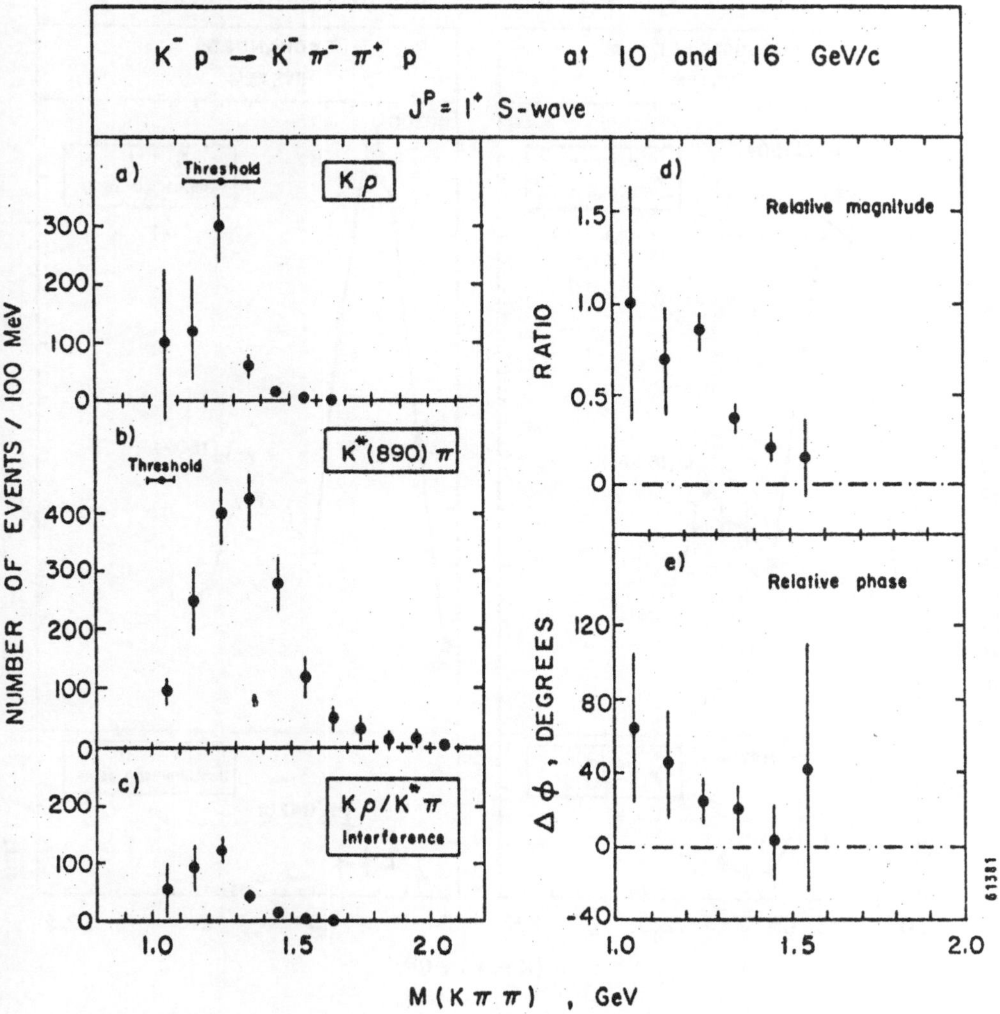

$K^- p \rightarrow K^- \pi^- \pi^+ p$ at 10 and 16 GeV/c
$J^P = 1^+$ S-wave
a)
Threshold
$K\rho$
b)
Threshold
$K^*(890)\pi$
c)
$K\rho / K^*\pi$
Interference
NUMBER OF EVENTS / 100 MeV
d)
Relative magnitude
RATIO
e)
Relative phase
$\Delta\phi$, DEGREES
M($K\pi\pi$), GeV
61381

Fig. 32

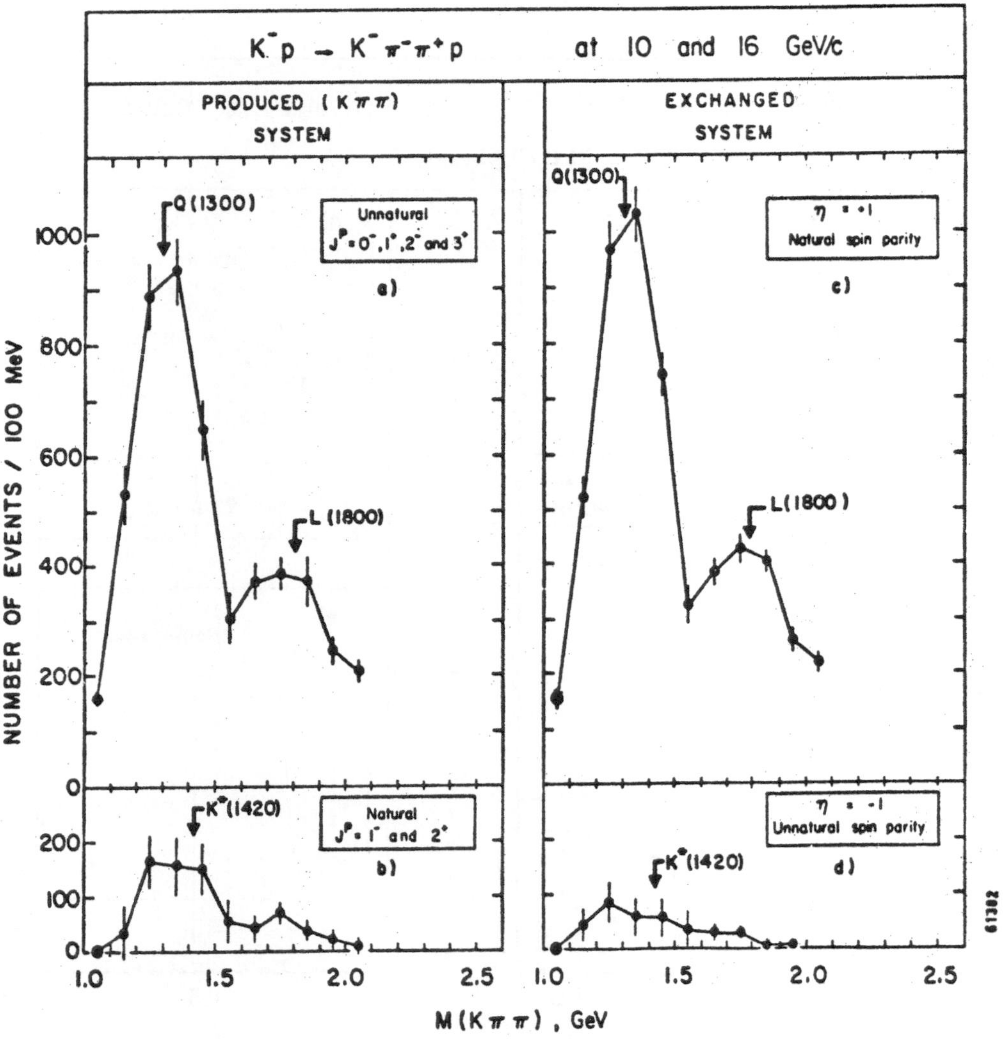

K⁻p → K⁻π⁻π⁺p at 10 and 16 GeV/c
PRODUCED (Kππ) SYSTEM
EXCHANGED SYSTEM
Q(1300)
Unnatural
J^P = 0⁻, 1⁺, 2⁻ and 3⁺
a)
L(1800)
η = +1
Natural spin parity
c)
K*(1420)
Natural
J^P = 1⁻ and 2⁺
b)
η = -1
Unnatural spin parity
d)
NUMBER OF EVENTS / 100 MeV
M(Kππ), GeV
1.0
1.5
2.0
2.5
0
100
200
400
600
800
1000
61382

Fig. 33

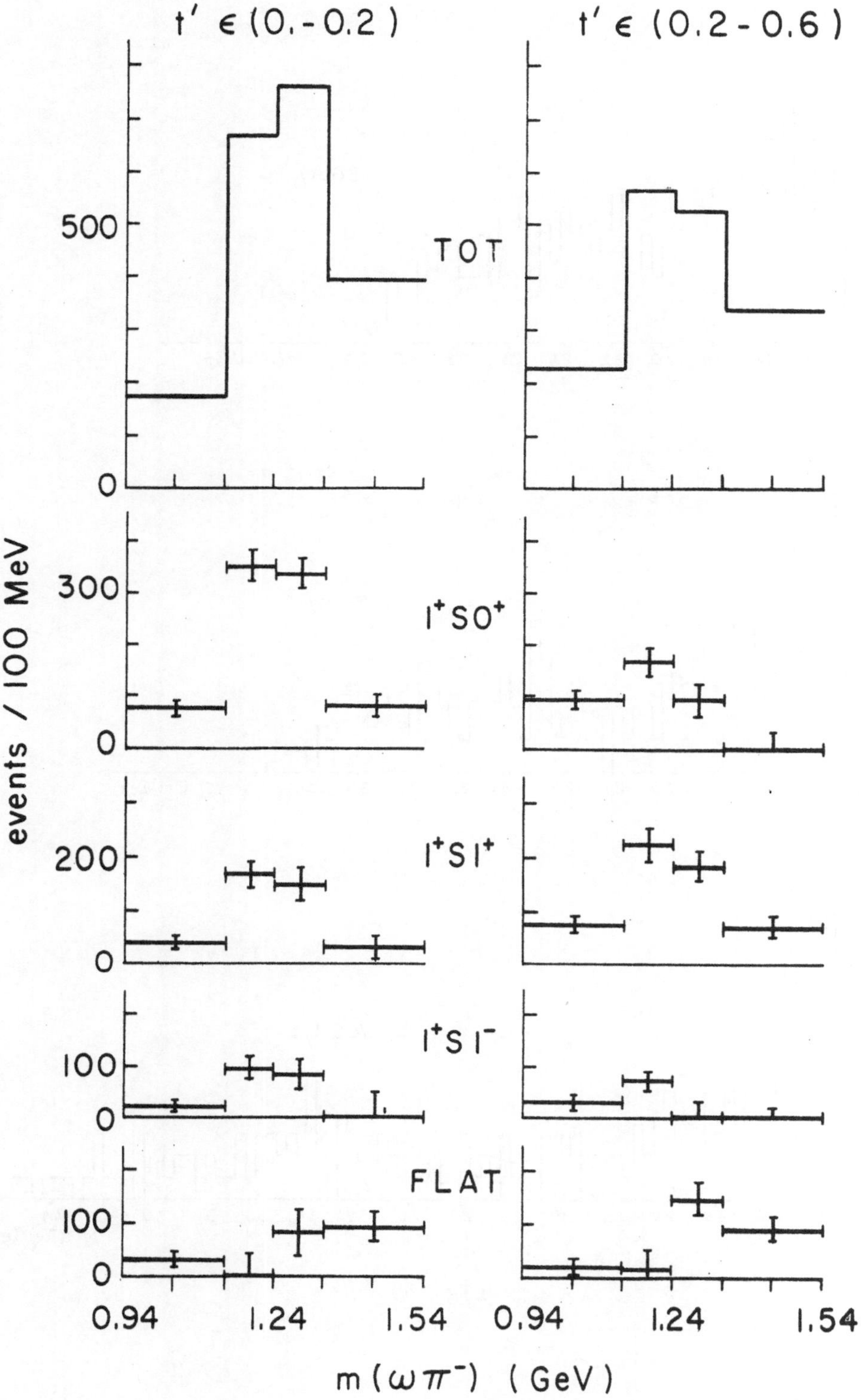
t' ∈ (0.-0.2)
t' ∈ (0.2-0.6)
500
0
TOT
300
0
1+S0+
200
0
1+S1+
100
0
1+S1-
FLAT
100
0
0.94
1.24
1.54
0.94
1.24
1.54
events / 100 MeV
m(ωπ-) (GeV)

Fig. 34

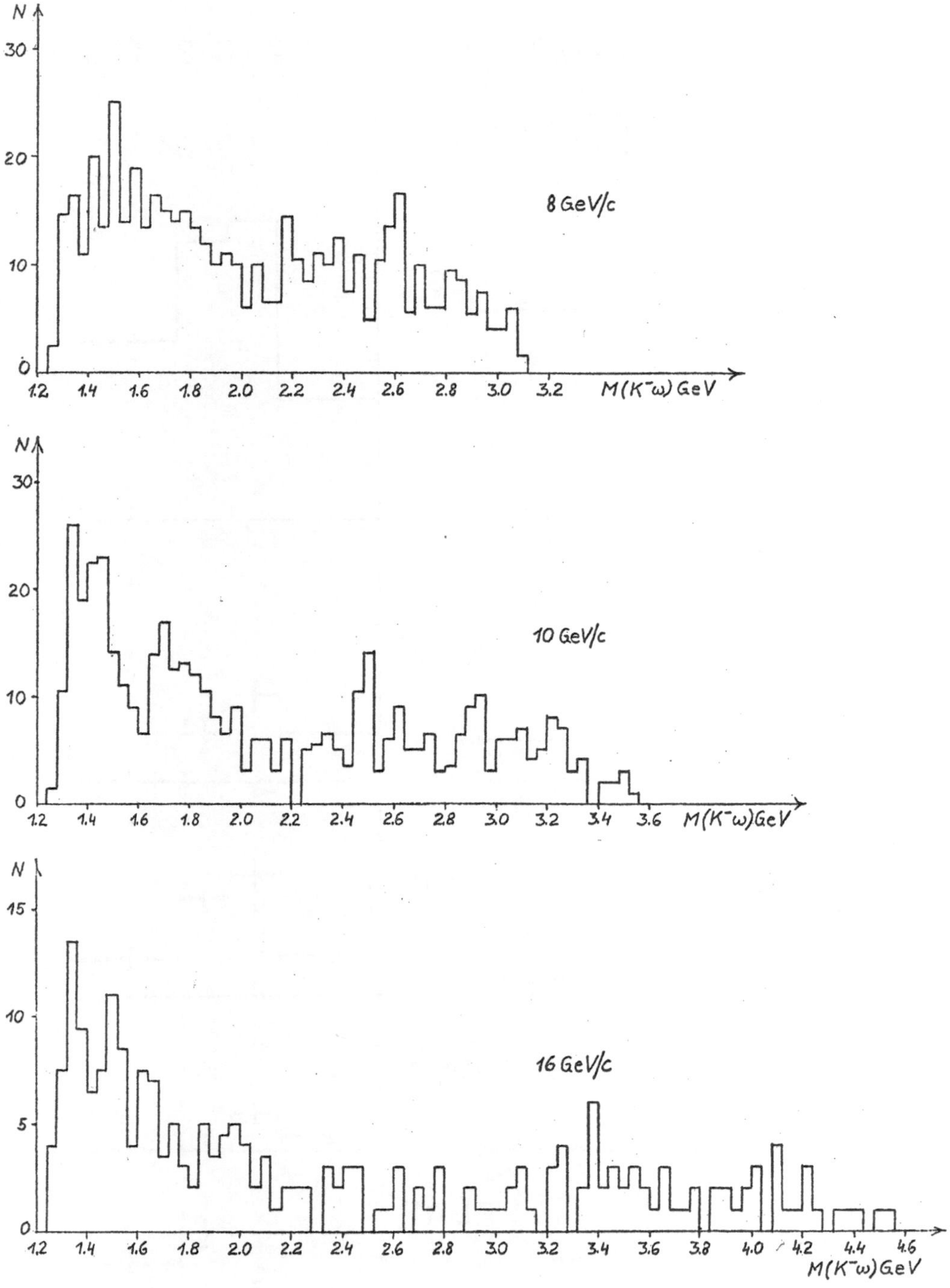
N
30
20
10
0
1.2 1.4 1.6 1.8 2.0 2.2 2.4 2.6 2.8 3.0 3.2
M(K⁻ω) GeV
8 GeV/c
N
30
20
10
0
1.2 1.4 1.6 1.8 2.0 2.2 2.4 2.6 2.8 3.0 3.2 3.4 3.6
M(K⁻ω)GeV
10 GeV/c
N
15
10
5
0
1.2 1.4 1.6 1.8 2.0 2.2 2.4 2.6 2.8 3.0 3.2 3.4 3.6 3.8 4.0 4.2 4.4 4.6
M(K⁻ω)GeV
16 GeV/c

Fig. 35

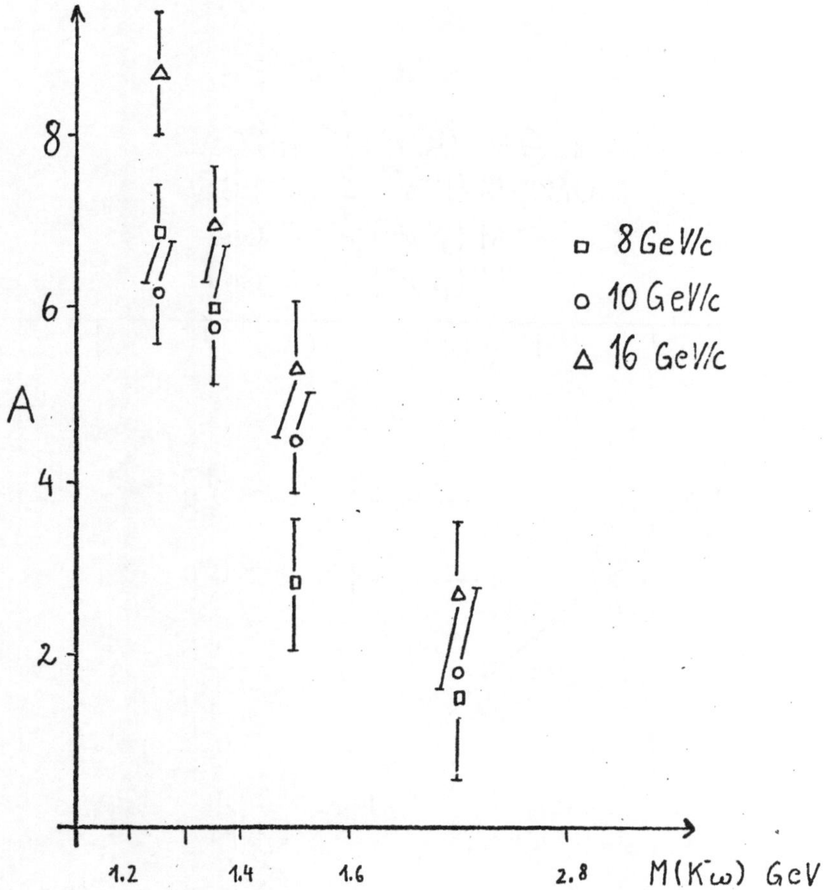
8
6
A
4
2
□ 8 GeV/c
○ 10 GeV/c
△ 16 GeV/c
1.2
1.4
1.6
2.8
M(K⁻ω) GeV

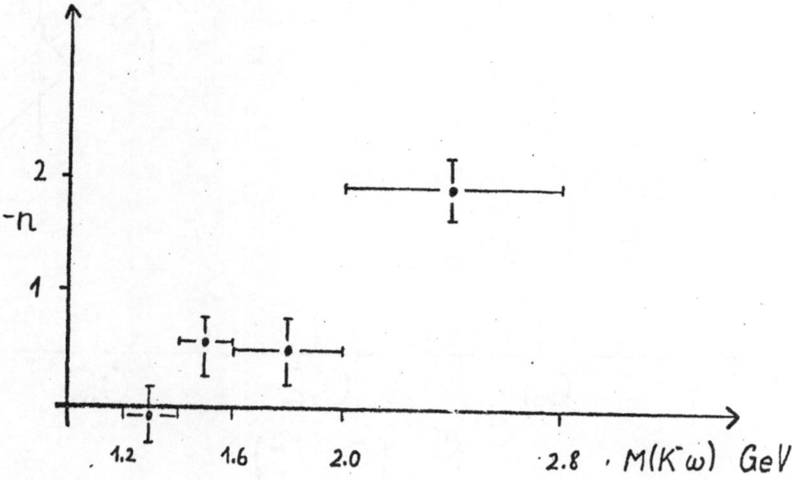
2
-n
1
1.2
1.6
2.0
2.8
M(K⁻ω) GeV

Fig. 36

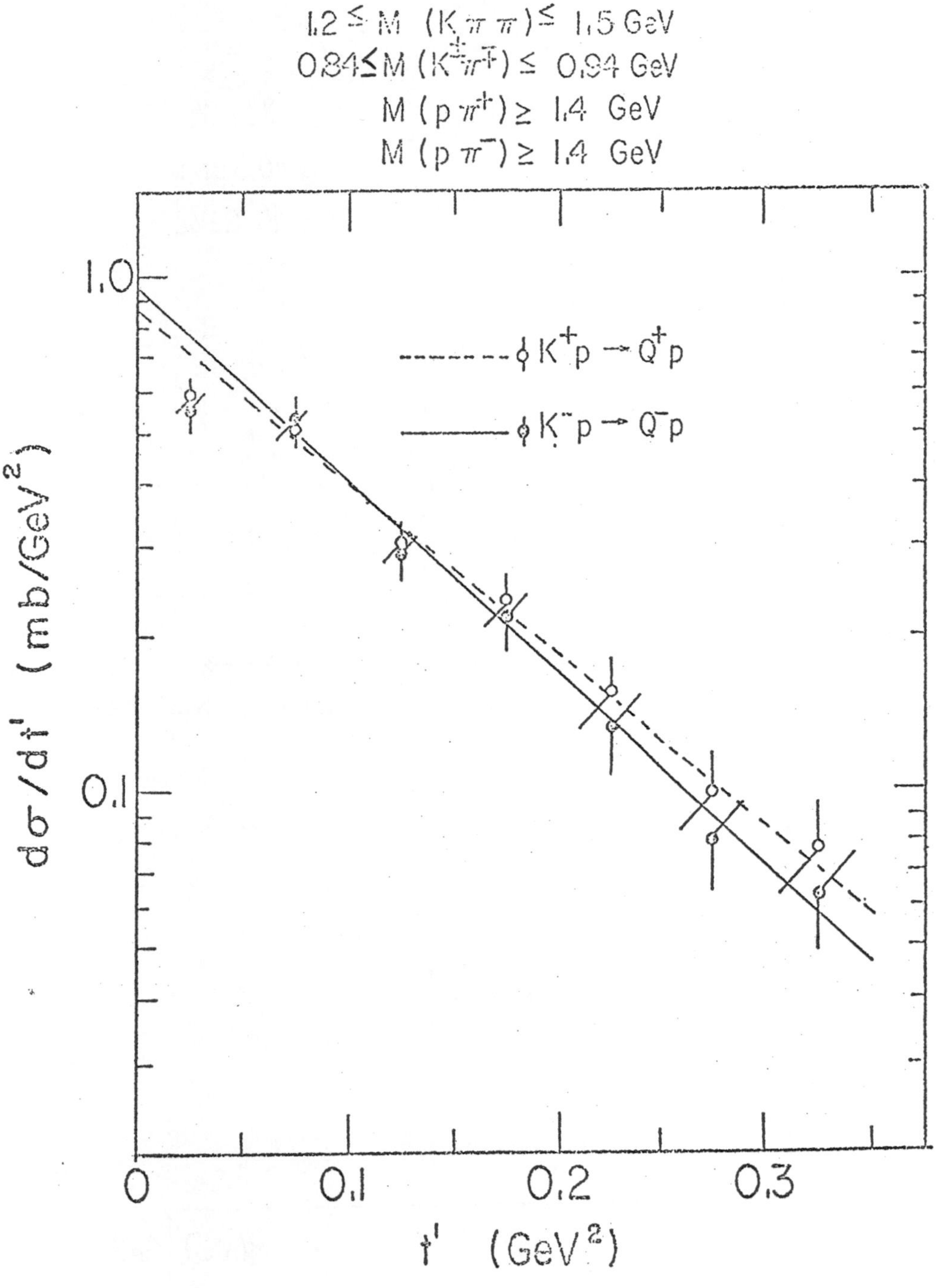
1.2 ≤ M (K π π) ≤ 1.5 GeV
0.84 ≤ M (K$^{\pm}$π$^{\mp}$) ≤ 0.94 GeV
M (p π^{+}) ≥ 1.4 GeV
M (p π^{-}) ≥ 1.4 GeV
K^{+}p → Q^{+}p
K^{-}p → Q^{-}p
dσ/dt' (mb/GeV^{2})
1.0
0.1
0
0.1
0.2
0.3
t' (GeV^{2})

Fig. 37

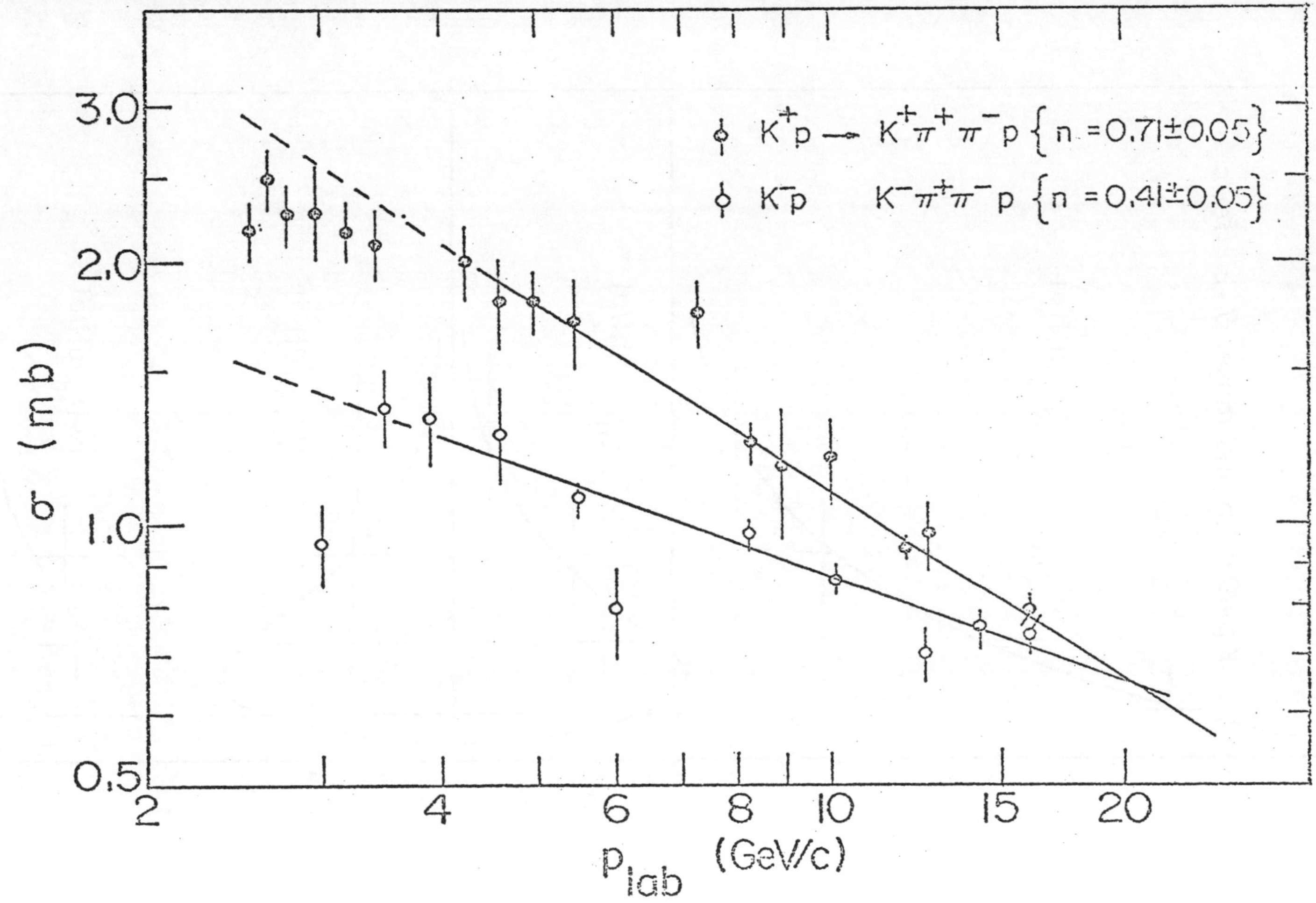
K+p → K+π+π−p {n = 0.71±0.05}
K−p K−π+π−p {n = 0.41±0.05}
3.0
2.0
1.0
0.5
σ (mb)
2
4
6
8
10
15
20
plab (GeV/c)

Fig. 38

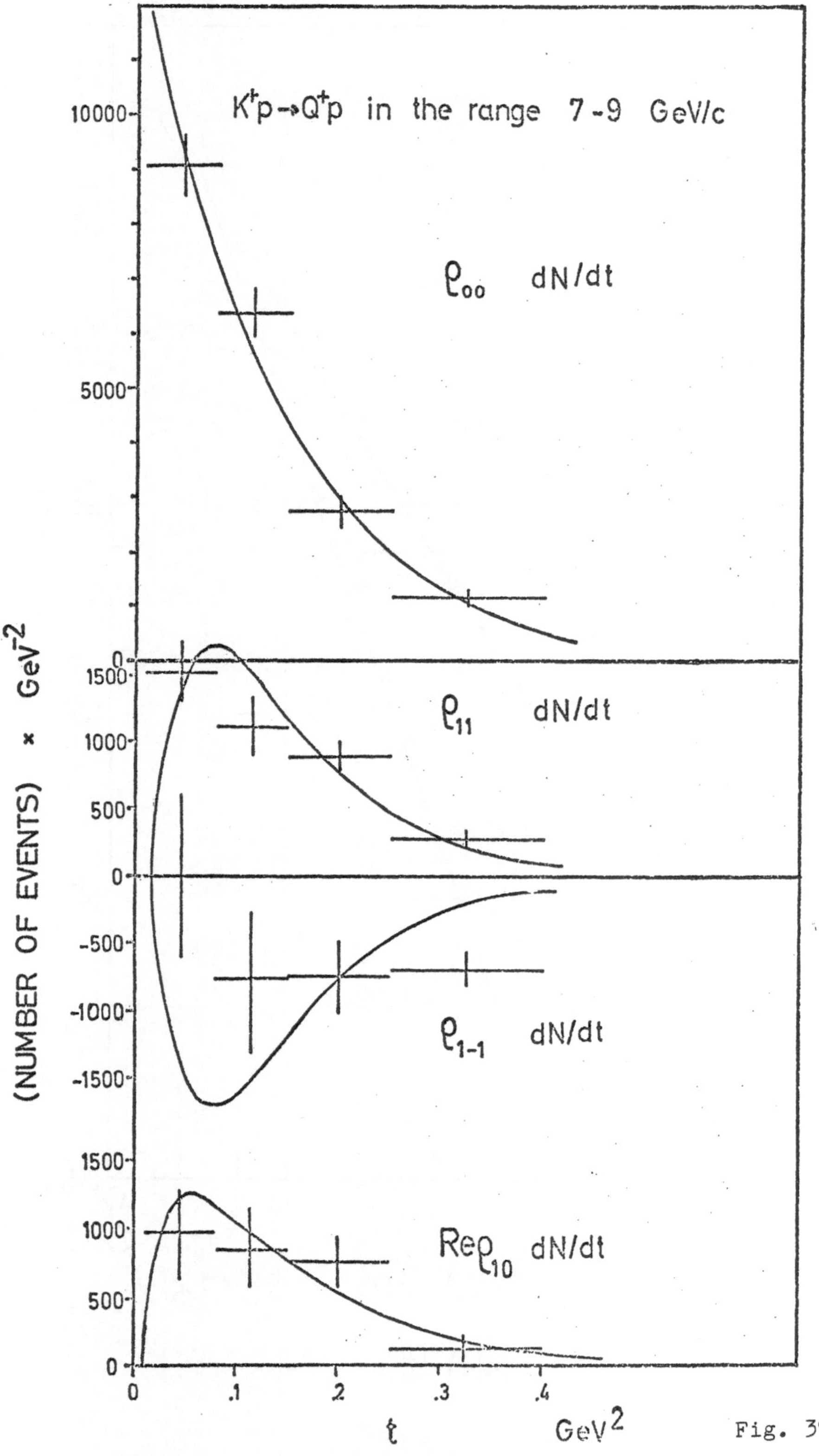
K+p→Q+p in the range 7-9 GeV/c
ρ00 dN/dt
ρ11 dN/dt
ρ1-1 dN/dt
Reρ10 dN/dt
(NUMBER OF EVENTS) × GeV^-2
10000
5000
0
1500
1000
500
0
-500
-1000
-1500
1500
1000
500
0
0
.1
.2
.3
.4
t
GeV^2

Fig. 39

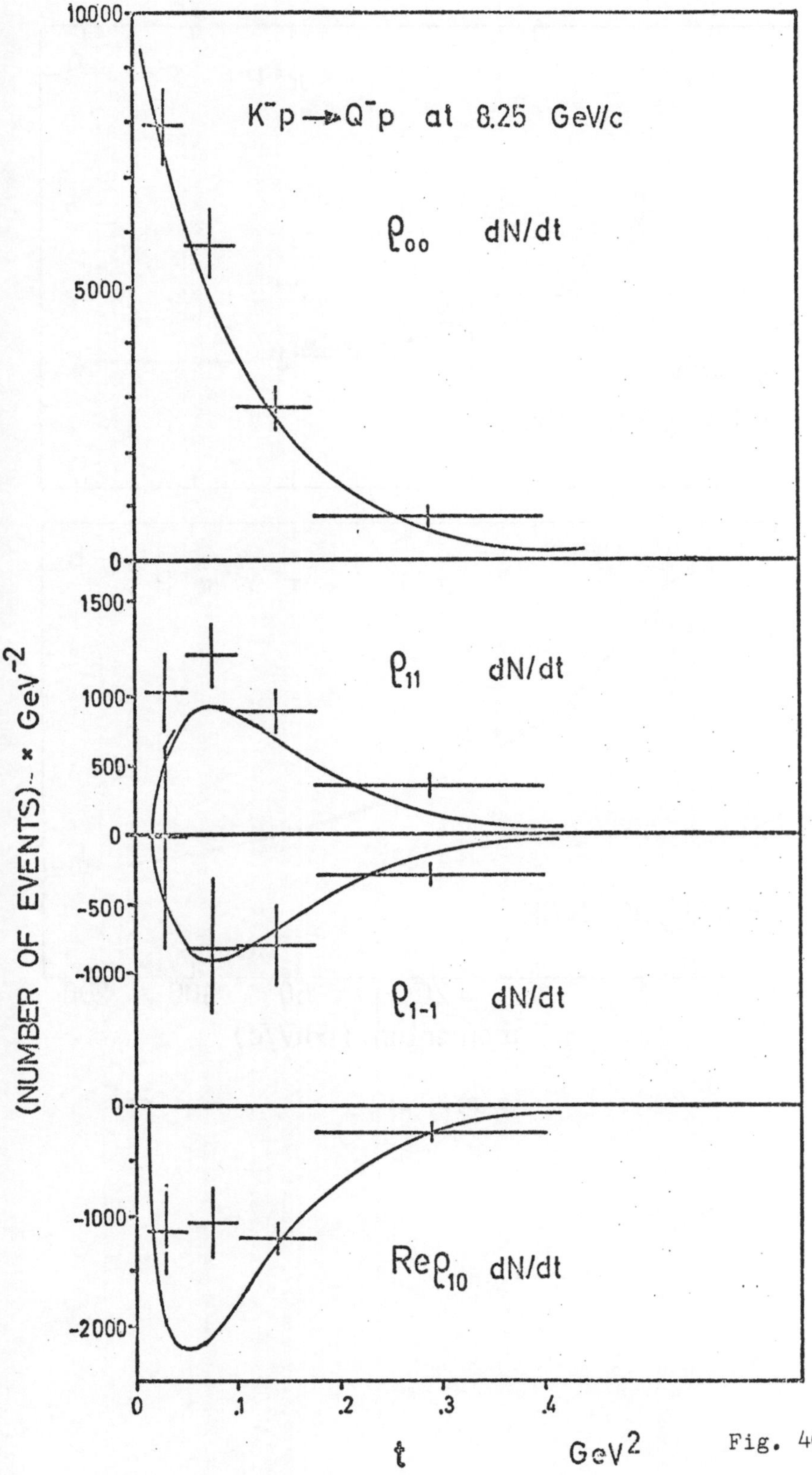
K⁻p → Q⁻p at 8.25 GeV/c
ρ00 dN/dt
ρ11 dN/dt
ρ1-1 dN/dt
Reρ10 dN/dt
(NUMBER OF EVENTS) × GeV-2
10000
5000
0
1500
1000
500
0
-500
-1000
0
-1000
-2000
0
.1
.2
.3
.4
t
GeV2

Fig. 40

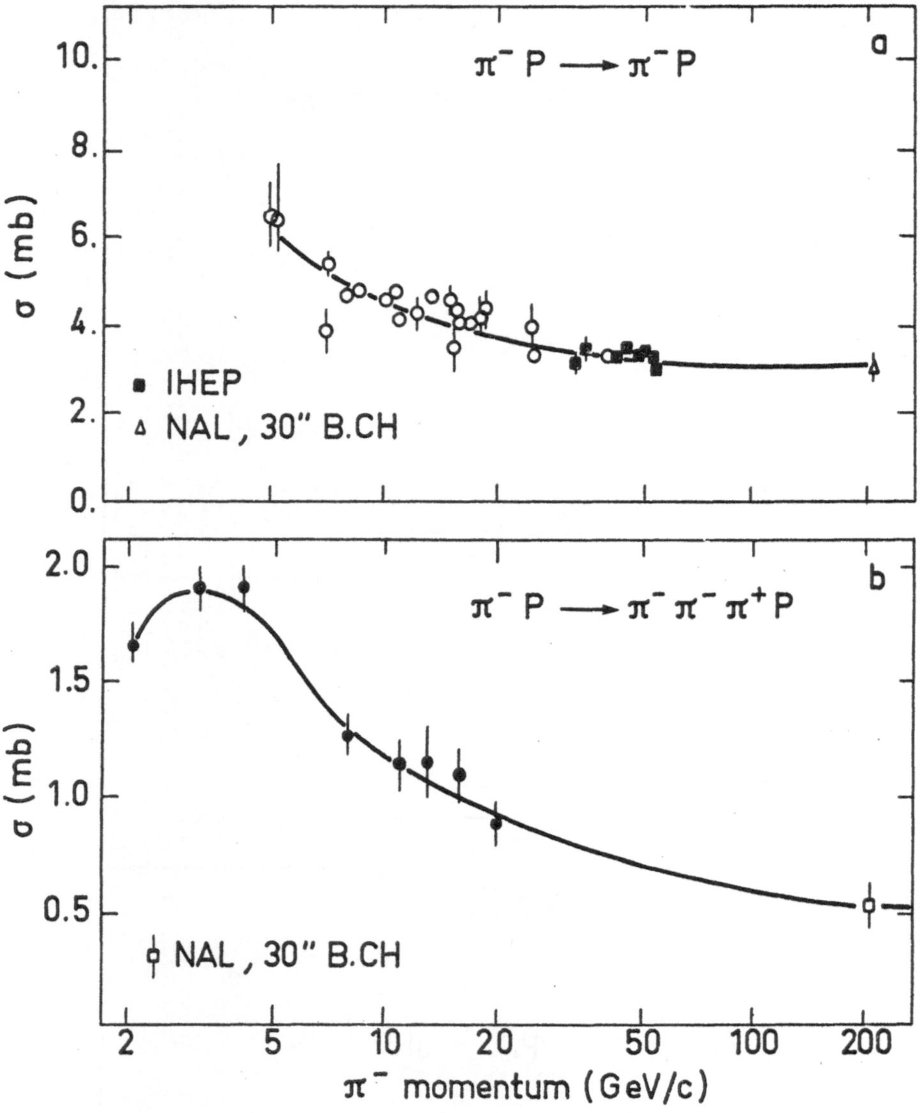
a
$\pi^- P \longrightarrow \pi^- P$
σ (mb)
10.
8.
6.
4.
2.
0.
IHEP
NAL , 30" B.CH
b
$\pi^- P \longrightarrow \pi^- \pi^- \pi^+ P$
σ (mb)
2.0
1.5
1.0
0.5
NAL , 30" B.CH
2
5
10
20
50
100
200
π^- momentum (GeV/c)

Fig. 41

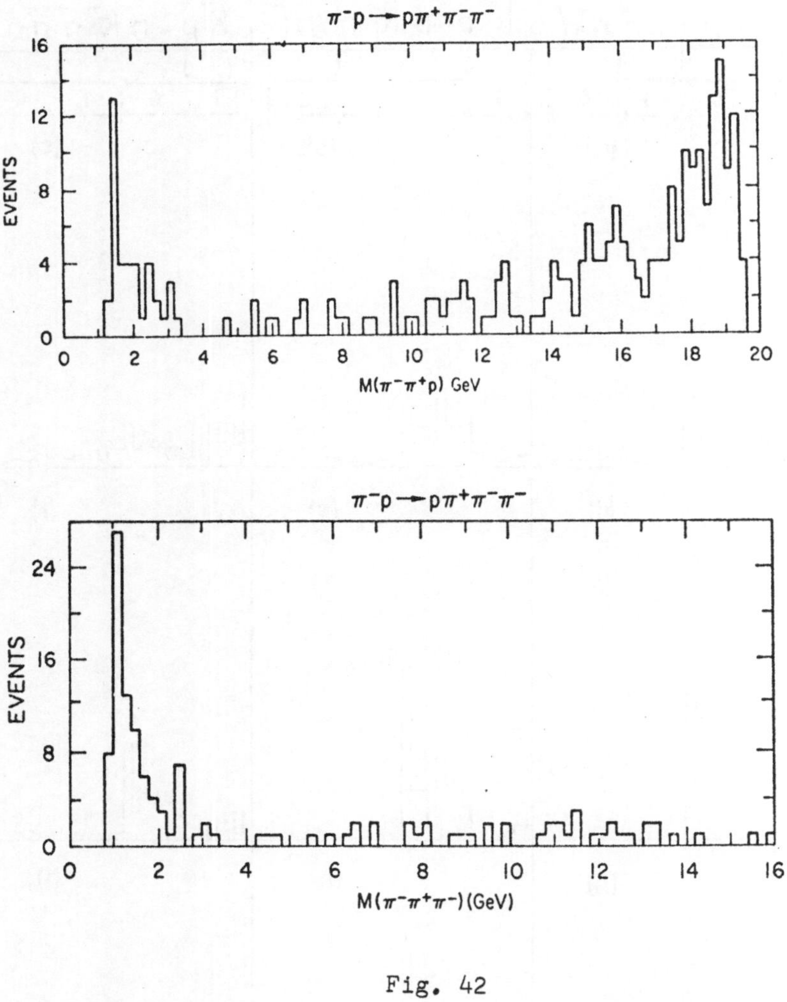

$\pi^- p \rightarrow p\pi^+\pi^-\pi^-$
EVENTS
0
4
8
12
16
0
2
4
6
8
10
12
14
16
18
20
$M(\pi^-\pi^+p)$ GeV
$\pi^- p \rightarrow p\pi^+\pi^-\pi^-$
EVENTS
0
8
16
24
0
2
4
6
8
10
12
14
16
$M(\pi^-\pi^+\pi^-)$(GeV)

Fig. 42

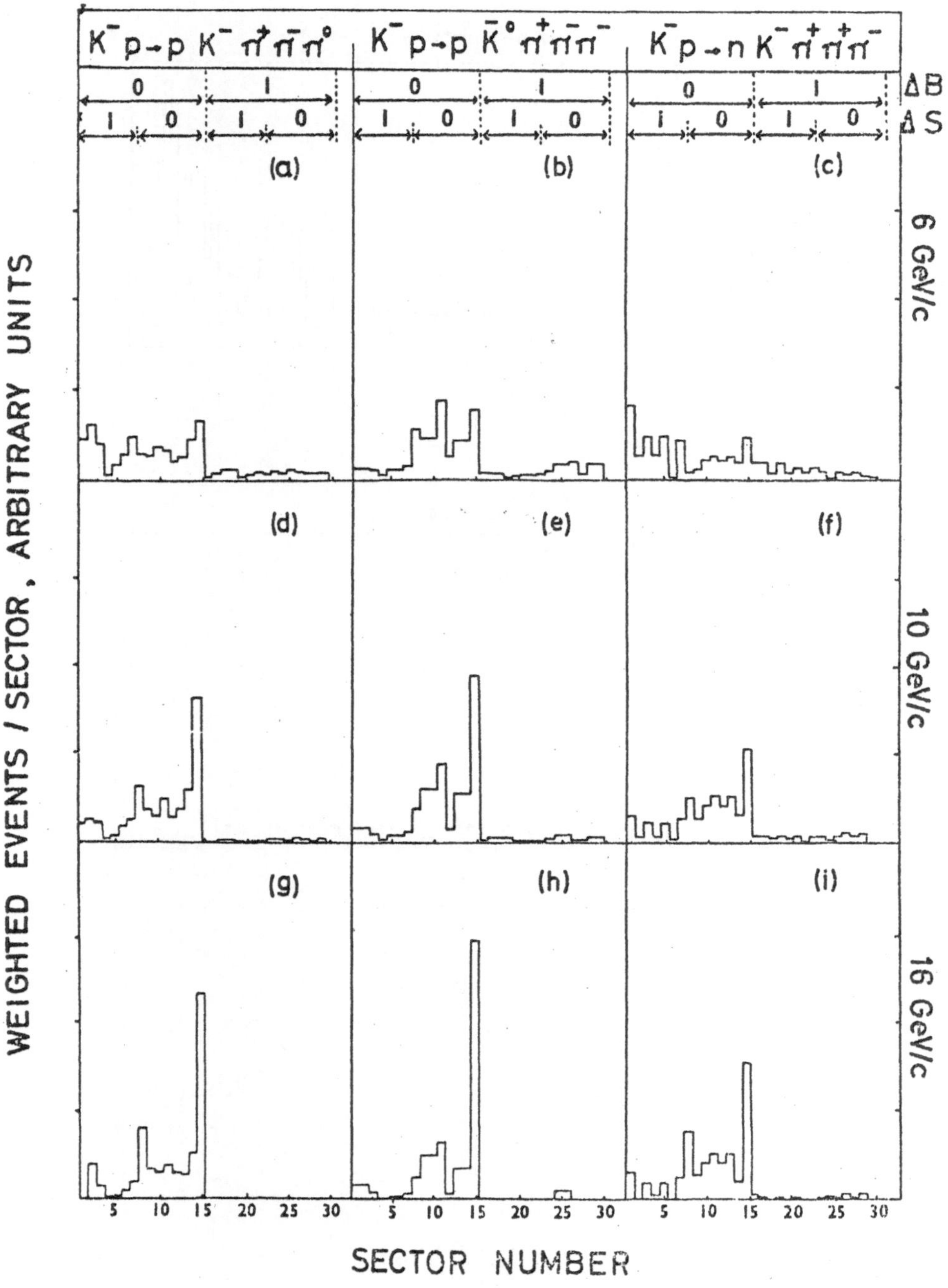

$K^- p \to p K^- \pi^+ \pi^- \pi^0$
$K^- p \to p \bar{K}^0 \pi^+ \pi^- \pi^-$
$K^- p \to n K^- \pi^+ \pi^+ \pi^-$
ΔB
ΔS
(a)
(b)
(c)
(d)
(e)
(f)
(g)
(h)
(i)
6 GeV/c
10 GeV/c
16 GeV/c
WEIGHTED EVENTS / SECTOR, ARBITRARY UNITS
5 10 15 20 25 30
SECTOR NUMBER

Fig. 43

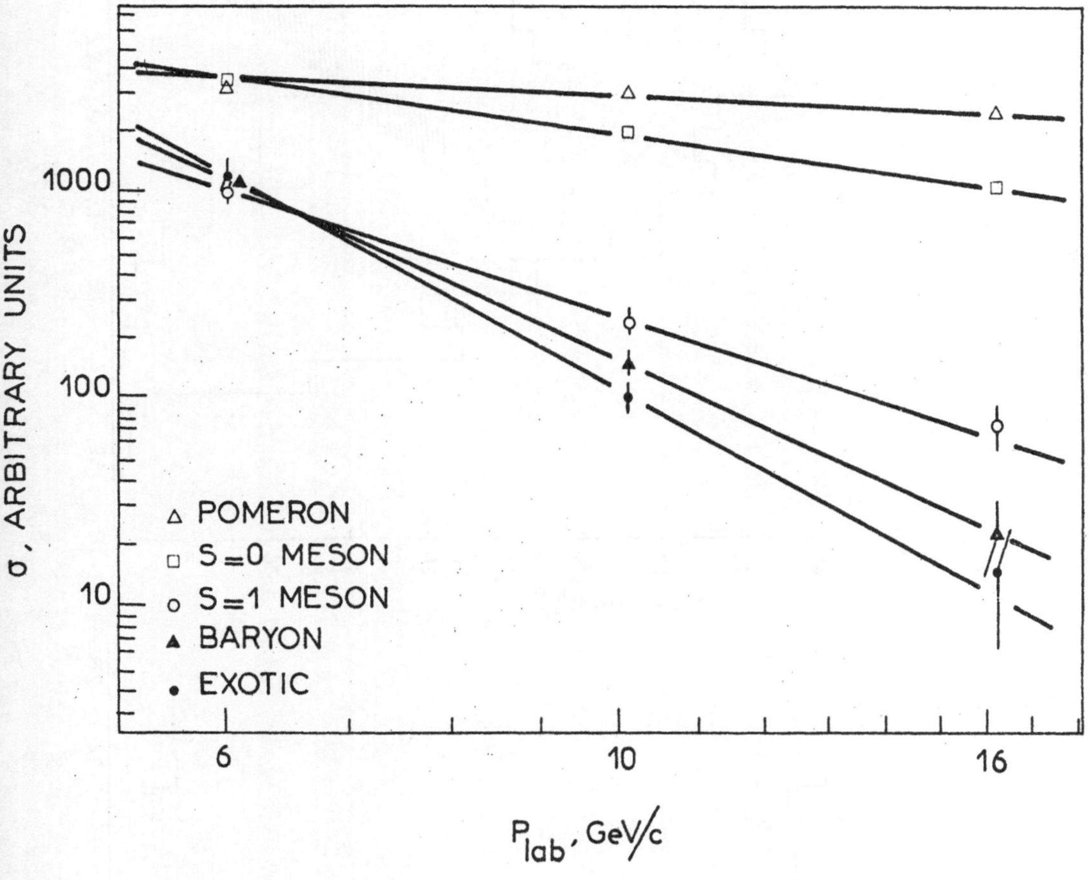
σ, ARBITRARY UNITS
1000
100
10
6
10
16
P_{lab}, GeV/c
△ POMERON
□ S=0 MESON
○ S=1 MESON
▲ BARYON
• EXOTIC

Fig. 44

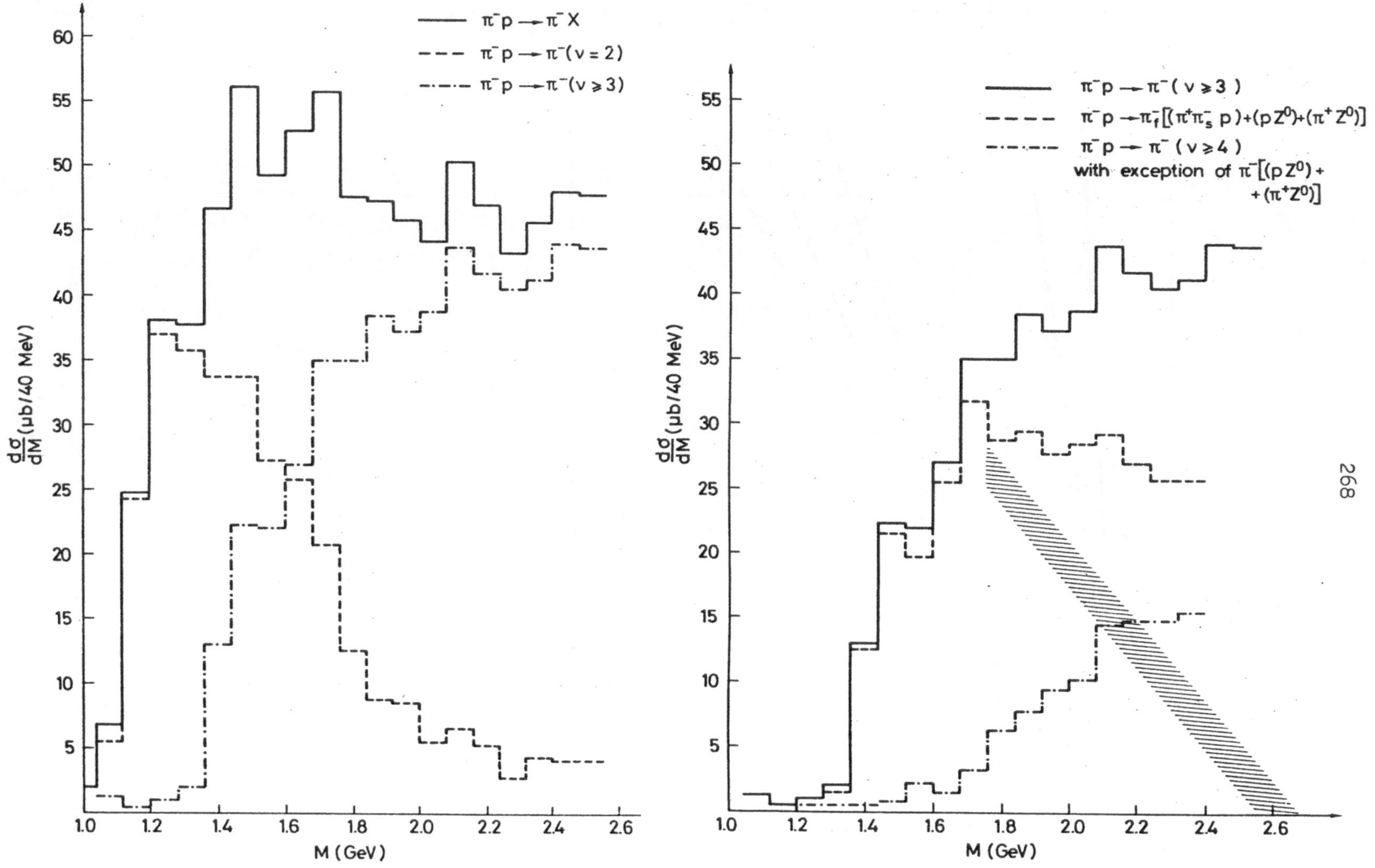
π⁻p → π⁻X
π⁻p → π⁻(ν = 2)
π⁻p → π⁻(ν ≥ 3)
dσ/dM (μb/40 MeV)
M (GeV)
π⁻p → π⁻(ν ≥ 3)
π⁻p → π⁻_f[(π⁺π⁻_s p)+(pZ⁰)+(π⁺Z⁰)]
π⁻p → π⁻(ν ≥ 4)
with exception of π⁻[(pZ⁰) + + (π⁺Z⁰)]
dσ/dM (μb/40 MeV)
M (GeV)

Fig. 45

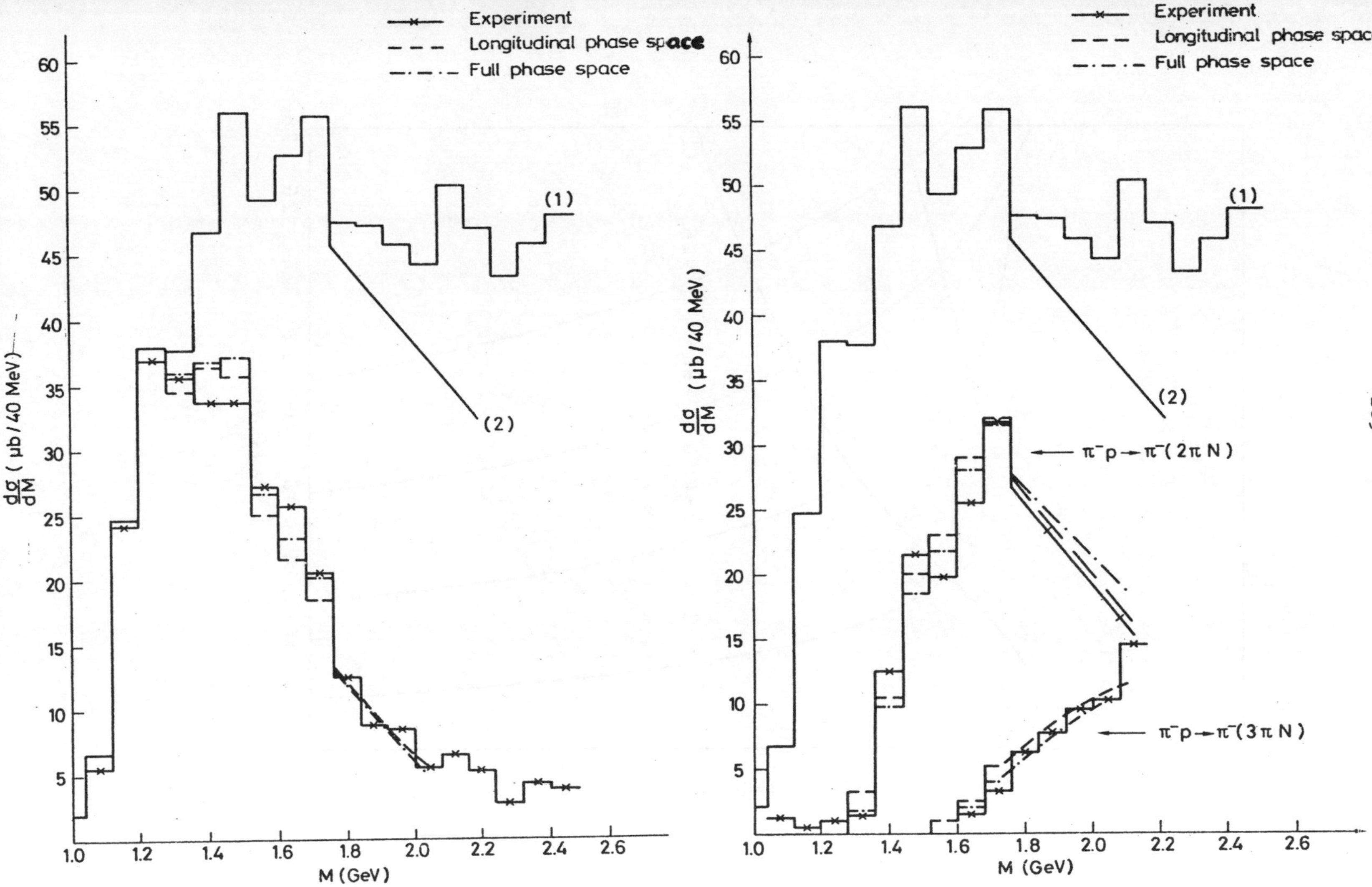
Experiment
Longitudinal phase space
Full phase space
$\frac{d\sigma}{dM}$ (μb/40 MeV)
M (GeV)
(1)
(2)
π⁻p → π⁻(2π N)
π⁻p → π⁻(3π N)

Fig. 46

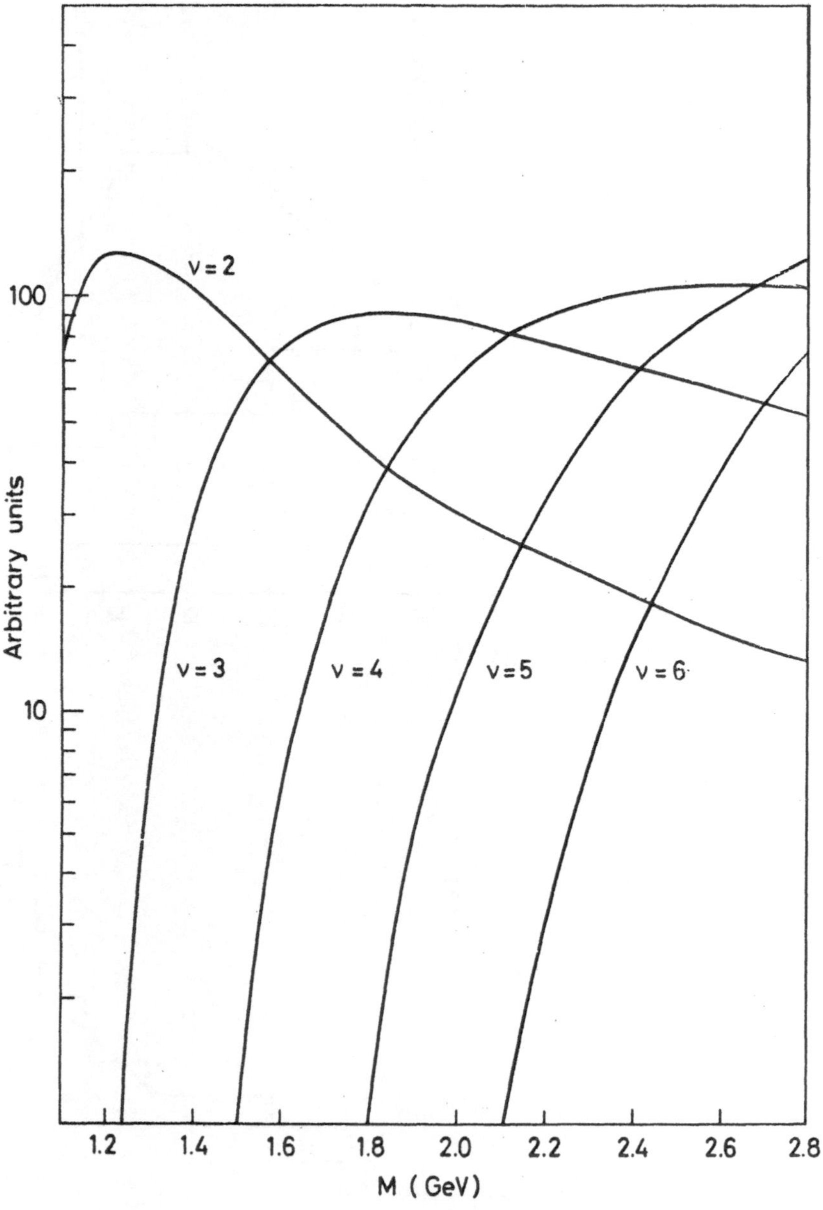
ν = 2
ν = 3
ν = 4
ν = 5
ν = 6
100
10
Arbitrary units
1.2
1.4
1.6
1.8
2.0
2.2
2.4
2.6
2.8
M (GeV)

Fig. 47

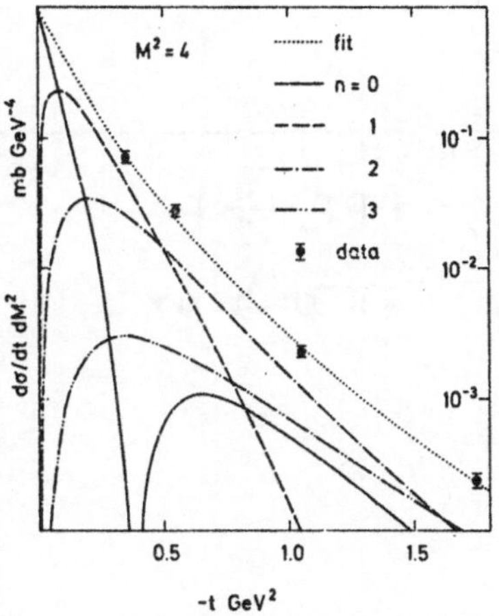

Contributions to $d^2\sigma/dt\,dM^2$, for $M^2 = 4$, of s-channel helicity amplitudes (labelled by net helicity flip n) in the case where the excitation vertex conserves t-channel helicity. Also shown is the best fit to the data.

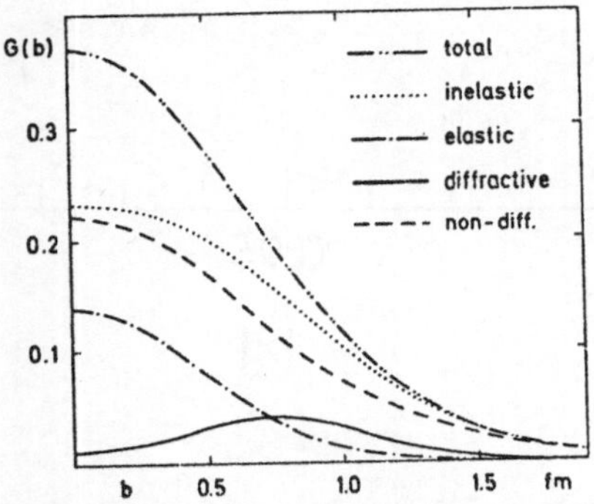

Overlap functions in b-space for t-channel helicity conservation. Normalization of $2\sigma_d$ is NAL value.

Fig. 48

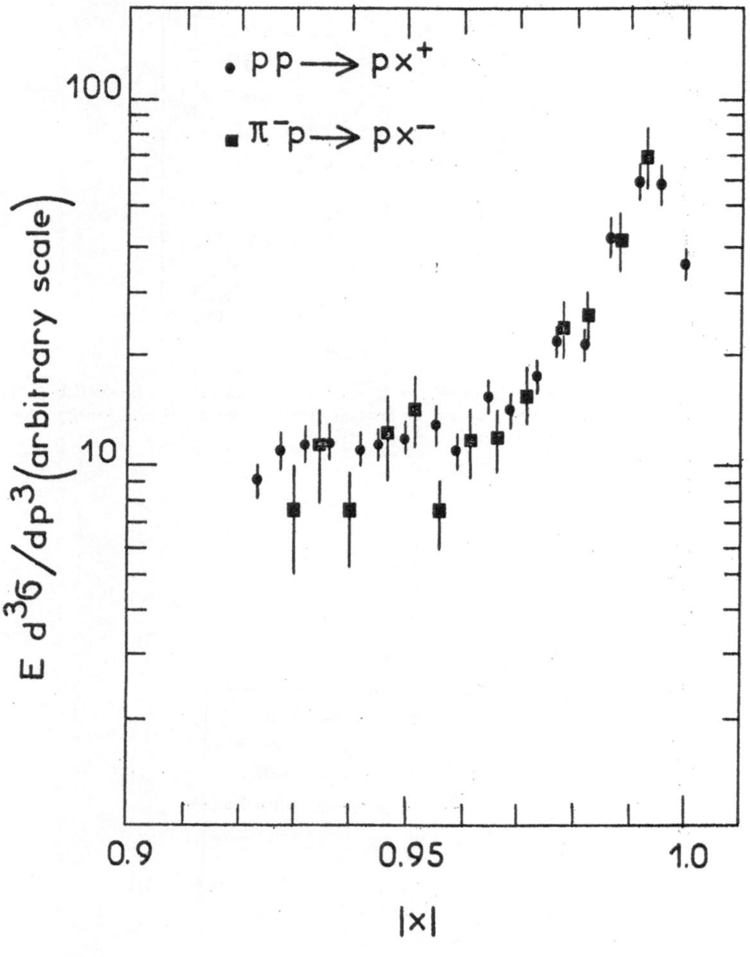

Fig. 49

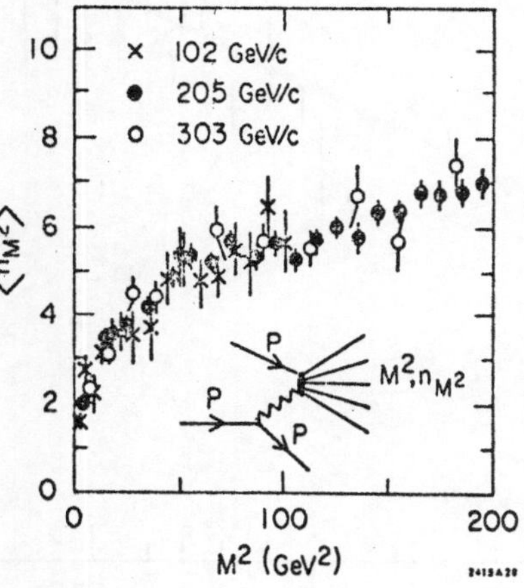

× 102 GeV/c
● 205 GeV/c
○ 303 GeV/c
⟨n$_{M^2}$⟩
P
P
P
M^2, n$_{M^2}$
0
2
4
6
8
10
0
100
200
M^2 (GeV^2)

Fig. 50

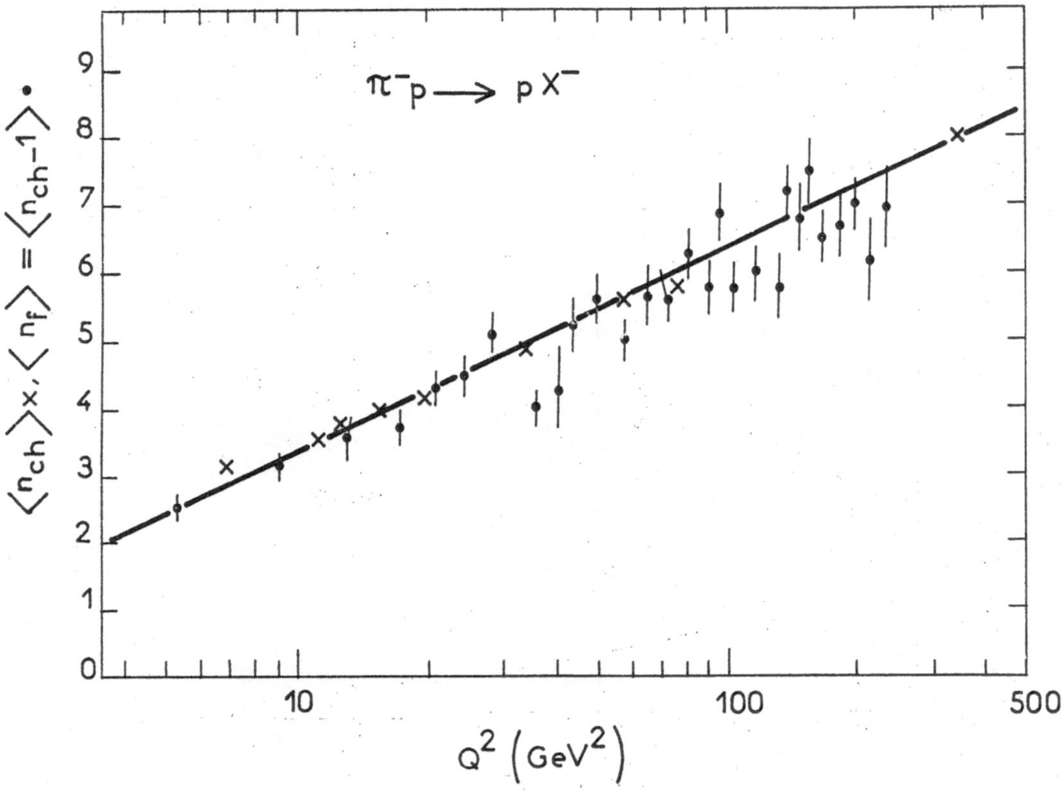
π⁻p ⟶ p X⁻
⟨n_ch⟩ ×, ⟨n_f⟩ = ⟨n_ch−1⟩ •
Q² (GeV²)
0
1
2
3
4
5
6
7
8
9
10
100
500

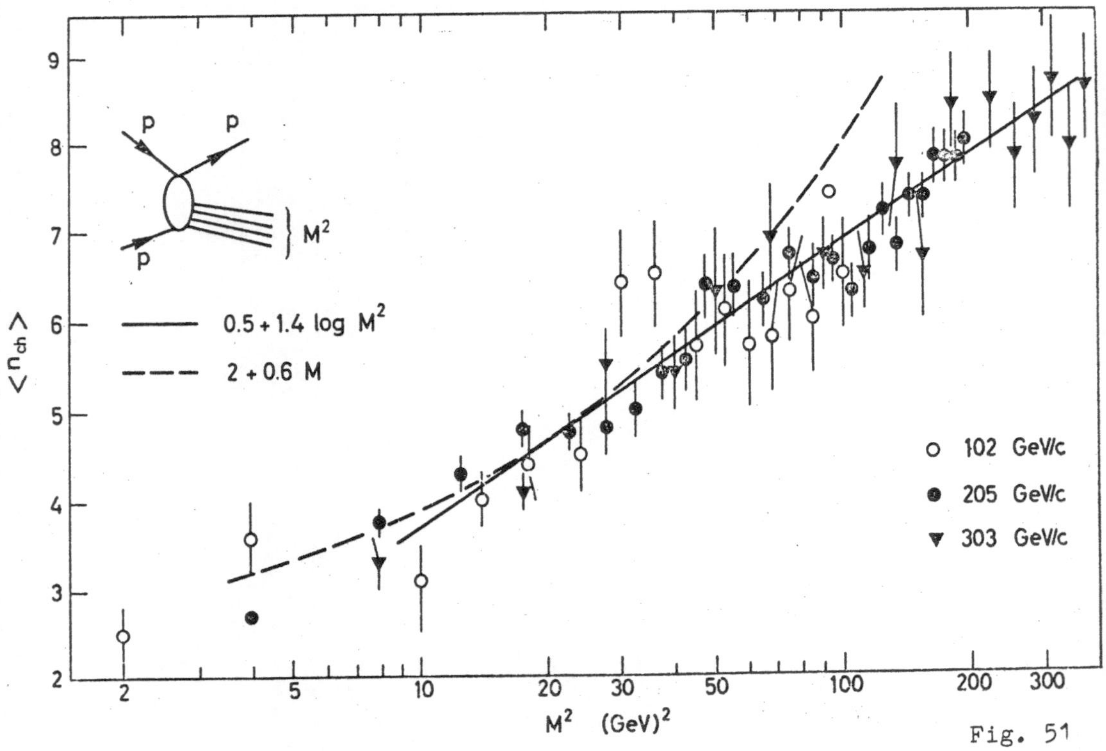
p
p
p
M²
0.5 + 1.4 log M²
2 + 0.6 M
<n_ch>
102 GeV/c
205 GeV/c
303 GeV/c
M² (GeV)²
2
3
4
5
6
7
8
9
2
5
10
20
30
50
100
200
300

Fig. 51

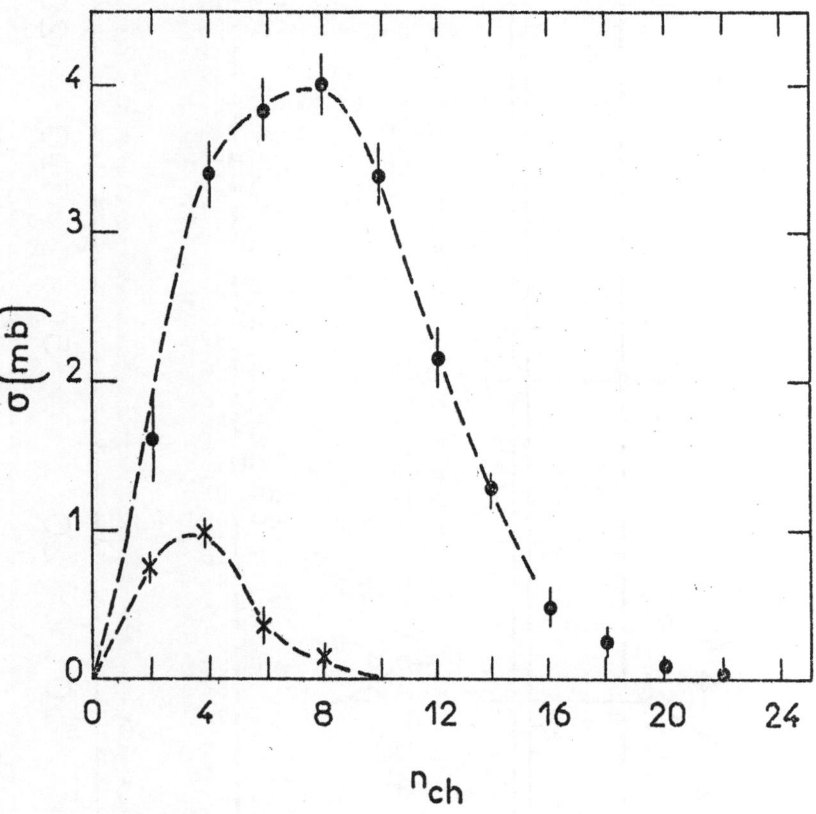

Fig. 52

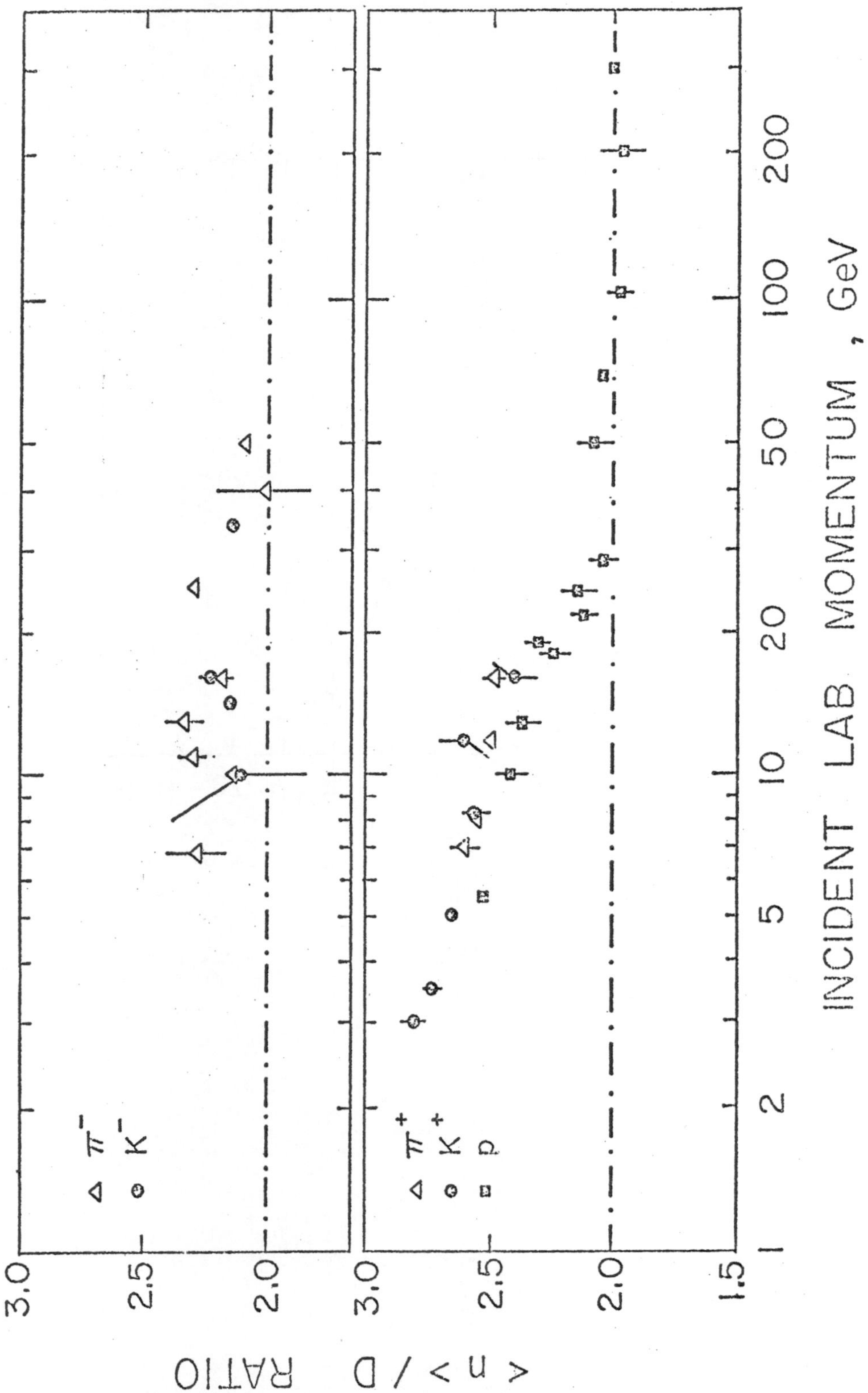

π⁻
K⁻
π⁺
K⁺
p
3.0
2.5
2.0
1.5
< n > / D RATIO
1
2
5
10
20
50
100
200
INCIDENT LAB MOMENTUM , GeV

Fig. 53

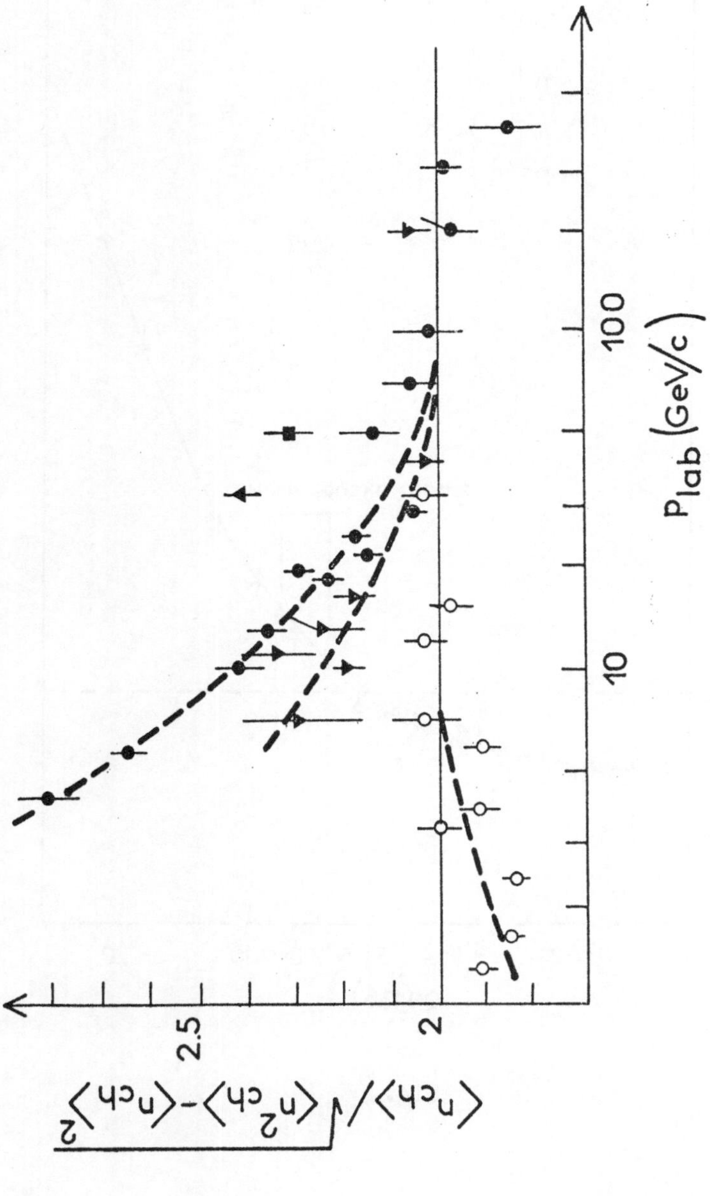

2.5
2
10
100
p_{lab} (GeV/c)
$\langle n_{ch}\rangle / \sqrt{\langle n^2_{ch}\rangle - \langle n_{ch}\rangle^2}$

Fig. 54

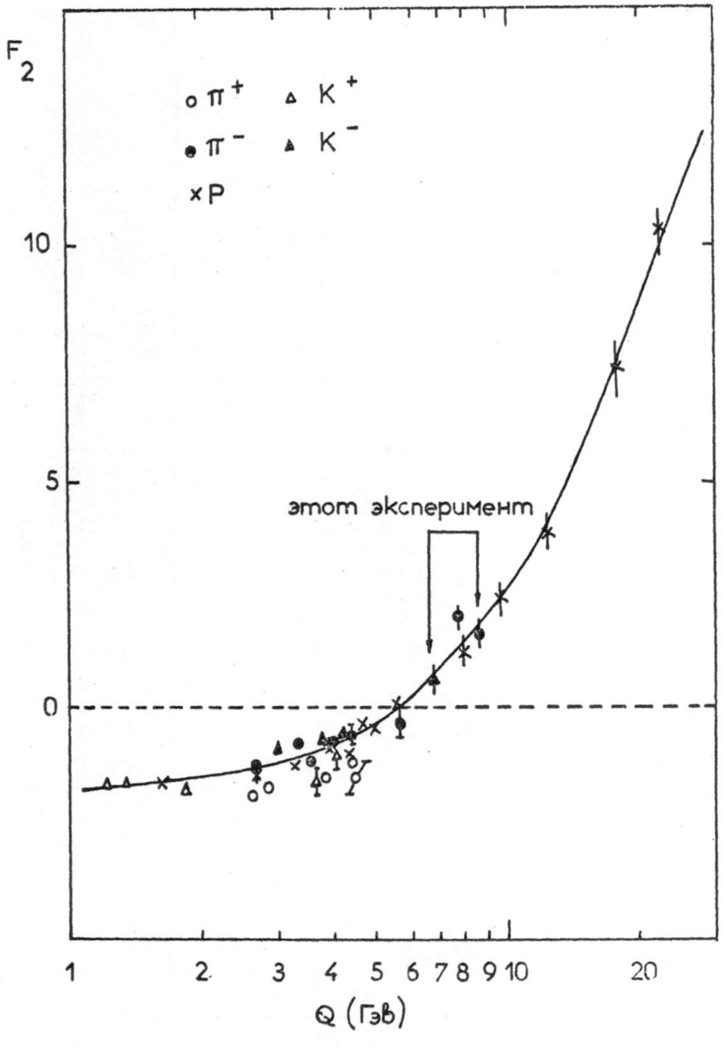

Fig. 55

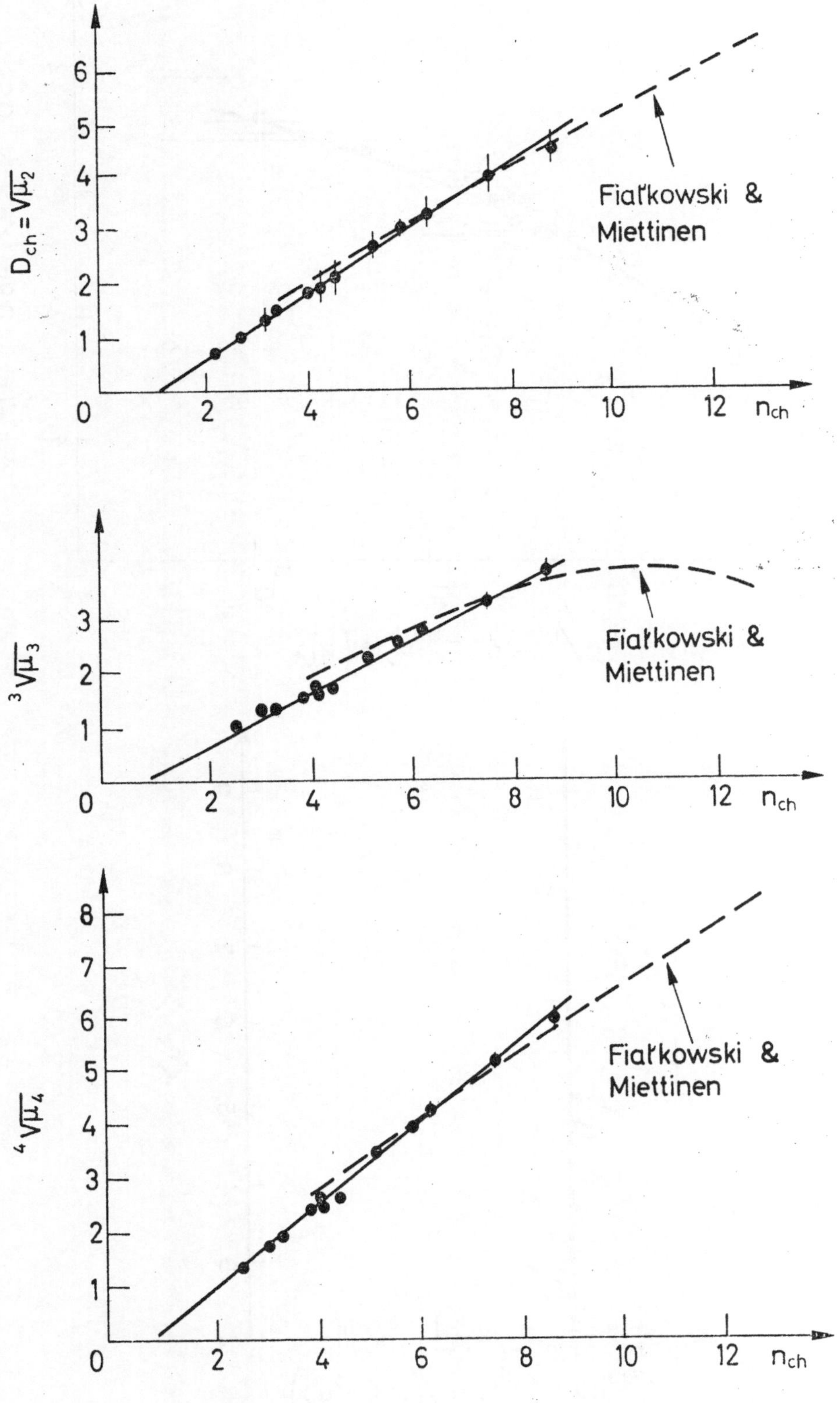

$D_{ch} = \sqrt{\mu_2}$
0
1
2
3
4
5
6
2
4
6
8
10
12
n_{ch}
Fiałkowski &
Miettinen
$\sqrt[3]{\mu_3}$
1
2
3
Fiałkowski &
Miettinen
$\sqrt[4]{\mu_4}$
1
2
3
4
5
6
7
8
Fiałkowski &
Miettinen

Fig. 56

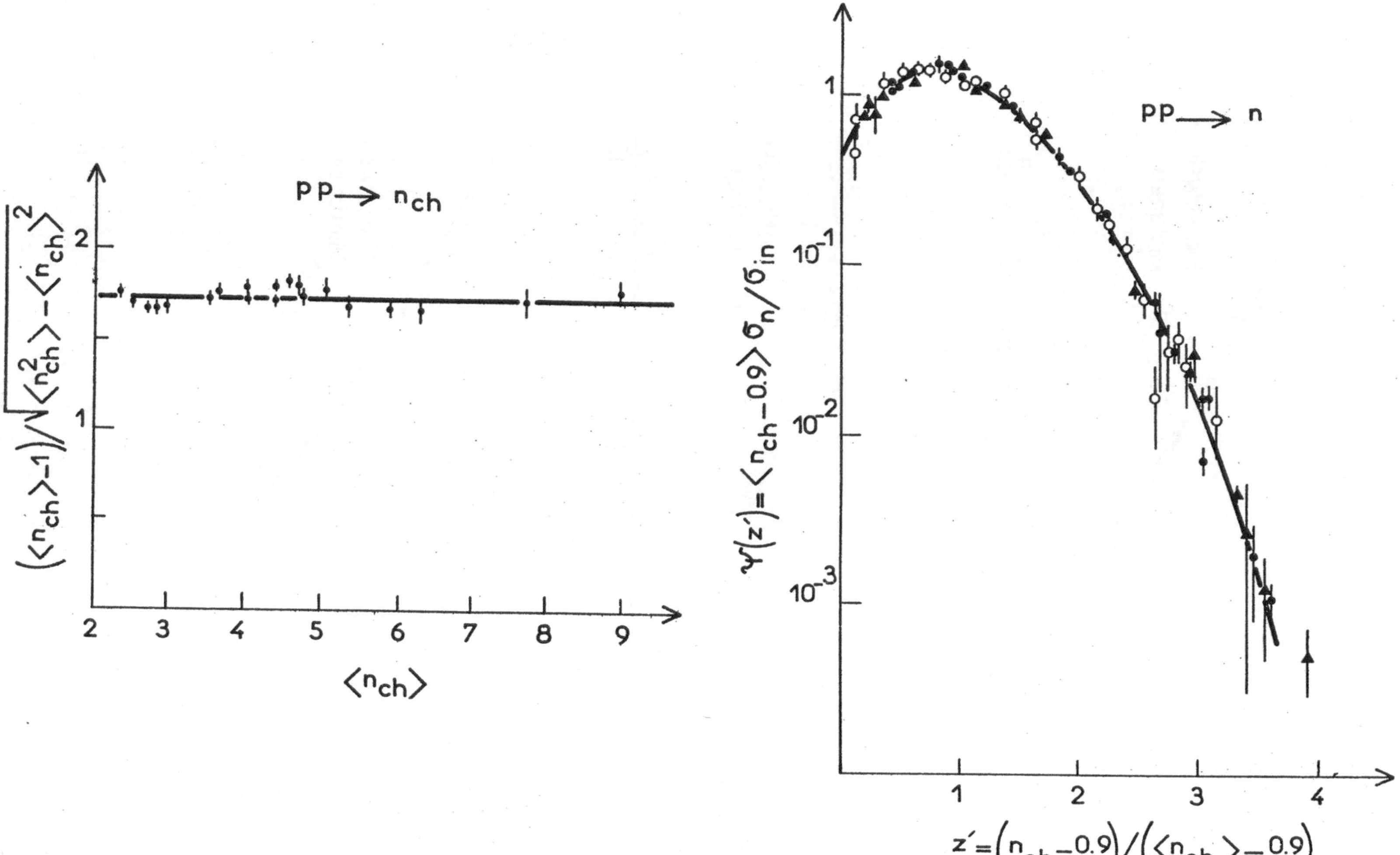
pp → n_ch
(⟨n_ch⟩−1)/√(⟨n_ch^2⟩−⟨n_ch⟩^2)
2
1
2 3 4 5 6 7 8 9
⟨n_ch⟩
pp → n
Ψ(z′)=⟨n_ch−0.9⟩σ_n/σ_in
1
10^-1
10^-2
10^-3
1 2 3 4
z′=(n_ch−0.9)/(⟨n_ch⟩−0.9)

Fig. 57

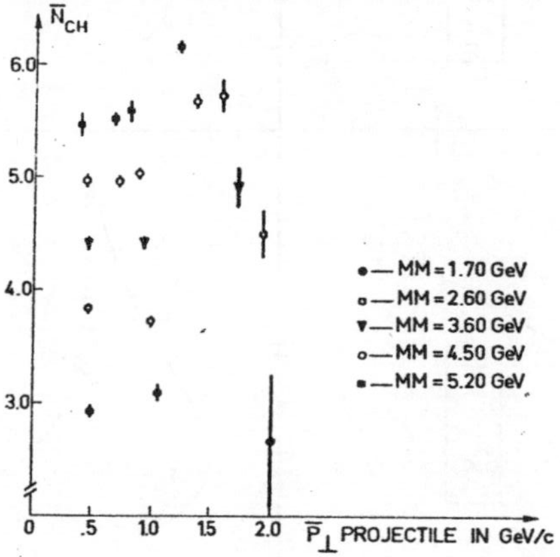

The variation of the average charged multiplicity $\bar{N}_{CH}$ with $p_{\perp}$ for five intervals of missing mass MM.

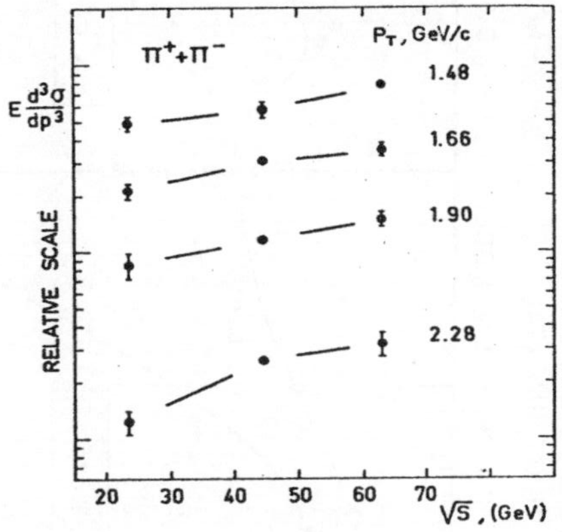

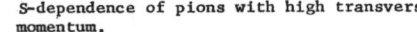

S-dependence of pions with high transverse momentum.

Fig. 58

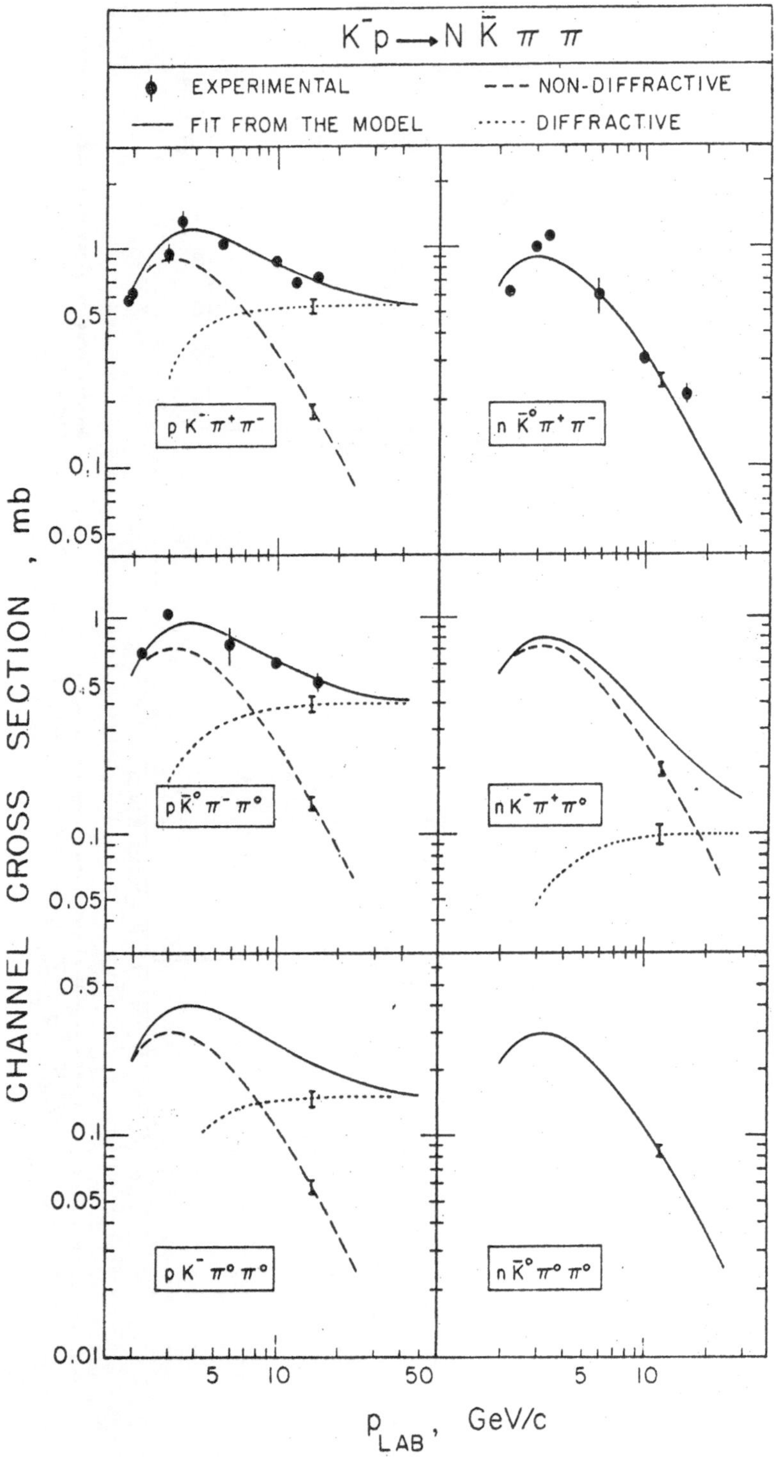
K⁻p → N K̄ π π
EXPERIMENTAL
FIT FROM THE MODEL
NON-DIFFRACTIVE
DIFFRACTIVE
p K⁻ π⁺ π⁻
n K̄° π⁺ π⁻
p K̄° π⁻ π°
n K⁻ π⁺ π°
p K⁻ π° π°
n K̄° π° π°
CHANNEL CROSS SECTION, mb
1
0.5
0.1
0.05
0.01
5
10
50
p_LAB, GeV/c

Fig. 59

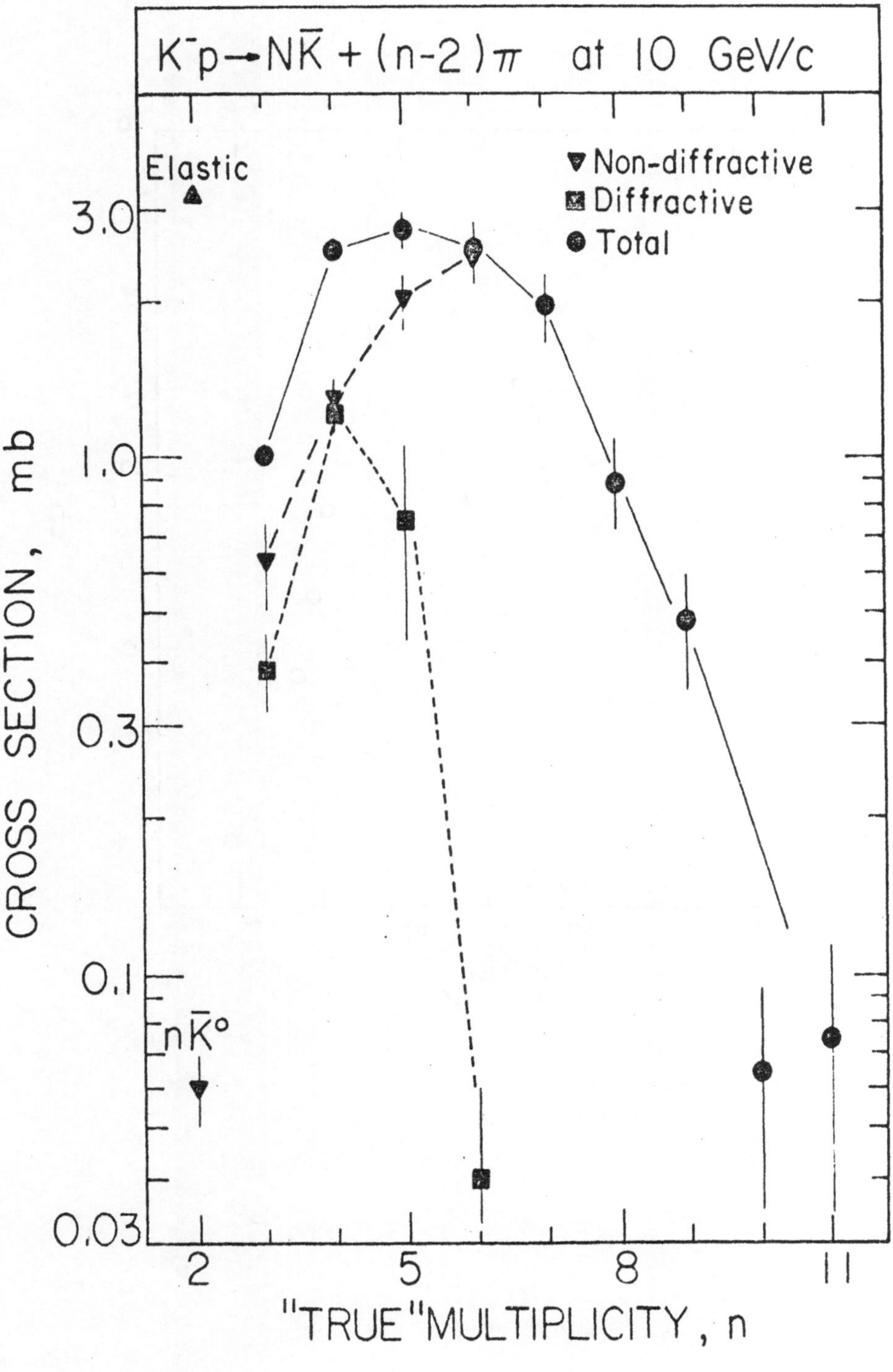
K⁻p → NK̄ + (n-2)π at 10 GeV/c
Elastic
Non-diffractive
Diffractive
Total
nK̄°
CROSS SECTION, mb
3.0
1.0
0.3
0.1
0.03
2
5
8
11
"TRUE" MULTIPLICITY, n

Fig. 60

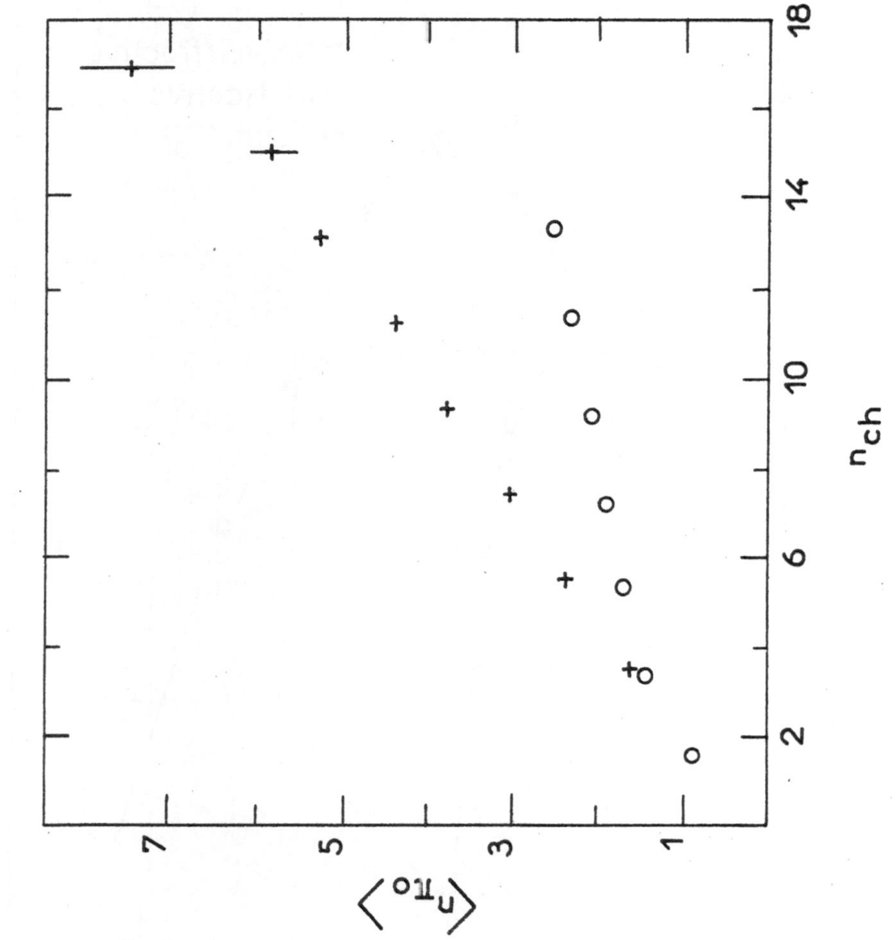

Fig. 61

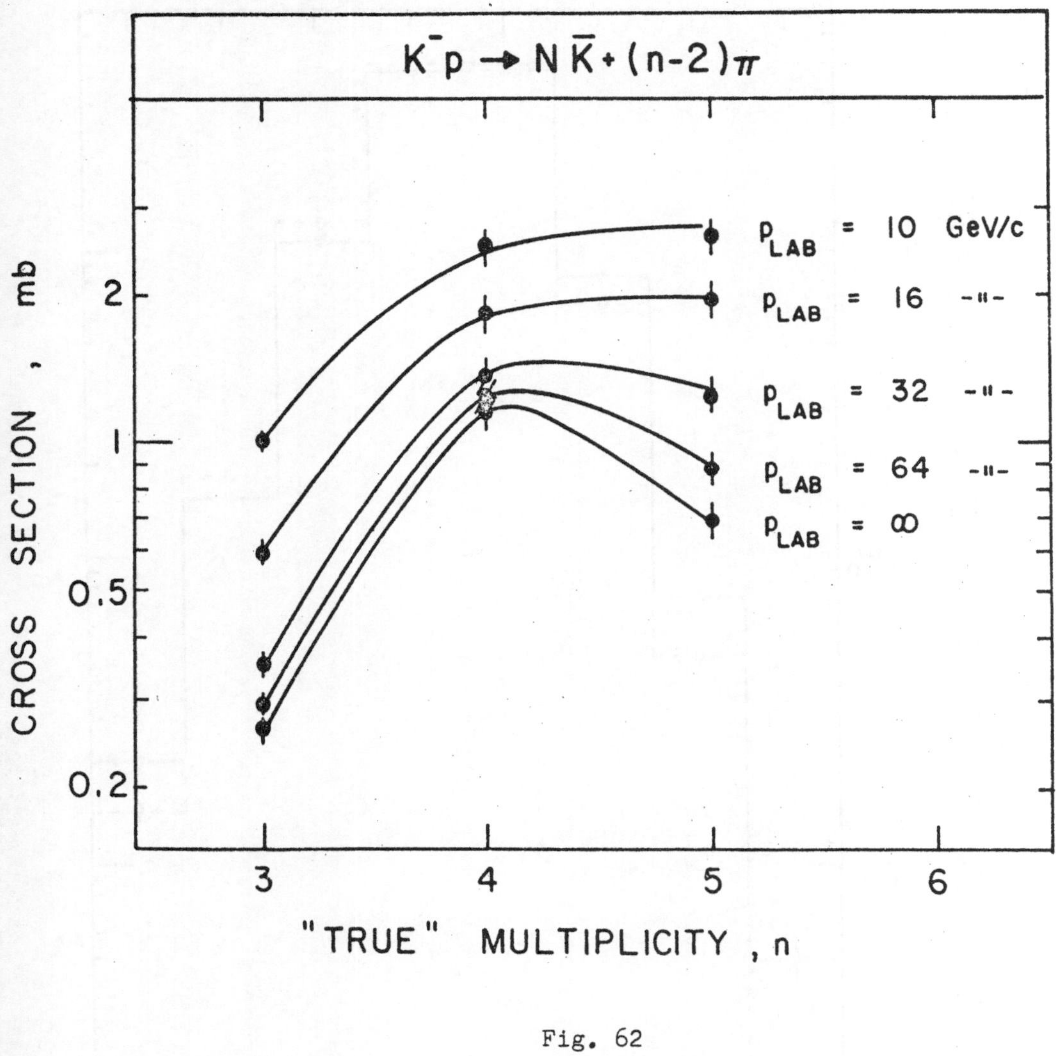
K⁻p → N K̄ + (n-2)π
CROSS SECTION , mb
2
1
0.5
0.2
3
4
5
6
"TRUE" MULTIPLICITY , n
p_LAB = 10 GeV/c
p_LAB = 16 -"-
p_LAB = 32 -"-
p_LAB = 64 -"-
p_LAB = ∞

Fig. 62

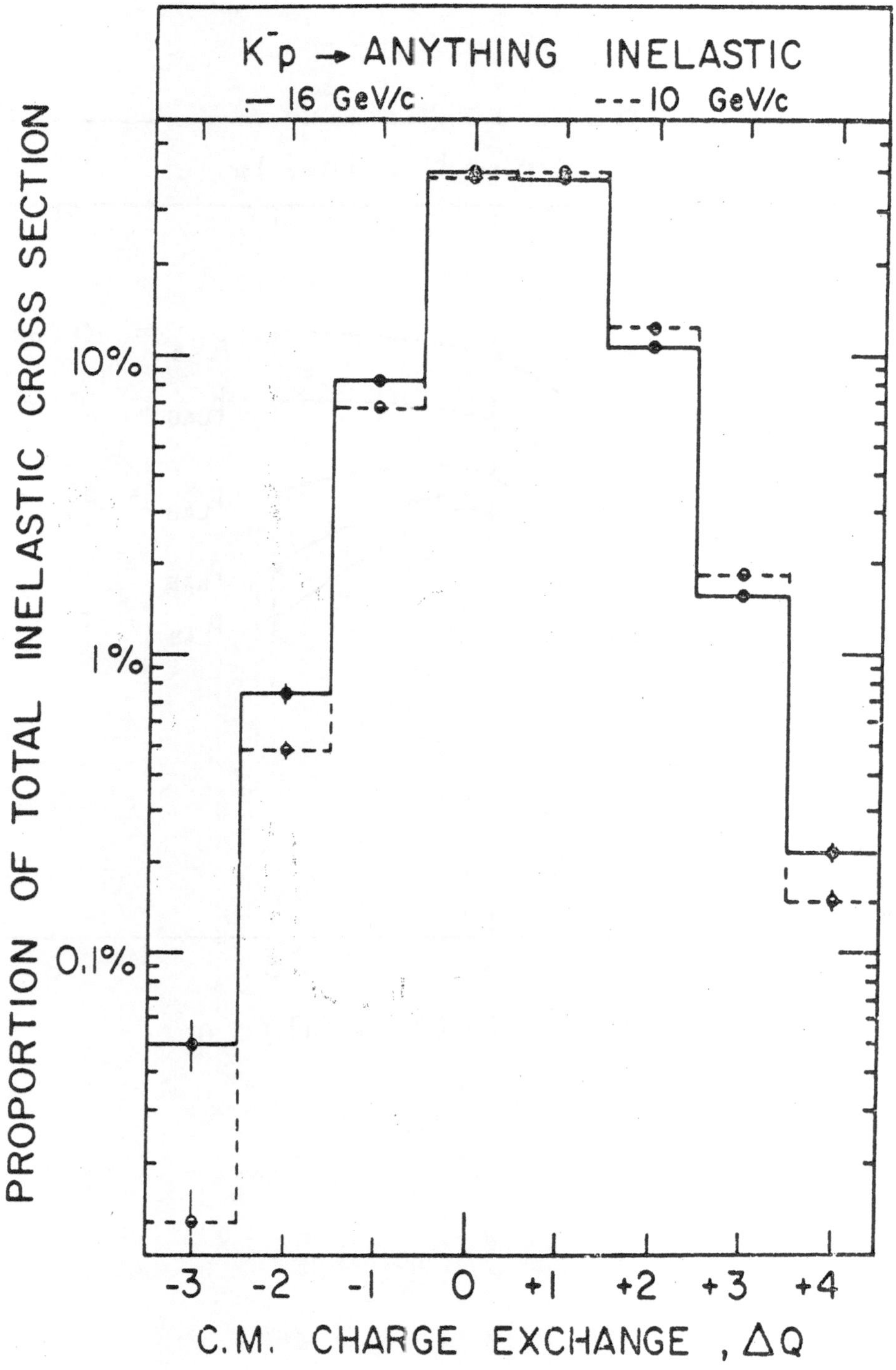
K⁻p → ANYTHING INELASTIC
— 16 GeV/c
--- 10 GeV/c
PROPORTION OF TOTAL INELASTIC CROSS SECTION
10%
1%
0.1%
-3
-2
-1
0
+1
+2
+3
+4
C.M. CHARGE EXCHANGE , ΔQ

Fig. 63

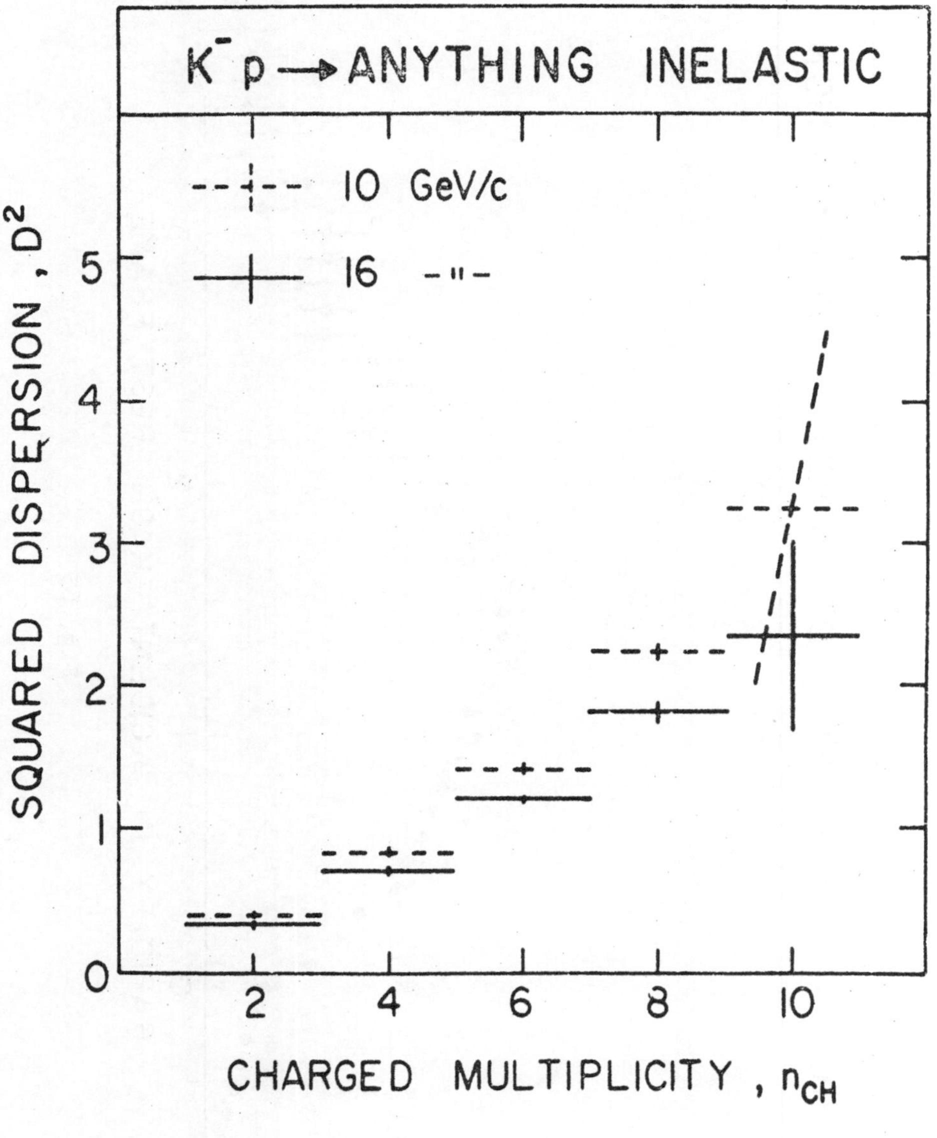
K⁻p → ANYTHING INELASTIC
10 GeV/c
16 -"-
SQUARED DISPERSION , D²
0
1
2
3
4
5
2
4
6
8
10
CHARGED MULTIPLICITY , n_CH

Fig. 64

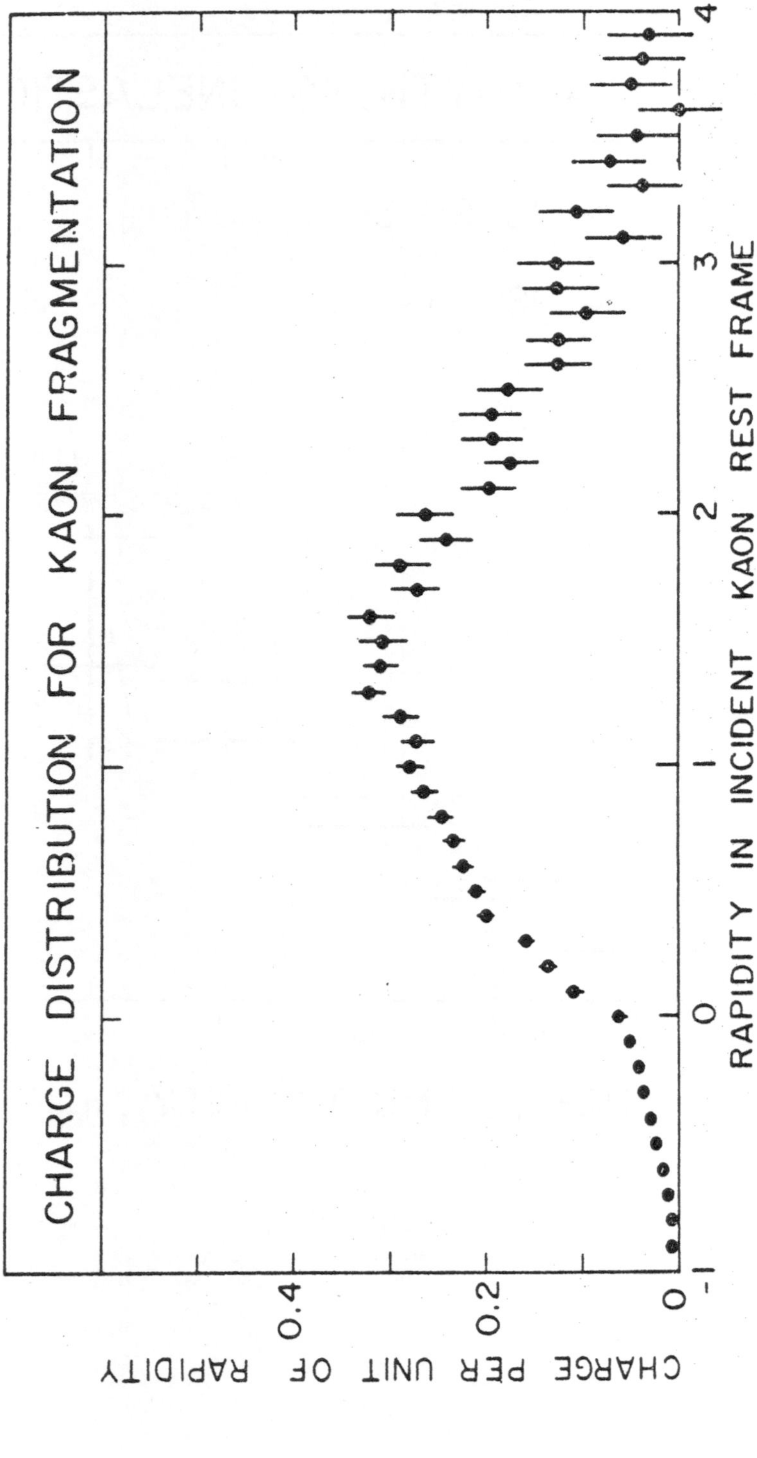
CHARGE DISTRIBUTION FOR KAON FRAGMENTATION
CHARGE PER UNIT OF RAPIDITY
0.4
0.2
0
-1
0
1
2
3
4
RAPIDITY IN INCIDENT KAON REST FRAME

Fig. 65

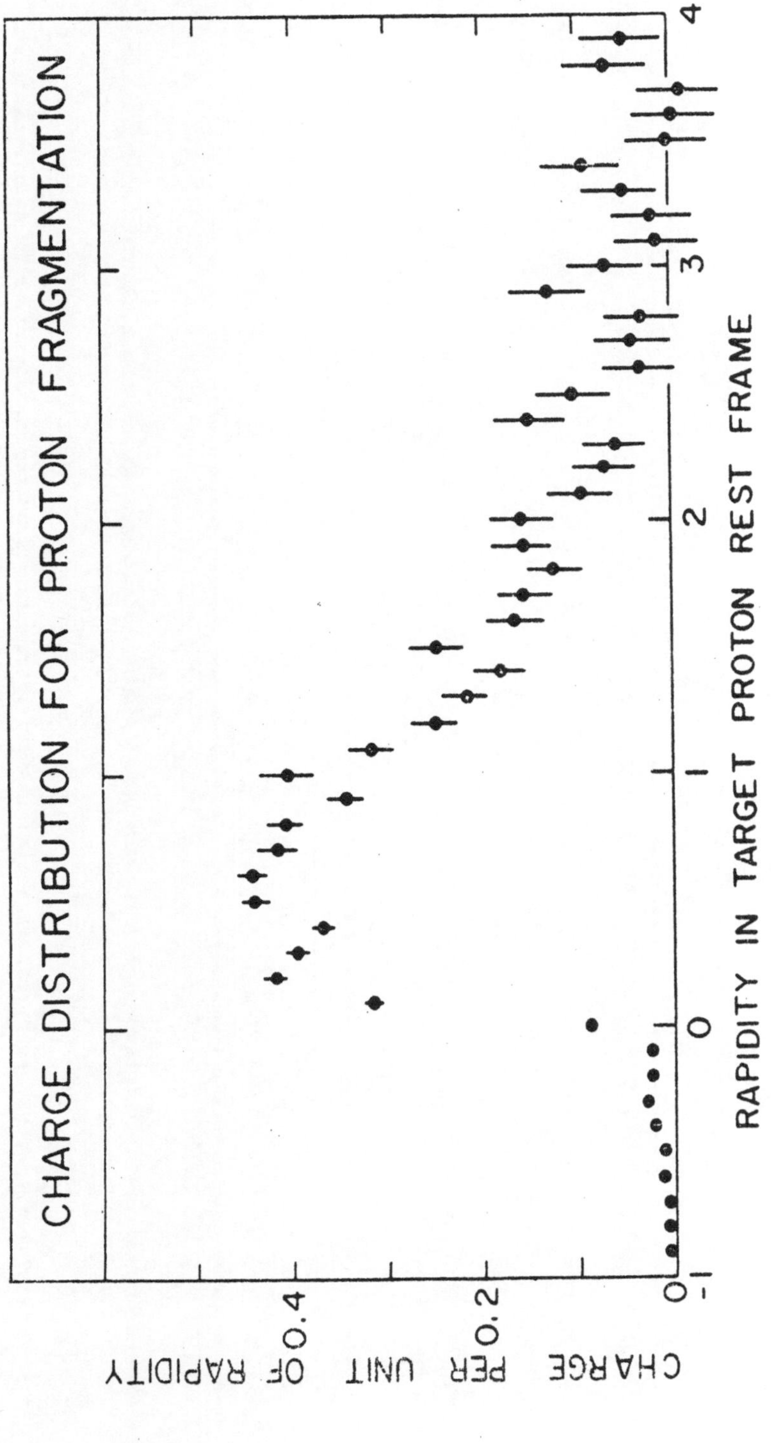
CHARGE DISTRIBUTION FOR PROTON FRAGMENTATION
CHARGE PER UNIT OF RAPIDITY
0
0.2
0.4
RAPIDITY IN TARGET PROTON REST FRAME
-1
0
1
2
3
4

Fig. 66

Acta Physica Austriaca, Suppl. XIII, 291–359 (1974)

STATUS OF MESON-MESON INTERACTION*)

by

J. L. PETERSEN
Nordita, Copenhagen

1. STATUS OF EXPERIMENTAL MESON-MESON PHASE SHIFT ANALYSES

The plan of these lectures is as follows: In 1. I shall briefly review the current experimental situation for $\pi\pi \to \pi\pi$, $\pi\pi \to K\bar{K}$ and $K\pi \to K\pi$ phase shift analysis; in 2. I shall give an account of some of the dispersion techniques which are evolving for the imposition of crossing symmetry and in 3. I shall present the results of some analyses which have been carried out in order to complete the fragmentary picture provided by the phase shift analyses and to enable one to compare with theoretical models.

1.1. $\pi\pi$ PHASE SHIFT ANALYSIS

At present the situation is dominated by two major

*) Lectures given at XIII. Internationale Universitätswochen für Kernphysik, Schladming, February 4-16, 1974.

experiments (for a summary on older ones cf. for example ref. [1]):

Berkeley [2]:

$\pi^+ p \to \pi^+\pi^-\Delta^{++}$ at 7.1 GeV/c (∿ 30.000 events) (1)

CERN-Munich [3] - [5]

$\pi^- p \to \pi^+\pi^- n$ at 17.2 GeV/c (∿ 300.000 events) (2)

The objective is to isolate the one pion exchange (OPE) contribution to these reactions following the idea of Chew and Low and of Goebel [6]. Thus if OPE dominates we have for reactions (1) and (2) respectively

$$\frac{\partial^5\sigma(s,t,M_{\pi\pi},\theta,\phi,M_{\pi p})}{\partial t\partial M_{\pi\pi}\partial^2\Omega(\theta,\phi)\partial M_{\pi p}} =$$

$$= \frac{1}{4\pi^3 p^2 s}\left(M_{\pi\pi}^2 q(t)\left(\frac{d\sigma}{d\Omega}\right)_{\pi\pi}\right)\frac{1}{(t-\mu^2)^2}\left(M_{\pi p}^2 Q(t)\sigma_{\pi p}(M_{\pi p})\right) \qquad (1')$$

$$\frac{\partial^4\sigma(s,t,M_{\pi\pi},\theta,\phi)}{\partial t\partial M_{\pi\pi}\partial^2\Omega(\theta,\phi)} =$$

$$= \frac{1}{4\pi p^2 s}\left(M_{\pi\pi}^2 q(t)\left(\frac{d\sigma}{d\Omega}\right)_{\pi\pi}\right)\frac{t}{(t-\mu^2)^2}\frac{G^2}{4\pi} \qquad (2')$$

Here μ is the pion mass, p the initial c.m. 3-momentum, q(t) and Q(t) the 3-momentum of the virtual pion in the $\pi^+\pi^-$ rest frame and the π^+p rest frame respectively, $M_{\pi p}$ is the mass of the Δ^{++} (1236), $\sigma_{\pi p}(M_{\pi p})$ is the π^+p cross section, $G^2/4\pi = 2 \times 14.6$ is the π^+pn coupling constant. The meaning of the remaining symbols follows from fig. 1 (we use units $\hbar = c = 1$).

In the physical regions for reactions (1) and (2) complications arise due to the presence of background terms, and for reaction (2) the vanishing of the OPE-signal at t = 0 implies that these are relatively more important for (2) than for (1). For (1) therefore a straightforward extrapolation procedure was adopted [2] and results for the p-wave and the I = 0 s-wave below 1 GeV are shown in figs. 2 and 3. To get results on δ_0^0 *) an assumption must be made for δ_0^2. Since from the isospin decomposition

$$F(\pi^+\pi^- \to \pi^+\pi^-) = \frac{1}{3}F^{(0)} + \frac{1}{2}F^{(1)} + \frac{1}{6}F^{(2)} \tag{3}$$

the weight of the I = 2 is small, no very large error should arise from uncertainties in δ_0^2.

The ρ-parameters corresponding to fig. 2 are

$$m_\rho = 775 \pm 4 \text{ MeV}, \qquad \Gamma_\rho = 160 \pm 10 \text{ MeV} \tag{4}$$

corresponding to a wider object than in the "educated guess" of PDG [7]:

---

*) A quantity A pertaining to angular momentum ℓ and isospin I is denoted by A_ℓ^I.

$$m_\rho = 770 \pm 5 \text{ MeV} \quad , \qquad \Gamma_\rho = 146 \pm 10 \text{ MeV} . \qquad (5)$$

A very significant result concerning the Berkeley experiment [2] is the observed absence of 4π inelasticity below 1 GeV (except for a small $\omega\pi$ signal in I = 1). This implies that partial wave amplitudes can be considered elastic down to the 1% level below 1 GeV.

The main interest centres around the behaviour of δ_0^0 (fig. 3). Below 700 MeV a "between" form is obtained in agreement with the consensus reached after 1970 (cf. ref. [1]). Above that energy a remarkable drop in the mass spectrum and in the sp interference at the $K\bar{K}$-threshold is succesfully correlated to the long known rapid rise of the $\pi\pi \rightarrow K\bar{K}$ cross section above threshold (cf. sect. 1.2 below) to resolve the well known "up-down" ambiguity in favour of the "down"-solution up to 900 MeV. Between 900 MeV and 1100 MeV I = 0 s-wave scattering is dominated by the $K\bar{K}$-threshold effect (or the S^*-effect). The characteristic feature is the sudden rise in δ_0^0 to a value close to 180^0 at the $K\bar{K}$-threshold followed by a sharp drop in η_0^0.

The insight gained as a result of the Berkeley experiment represented a qualitative breakthrough in low energy $\pi\pi$ scattering. For information on the behaviour of the amplitude above 1 GeV and for quantitative details below 1 GeV workers in the field have been looking forward for a long time to the analysis of the CERN-Munich experiment on reaction (2) [3-5]. Three different analyses of the data have been published.

Grayer et al. [3] used (a) a simple Chew-Low extrapolation and (b) a t-channel amplitude analysis scheme suggested by Froggatt and Morgan [8]. Results of the two agreed with one another and with the Berkeley-results (cf. fig. 3) which was very encouraging.

Estabrooks et al. [4] argued in favour of an s-channel analysis which surprisingly produced two different solutions (1) and (2) of which (2) was excluded by appealing to $\pi^o\pi^o$ experiments or by using the S^*-argument of ref. [2]. δ_o^o for solution (1), however, was 10° above the data of refs. [2] and [3] (fig. 3).

Finally Hyams et al. [5] have carried out the only analysis so far which covers the complete region of the experiment i.e. $M_{\pi\pi}$ up to 1800 MeV. The results, based on generalized absorption model calculations, agree with refs. [2] and [3] and disagree with Estabrooks et al.[4] (fig. 3).

The controversy between refs. [4] and [5], say, is one of the main issues to be settled in $\pi\pi$ scattering below 1 GeV. In sect. 3 I shall present reasons to believe that the Estabrook et al. [4] results are inconsistent at least at very low energies ($M_{\pi\pi}$ < 600-700 MeV)

Let us briefly look at the current status of pion production amplitude analysis.*)

We write the differential production cross section in terms of helicity amplitudes as follows

$$I(s,t,M_{\pi\pi},\theta,\phi) = \frac{1}{2} \sum_{\lambda\lambda'} | H_{\lambda\lambda'}|^2$$

$$= N \sum_{L=0}^{\infty} \sum_{M=-L}^{L} \langle Y_L^M \rangle Y_L^M(\theta,\phi) \qquad (6)$$

*) I am much indebted to C.D. Froggatt for many clarifying conversations on this subject.

where

$$H_{\lambda\lambda'} = \sum_{j=0}^{\infty} \sum_{m=-j}^{j} (2j+1)^{1/2} H^{j}_{\lambda\lambda';m} d^{j}_{m0}(\theta)e^{im\phi} \tag{7}$$

$$N \equiv \frac{\partial^2\sigma}{\partial M_{\pi\pi}\,\partial t} \tag{8}$$

Here λ,λ' are initial and final nucleon helicities (not observed) and m is the dipion helicity. For simplicity we consider the s-p approximation and define

$$S \equiv (H^{0}_{++;0} \; ; \; H^{0}_{+-;0})$$

$$P_m \equiv (H^{1}_{++;m} \; ; \; H^{1}_{+-;m}) \qquad m = 0,\pm 1 \; . \tag{9}$$

The dipion polar angles (θ,ϕ) and the helicities are normally described either in the s-channel (or helicity) frame or in the t-channel (or GJ [9] frame). In both $\hat{y}$ is normal to the production plane (beam x outgoing nucleon). In the s-channel frame $\hat{z}$ is opposite to the direction of the outgoing neutron as seen in the dipion rest frame. In the t-channel frame $\hat{z}$ is the direction of the incoming π^- as seen in the dipion rest frame.

In the t-channel frame OPE takes the following simple form (omitting the nucleon spin factors and using the high energy approximation $t_{min} \approx 0$):

$$S^{(t)}(\text{OPE}) = \frac{M_{\pi\pi}\sqrt{-t}}{t-\mu^2}\, f^{S}_{\pi\pi}$$

$$P_o^{(t)}\,(\mathrm{OPE}) = \frac{M_{\pi\pi}\sqrt{-t}}{t-\mu^2}\, f^P_{\pi\pi} \tag{10}$$

$$P^{(t)}_{\pm 1}\,(\mathrm{OPE}) \equiv 0 \,.$$

Here $f^L_{\pi\pi}$ is the $\pi\pi$ partial wave amplitude. It follows that only the $M = 0$ moments $\langle Y^O_L \rangle$ contain the OPE in the t-channel frame. These are therefore the only ones which need be considered in a Chew-Low extrapolation. However, as has been emphasized by many people [8], [10], [11] the result of the extrapolation will depend sensitively on details of the background for data with finite accuracy due to the vanishing of OPE at $t = 0$. One therefore attempts to make use of the $M \neq 0$ moments to get information on the background. That, however, cannot be done in a model-independent way for an unpolarized target experiment. Thus in the s-p approximation 6 numbers ($\langle Y^M_L \rangle$ for $L \leq 2$) can be measured whereas the process is described by the 8 complex amplitudes eq. (9). In practice therefore amplitudes are parametrized in a way suggested by absorption models:

First transform eqs. (10) to the s-channel frame (where absorption is most easily described)

$$S^{(s)}\,(\mathrm{OPE}) = \frac{M_{\pi\pi}\sqrt{-t}}{t-\mu^2}\, f^S_{\pi\pi} \qquad (n = 1)$$

$$P_o^{(s)}\,(\mathrm{OPE}) = \frac{M_{\pi\pi}\sqrt{-t}}{t-\mu^2}\, f^P_{\pi\pi} \qquad (n = 1) \tag{11}$$

$$P_1^{(s)}\,(\mathrm{OPE}) = \frac{\sqrt{2}t}{t-\mu^2}\, f^P_{\pi\pi} \qquad (n = 0)$$

$$P_{-1}^{(s)}\,(\mathrm{OPE}) = -\frac{\sqrt{2}t}{t-\mu^2}\, f^P_{\pi\pi} \qquad (n = 2) \quad .$$

Numbers in parenthesis denote net helicity flip. Apart from $P_1^{(s)}$ the behaviour at $t = 0$ is the one required by angular momentum conservation: $(-t)^{n/2}$. Hence $P_1^{(s)}$ is believed to acquire the main modification as a result of absorption, the other amplitudes mainly being modified by an overall t-dependent function. Defining the amplitudes

$$P_{\pm} \quad \frac{1}{\sqrt{2}}\,(P_1 \pm P_{-1}) \tag{12}$$

we thus arrive at the model

$$\begin{aligned}
S^{(s)} &= \frac{M_{\pi\pi}\sqrt{-t}}{t-\mu^2}\, f^S_{\pi\pi}\,\Phi(t) \\
P_o^{(s)} &= \frac{M_{\pi\pi}\sqrt{-t}}{t-\mu^2}\, f^P_{\pi\pi}\,\Phi(t) \\
P_-^{(s)} &= \left(\frac{2t}{t-\mu^2} - C\right) f^P_{\pi\pi}\,\Phi(t) \\
P_+^{(s)} &= C\, f^P_{\pi\pi}\,\Phi(t)
\end{aligned} \tag{13}$$

where for $\Phi(t)$ something like

$$\Phi(t) = e^{b(t-\mu^2)} \tag{14}$$

is taken. Finally we can rotate eqs. (13) back to the t-channel frame:

$$S^{(t)} = \frac{M_{\pi\pi}\sqrt{-t}}{t-\mu^2} f^S_{\pi\pi} \Phi(t)$$

$$P_o^{(t)} = \left(\frac{M_{\pi\pi}\sqrt{-t}}{t-\mu^2} + c \sin\chi\right) f^P_{\pi\pi} \Phi(t) \tag{15}$$

$$P_-^{(t)} = c \cos\chi$$

$$P_+^{(t)} = c$$

where χ is the crossing angle

$$\cos\chi \approx \frac{M^2_{\pi\pi}+t}{M^2_{\pi\pi}-t} \qquad \sin\chi \approx \frac{2M_{\pi\pi}\sqrt{-t}}{M^2_{\pi\pi}-t} \ . \tag{16}$$

To complete the model the relative direction in spin-space (eq. (9)) has to be specified. All analyses are based on the spin-coherence assumption according to which all amplitudes have the same relative amounts of flip to non-flip. This assumption is justified if A_1-type exchange can be ignored as is currently believed. This model (eqs. (13) or (15)) seems to account very well for all the observed features in the data (4), (12).

In the simplest versions of the absorption model c is real (phase coherence), in fact c = 1 in the Williams model [11]. In general, however, c may be complex. This point was emphasized by Estabrooks et al. [4] who argued that for complex c only the s-channel amplitudes provide a "clean" way of obtaining the p-wave (P_o) and the s-wave from the interference

$$\sqrt{\pi}\ N \langle Y_1^o \rangle = \mathrm{Re}\ (sP_o^*) \qquad (17)$$

(also they add an A_2-exchange term to $P_+^{(s)}$).

However, the preferred solution (1) of ref. (4) is compatible with c real (!) and so this cannot account for the difference to the results of ref. [5] which are based on the phase coherence assumption from the beginning and the analysis of t-channel moments. In reality the s-p approximation is not used and indeed d-waves are treated rather differently by the two groups. So far, however, it is not possible to point clearly to the origin of the discrepancy.

In fig. 4 we show the Argand diagrams of ref. [5]. The ρ-parameters are given as

$$m_\rho = 778 \pm 2\ \mathrm{MeV} \qquad \Gamma_\rho = 152 \pm 2\ \mathrm{MeV} \qquad (18)$$

also wider than the PDG average eq. (5) although not so much so as Berkeley's eq. (4).

Above 1 GeV phase shift ambiguities start to become a problem. In ref. [5] they are dealt with by using an energy dependent fit to stabilize the results. Of particular interest is the appearance of a nearly elastic ε' with

$$m_{\varepsilon'} = 1537 \text{ MeV} , \qquad \Gamma_{\varepsilon'} = 233 \text{ MeV}$$

and a 25% ± 10% elastic ρ' with

$$m_{\rho'} = 1590 \pm 20 \text{ MeV}, \qquad \Gamma_{\rho'} = 180 \pm 50 \text{ MeV}.$$

Finally in fig. 5 we show results for δ_0^2 coming from the reaction

$\pi^+ p \rightarrow \pi^+ \pi^+ n$ at 12.5 GeV/c [13] [14].

Results are compared with several other data. The picture looks rather confusing. As we shall see in sect. 3 it is crucial to improve the experimental situation regarding δ_0^2.

1.2. $\pi\pi \rightarrow K\bar{K}$

The most detailed information comes from Grayer et al. [14] who analyze the data on reaction (2) together with data on

$$\pi^- p \rightarrow K^+ K^- n \qquad \text{at } 9.8 \text{ GeV/c}$$

in a coupled channel formalism around the $K\bar{K}$-threshold. In fig. 6 we give their results on η_0^0. Also shown are the results of Protopopescu et al. [2] using data on

$$\pi^+ p \to K^+K^- \Delta^{++} \qquad \text{at 7.1 GeV/c}$$

not measuring the angular distribution but assuming the $\pi^+\pi^- \to K^+K^-$ cross section to be pure s-wave. Further in fig. 6 are shown the predictions of refs. [4]and [5] based on the $\pi^+\pi^-$ angular distribution alone.

1.3. Inelastic Resonances. ε and S^*.

Let us consider the S-matrix element

$$S_0^0 = \eta_0^0 \, e^{2i\delta_0^0} \quad . \tag{19}$$

A simple elastic s-wave resonance is conveniently described in the q-plane by the following Breit-Wigner expression

$$S_{BW}^{el}(q) = \frac{q-q_R^*}{q-q_R} \cdot \frac{q+q_R}{q+q_R^*} \equiv R(q;q_R) \tag{20}$$

where $q = q_R$ is the 2nd sheet pole position

$$q_R^2 = \frac{1}{4}(s_R-4\mu^2)$$

$$S_R = M^2_{\pi\pi,R} = (M_R - i\,\frac{\Gamma_R}{2})^2 \quad . \tag{21}$$

We must now bring in the $K\bar{K}$-threshold and we find it convenient to bring in at the same time the left-hand cut threshold s = 0. We do this by introducing a new variable z similar to q near q = 0 but unfolding the 4 relevant Riemann sheets [15]

$$q^2 = \frac{R^2 z^2}{(z^2-z_o^2)(z^2-z_o^{*2})}$$

$$R = 2\sqrt{1-\frac{\mu}{m_k}} \qquad z_o = e^{i\theta} \qquad \tan\theta = \frac{\mu}{\sqrt{m_k^2-\mu^2}} \tag{22}$$

(cf. fig. 7). One easily proves for the elastic Breit Wigner

$$R(q;q_R) \equiv R(z;z_R)R(z;z_R') \tag{23}$$

where z_R and z_R' are the images of $s = s_R$ on the 2nd and 3rd sheets respectively. In fact $z_R' = z_R^{-1}$. For an inelastic resonance, however, we expect $z_R' \neq z_R^{-1}$. Actually the experimental δ_o^o of fig. 3 is qualitatively accounted for by the elastic form (20) below 900 MeV. If we interpret the pole as an ε-resonance we find [15]

$$z_\varepsilon \approx 0.7 - 20.3; \qquad z_\varepsilon' \approx z_\varepsilon^{-1}$$

$$m_\varepsilon \approx 300\text{-}500 \text{ MeV}; \qquad \Gamma_\varepsilon \approx 500\text{-}900 \text{ MeV} \tag{24}$$

but of course for such a wide object a resonance interpretation is doubtful.

For the description of (δ_o^o, η_o^o) near the $K\bar{K}$-threshold elastic unitarity needs to be strongly broken. We can easily convince ourselves that something like

$$S_0^0 \approx R(q;q_\varepsilon)\ R(z;z_{S^*}) \tag{25}$$

will work. Here z_{S^*} is shown on fig. 7 and for the solution of ref.[5]

$$z_{S^*} \approx 0.98 - i0.05 \quad . \tag{26}$$

The factor $R(z;z_{S^*})$ in eq. (25) gives the right shape of η_0^0, in particular the minimum of η_0^0 occurs for a value of s above the $K\bar{K}$-threshold such that Im $z(s) \approx$ -Im z_{S^*} (notice, $|z(s)| = 1$ for $s > 4m_K^2$). Below about 900 MeV $R(z;z_{S^*}) \approx 1$ since Im z_{S^*} is so small (the S^* is so narrow) but the contribution to δ_0^0 from $R(z;z_{S^*})$ rises quickly above 900 MeV to become $\approx 90^0$ at $s = 4m_K^2$ ($z = 1$). That is just what we want. But for this argument it was crucial <u>not</u> to have a factor $R(z;z'_{S^*})$ with z'_{S^*} anywhere near $z_{S^*}^{-1}$ since that would about cancel the increase in δ_0^0 produced by $R(z;\ z_{S^*})$. So we are led to the following important dynamical identification for the S^*:

<u>The S^* is described as a pole on the 2nd sheet with no companion on the 3rd sheet.</u>

It is instructive to see how far removed the S^* is from an "ordinary" inelastic resonance by comparison with a conventional inelastic Breit-Wigner

$$S_{BW}^{inel.}(s) = \frac{s-m_{S^*}^2 +2i\ \Gamma_K p-2i\Gamma_\pi q}{s-m_{S^*}^2 +2i\ \Gamma_K p+2i\Gamma_\pi q} \quad . \tag{27}$$

Here p is the kaon c.m. momentum and Γ_K and Γ_π are the partial widths to the channels $K\bar{K}$ and $\pi\pi$ respectively. For simplicity let us work in the following (very good) approximation

$$\Gamma_\pi/m_{S*} << 1 \; , \qquad \Gamma_K/m_{S*} << 1$$

$$m_{S*} = 2m_K \; .$$

The form (27) has $\delta_0^0(4m_K^2) = 90^0$ but we can correct for that by multiplying with a slowly varying background phase like $R(q;q_\varepsilon)$ as in eq. (25). The real trouble concerns the singularity structure shown in fig. 8 in a small region around z = 1. We see that the position of z'_{S*} is correlated to z_{S*} in a way which is against what the data require according to the above discussion. We conclude that phenomenology of the S^* based on an inelastic Breit-Wigner can lead to a serious bias.

Most workers now use a 2-channel K-matrix description which should work well in a limited region around 1 GeV (cf. also the discussion in sect. 3.2). Thus in the 2-channel study of ref. [14] the following mass and width values were obtained

$$m_{S*} = 1012 \pm 6 \text{ MeV} \; , \qquad \Gamma_{S*} = 34 \pm 10 \text{ MeV} \qquad (28)$$

1.4. $K\pi$ Phase Shift Analysis

Chew Low extrapolation studies of kaon initiated pion-production experiments are necessarily far less sophisticated than in the case of $\pi\pi$ scattering due to the smaller cross section. Also, data from several different groups are often combined when phase shift analyses are carried out (for reviews see Trippe [16] and Schlein [17]). As a consequence the level of accuracy is 10° - 15° so that many of the subtleties discussed in $\pi\pi$ studies become irrelevant.

The $I = \frac{1}{2}$ p-wave is found to be very well described by an elastic Breit-Wigner with mass and width values as given by PDG [7] for the K^{*} (890). Figs. 9 and 10 give an idea about the exotic s- and p-waves. Analogous to the $\pi\pi$ situation the main interest centres around the $\delta_{0}^{1/2}$ wave. Fig. 11 shows the compilation made by Trippe [16] with "down" and "up" branches above the K^{*}(890). Matison et al. [18] recently claimed to be able to rule out the "up"-branch on the basis of improved experimental resolution (fig. 12).

2. NEW METHODS IN DISPERSION THEORY

To see the importance of imposing crossing in phenomenological analyses of data let us consider the problem of obtaining low energy parameters. Fig. 13 is a plot of the real part of the forward $\pi\pi \to \pi\pi$-amplitude*) for I = 0 which at low energies is very close to the real part of the s-wave. To obtain the scattering length by extrapolating the data in the physical region to threshold is clearly very hard. However, using crossing we can relate the forward I = 0 amplitude for unphysical energies to forward scattering

*) The illustrative power of the forward amplitude plot was emphasized by D. Morgan.

in the crossed channel. Obviously comparison for example with the soft meson prediction [19] (also shown) is greatly helped thereby, in particular the prediction of a zero in the low energy region is immediately confirmed without any further work. Fig. 14 shows the analogous figure for $I = \frac{1}{2} K\pi$ scattering. Note that in this case the larger mass of the kaon helps effectively to get experimental points relatively closer to threshold.

The existence of branch points at the physical and at the crossed thresholds imply that dispersion techniques must be used in quantitative work. To these we now turn.

2.1. The Roy Equations for $\pi\pi$ Scattering [20]

We normalize [1] $\pi\pi$ amplitudes as follows

$$F^I(s,t,u) = \sum_{\ell=0}^{\infty} (2\ell+1) f_\ell^I(s) P_\ell(\cos\theta) \tag{29}$$

where

$$f_\ell^I(s) = \sqrt{\frac{s}{s-4}}\,(\eta_\ell^I(s) e^{2i\delta_\ell^I(s)} - 1)/2i \tag{30}$$

(from now on we put $\mu = 1$).

For simplicity consider first the fully symmetric $\pi^0\pi^0$ amplitude

$$\begin{aligned} F(s,t,u) &= F(s,u,t) = F(u,s,t) \\ &= \frac{1}{3}(F^{(0)}(s,t,u) + 2\,F^{(2)}(s,t,u)) \end{aligned} \tag{31}$$

A fixed-t dispersion relation with 2 subtractions is expected to exist [21] and following Roy [20] we write it as

$$F(s,t) = \frac{1}{\pi}\int_4^\infty ds' \frac{\mathrm{Im}F(s',t)}{s'^2}\left(\frac{s^2}{s'-s} + \frac{u^2}{s'-u}\right) + C(t) \qquad (32)$$

where $u = 4 - s - t$.

The basic remark of Roy was that $C(t)$ could be expressed by a sum rule using crossing symmetry:

$$F(0,t) = F(t,0) \qquad u \equiv 4 - t \qquad (33)$$

This gives

$$F(0,t) = \frac{1}{\pi}\int_4^\infty \frac{ds'}{s'^2}\mathrm{Im}F(s',t)\,\frac{(4-t)^2}{s'-4+t} + C(t)$$

(34)

$$= F(t,0) = \frac{1}{\pi}\int_4^\infty \frac{ds'}{s'^2}\,\mathrm{Im}F(s',0)\left(\frac{t^2}{s'-t} + \frac{(4-t)^2}{s'-4+t}\right) + C(0)$$

Also $$F(4,0) = a_0 = \frac{1}{3}(a_0^0 + 2a_0^2) \qquad (35)$$

where a_0^I is the s-wave scattering length for isospin I. Thus

$$a_0 = C(0) + \frac{1}{\pi}\int_4^\infty ds' \frac{\mathrm{Im}F(s',0)}{s'^2}\,\frac{16}{s'-4} \qquad (36)$$

Now we can eliminate C(t) and obtain

$$F(s,t) = a_o - \frac{16}{\pi}\int_4^\infty ds' \frac{\mathrm{Im}F(s',0)}{s'^2(s'-4)}$$

$$+ \frac{1}{\pi}\int_4^\infty \frac{ds'}{s'^2}\,\mathrm{Im}F(s',t)\left(\frac{s^2}{s'-s} + \frac{u^2}{s'-u}\right) \qquad (37)$$

$$+ \frac{1}{\pi}\int_4^\infty \frac{ds'}{s'^2}\left(\mathrm{Im}F(s',0)\frac{t^2}{s'-t} + [\mathrm{Im}F(s',0)-\mathrm{Im}F(s',t)]\frac{(4-t)^2}{s'+t-4}\right).$$

This equation is manifestly $s \leftrightarrow u$ symmetric as it should be from the derivation. In addition of course $F(s,t,u) = F(s,u,t)$ but that symmetry is not manifest. That is, $\mathrm{Im}F(s',t)$ will have to satisfy certain "supplementary conditions" in order for this extra symmetry to be fulfilled. However, consider approximating $\mathrm{Im}F(s',t)$ by the s-wave (the p-wave is zero for $\pi^o\pi^o \rightarrow \pi^o\pi^o$) i.e.

$$\mathrm{Im}F(s',t) \approx \mathrm{Im}F(s',0) \approx \mathrm{Im}f_o(s') \;. \qquad (38)$$

Then eq. (37) becomes manifestly fully crossing symmetric.

For the pure isospin amplitudes the equations become rather lengthy ones and we refer to the original papers for details [20]. In order to be able to impose unitarity it is convenient to project on partial waves. Interchanging summations and integrations then lead to partial wave relations (PWRs) of the form

$$f_\ell^I(s) = \begin{pmatrix} a_0^0 \\ 0 \\ a_0^2 \end{pmatrix} \delta_{\ell 0} + \frac{1}{3}(2a_0^0 - 5a_0^2) q^2 \begin{pmatrix} \delta_{\ell 0} \\ \frac{1}{6}\delta_{\ell 1} \\ -\frac{1}{2}\delta_\ell \end{pmatrix}$$

$$+ \sum_{I'=0}^{2} \sum_{\ell'=0}^{\infty} \int_4^\infty ds' \, K_{\ell,I}^{\ell',I'}(s,s') \, \mathrm{Im} f_{\ell'}^{I'}(s') \qquad (39)$$

where

$$f_\ell^I(s) = \frac{1}{2}\int_{-1}^{1} d\cos\theta \; F^I(s,\cos\theta) P_\ell(\cos\theta) \qquad (40a)$$

$$= \int_0^1 d\cos\theta \; F^I(s,\cos\theta) P_\ell(\cos\theta). \qquad (40b)$$

In eq. (40b) Bose symmetry or $t \leftrightarrow u$ crossing has been used. For eqs. (40a) and (40b) two different sets of kernel-functions are obtained. The input values for $\mathrm{Im} f_{\ell'}^{I'}(s')$ will thus have to satisfy supplementary conditions ensuring that the results are independent of the form chosen. However, analogous to what was described for $\pi^0\pi^0 \to \pi^0\pi^0$ the twice subtracted nature of the original fixed-t dispersion relations imply that the kernel-functions pertaining to s- and p-waves are invariant under going from eq. (40a) to eq. (40b). Thus we can separate in a crossing symmetric way the s-p contribution from the imaginary parts and write

$$f_\ell^I(s) = \hat{f}_\ell^I(s) + \phi_\ell^I(s) \qquad (41)$$

where $\hat{f}_\ell^I$ is the s-p contribution and ϕ_ℓ^I is called the driving term. For s- and p-waves we get [20]

$$\hat{f}_o^o(s) = a_o^o + \frac{1}{3}(2a_o^o-5a_o^2)q^2 + \frac{s-4}{\pi}\int_4^N \frac{\mathrm{Im}f_o^o(s')}{(s'-s)(s'-4)}\,ds'$$

$$+ \frac{2}{\pi}\int_4^N\frac{ds'}{s'}[Q_1(z')-Q_o(z')][\tfrac{1}{3}\mathrm{Im}f_o^o(s')+3\chi'\mathrm{Im}f_1^1(s')+\tfrac{5}{3}\mathrm{Im}f_o^2(s')]$$

$$- \frac{4}{3}q^2\frac{1}{\pi}\int_4^N \frac{ds'}{s'(s'-4)}[2\mathrm{Im}f_o^o(s')-9\mathrm{Im}f_1^1(s')-5\mathrm{Im}f_o^2(s')] \quad (42a)$$

$$\hat{f}_1^1(s) = \frac{1}{18}(2a_o^o-5a_o^2)q^2 + \frac{s-4}{\pi}\int_4^N \frac{\mathrm{Im}f_1^1(s')}{(s'-s)(s'-4)}\,ds' \quad (42b)$$

$$+ \frac{4}{\pi(s-4)}\int_4^N ds'Q_1(z')[\tfrac{1}{3}\mathrm{Im}f_o^o(s')+\tfrac{3}{2}\chi'\mathrm{Im}f_1^1(s')-\tfrac{5}{6}\mathrm{Im}f_o^2(s')]$$

$$- \frac{2}{9}q^2\frac{1}{\pi}\int_4^N \frac{ds'}{s'(s'-4)}[2\mathrm{Im}f_o^o(s') + 27\,\mathrm{Im}f_1^1(s')-5\,\mathrm{Im}f_o^2(s')]$$

$$\hat{f}_o^2(s) = a_2-\frac{1}{6}(2a_o^o-5a_o^2)q^2 + \frac{(s-4)}{\pi}\int_4^N \frac{\mathrm{Im}f_o^2(s')}{(s'-s)(s'-4)}\,ds'$$

$$+ \frac{2}{\pi}\int_4^N \frac{ds'}{s'}[Q_1(z')-Q_o(z')][\tfrac{1}{3}\mathrm{Im}f_o^o(s')-\tfrac{3}{2}\chi'\mathrm{Im}f_1^1(s')+\tfrac{1}{6}\mathrm{Im}f_o^2(s')]$$

$$+ \frac{2}{3}q^2\frac{1}{\pi}\int_4^N \frac{ds'}{s'(s'-4)}[2\mathrm{Im}f_o^o(s')-9\mathrm{Im}f_1^1(s')-5\mathrm{Im}f_o^2(s')] \quad (42c)$$

where $\chi' = 1 + 2s/(s'-4)$, $z' = 1+2s'/(s-4)$ and where N may be ∞ or any other cut off. In practice $N \approx \frac{1}{2}(m^2_{f^o(1260)} + m^2_{g(168o)})$. Clearly

$$\mathrm{Im}\ \hat{f}^I_{o,1}(s) = \mathrm{Im}\ f^I_{o,1}(s), \quad \mathrm{Im}\ \hat{f}^I_{\ell\geq 2}(s) \equiv 0 \qquad (43)$$

The functions $\hat{f}^I_\ell$ themselves satisfy all requirements of crossing and positivity [20]. So the supplementary conditions apply to the driving terms only.

In the BFP [15] analyses which I shall describe in sect. 3 ϕ^I_ℓ was considered a fixed input evaluated from phenomenological assumptions by (a) for $s < N$ ignoring the imaginary part of all higher waves except $\mathrm{Im}\ f^o_2(s)$ which was approximated by an inelastic Breit-Wigner for the $f^o(1260)$, (b) for $s > N$ using for $I_t = 1$ the modified B_4-model [22] for ρ-exchange and for $I_t = 0$ using the same model for f^o-exchange added to the result of a simple pomeron exchange model and finally ignoring $I_t = 2$ exchange. The results [15] turned out to be rather small below 1 GeV and to not change a great deal going from eq. (40a) to eq. (40b) in fact experimental uncertainties (such as on the asymptotic cross section) are more important than lack of full crossing symmetry of the driving terms.

The presence of the double spectral function prevents eqs. (39) from converging for all s. Thus assuming Mandelstam analyticity one has convergence in the following intervals

$$\begin{aligned} -4 < s < 36 \quad & \text{for eq. (40a)} \\ -4 < s < 68 \quad & \text{for eq. (40b)}\ . \end{aligned} \qquad (44)$$

Recently Mahoux, Roy and Wanders [23] have generalized the equations basing them on curved dispersion paths described by the variables

$$\chi = st + tu + us \qquad \text{and} \qquad y = stu.$$

This way they obtain a whole family of equations each valid in its own separate interval. The union of all intervals is

$$-28 < s < 125 \tag{45}$$

However, the generalization leaves the s-p contribution $\hat{f}^I_\ell$ unchanged. The authors calculate the $f^o(1260)$ contribution to ϕ^o_o and find it comfortingly close to the result of the original eqs. for $s < 60$.

2.2 Hyperbolic Dispersion Relations for Kπ Scattering

For the $K\pi$ system we have two invariant amplitudes

$$F^{(+)}(s,t,u) = F^{(+)}(u,t,s) = \frac{1}{\sqrt{6}} F_t^{(0)} = \frac{1}{3}(2F_s^{(3/2)} + F_s^{(1/2)})$$

$$F^{(-)}(s,t,u) = -F^{(-)}(u,t,s) = \frac{1}{2}F_t^{(1)} = \frac{1}{3}(F_s^{(1/2)} - F_s^{(3/2)}). \tag{46}$$

We use a normalization [1] so that the s-channel partial wave expansion takes the form

$$F^I_s(s,\cos\theta) = -8\pi\sqrt{s}\sum_\ell f^I_\ell(s)\, P_\ell(\cos\theta)\,(2\ell+1) \tag{47}$$

where $$f_\ell^I(s) = \frac{\eta_\ell^I(s)e^{2i\delta_\ell^I(s)}-1}{2\,i\,q(s)}$$

q being the c.m. 3-momentum. We then define partial wave amplitudes in the t-channel by

$$F^{(\pm)}(t,\cos\theta_t) = \sum_J (2J+1)(p_\pi p_K)^J g_J^{(\pm)}(t)P_J(\cos\theta_t) \tag{48}$$

with $$\cos\theta_t = (s-u)/4p_\pi p_K \;.$$

Consider the s-u even amplitude $F^{(+)}(s,t,u)$. A twice subtracted fixed-t dispersion relation may be written

$$F^{(+)}(s,t,u) = \frac{1}{\pi}\int_{(m_K+\mu)^2}^{\infty} ds' \frac{\mathrm{Im}F^{(+)}(s',t)}{s'^2}\left(\frac{s^2}{s'-s}+\frac{u^2}{s'-u}\right)+C(t) \tag{49}$$

Now it is not very useful to proceed in complete analogy to the $\pi\pi$-case and use a fixed-s dispersion relation to make explicit the t-dependence of the subtraction term. The reason is that the resulting PWRs would converge only in a very small interval. The trouble, however, is cured by working with hyperbolic dispersion relations [24] as discussed in detail by Hite et al. in the case of $\pi N\to\pi N$. They consider hyperbolas of the form

$$(s-a)(u-a) = b \tag{50}$$

which generalize the fixed-s (and -u) lines (b = 0). We find it useful to put a = 0 and let b vary. Consider a fixed point (s,t,u) where eq. (49) is evaluated, i.e. take b = s.u. A once subtracted dispersion relation along that hyperbola (which will converge when K^*-Regge exchange dominates asymptotically) can then be written as

$$F^{(+)}(s,t,u) = \frac{t}{\pi}\int_4^\infty dt' \frac{\mathrm{Im}F^{(+)}(t',\cos\theta_t')}{t'(t'-t)}$$
$$+ \frac{1}{\pi}\int_{(m_K+1)^2}^\infty ds' \frac{\mathrm{Im}F^{(+)}(s',u')}{s'}\left(\frac{s}{s'-s} + \frac{u}{s'-u}\right) + h(\nu) \tag{51}$$

where

$$\cos\theta_t' = 4p_\pi(t')p_K(t') \sqrt{(t'-\Sigma)^2-4su}$$

$$u' = su/s', \qquad \Sigma \equiv 2m_K^2+2 \qquad \nu=(s-u)/4m_K \quad .$$

The two representations (49) and (51) should agree everywhere. By identifying them at $t = 0$ we get a sum rule representation for the ν-dependence of $h(\nu)$ and by identifying them at backward angles, $s.u = (m_K^2-1)^2$, we get a representation of the t-dependence of $C(t)$. This way $C(t)$ and $h(\nu)$ may be eliminated in analogy to what was done to derive the Roy dispersion relation eq. (37).

Two rather lengthy expressions for $F^{(+)}$ follow and these might be used to construct PWRs. Imaginary parts would have to satisfy supplementary conditions ensuring that the two sets of relations are equivalent. One is reluctant to do this if only because the computation of kernel functions is time consuming first for human beings and later for the computer (this is true already for eqs.

(42) and certainly for the PWRs of ref. [23]). If, however, an s-p approximation is used for the imaginary part and an s-p-d approximation for the real part of the amplitudes in the low energy region (which should be very reasonable) a drastic saving could be obtained by replacing the integration involved in the partial wave projection by a simple <u>evaluation</u> at the energy considered (be it in the s- or in the t-channel) at the values of $\cos\theta$ for which $P_2(\cos\theta) = 0$ or $P_3(\cos\theta) = 0$. In general if L partial waves are considered the principle of Gaussian integration would require evaluations at the points where $P_L(\cos\theta) = 0$ or $P_{L+1}(\cos\theta) = 0$. Such a scheme was proposed by Burkhardt and McCauley and by Lang [25]. These authors proceed to write down dispersior relations for fixed angle. As emphasized by Hite et al. [24]this leads to cumbersome kinematical singularities. Also one has the problem of not being able to evaluate the left hand cut imaginary part below a finite energy. Both of these troubles are cured by fixed-t dispersion relations and hyperbolic dispersion relations.

Such a combined analysis is for the future. The ones I shall describe are based on simplified versions of <u>either</u> eq. (49) <u>or</u> eq. (51). Thus Nielsen and Oades [26] rewrite eq. (49) in the following finite contour relation form

$$F^{(\pm)}(s,t) = \frac{1}{\pi} \int_{(m_K+1)^2}^{N} ds' \quad \mathrm{Im}F^{(\pm)}(s',t)\left(\frac{1}{s'-s} \pm \frac{1}{s'-u}\right)$$

$$+ \Delta^{(\pm)}(s,t) \qquad (52)$$

Here $\Delta^{(\pm)}(s,t)$ can be evaluated in the experimental region from the $K\pi$ phase shifts using the equations. Since it is regular (for fixed t) in the s-plane from $s = 2m_K^2 + 2 - N - t$ to $s = N$ (and therefore smoother in that interval than $F^{(\pm)}(s,t)$ itself) and is known in part of that interval, it can be extrapolated with relative ease to the threshold region in order to obtain low energy parameters.

Johannesson and I [27] used eq. (51) in our study of the $\pi\pi \rightarrow K\bar{K}$ s-wave. Projecting out $g_0^{(+)}$ gives

$$g_0^{(+)}(t) = S_0^{(+)}(t) + \frac{t}{\pi}\int_4^{N_t} dt' \frac{\text{Im}\,g_0^{(+)}(t')}{t'(t'-t)}$$
$$+ \sum_{\ell=0}^{1} \int_{(m_K+1)^2}^{N_s} ds'\, K_{\ell 0}^{(+)}(t,s')\,\text{Im}\,f_\ell^{(+)}(s') + \phi_0^{(+)}(t) \tag{53}$$

with

$$S_0^{(+)}(t) = \int_0^1 d\chi\ \text{Re}\ F^{(+)}(t=0,\ s = s(\chi,t)) \tag{54}$$

where

$$s(\chi,t) = \frac{1}{2}\left(\Sigma + [t(2\Sigma - t) + 4p_\pi^2(t)\,p_K^2(t)\,\chi^2]^{1/2}\right) . \tag{55}$$

Thus $S_0^{(+)}(t)$ is the average in a small t range near threshold over the real part of the forward $K\pi$ amplitude. Therefore we have replaced it by the constant $g_0^{(+)}(0)$. Hedegaard-Jensen [28] recently showed that this may not be a good approximation for more accurate studies. For details concerning the form of the kernel functions and for evaluation of the driving term from high energy information I refer to our paper [27].

3. RESULTS FROM DISPERSION PHENOMENOLOGY

3.1. Construction of Phenomenological $\pi\pi$-Amplitudes

I shall essentially restrict myself to a description of the work by Basdevant, Forggatt and myself (BFP) [15]. Somewhat similar analyses using the Roy eqs. have been carried out by other groups [29] [30] [31]. One concentrates on the low energy region below the $K\bar{K}$-threshold where the Roy eqs. are applicable, where the driving term is a minor contribution and where the imaginary parts can be approximated by s- and p-waves. Eqs. (42) are then used to impose crossing symmetry. To further impose unitarity we parametrize s- and p-waves by writing the corresponding S-matrix elements as rational functions of suitably chosen conformal variables z of the type eq. (22) as discussed in the first lecture, however, allowing for various background singularities in addition to the resonance poles [15]. By requiring the Roy eqs. for s- and p-waves to be self reproducible to some accuracy (typically 1%) we obtain what we call an approximately crossing symmetric and unitary low energy model. In this, higher partial waves are purely real below 1 GeV.

To further obtain phenomenological models we require fit to experimental phase shifts. More precisely, we fix the input mass and width of the ρ-meson but leave the p-wave scattering length and the details of the phase near 900 MeV to be determined by the equations. For δ_0^0 we require a χ^2 of better than 2 per point to the particular set of phase shifts below 1 GeV which we are considering. Also we require the model to reproduce in detail the characteristic behaviour of δ_0^0 and η_0^0 in the S^* region as discussed in sect. 1. Finally the imaginary parts $\mathrm{Im}f_0^0$ and $\mathrm{Im}f_1^1$ are kept fixed from 1 GeV to 1.4 GeV (the cut-off energy) in a way which agrees with the experimental situation. Available data on δ_0^2 are found not to impose a useful constraint (but see below). These various pieces of experimental input are varied by reasonable amounts to get estimates of the uncertainties. For simplicity of presentation I shall for the most part only show results based on the Hyams et al. [5] analysis.

Fig. 15 shows the s-wave phases for 3 typical solutions corresponding to 3 different fixed I = 0 s-wave scattering lengths a_0^0. These solutions are members of a one dimensional continuum. In the (a_0^0,a_0^2)-plane the allowed region form a curve of finite extent called the universal curve after Morgan and Shaw [32], see fig. 16. The name at the time of Morgan and Shaw's work was chosen to emphasize the fact that in spite of the considerable ambiguity in the information then available, the scattering lengths found (using forward dispersion relations) lined up along a narrow band in the (a_0^0,a_0^2)-plane. Today the remarkable fact is that despite the much improved degree of sophistication concerning the imposition of crossing and analyticity and despite the large increase in experimental information imposed, we still <u>do find solutions</u> for all points along such a curve.

First let us compare with the soft meson predictions [19]. These pertain to a linear approximation of the amplitude on-shell and off-shell of the external pions. In the approximation we have on-shell

$$f_0^0(s) = a_0^0 + \frac{1}{3}q^2(2a_0^0 - 5a_0^2)$$

$$f_0^2(s) = a_0^2 - \frac{1}{6}q^2(2a_0^0 - 5a_0^2) \qquad (56)$$

$$f_1^1(s) = \frac{1}{18}(2a_0^0 - 5a_0^2)q^2$$

so that we have the sum rule

$$2a_0^0 - 5a_0^2 = 18a_1^1 \quad . \qquad (57)$$

The Gell-Mann current algebra then gives the Adler-Weissberger relation

$$L \equiv 2a_0^0 - 5a_0^2 = 18a_1^1 = \frac{3\mu}{4\pi F_\pi^2} = (0.60 \pm 0.06)\mu^{-1} \quad . \qquad (58)$$

Here $F_\pi = (89 \pm 4)$ MeV is the pion decay or PCAC constant. Further assuming the $\Delta I = 2\sigma$-term to vanish, Weinberg gets

$$a_0^0/a_0^2 = -\ 7/2 \ . \qquad (59)$$

The lines (58) and (59) are shown on fig. 16 for comparison. It is seen that whereas eq. (58) is qualtiatively verified by the data, so far it has been impossible to get much information concerning the value of the $\Delta I = 2\sigma$-term (eq. (59)). More precisely the universal curve may be approximately represented by the form

$$2a_0^0 - 5a_0^2 = C + 0.5\ a_0 + 0.6(a_0^0)^2 \ , \qquad (60)$$

where the constant c for figs. 15 and 16 is c = 0.45 ± 0.04, but depending on which set of experimental δ_0^0-

values, ρ-widths etc. is taken c may vary from 0.38 to 0.52. We further find that whereas the value of $2a_0^0-5a_0^2$ varies by about a factor of 2 going from one end to the other of the universal curve the predicted value of a_1^1 is much more stable varying by less than 30 % for a given set of inputs. A realistic estimate taking the spread in inputs into account is

$$a_1^1 = 0.038 \pm 0.006 \qquad (61)$$

in nice agreement with the soft meson result.

This different behaviour of $2a_0^0-5a_0^2$ and $18a_1^1$ (cf. eq. (57)) gives an idea about the accuracy to which the predictions (58) and (59) may at all be tested.

Now let us see if we can understand in qualitative terms (a) how it is possible to have solutions for all points on the curve in fig. 16 and (b) that the dependence of $2a_0^0-5a_0^2$ on a_0^0 has to be of the form eq. (60). Let us look back at eqs. (41) and (42). We first notice that to a very good approximation $\mathrm{Im}f_1^1$ is independent of a_0^0. Further the values of δ_0^2 are so small that we can almost ignore $\mathrm{Im}f_0^2$ and even better the <u>dependence</u> of $\mathrm{Im}f_0^2$ on a_0^0. Now consider a value of a_0^0 and a corresponding solution (f_0^0, f_0^2, f_1^1). We want to understand how a new solution is obtained when a_0^0 is changed into $\tilde{a}_0^0 = a_0^0 + \Delta a_0^0$. First suppose we have somehow managed to satisfy the eqs. for f_0^0 and f_1^1. Ignoring $\mathrm{Im}f_0^2$ the right hand side of eq. (42c) then gives a prediction for $\mathrm{Re}f_0^2$. As unitarity is a weak constraint for f_0^2 a simple iteration procedure would quickly generate a solution for f_0^2 as well. The real problem is to understand how to satisfy eqs. (42a,b). For simplicity of presentation let us say we have to do this by varying the following 3 parameters: (1) $2a_0^0-5a_0^2$, (2) a_1^1 and (3) b_0^0 defined by

$$\mathrm{Re}f_0^0(s) \approx a_0^0 + b_0^0 q^2 \quad \text{near} \quad s = 4 \tag{62}$$

so

$$\mathrm{Im}f_0^0(s) \approx q(a_0^0)^2 + 2q^3 a_0^0 b_0^0 + \ldots \text{ near } s = 4 \quad .$$

The terms

$$\frac{s-4}{\pi} \int_4^N \frac{\mathrm{Im}f_\ell^I(s')}{(s'-s)(s'-4)} ds'$$

in eqs. (42a,b) will ensure that the discrepancy is a smooth function of s. Therefore let us for simplicity only look at eq. (42a) at the point $s = s_0$ where $\mathrm{Re}f_0^0(s_0) = 0$ (by data $\sqrt{s_0} = M_{\pi\pi} \approx 900$ MeV) and at eq. (42b) at $s = s_1 = m_\rho^2$ where $\mathrm{Re}f_1^1(s_1) = 0$. At these s-values we have from eqs. (42) using

$$\frac{\partial}{\partial a_0^0} (\mathrm{Im}f_1^1, \mathrm{Im}f_0^2) \equiv 0$$

$$\frac{1}{q^2} \frac{d}{da_0^0}(\mathrm{Re}f_0^0(s_0))=0 = \frac{1}{q_0^2} + \frac{1}{3} \frac{d}{da_0^0}(2a_0^0-5a_0^2)$$

$$+ \frac{s_0-4}{\pi q_0^2} \int_4^{s_L} ds' \frac{\frac{\partial}{\partial a_0^0} \mathrm{Im}f_0^0(s')}{(s'-s)(s'-4)} + \frac{2}{3\pi q_0^2} \int_4^{s_L} \frac{ds'}{s'}(Q_1(z')-Q_0(z')) \cdot$$

$$\cdot \frac{\partial}{\partial a_0^0} \mathrm{Im}f_0^0(s') - \frac{8}{3} \frac{1}{\pi} \int_4^{s_L} \frac{ds'}{s'(s'-4)} \frac{\partial}{\partial a_0^0} \mathrm{Im}f_0^0(s') \tag{63}$$

$$\frac{6}{q_1^2}\frac{d}{da_o^o}(\mathrm{Ref}_1^1(s_1)) = 0 = \frac{1}{3}\frac{d}{da_o^o}(2a_o^o - 5a_o^2)$$

$$+ \frac{2}{\pi q_1^4}\int_4^{s_L} ds'\, Q_1(z')\frac{\partial}{\partial a_o^o}\mathrm{Imf}_o^o(s') - \frac{81}{3\pi}\int_4^{s_L}\frac{ds'}{s'(s'-4)}\frac{\partial}{\partial a_o^o}\mathrm{Imf}_o^o(s').$$

Here $\sqrt{s_L} \approx$ 600-700 MeV above which energy Im f_o^o is essentially fixed by data so that $\frac{\partial}{\partial a_o^o}$ Im $f_o^o(s) \approx 0$ for $s > s_L$. To satisfy even only these two eqs. we really have only one parameter to play with: $2a_o^o - 5a_o^2$. This is because first a_1^1 does not enter eqs. (63) at all, it is used afterwards only to fix up the f_1^1 eq. (42b) in detail near $s = 4$, and second b_o^o is (to a first approximation) fixed already by our choice of a_o^o since we must fit the lowest data-points for Ref$_o^o$. However, we can see from eqs. (63) that the dominant terms for large s have the required degeneracy. Ignoring the effect of all others we thus get

$$\frac{d}{da_o^o}(2a_o^o-5a_o^2) = \frac{8}{\pi}\int_4^{s_L}\frac{ds'}{s'(s'-4)}\frac{\partial}{\partial a_o^o}\mathrm{Imf}_o^o(s') \qquad (64)$$

This way we can immediately understand the sign of $\frac{d}{da_o^o}$ $(2a_o^o-5a_o^2)$ and using eqs. (62) one may even from eq. (64) derive eq. (60) in the form

$$\frac{d}{da_o^o}(2a_o^o-5a_o^2) \approx 0.5 + 1.2a_o^o \qquad (65)$$

The above discussion also explains why a_1^1 is more slowly varying with a_0^0 than would be expected from eq. (57): the variation of the first term in eq. (42b) (the linear approximation term) is cancelled by the variation of the last term. To account for all the smaller effects ignored is a task for the computer.

An interesting result is found for b_0^0. Naively one might expect b_0^0 to decrease with increasing a_0^0 since $\mathrm{Re}f_0^0$ is fixed at the lowest energy for which data exist. On the other hand in the linear approximation eq. (56) crossing requires

$$b_0^0 = \frac{1}{3}(2a_0^0 - 5a_0^2)$$

which we have seen <u>increases</u> with increasing a_0^0. The Roy eqs. decide on a compromise and let b_0^0 increase for small values of a_0^0 to reach a maximum and finally decrease to 0 at $a_0^0 \simeq 0.6$.

How can the remaining ambiguity at low energies be resolved? Ke_4-experiments [33] are not yet accurate enough but seem to rule out $a_0^0 < 0$. For future analysis of high precision Ke_4 data our predicted (a_0^0, b_0^0) correlation should be used.

$\pi^0\pi^0$ production experiments [34] often favour large values for a_0^0 but results from different groups are too conflicting for firm conclusions to be drawn.

High precision dispersion theory studies of πN d-waves [35] seem to require $a_0^0 < 0.4$.

From fig. 15 one notices a correlation between a_0^0 and our prediction for δ_0^2 in the 800 MeV region, say, so that measuring δ_0^2 with good accuracy here would be helpful. Thus the results of Hoogland et al. [13] [14] (fig.5) would select in favour of $a_0^0 > 0.1$. However, the details of the correlation sensitively depend on the inputs for $\delta_0^0, \Gamma_\rho, \sigma_{\pi\pi}^\infty$ etc. and at present we cannot use the effect in a decisive way.

Perhaps the most obvious way is to get the Chew-Low extrapolation to work at low energies (this kind of argument was used in ref.[3]). Thus the analysis of Estabrooks et al. [4] which give accurate values of δ_0^0 at quite low energies favour a large value of δ_0^0. However, constructing $\pi\pi$ models on the basis of their phase shifts leads to a serious inconsistency just at these low energies. Fig. 17 shows our predictions for the low energy part of the p-wave based on the δ_0^0-values of ref. [4]. Curves (1) and (3) represent extreme possibilities. Clearly the low energy δ_1^1 values of ref. [4] are much too large. Since furthermore the δ_0^0-points were obtained from the δ_1^1-points using the $\langle Y_1^0 \rangle$-moment (cf. eq. (17)) we conclude that the whole analysis must be unreliable at least at low energies. In reality d-waves enter the problem. In ref. [4] δ_2^0 was obtained from $\langle Y_4^0 \rangle$ and the main p-d interference effect was subtracted from $\langle Y_1^0 \rangle$ using $\langle Y_3^0 \rangle$. Again, the low energy d-waves came out to be very much bigger than can be accepted by crossing (table 1), which we believe at present to predict higher waves much more accurately than they can be measured. We recommend table 1 be used in future analyses as input.

3.2. Dispersion Phenomenology of the $K\pi$ System

This field is far less matured than the field of $\pi\pi$-physics and results have a preliminary character.

Nevertheless some interesting structures are emerging and qualified questions are being formulated.

Ader et al. [36] have analyzed the complete system using PWRs based on fixed-t and fixed-s dispersion relations and thus ignoring the convergence troubles mentioned in sect. 2.2. The consequence of this is hard to assess without having the details of their work. I shall concentrate here on the works of Nielsen and Oades [37] and of Johannesson and myself [27].

First consider the problem of determining low energy parameters. Let us start by summarizing the soft meson predictions [19]:

Linearity of $F^{(\pm)}$ (eqs. (46)) gives a sum rule analogous to eq. (57)

$$a_0^{(-)} \equiv \frac{1}{3}(a_0^{(1/2)} - a_0^{(3/2)}) = 6m_K\mu a_1^{(-)} \equiv 2m_K\mu(a_1^{(1/2)} - a_1^{(3/2)}) . \quad (66)$$

The Adler-Weissberger relation is

$$a_0^{(-)} = \frac{m_K}{m_K + \mu} \frac{\mu}{8\pi F_\pi^2} \approx 0.08 \pm 0.01 \quad (67)$$

whereas no results follow for the (+) amplitude without a non-exoticity assumption. Since the σ-term information apply rather far off the mass shell it is of interest to work out the consequence of the experimentally well verified assumption (cf. fig. 10)

$$a_1^{(3/2)} = 0 . \quad (68)$$

Writing in the normalization of eq. (46)

$$F^{(+)}(s,t,u) = - 8\pi(m_K+\mu)a_o^{(+)} + b_o^{(+)}\cdot t \qquad (69)$$

then gives

$$b_o^{(+)} = - 6(1 + \mu/m_K)a_o^{(-)} \approx - \frac{3\mu}{4\pi F_\pi^2} \approx -0.60\pm0.06. \qquad (70)$$

Finally assuming absence of both $\Delta I = 3/2$ σ-terms gives back eqs. (68), (70) together with

$$a_o^{(+)} = 0 \qquad (71)$$

Nielsen and Oades [37] basing their analysis on the finite contour relation eq. (52) found the following parameter values:

$$\begin{aligned} a_o^{(-)} &\approx 0.06 \pm 0.02 \\ b_o^{(+)} &\approx 0.40 \pm 0.15 \\ a_o^{(+)} &\approx 0.00 \pm 0.02 . \end{aligned} \qquad (72)$$

Also eq. (68) was found to be well satisfied.

Eqs. (72) actually correspond to a more accurate determination of low energy parameters than what we obtained for $\pi\pi$ scattering. That seems remarkable and deserves further study (it should be noticed that a large range of $a_o^{(+)}$ values are claimed possible in ref. [36]).

Nielsen and Oades also attempted to calculate the t-channel partial wave amplitudes $g_o^{(+)}$, $g_1^{(-)}$ of eq. (48) in the pseudo-physical range $4 < t < 4m_K^2$ to get information on ρ, ε and S^*-exchange which is of interest also for KN dynamics. They used an Omnès function technique to write

$$g_J^{(I)}(t) = V_J^{(I)}(t) \cdot F_J^{(I)}(t) \tag{73}$$

where

$$F_J^{(I)}(t) \propto \exp\{\frac{t}{\pi} \int_4^{4m_K^2} \frac{\delta_J^{(I)}(t')}{t'(t'-t)} dt'\} \tag{74}$$

and $\delta_J^{(I)}(t)$ is the elastic $\pi\pi$ phase shift, so that $V_J^{(I)}(t)$ is real on $4 < t < 4m_K^2$ by virtue of the generalized unitarity condition

$$\arg g_J^{(I)}(t) \equiv \delta_J^{(I)}(t) \pmod{\pi} . \tag{75}$$

Evaluating $g_J^{(I)}(t)$ and thus $V_J^{(I)}(t)$ for $t < 0$ from eq.(48) one can extrapolate $V_J^{(I)}(t)$ (which is smoother than $g_J^{(I)}$ itself) to the pseudophysical range and reconstruct $g_J^{(I)}$ there.

This scheme turned out to work well for $g_1^{(-)}$ and universality for $g_{\rho K\bar{K}}/g_{\rho\pi\pi}$ was found to work within 30%. For $g_o^{(+)}$ the results appeared very unstable.

Johannesson and I [27] then proposed to make use of 2-channel unitarity (ignoring 4π production) by using data on $\sigma_s^{I=0}(\pi\pi \to K\bar{K})$ at and above the $K\bar{K}$-threshold via the relation

$$|g_o^{(+)}(t)|^2 = \frac{8\pi}{3}\frac{p_\pi}{p_K}\sigma_s^{I=0}(\pi\pi \to K\bar{K})$$

$$= \frac{16}{3}\pi^2 \frac{1-(\eta_o^o(t))^2}{p_\pi p_K} \qquad (76)$$

(notice the equal particle factor [1]).

The usual way of attacking such a coupled channel problem is by the K-matrix method. However, a number of difficulties appear. For example, whereas the partial wave amplitudes for $\pi\pi \to \pi\pi$ and $\pi\pi \to K\bar{K}$ have the same singularity structure, $K\bar{K} \to K\bar{K}$ partial waves have an "extra" left hand cut overlapping the right hand cut and ending at $t = 4(m_K^2 - \mu^2)$. This cut must appear in <u>all</u> the K-matrix elements although of course in reconstructing the partial waves for $\pi\pi \to \pi\pi$ and $\pi\pi \to K\bar{K}$ it cancels out. That suggests that K-matrix elements are messy ones to parametrize for studies where the whole interval $0 < t < 4m_K^2$ is important such as will be the case here. Also, since we have already the analytic BFP models [15] for $\pi\pi$ scattering at our disposal we find it convenient to generalize their parametrization to include the 2-channel description. In particular it is attractive to replace the Omnès function eq. (74) by a rational function of z. Writing the elastic $\pi\pi$ S-matrix element as

$$S_o^o(z) = \prod_i \frac{(z-z_i^*)}{(z-z_i)} \qquad (77)$$

we define

$$\bar{F}_o^{(+)}(t) = k \prod_i \frac{1}{(z-z_i)} \tag{78}$$

(we ignore for simplicity of presentation a small kinematical correction, cf. refs. [15], [27]).

Clearly $\bar{F}_o^{(+)}(t)$ defined by eq. (78) have the following properties in common with the Omnès function eq. (74):

(a) $\arg \bar{F}_o^{(+)}(t) = \delta_o^o(t)$ for $4 < t < 4m_K^2$

(b) $\bar{F}_o^{(+)}(t)$ has no zeros on the physical sheet.

(c) k may be chosen so that $\bar{F}_o^{(+)}(0) = 1$.

In addition $\bar{F}_o^{(+)}(t)$ has a convenient behaviour at $t = 4m_K^2$: as a function of z it is regular at z = 1. $F_o^{(+)}(t)$ on the other hand has a logarithmic singularity which makes the use of eq. (76) cumbersome. To that end it is further convenient to use for $\eta_o^o(t)$ the analytic representation of BFP eq. (77).

The complete 2-channel formalism may now be set up. I refer to our article for details [27]. The most natural parametrization for $V_o^{(+)}(z)$ is currently being investigated. For obtaining $g_o^{(+)}(t)$ for $0 < t < 4m_K^2$ a simple form such as the one in ref. [27] should do.

Now knowing the value of $|V_o^{(+)}(4m_K^2)|$ by eq. (76) and insisting on a behaviour near t = 0 consistent with eqs. (72) we then know that $V_o^{(+)}(t)$ must qualitatively behave as shown in fig. 18 corresponding to $V_o^{(+)}(4m_K^2)$ <0 (1) or >0 (2). To decide among these two classes and on the details of the shape of $V_o^{(+)}(t)$ we use crossing as expressed by the hyperbolic PWR eq. (53), using the simplified approximation $S_o^{(+)}(t) = g_o^{(+)}(0)$. The individual contributions to the last two terms of that eq. are shown in fig. (19). They are seen to cancel to a good approximation.

We thus have the following three different types of input.

(a) low energy behaviour of $g_o^{(+)}(t)$ must be consistent with low energy $K\pi$ amplitudes,

(b) 2-channel unitarity

(c) crossing as expressed by eq. (53).

The shape of $V_o^{(+)}(t)$ is now varied by means of the parametrization to get consistency with these. To get a feeling for how (a), (b) and (c) work together to determine a solution for $g_o^{(+)}(t)$ and in particular decide between the possibilities (1) and (2) of fig. 17, consider the case of $S_o^{(+)}(t)$ in eq. (53) constant and of the sum of the last two terms vanishing. If $g_o^{(+)}(t)$ is a solution to eq. (53) then so is $-g_o^{(+)}(t)$ in that approximation and both of course will satisfy the unitarity condition eq. (76). However, both cannot have the same low energy behaviour. In fact consider the derivative at $t = 0$. In the above approximation we get from eq. (53)

$$\frac{d}{dt}g_o^{(+)}(0) = \frac{1}{\pi}\int_4^{N_t} dt' \frac{\mathrm{Im}g_o^{(+)}(t')}{t'^2} \quad . \tag{79}$$

Since from eqs. (72) and (69) the left hand side is known

to be negative (input (a) above) one can understand why eq. (53) selects the class (1) of fig. 17 for $V_0^{(+)}(t)$.

Finally fig. 20 shows solutions for $g_0^{(+)}$ based on extreme BFP solutions ((1) and (3)) for δ_0^0 (cf. fig. 15). We conclude that ε-exchange is an attractive force, in agreement for example with the modified B_4-model [22]. I emphasize, however, that the spread in solutions in fig. 20 is in no way representative of the true error. A more careful evaluation of the effect of varying the input by reasonable amounts is needed. These and other aspects of $K\pi$ physics are currently being investigated.

Table 1

$M_{\pi\pi}$ (MeV)	δ_2^0	δ_2^2	δ_3^1
400	$0^\circ.07 \pm 0^\circ.01$	$0^\circ.00 \pm 0^\circ.01$	$0^\circ.002 \pm 0^\circ.001$
600	$0^\circ.9 \pm 0^\circ.1$	$-0^\circ.1 \pm 0^\circ.1$	$0^\circ.04 \pm 0^\circ.02$
800	$3^\circ.5 \pm 0^\circ.5$	$-0^\circ.5 \pm 0^\circ.2$	$0^\circ.2 \pm 0^\circ.1$
1000	$11^\circ \pm 2^\circ$	$-1^\circ.0 \pm 0^\circ.6$	$0^\circ.8 \pm 0^\circ.2$

BFP [15] predictions for $\pi\pi$ d- and f-waves below 1 GeV.

FIGURE CAPTIONS

Fig. 1. Pion production. Kinematical notation and one pion exchange (OPE) contribution.

Fig. 2. The $\pi\pi$ p-wave phase shift from ref. [2] (squares) ref. [4] (dots) and ref. [5] (crosses).

Fig. 3. The $\pi\pi$ I = 0 s-wave phase shift below 1 GeV from ref. [2] (■), ref. [3] (✕), ref. [4] (●) and ref. [5] (□).

Fig. 4. Argand diagrams from the $\pi\pi$ phase shift analysis of Hyams et al. (5).

Fig. 5. Compilation of I = 2 s-wave phase shifts. Data points are from:

■ J. P. Baton et al., Phys. Letters 33b, 525, 528 (1970).

□ W. M. Katz et al., Proceedings Argonne Conf. on $\pi\pi$ and $K\pi$ Interactions (May 1969), Eds. F. Loeffler and E. Malamud.

○ ● E. Colton et al., Phys. Rev. D3, 2028 (1971).

△ W. D. Walker et al., Phys. Rev. Letters 38B, 555 (1972).

D. Cohen et al., Phys. Rev. D7, 661 (1973).

M. Baubillier et al., Contr. to the 16th Int. Conf. on High-Energy Physics, Batavia 1972.

◆ refs. [13] [14].

Fig. 6. I = 0 $\pi\pi$ s-wave inelasticities above the $K\bar{K}$-threshold. Curve is from ref. [14], squares are from ref. [2], dots are from ref. [4] and crosses are from ref. [5].

Fig. 7. z-plane image of the 4-sheeted Riemann-surface for s corresponding to eq. (22). Encircled numbers denote sheet number in s-plane. The physical region is $0 \leq z \leq 1$ (Im z = 0) for $4\mu^2 \leq \text{Re}s \leq 4\, m_K^2$ (Ims = $+\varepsilon$) and $|z| = 1$ ($0 \leq \text{Im}z \leq \text{Im}z_0$) for Res $\geq 4m_K^2$ (Ims = $+\ \varepsilon$). Resonance singularities corresponding to a typical BFP [15] solution are shown. Poles are denoted by crosses (x), zeros by circles (o).

Fig. 8. Small section of the z-plane around the point z = 1. Singularities (poles denoted by x, zeros by o) of the inelastic Breit-Wigner eq. (27) are shown for $m_{S*} = 2\, m_K$, Γ_π, $\Gamma_K/m_{S*} << 1$. Encircled numbers are s-plane sheet numbers.

Fig. 9. Trippe's [16] compilation of I = $^3/2$ $K\pi$ cross-sections.

Fig.10. I = $^3/2$ s- and p-wave $K\pi$ phase shifts from ref. [16]. Data are from the SABRE collaboration (●) and from the Amsterdam-Nijmwegen collaboration (○). Dashed curve corresponds to a constant I = $^3/2$ s-wave cross section of 1.8 mb (cf. fig. 9).

Fig.11. Trippe's [16] compilation of I = $\frac{1}{2}$ $K\pi$ s-wave phase shifts.

Fig.12. I = $\frac{1}{2}$ $K\pi$ s-wave phase shift from ref. [18].

Fig.13. Real part of the I = 0 $\pi\pi$ forward amplitude from data on δ_0^0 and δ_1^1 from ref. [5] and on δ_0^2 from refs. [13] [14]. ω_L is pion lab energy. Straight line is soft-meson prediction [19] (eq. (56)).

Fig.14. Real part of the I = $\frac{1}{2}$ $K\pi$ forward amplitude. Straight line is soft meson prediction [19] (eqs. (67)-(71)).

Fig.15. BFP [15] solutions based on the energy-dependent

phase-shift analysis results on δ_0^0 from ref. [5]. Data on δ_0^0 from refs. [13], [14] are shown for comparison.

Fig. 16. Universal curve corresponding to BFP solutions [1], [2], [3] of fig. 15 together with intermediate ones. The soft meson lines eqs. (58), (59) are shown for comparison.

Fig. 17. BFP [15]-prediction for the low energy behaviour of the $\pi\pi$ p-wave based on δ_0^0-data from ref. [4]. Solutions (1) and (3) represent extreme possibilities. Also shown are solutions (1) (●) and (2) (▲) of ref. [4] for the p-wave itself.

Fig. 18. Typical expected [27] behaviour for $V_0^{(+)}$ compatible with eqs. (72) and corresponding to $V_0^{(+)}(4m_K^2)>0$ (class (2)) or <0 (class (1)).

Fig. 19. Contributions [27] to $\Delta(t)$ defined as the last two terms in eq. (53).

Fig. 20. The prediction for the $\pi\pi \to K\bar{K}$ $I = 0$ s-wave $g_0^{(+)}(t)$ from ref. [27].

REFERENCES

1. J. L. Petersen, Physics Reports 2C, 155 (1971).

2. S. D. Protopopescu et al., Proc. Int. Conf. on $\pi\pi$ Scattering and Associated Topics. Tallahassee (1973).

3. G. Grayer et al., in "Experimental Meson Spectroscopy 1972", Proc. Third Int. Conf. Philadelphia (1972), eds. F. W. Lai and A. H. Rosenfeld.

G. Grayer et al., Contribution to the XVI Int. Conf. on High Energy Physics, Batavia Sept. 1972.

4. P. Estabrooks et al., Proc. Int. Conf. on $\pi\pi$ Scattering and Associated Topics, Tallahassee (1973).

5. B. Hyams et al., Proc. Int. Conf. on $\pi\pi$ Scattering and Associated Topics, Tallahassee (1973).

6. C. Goebel, Phys. Rev. Letters 1, 337 (1958).
G. F. Chew, F. E. Low, Phys. Rev. 113, 1640 (1959).

7. Particle Data Group. Rev. Mod. Phys. 45, 1 (1973).

8. C. D. Froggatt, D. Morgan, Phys. Rev. 187, 2044 (1969).

9. K. Gottfried, J. D. Jackson, Nuovo Cim. 33, 309 (1964).

10. G. L. Kane, M. Ross, Phys. Rev. 177, 2353 (1969).

11. P. K. Williams, Phys. Rev. D1, 1312 (1970).

12. P. Estabrooks, A. D. Martin, Phys. Letters 41B, 350 (1972).

13. W. Hoogland et al. CERN preprint July (1973).

14. G. Grayer et al., Proc. Int. Conf. on $\pi\pi$ Scattering and Associated Topics, Tallahassee (1973).

15. J. L. Basdevant, C.D. Froggatt, J. L. Petersen, Phys. Letters 41B, 173, 178 (1972); Proc. Int. Conf. on $\pi\pi$ Scattering and Associated Topics, Tallahassee (1973); Nucl. Phys. B, to be published. J. L. Basdevant et al., Comm. to the IInd Int. Aix-en-Provence Conf. on Elementary Particles (1973).

16. T. G. Trippe, Talk at the Summer Workshop No. 1 "$\pi\pi$ and $K\pi$ Interactions" ANL June (1971), LBL-763.

17. P. E. Schlein, Proc. Int. School of Subnuclear Phys., Erice (1970).

18. M. J. Matison et al., LBL-1537 preprint (1973).

19. S. Weinberg, Phys. Rev. Letters 17,616 (1966).
R. W. Griffith, Phys. Rev. 176, 1705 (1968).

20. S. M. Roy, Phys. Letters 36B, 353 (1971).
J. L. Basdevant, J.C. Le Guillou, H. Navelet, Nuovo Cim. 7A, 363 (1972).

21. Y. S. Jin, A. Martin, Phys. Rev. 135, B1375 (1964).

22. C. Lovelace, Phys. Letters 28B, 264 (1968); Proc. of Conf. on $\pi\pi$ and $K\pi$ Int., ANL (1969).
J. A. Shapiro, Phys. Rev. 179, 1345 (1969).
K. Kawarabayashi et al., Phys. Letters 28B, 432 (1969).

23. G. Mahoux, S. M. Roy, G. Wanders, Saclay preprint (1973).

24. P. W. Greenberg, J. C. Sanudsky, Nuovo Cim. 6A, 617 (1971).
J. C. Sanudsky, Nuovo Cim. 6A, 627 (1971).
G. E. Hite, R. Jacob, F. Steiner, Phys. Rev. D6, 3333 (1972).
G. E. Hite, F. Steiner, CERN-preprint TH-1590 (1972).

25. H. Burkhardt, G. McCauley, Nuovo Cim. 38, 872 (1965).
C. B. Lang, Acta Phys. Austr. 38, 367 (1973).

26. H. Nielsen, G. C. Oades, Nucl. Phys. B55, 3ol (1973).

27. N. O. Johannesson, J. L. Petersen, Nucl. Phys. B to be published.

28. N. Hedegaard-Jensen, University of Aarkus preprint, Jan. 1974.

29. B. Bonnier, P. Gauron, Nucl. Phys. B52, 506 (1973).

30. M. R. Pennington, S. D. Protopopescu, Phys. Rev. D7, 1429 (1973), Proc. Conf. on $\pi\pi$ Scattering and Associated Topics (1973).

31. J. C. Le Guillou, J.L. Basdevant, Ann. Inst. Henri-Poincare 17, 221 (1972).

32. D. Morgan, G. Shaw, Phys. Rev. D2, 520 (1970); Nucl. Phys. B10, 261 (1969).

33. A. Zylbersztejn et al., Phys. Letters 38B, 457 (1972). E. W. Beier et al., Phys. Rev. Letters 29, 511 (1972); 30, 399 (1973).

34. P. Sonderegger, P. Bonamy, Contr. to the Fifth Int. Conf. on Elementary Particles, Lund (1969).
J. R. Bensinger et al., Phys. Letters 36B, 134 (1971).
W. Deinet et al., Phys. Letters 30B, 359 (1969).
E. I. Shibata et al. Phys. Rev. Letters 25, 1227 (1970).
G. Villet et al., Proc. Int. Conf. on $\pi\pi$ Scattering and Associated Topics, Tallahassee (1973).

35. H. Nielsen, G. C. Oades, Comm. to the IInd Int. Aix-en-Provence Conf. on Elementary Particles (1973).

36. J. P. Ader, C. Meyers, B. Bonnier, Phys. Letters 46B, 403 (1973).

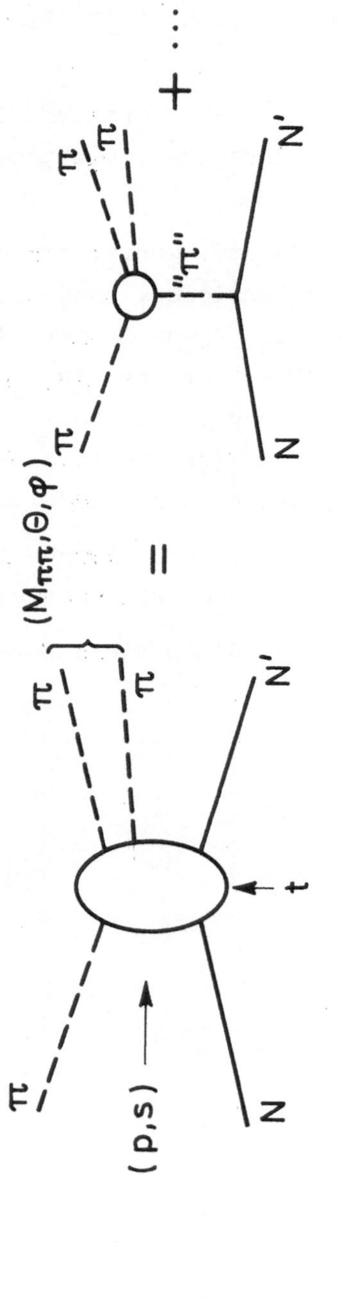

FIG. 1

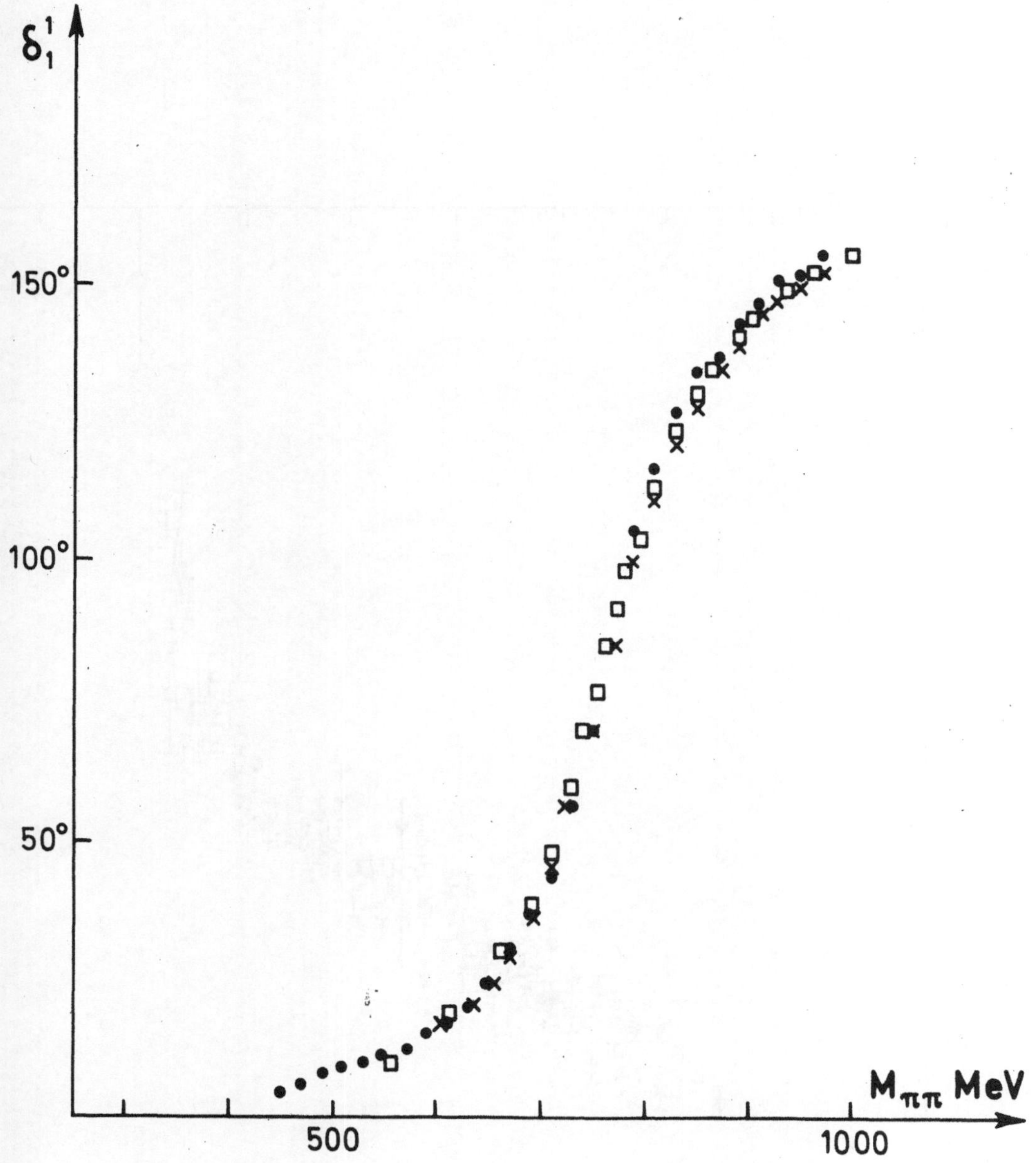

Fig. 2

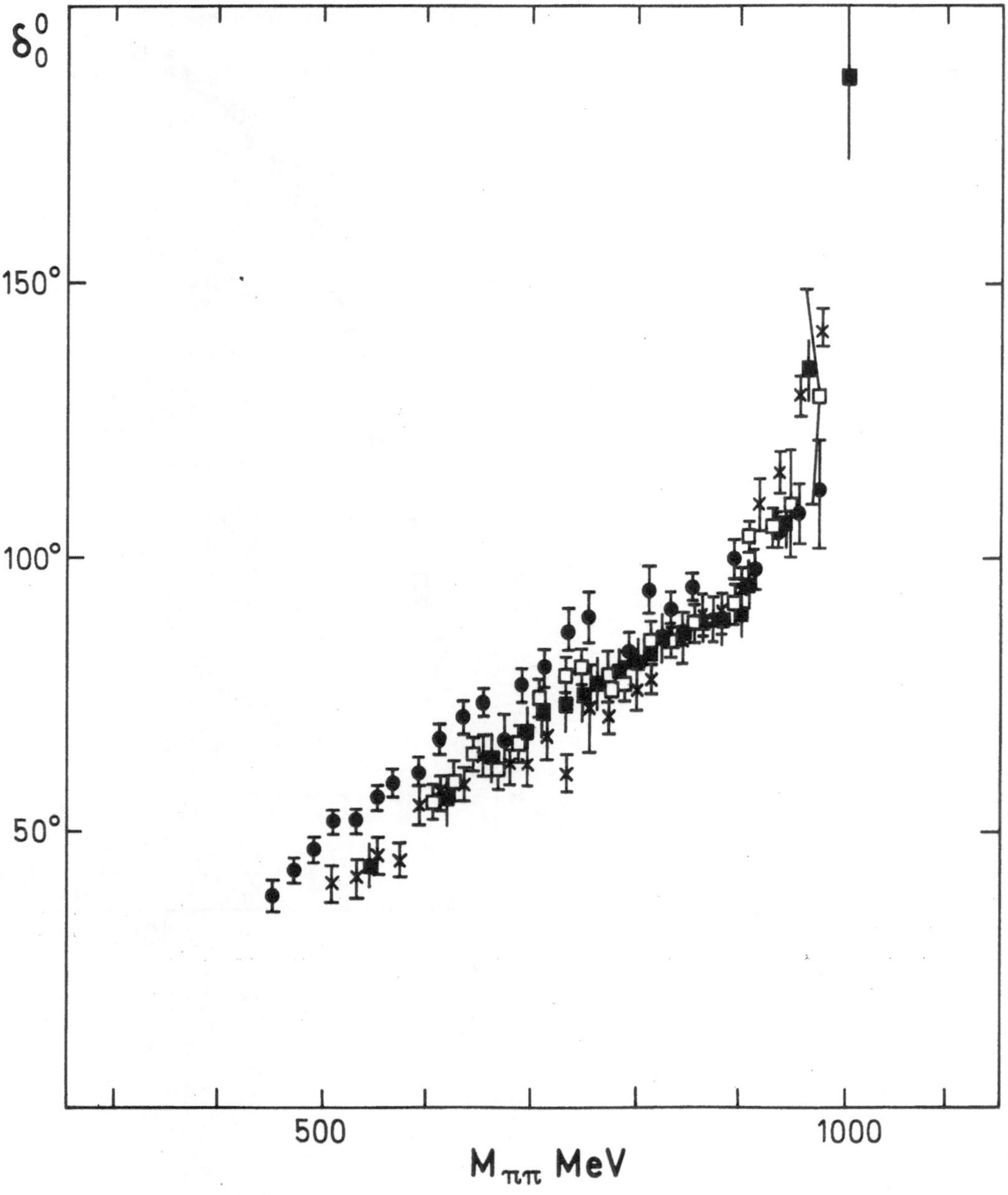

Fig. 3

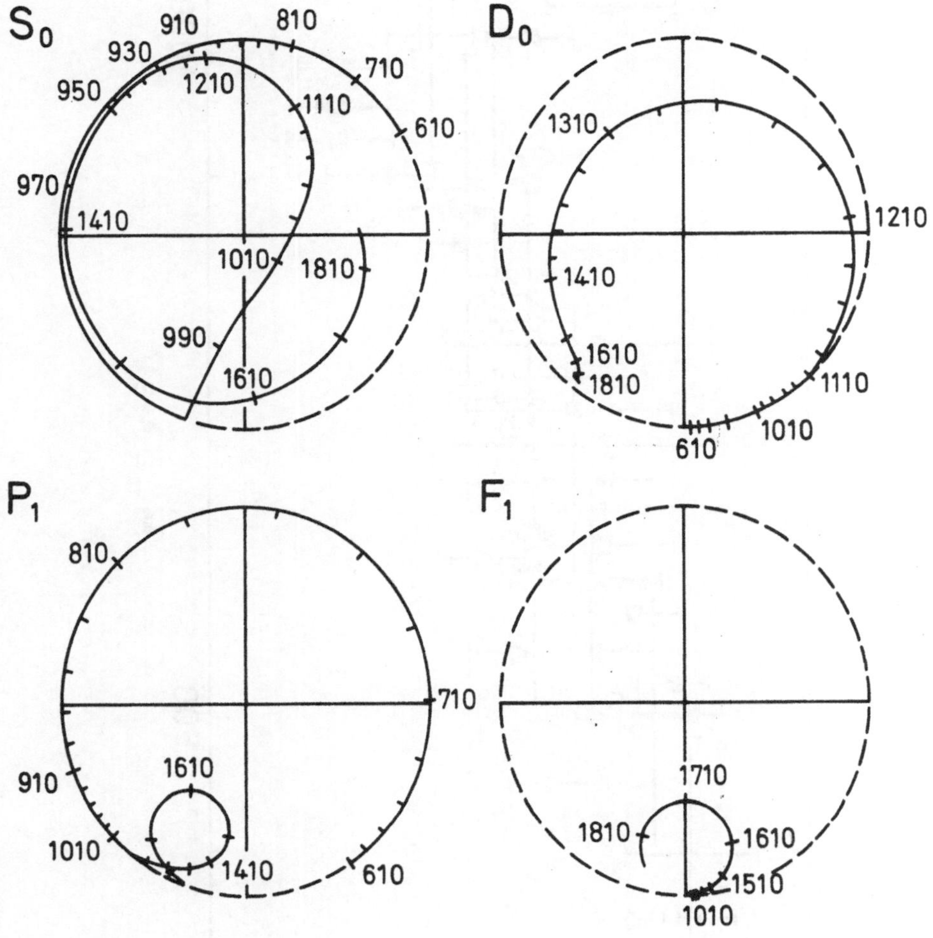

S0
910
810
930
710
950
1210
1110
610
970
1410
1010
1810
990
1610
D0
1310
1210
1410
1610
1810
1110
1010
610
P1
810
710
910
1610
1010
1410
610
F1
1710
1810
1610
1510
1010

Fig. 4

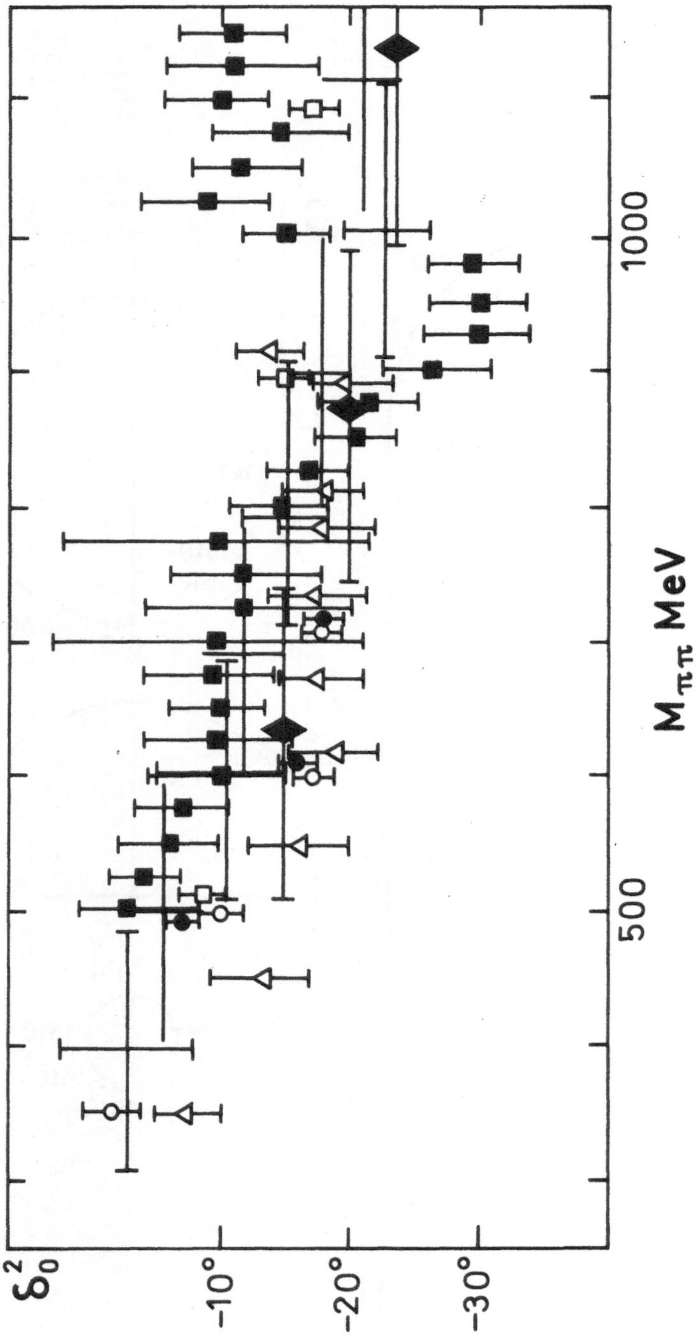

Fig. 5

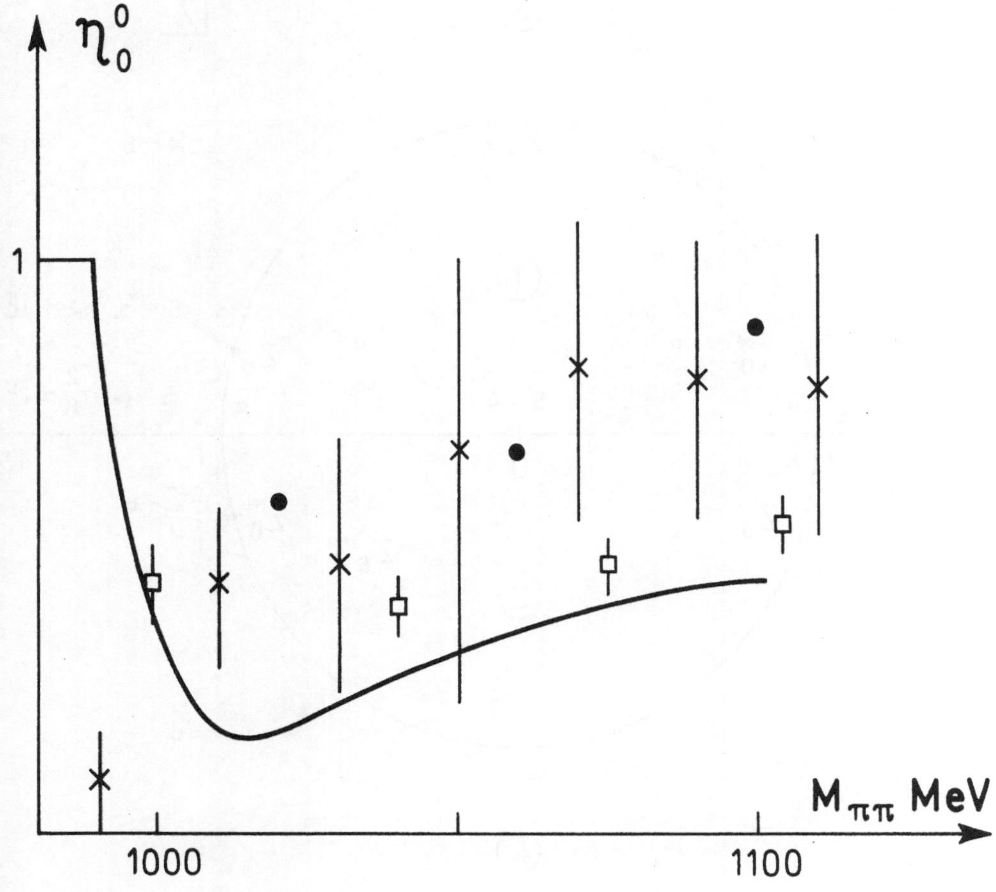

Fig. 6

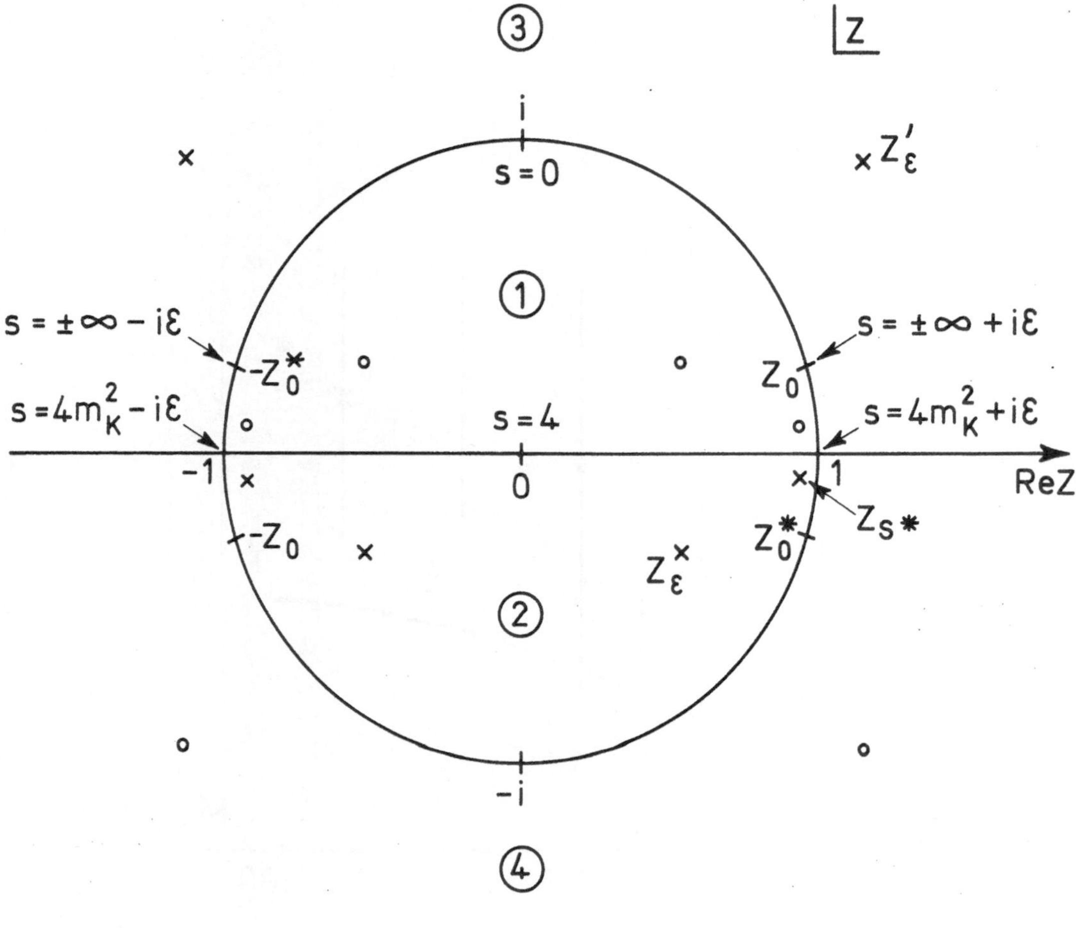
③
Z
i
s = 0
x Z'_ε
①
s = ±∞ − iε
−Z_0*
s = ±∞ + iε
Z_0
s = 4m_K^2 − iε
s = 4
s = 4m_K^2 + iε
−1
0
1
ReZ
Z_s*
−Z_0
Z_0*
Z_ε
②
−i
④

Fig. 7

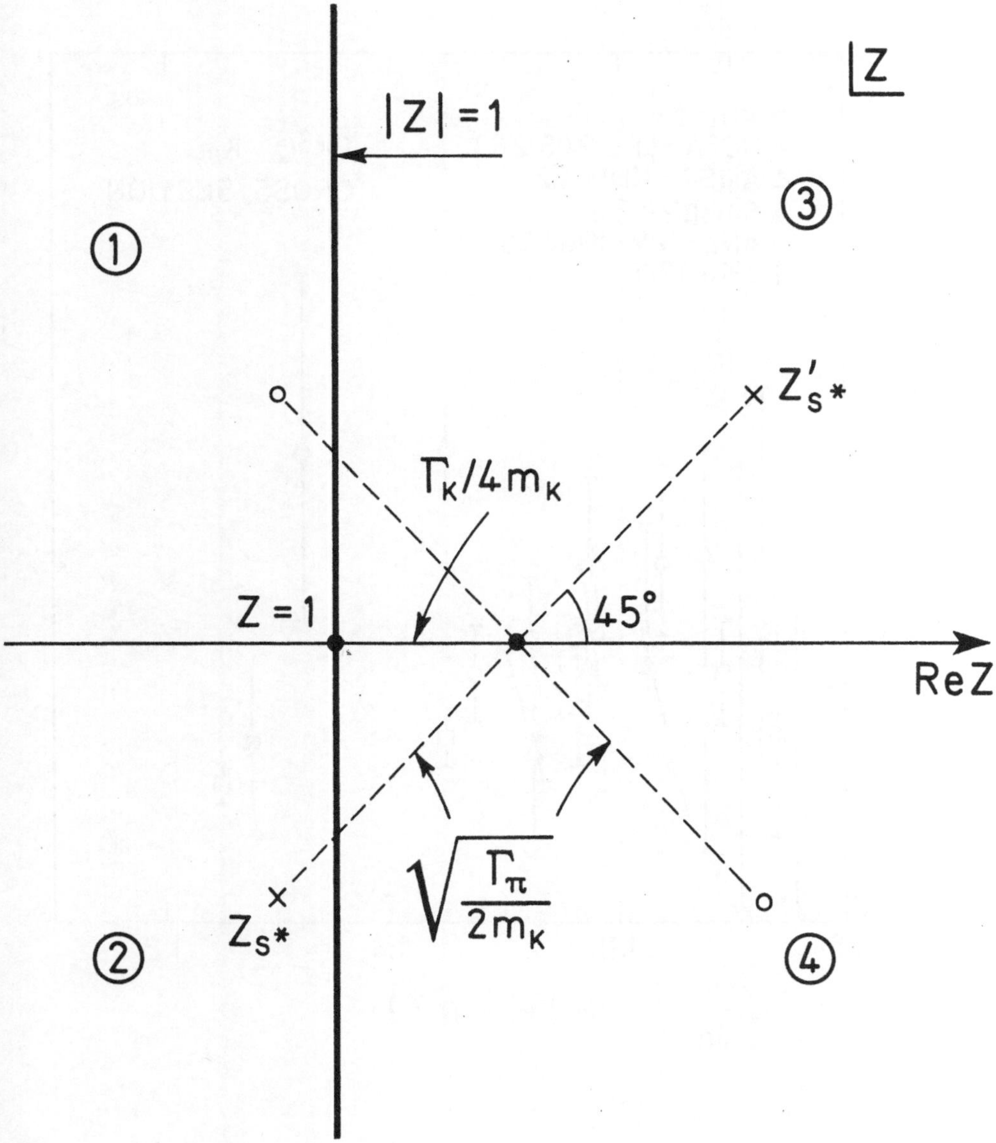

Fig. 8

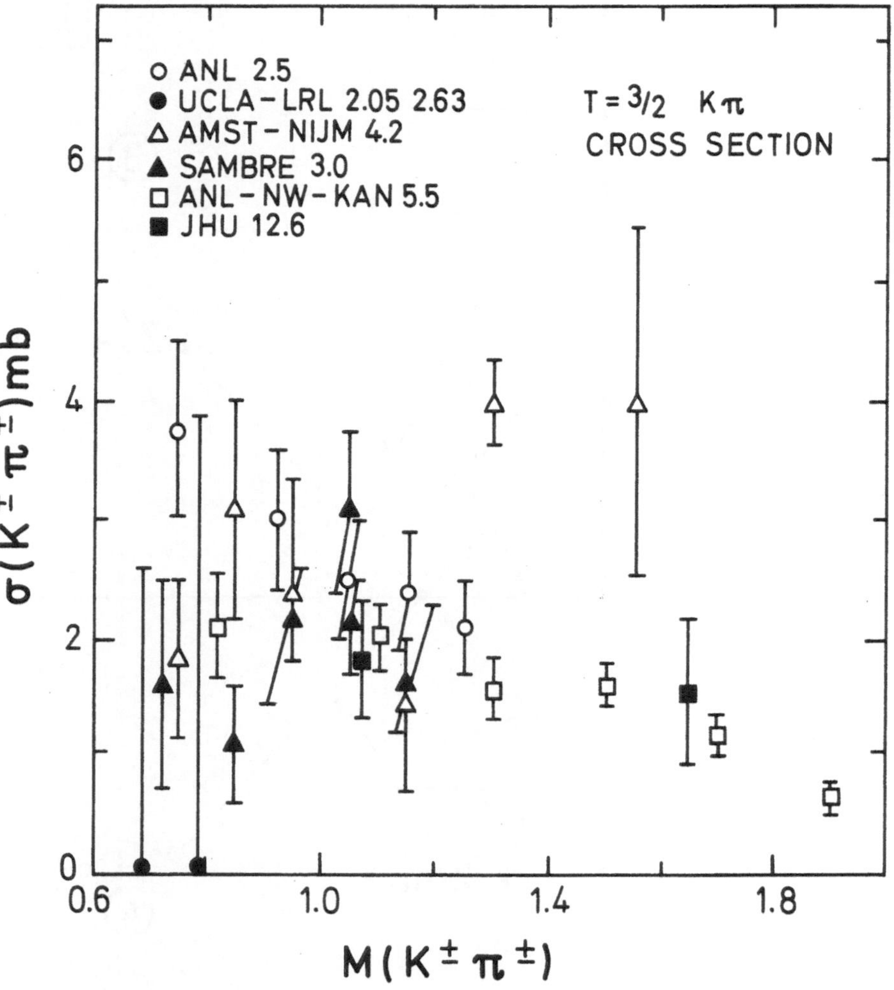
○ ANL 2.5
● UCLA-LRL 2.05 2.63
△ AMST-NIJM 4.2
▲ SAMBRE 3.0
□ ANL-NW-KAN 5.5
■ JHU 12.6
T = 3/2 Kπ
CROSS SECTION
σ(K± π±)mb
0
2
4
6
0.6
1.0
1.4
1.8
M(K± π±)

Fig. 9

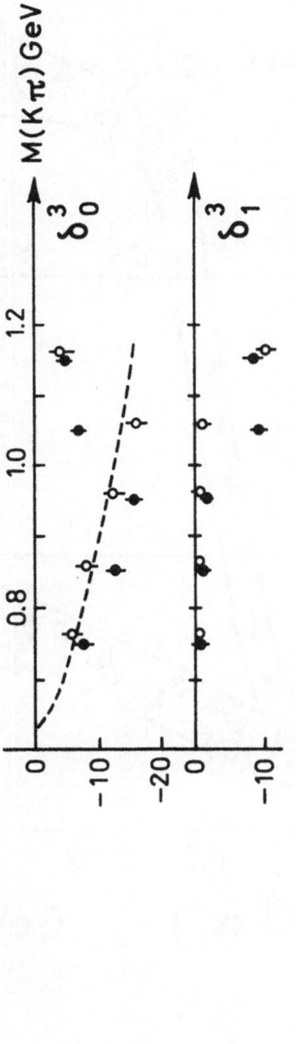

Fig. 10

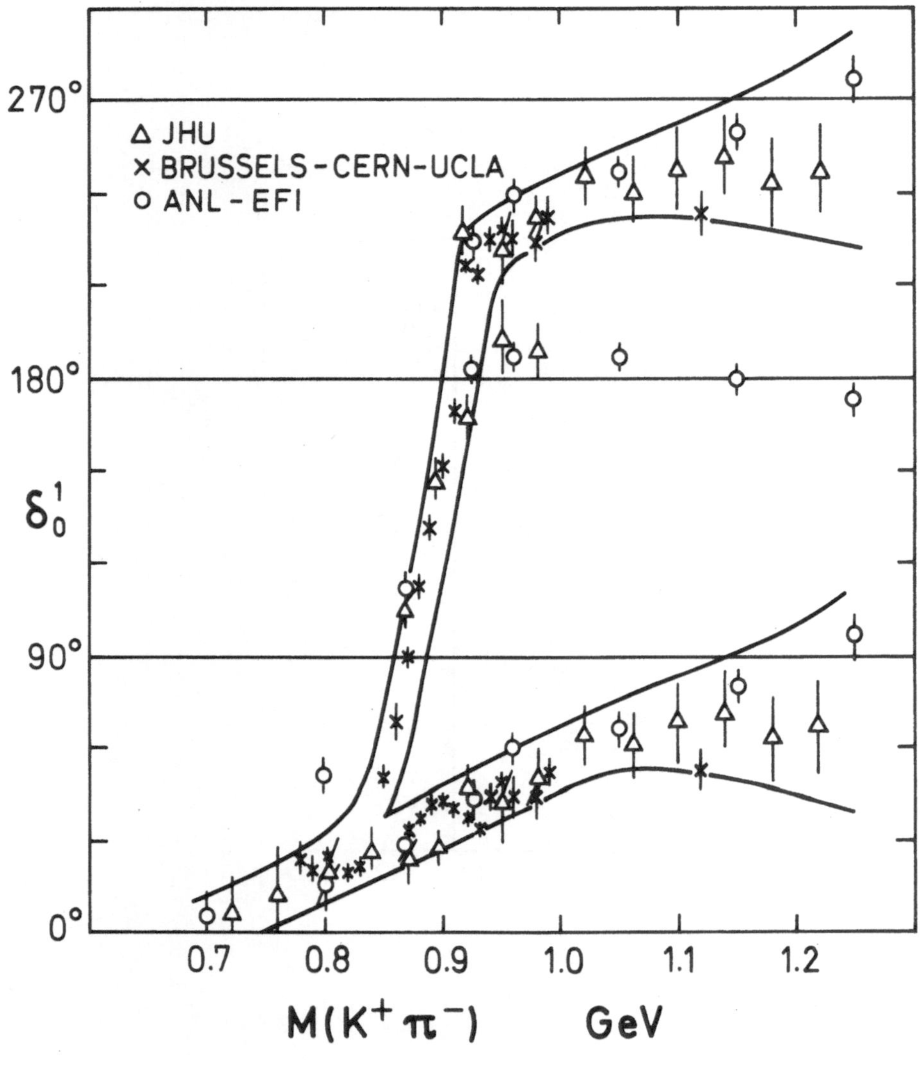
270°
180°
90°
0°
Δ JHU
× BRUSSELS-CERN-UCLA
○ ANL-EFI
δ_0^1
0.7
0.8
0.9
1.0
1.1
1.2
M(K⁺ π⁻) GeV

Fig. 11

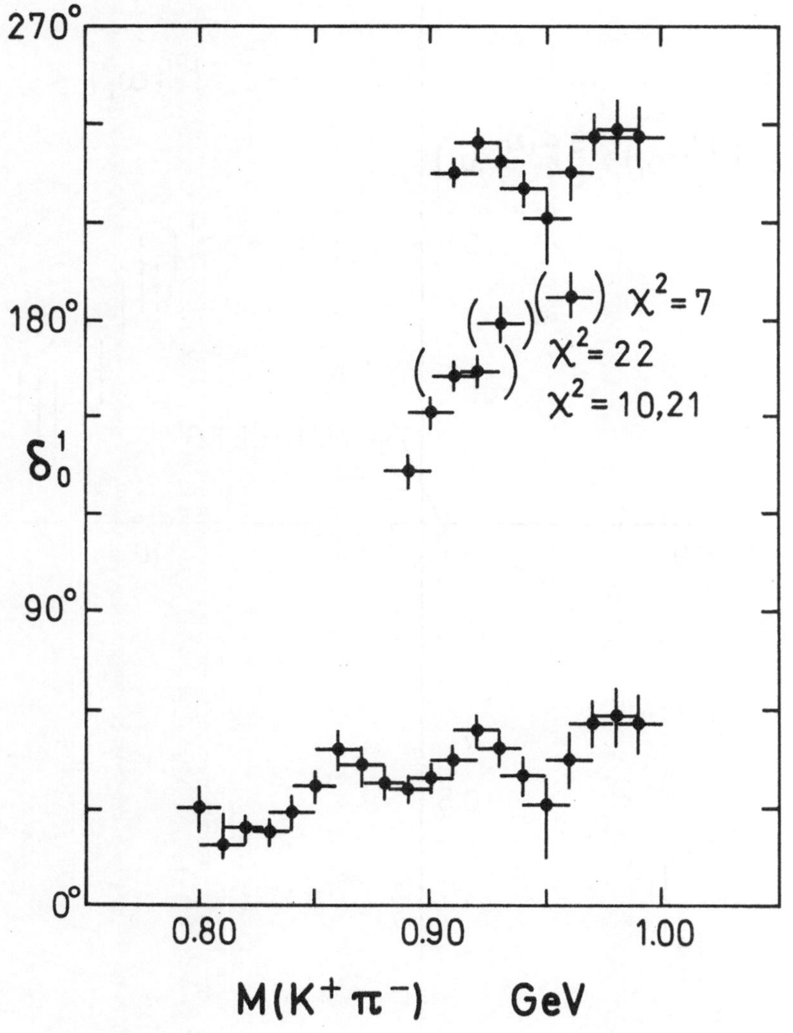

270°
180°
90°
0°
δ_0^1
$\chi^2 = 7$
$\chi^2 = 22$
$\chi^2 = 10,21$
0.80
0.90
1.00
M(K$^+\pi^-$) GeV

Fig. 12

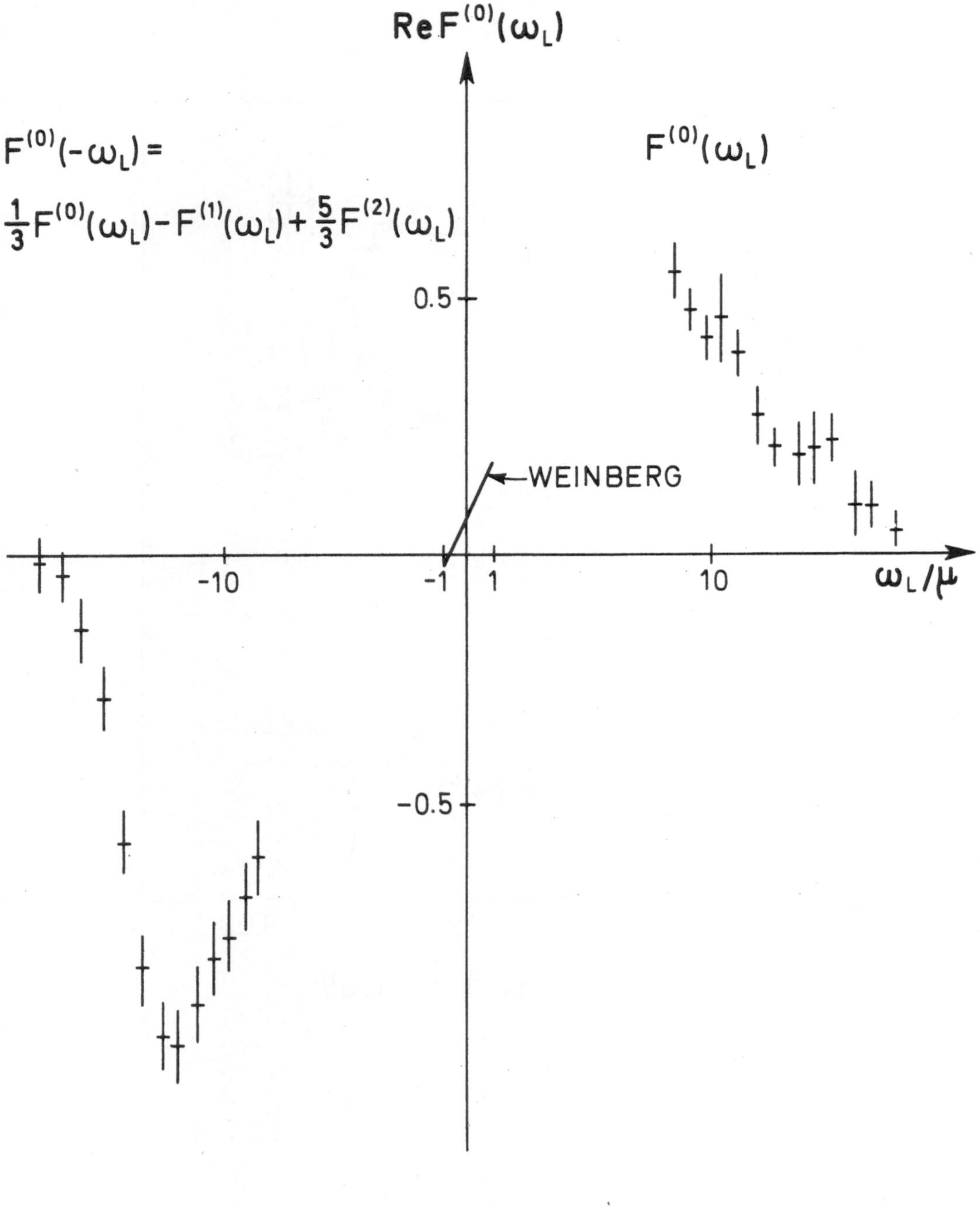
Re $F^{(0)}(\omega_L)$
$F^{(0)}(-\omega_L) =$
$\frac{1}{3}F^{(0)}(\omega_L) - F^{(1)}(\omega_L) + \frac{5}{3}F^{(2)}(\omega_L)$
$F^{(0)}(\omega_L)$
0.5
WEINBERG
-10
-1
1
10
ω_L/μ
-0.5

Fig. 13

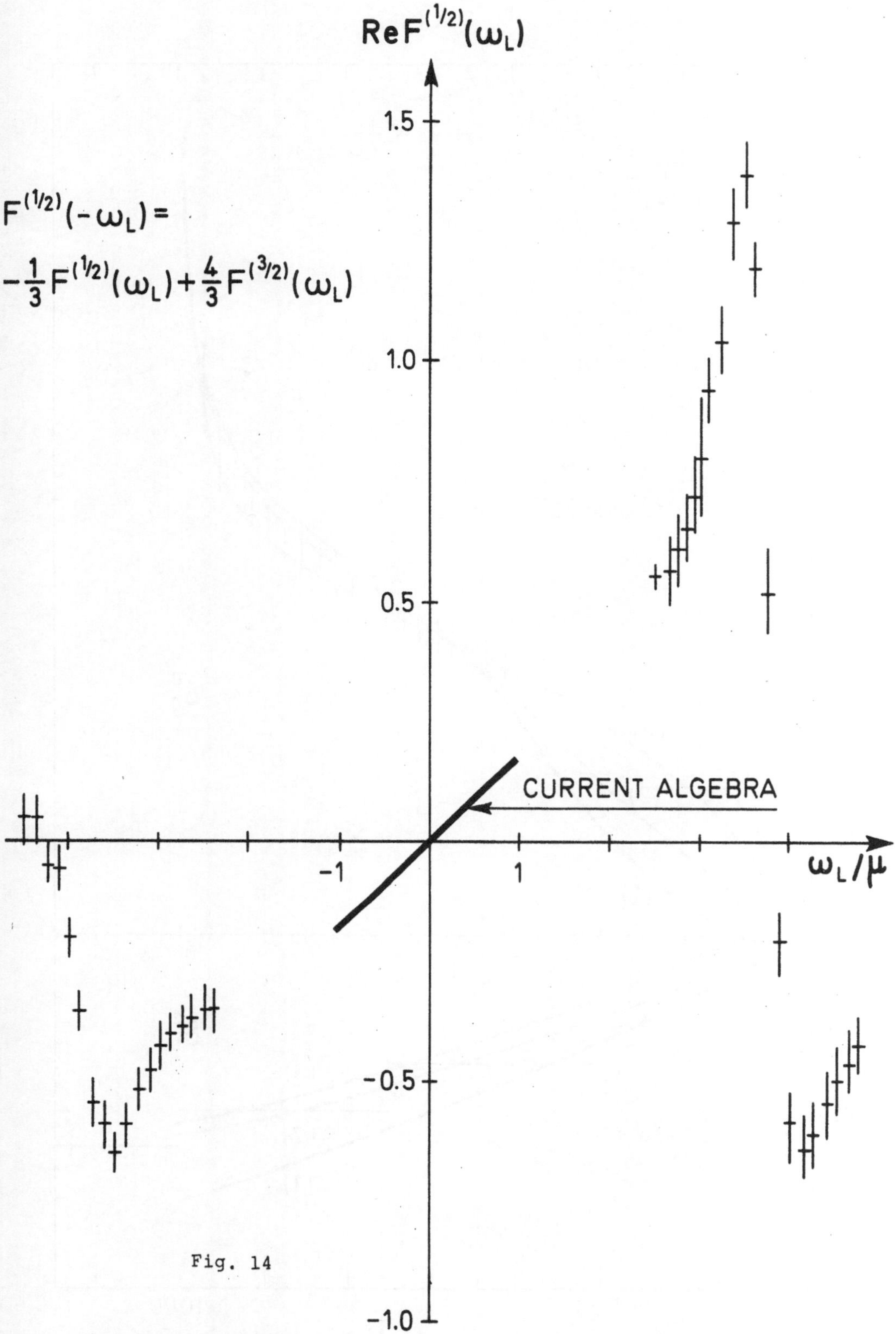

$\mathrm{Re}\,F^{(1/2)}(\omega_L)$
1.5
1.0
0.5
-0.5
-1.0
-1
1
ω_L/μ
$F^{(1/2)}(-\omega_L) =$
$-\frac{1}{3}F^{(1/2)}(\omega_L)+\frac{4}{3}F^{(3/2)}(\omega_L)$
CURRENT ALGEBRA

Fig. 14

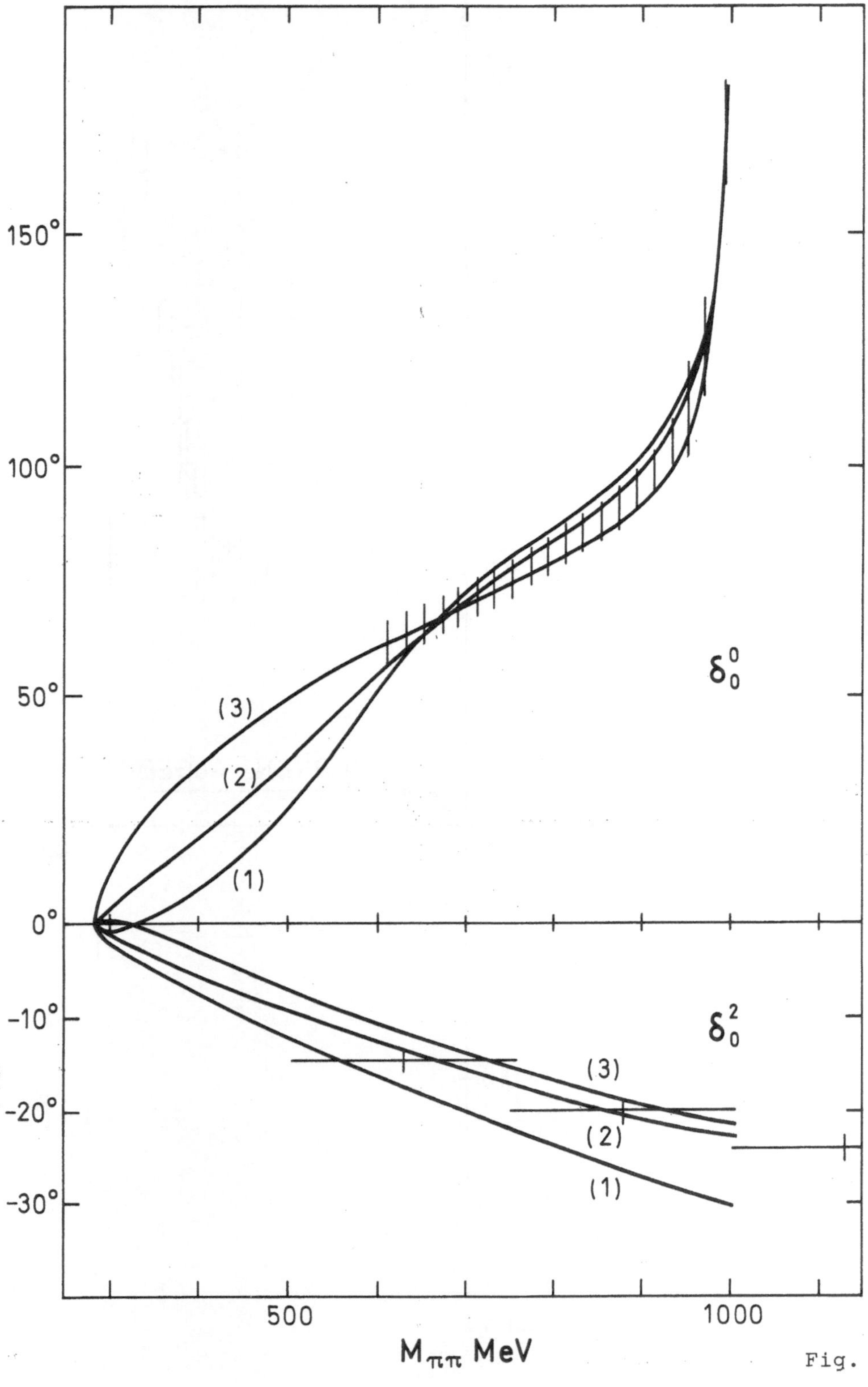

150°
100°
50°
0°
-10°
-20°
-30°
(3)
(2)
(1)
δ_0^0
δ_0^2
(3)
(2)
(1)
500
1000
$M_{\pi\pi}$ MeV

Fig. 15

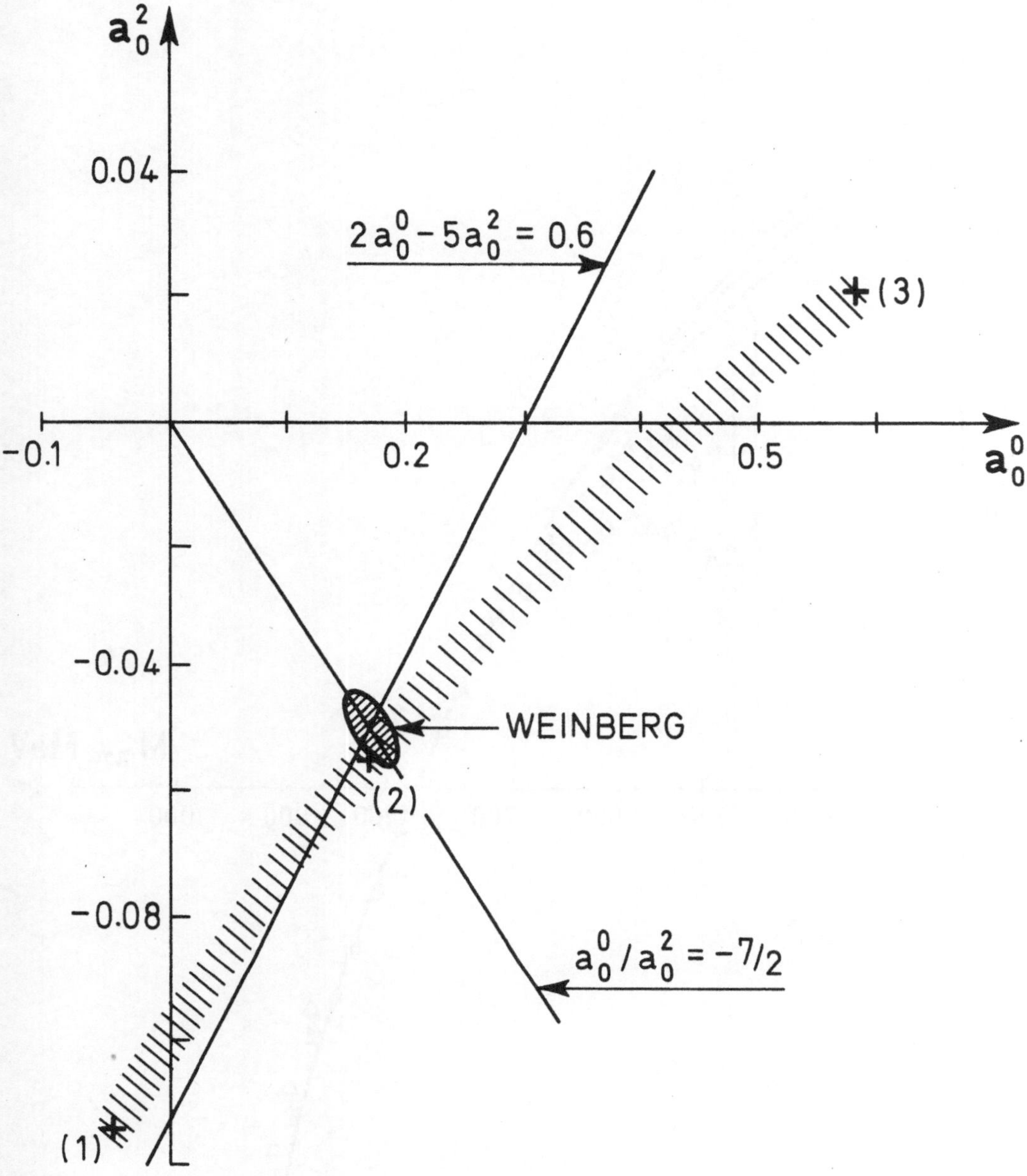

a_0^2
0.04
$2a_0^0 - 5a_0^2 = 0.6$
(3)
-0.1
0.2
0.5
a_0^0
-0.04
WEINBERG
(2)
-0.08
$a_0^0 / a_0^2 = -7/2$
(1)

Fig. 16

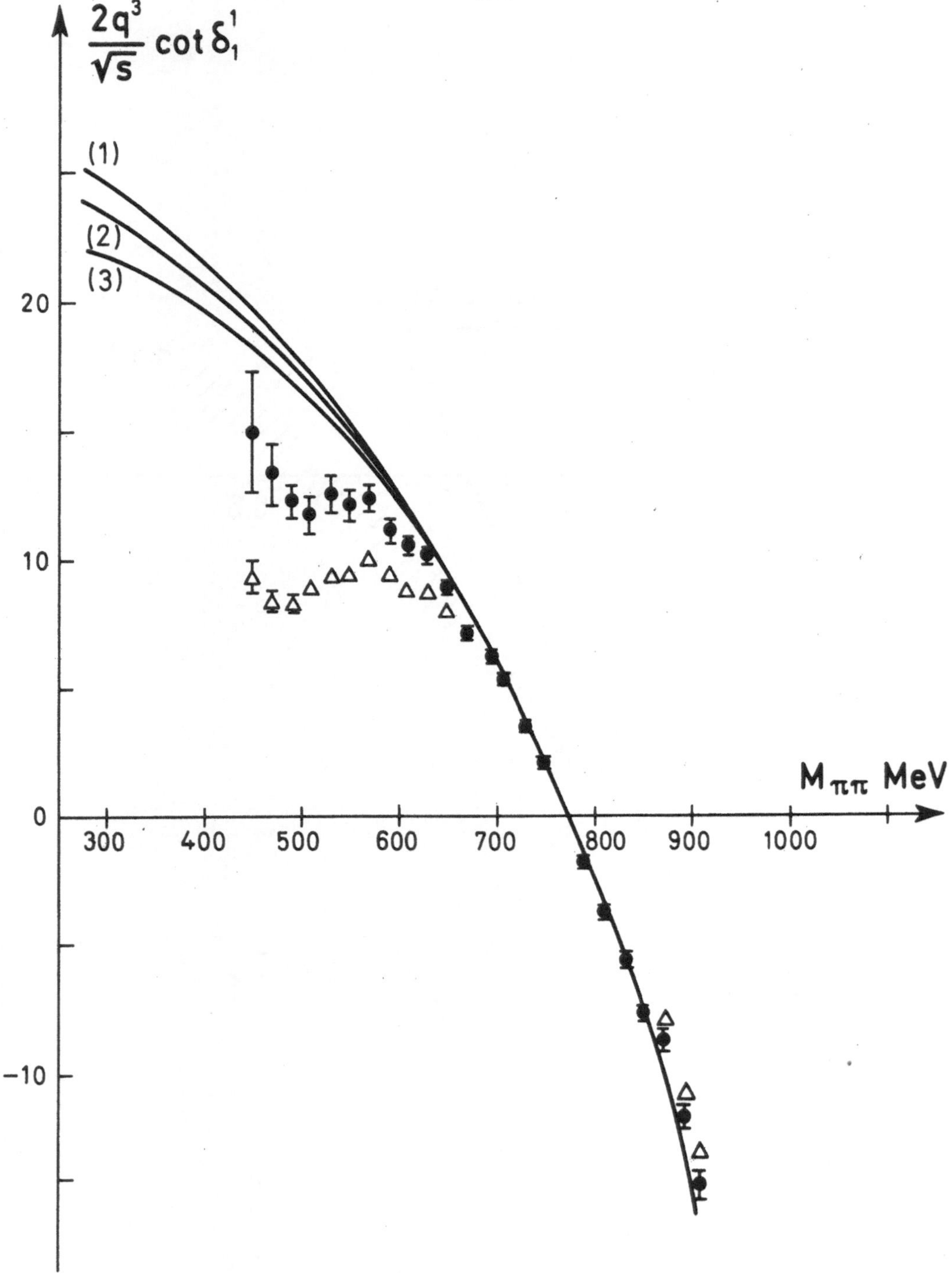
$\frac{2q^3}{\sqrt{s}} \cot \delta_1^1$
(1)
(2)
(3)
20
10
0
-10
$M_{\pi\pi}$ MeV
300
400
500
600
700
800
900
1000

Fig. 17

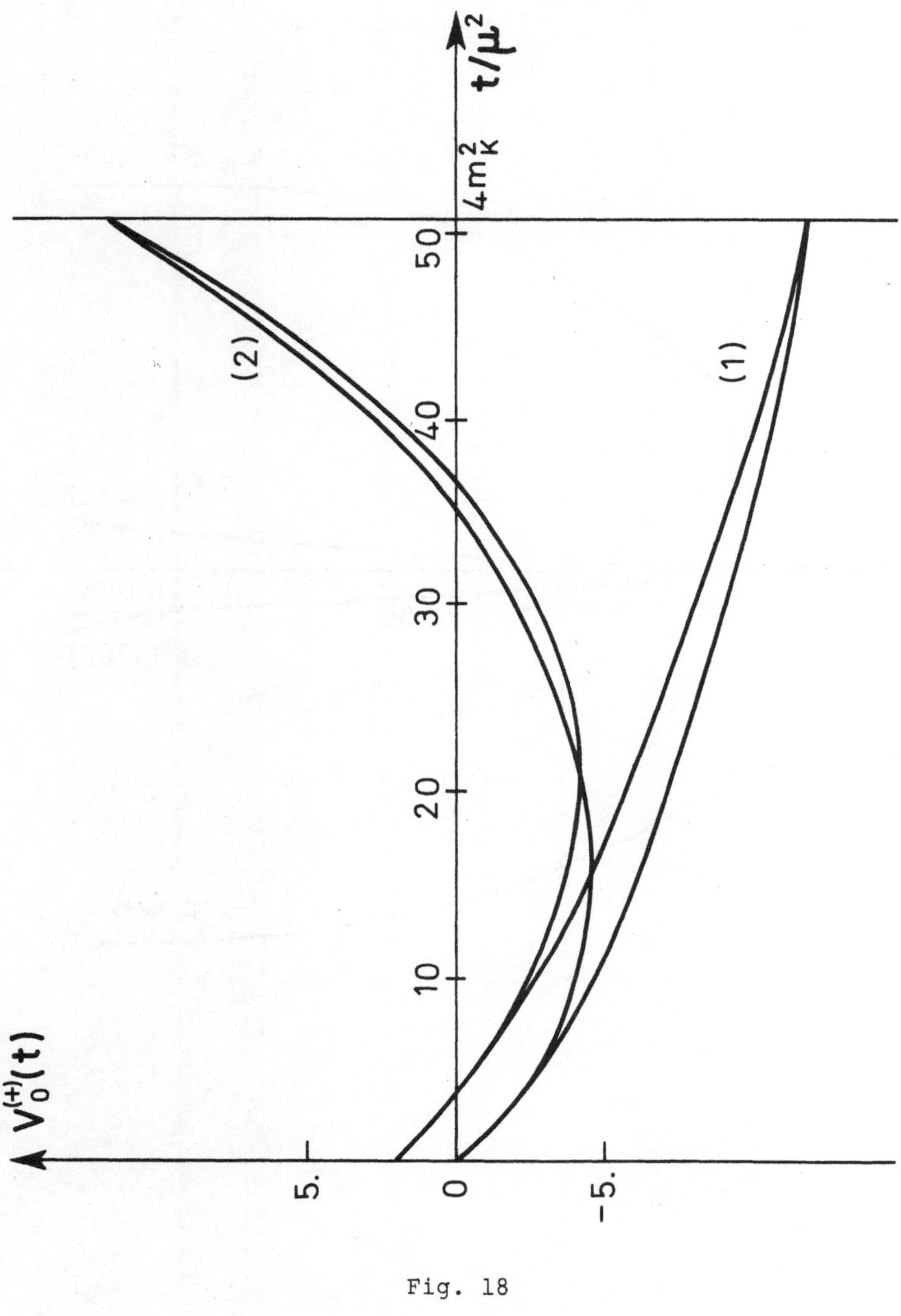

Fig. 18

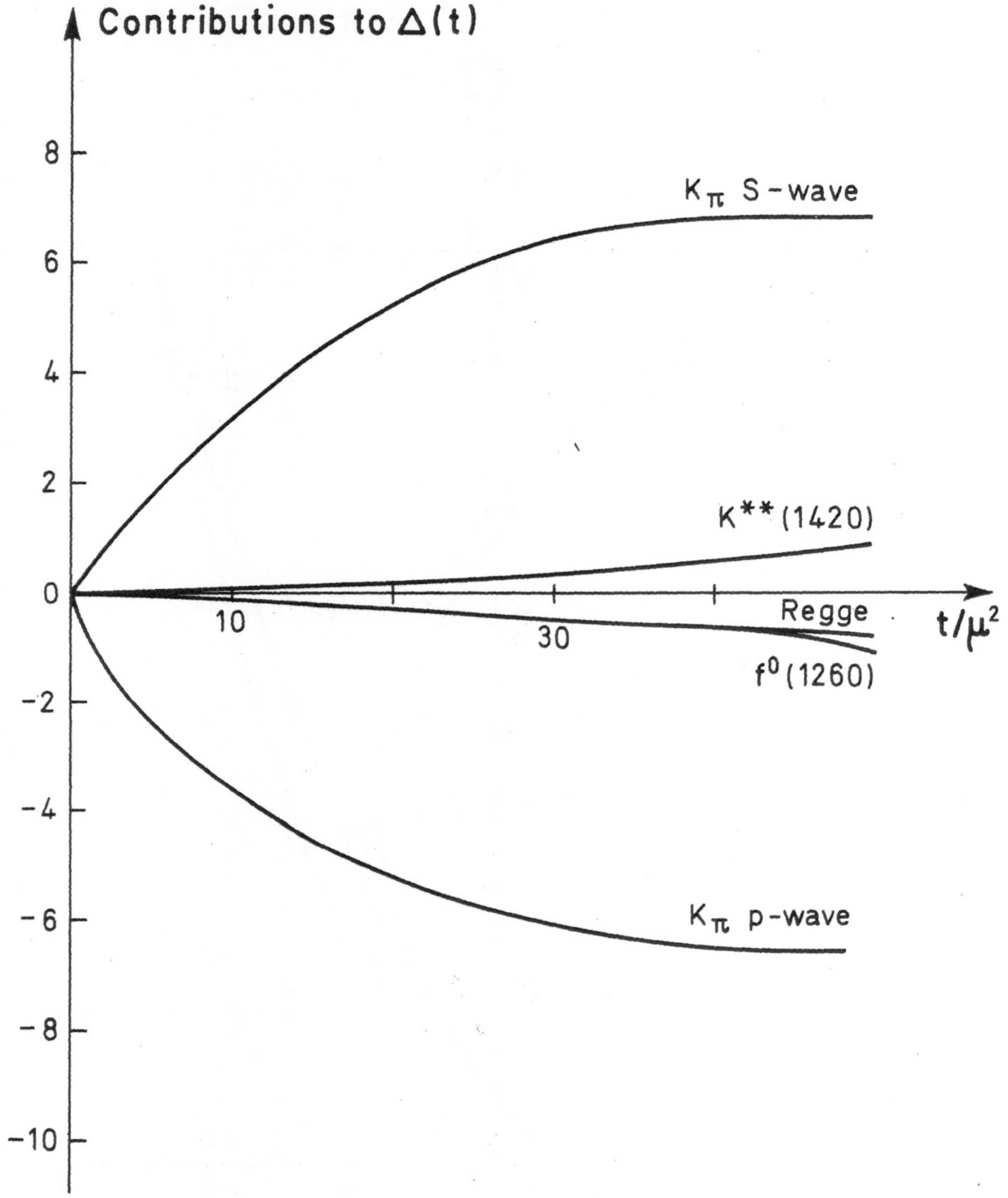

Contributions to Δ(t)
8
6
4
2
0
-2
-4
-6
-8
-10
10
30
t/μ²
Kπ S-wave
K** (1420)
Regge
f⁰ (1260)
Kπ p-wave

Fig. 19

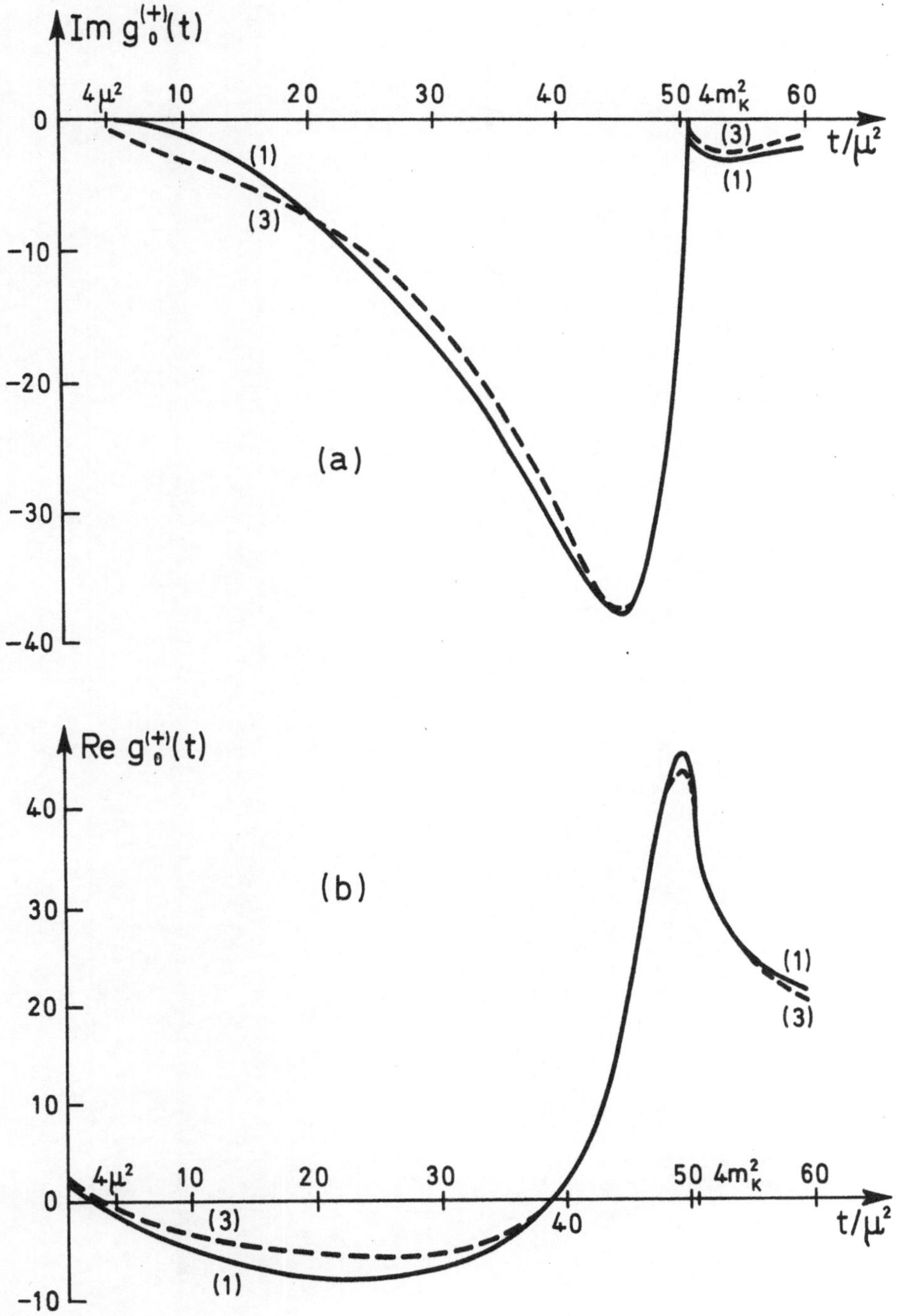
Im $g_0^{(+)}(t)$
$4\mu^2$
10
20
30
40
50
$4m_K^2$
60
t/μ^2
0
−10
−20
−30
−40
(1)
(3)
(a)
Re $g_0^{(+)}(t)$
40
30
20
10
0
−10
(b)

Fig. 20

Acta Physica Austriaca, Suppl. XIII, 361–394 (1974)

ASYMPTOTIC FREEDOM AND ALMOST-FREEDOM

by

W. KUMMER*)
Inst. f. Theoret. Physik, Techn. Hochschule Wien
Wien, Austria

1. INTRODUCTION

One of the classical ideas in theoretical physics is the generalization of the gauge invariance of the electromagnetic field to vector fields with nonabelian gauge groups [1]. However, it took a long time until the quantization of such fields was carried through consistently for the massless case [2] and until the renormalizability was shown for massive versions [3] if they result from spontaneous symmetry breaking via the Higgs-mechanism [4]. At first this stimulated new interest for models in which a unification of weak and electromagnetic interactions was proposed in terms of nonabelian gauge theories of the photon and to the hypothetical massive vector bosons which mediate weak interactions [5].

---

*) Lecture given at XIII. Internationale Universitätswochen für Kernphysik, Schladming, Austria, February 4-15, 1974.

To the pleasant surprise of particle theorists during 1973 also some evidence accumulated in favor of a basic interaction of the same kind for hadronic phenomena [6,7,8]. This will be the topic of these lectures.

2. SCALING IN DEEP INELASTIC LEPTON-HADRON SCATTERING

2a) Basic Features

Asymptotic "freedom" is motivated by the experimental results of deep inelastic e-p-scattering which is therefore reviewed shortly. If in the process of fig. 1 no particle from the lower vertex is detected experimentally its inclusive cross-section reads

$$\frac{d^2\sigma}{dq^2 d\nu} = W_{\mu\nu} \quad \ell^{\mu\nu} \quad . \tag{2.1}$$

where $q^2 < 0$ is the virtual momentum of the exchanged photon, $\nu = pq = m(E_\ell - E_\ell')$ is the other invariant variable and

$$W_{\mu\nu} = \int \langle p|[j_\mu(x), j_\nu(0)]|p\rangle \, e^{iqx} \, d^4x \tag{2.2}$$

$\ell^{\mu\nu}$ is proportional to the spin-averaged absolute square of the electron current. It will be of no interest here. The hadronic part $W_{\mu\nu}$ can be decomposed into

$$W_{\mu\nu} = (g_{\mu\nu} - q_\mu q_\nu/q^2) V_1(q^2,\nu) +$$

$$+ 2q^{-4}[p_\mu p_\nu q^2+g_{\mu\nu}\nu^2-(p_\mu q_\nu+p_\nu q_\mu)\nu]V_2(q^2,\nu) \qquad (2.3)$$

Again a spin-average is assumed.

$W_{\mu\nu}$ represents the absorptive part of the analytic continuation $T^{an}_{\mu\nu}$ of

$$T_{\mu\nu} = i\int \langle p|T(j_\mu(x), j_\nu(0))|p\rangle e^{iqx} d^4x \qquad (2.4)$$

which we may express in a similar fashion as (2.3) with invariant functions T_1 and T_2:

$$2iV_j = \text{abs } T_j^{an} \quad . \qquad (2.5)$$

Our normalization is such that the V_j (and T_j) are dimensionless functions

$$V_j = V_j(\omega, q^2/m^2, m_i/m_j\ldots, g\ldots.)$$

where

$$\omega = x^{-1} = - 2\nu/q^2 \qquad (2.6)$$

and where a possible dependence on mass ratios and dimensionless couplings of basic fields is indicated. If the limit $q^2,\nu \to \infty$ at fixed ω exists then the dependence on q^2 in (2.6) disappears. This naive scaling seems to be the case experimentally; one finds also that $V_1 \simeq 0$ in this region [9]. It should be remarked that the latter quantity is usually called $(\omega/2)\, F_L$ whereas

$$V_2 = - (\nu W_2)/\omega \qquad (2.7)$$

2b) Light Cone Expansion

A composite model of the nucleon with such a behaviour in the deep inelastic region is represented by the picture of a cloud of quasifree pointlike "partons" with spin 1/2 [10]. For our purpose it is preferable to concentrate on the alternative more formal description, by means of a light cone expansion.

If we introduce "light-like" variables for x and q

$$x_\pm = (x_o \pm x_3)/\sqrt{2} \qquad q_\pm = (q_o \pm q_3)/\sqrt{2}$$

$$\vec{x}_\perp \qquad \vec{q}_\perp$$

in

$$q^2 = 2q_+q_- - \vec{q}_\perp^{\,2}$$

$$\nu = p_+q_+ + p_-q_- - \vec{p}_\perp \vec{q}_\perp$$

one observes immediately that the Bjorken limit $q^2 \to \infty$, $\nu \to \infty$ [11] referred to above is realized by, say $q_+ \to \infty$.

In the oscillating exponential from

$$qx = q_+x_- + q_-x_+ - \vec{q}_\perp \vec{x}_\perp$$

a noticable contribution arises then only for $x_- \sim 0$ or at

$$x^2 = 2x_+ x_- - \vec{x}_\perp^{\,2} \sim - \vec{x}_\perp^{\,2}$$

The commutator in (2.2) vanishes, however, for $x^2 < 0$ (causality!). The contribution of $j_\mu(x)j_\nu(0)$ is therefore important near $x^2 \sim 0$. This argument works only for the commutator. Taking $\int d^4q' \delta(q^2-q'^2)\delta(\nu-pq')T_{\mu\nu}(q',p)$ for $q^2,\nu \to \infty$ at fixed ω can be shown to select $x^2 \sim 0$ also in $T_{\mu\nu}$ and in other matrix-elements [12,13].

It is possible to write for two currents separated by $z^2 = (x-y)^2 \sim 0$ a light cone expansion

$$j_\mu(x)j_\nu(y) \underset{z^2\sim 0}{=} P_{\mu\nu\alpha\beta} R^{\alpha\beta} \quad . \tag{2.8}$$

The differential operator

$$P_{\mu\nu\alpha\beta} = g_{\mu\nu}\partial_\alpha^{(x)}\partial_\beta^{(y)} - g_{\nu\alpha}\partial_\mu^{(y)}\partial_\beta^{(x)} - g_{\mu\beta}\partial_\alpha^{(y)}\partial_\nu^{(x)} +$$

$$+ g_{\mu\beta}\, g_{\nu\alpha}(\partial^{(x)}\partial^{(y)}) \tag{2.9}$$

guarantees current conservation $\partial^\mu j_\mu = 0$, whereas the quantity [14]

$$R^{\alpha\beta} = z^\gamma(z^2-i\varepsilon z_o)^{-1} j_\gamma(z;y)g^{\alpha\beta} + j^{\alpha\beta}(z;y) + \ldots \tag{2.10}$$

contains bilocal operators depending on local operators $O^{(n,i)}(y)$, e. g.

$$j_{\rho\sigma}(z;y) = \sum_{n,i} c_{n,i}\, O^{(n,i)}_{\rho\sigma\alpha_1\ldots\alpha_n}(y)\, z^{\alpha_1}\ldots . z^{\alpha_n} \tag{2.11}$$

in many different combinations from which only two among the leading ones have been selected in (2.10).

In a theory with, say, only free fermions (free

partons!), the operators $O^{(n,i)}$ can be

$$O_{\rho}^{(o)} = \bar{\psi}\,\gamma_{\rho}\psi$$

$$O_{\alpha_\rho}^{(1,1)} = (\partial_{\alpha}\,\bar{\psi})\gamma_{\rho}\psi \qquad\qquad O_{\alpha\rho}^{(1,2)} = \bar{\psi}\,\gamma_{\rho}\,\partial_{\alpha}\,\psi$$

....

They can be characterized by a fixed twist = dimension - spin = 2 [15]. Simple counting of dimensions in (2.8) with (2.9), (2.10) shows that the $C_{n,i}$ in (2.11) are dimensionless. At least in a theory with free fields one can expect that masses are negligible at large momentum variables or small z^2; the $C_{n,i}$ which could be otherwise functions $C_{n,i}(m^2(z^2-i\varepsilon z_o))$ are then in general constants. In the second term of (2.10) also a logarithmic $C_{n,i}$ = $\log M^2(z^2-i\varepsilon z_o)$ fulfills the same criterion because the mass M^2 drops out during differentiation by P.

It is not difficult to insert (2.8) into (2.4) [16]. Defining

$$<p|O^{(n,i)}_{\rho\sigma\alpha_1\cdots\alpha_n}(0)|p> = i^n\,c_n^{(i)}p_{\rho}p_{\sigma}p_{\alpha_1}\cdots p_{\alpha_n} +$$

$$+ \text{ terms with } g_{\rho\sigma} \text{ etc.}$$

one notices that any $g_{\rho\sigma}$, $g_{\alpha_1\alpha_2}$ etc. in the latter expression yields a reduced singularity at $z^2 = x^2 \sim 0$. The contributions to e.g. T_2 originates from the second term in (2.10) only:

$$T_2 = \frac{g^4}{2}\int e^{iqx}d^4x \sum_{n,i} C_n^{(i)}(ipx)^n\,c_n^{(i)} + O(\frac{1}{\nu}) =$$

$$= \sum_n \omega^n E_n^{[2]} + O(\nu^{-1}) \tag{2.12}$$

$$E_n^{[2]} = \frac{1}{2} \sum_i c_n^{(i)} (-q^2)^{n+2} \left(\frac{\partial}{\partial q^2}\right)^n \int e^{iqx} d^4x \, C_n^{(i)} \, . \tag{2.13}$$

A similar relation for the $E_n^{[1]}$ of T_1 can be derived in a completely analogous way.

In the "free-field" case

$$\int e^{iqx} d^4x \log (x^2 - i\varepsilon x_o) \underset{q^2 \to \infty}{\propto} q^{-4}$$

and the $E_n^{[2]}$ in (2.13) are independent of q^2.

Note that (2.13) connects certain expansion coefficients of the amplitude T for forward virtual Compton scattering with the Fourier transform of the c-number coefficient function in the light-cone expansion. These coefficients coincide, however, also with the moments of the experimentally observable structure functions V_1 and V_2 in (2.3) for the commutator. This can be seen as follows [17], e.g. again for T_2.

The physical T_2 is the boundary value of an analytic function T_2^{an} which obeys the crossing relation

$$\text{abs } T_2^{an}(\omega, q^2) = - \text{ abs } T_2^{an}(-\omega, q^2) \tag{2.14}$$

whereas from (2.4) T_2 must be even in ω. Physical values of T_2 start at the one nucleon intermediate state $\omega = 1$. Near $\omega = 0$ $T_2 = T_2^{an}$ so that (2.12) can be inverted

$$E_n^{[2]} = (2\pi i)^{-1} \oint d\omega\ \omega^{-n-1}\ T_2^{an}(\omega, q^2)\ . \tag{2.15}$$

The integral is performed along a small circle around $\omega = 0$. We extend this curve as in fig. 2. Neglecting the infinite half-circles - which is equivalent to a no-subtraction hypothesis - one arrives with

$$T_2^{an}(\omega + i\varepsilon) - T_2^{an}(\omega - i\varepsilon) = \text{abs}\ T_2^{an} = 2iV_2$$

and equ. (2.14) in (2.15) at ($\omega = x^{-1}$, cf. (2.7))

$$E_n^{[2]} = -\pi^{-1} \int_0^1 dx\ x^n\ (\nu W_2)\ . \tag{2.16}$$

This establishes the connection between experimentally accessible [9] quantities and the Fourier transform of coefficient functions on the light cone.

2c) Scaling and Perturbation Theory

Everybody who has ever done a perturbation calculation in field theory knows that - at least order by order in perturbation theory - the Feynman diagrams never exhibit the simple property of naive scaling, i.e. that at high q^2, ν the dependence on the masses never disappears completely. The "scaling" properties assumed via the light cone expansion for the "free" case are therefore certainly not true in higher orders. They hold only without internal interactions. To order g^o in some coupling constant g the trivial evaluation of the graphs of fig. 3 yields

$$T_1 = 0, \ \omega \ T_2 \propto (\omega-1)^{-1} + (\omega+1)^{-1} \qquad (2.17)$$

and hence

$$\nu W_2 = - \ \omega \ V_2 \propto \ \delta(\omega-1)$$

in the physical range $\omega \geq 1$. This obeys scaling but is a trivial function. The trouble for scaling can be noticed already in a contribution of the type of fig. 4 to $O(g^2)$ where an additional "pion" is exchanged internally. This graph is perfectly finite so that no problems of renormalization arise. Nevertheless, if one tries to neglect all masses as compared to the large external momentum q^2 the remaining integral looks like

$$\int d^4k/k^4$$

at $k = 0$. We encounter therefore a logarithmic (infrared, IR) singularity so that the integral will behave like $\log \mu^2$ for some mass parameter $\mu \to 0$. For dimensional reasons a dependence like $\log q^2/\mu^2$ for $q^2 \to \infty$ is therefore plausible.

More generally the argument can also be represented as follows:

Large momenta seem to permit neglect of all masses. Such a theory is then ill-defined, however, except we renormalize (i.e. fix certain coupling constants etc.) at some arbitrary momentum, measured by μ^2. In such a way at least one mass parameter sneaks back into the game.

In fact in higher orders the powers of the log's increase. To reconcile this behaviour of $T_{\mu\nu}$ or of the moments

$$E_n \sim \sum_m (\log q^2/\mu^2)^m \, c_{nm} \, g^{2m} + \ldots$$

with the scaling result of experiments a widespread belief among optimistic theorists was for some time that somehow all orders of perturbation theory sum up so that e.g. the leading log's combine to a decreasing power like ($c > 0$)

$$\sum_{m=0}^{\infty} \frac{(-g^2c)^m}{m!} \log^m (q^2/\mu^2) = (q^2/\mu^2)^{-cg^2} \underset{q^2 \to \infty}{\to} 0$$

and that a similar effect takes place also for the non-leading terms. Such opinions could be maintained, of course, only as long as there was no "danger" of a computation of the asymptotic behaviour to all orders in perturbation theory.

3. CALLAN-SYMANZIK (C.-S.) EQUATION

We describe now the new very powerful method which under certain conditions - is able to cope with the asymptotic regime without requiring a perturbation calculation to all orders [19].

3a) C.-S. Equation for Green's Functions

For simplicity we consider at first only one scalar field with $g^2\phi^4$ - coupling. The Fourier transform of the "amputated" [20] renormalized function of the time-ordered Heisenberg fields ϕ

$$\Gamma^{(N)}(x_1 \ldots x_N; g, m) = \langle 0|T\phi(x_1)\ldots\phi(x_N)|0\rangle_{amp} \tag{3.1}$$

is the Feynman amplitude $\tilde{\Gamma}^{(N)}$ for the N-particle process

$$\int \prod_{i=1}^{N} (e^{iq_i x_i} d^4x_i)\,\Gamma^{(N)} = (2\pi)^4 \delta^4(\Sigma q_i)\,\tilde{\Gamma}^{(N)}(q_1,\ldots q_N). \tag{3.2}$$

It is related to the unrenormalized quantity $\Gamma_0^{(N)}$ by the wave function renormalization $\sqrt{Z_\phi}$ for N external lines

$$\Gamma^{(N)} = \Gamma_0^{(N)} Z_\phi^{N/2} \quad . \tag{3.3}$$

The physical mass and Z_ϕ are fixed by the requirement that the renormalized propagator $[\Gamma^{(2)}]^{-1}$ has a pole at m with residue 1.

We regularize with a cutoff Λ so that in terms of bare quantities

$$\begin{aligned} m &= m\,(m_0, g_0, \Lambda) \\ Z_\phi &= Z_\phi\,(m_0, g_0, \Lambda) \end{aligned} \tag{3.4}$$

Requiring the "physical" coupling to be g for some prescribed external momenta in $\Gamma^{(N)}$ leads to a relation

$$g = g\ (m_o,\ g_o,\ \Lambda) \tag{3.5}$$

We compute next the variation of $\Gamma_o^{(N)}$ with respect to m_o at fixed g_o and Λ in the form

$$Z^{N/2}\ m_o^1\ \frac{\partial}{\partial m_o^2}\ \Gamma_o^{(N)}\ =$$

$$= m_o^2\ [\frac{\partial m^2}{\partial m_o^2}\frac{\partial}{\partial m^2} + \frac{\partial g}{\partial m_o^2}\frac{\partial}{\partial g} - \frac{N}{2}\,\frac{\partial(\log Z_\phi)}{\partial m_o^2}]\ \Gamma^{(N)}\quad . \tag{3.6}$$

For our unrenormalized $\Gamma_o^{(N)}$ the relation to interaction picture operators $\bar{\phi}$ reads

$$\Gamma_o^{(N)} = \langle 0|:T(\bar{\phi}_o(x_1)\ldots\bar{\phi}_o(x_N))\ \exp\{i\int L_{int}(\bar{\phi})\,d^4x\}:|0\rangle_{amp} \tag{3.7}$$

For convenience we make the free Lagrangian depend on some mass m'^2 to get m_o^2 into L_{int}. Therefore the only m_o^2 in (3.7) occurs in

$$L_{int}(\bar{\phi}) = -\ g_o^2\ \bar{\phi}^4 - \frac{m_o^2 - m'^2}{2}\ \bar{\phi}^2\quad .$$

Going back to the Heisenberg picture after performing the differentiation in (3.7) the left hand side of

(3.6) is therefore

$$m_o^2 \frac{\partial}{\partial m_o^2} \Gamma_o^{(N)} = i\langle 0|T(- \frac{m_o^2}{2} \int \phi_o^2(x) d^4x, \phi_o(x_1) \ldots)|0\rangle_{amp} \equiv$$

$$\equiv - \frac{i m_o^2}{2} \Gamma^{(N)}_{o\phi^2} \quad . \tag{3.8}$$

This represents the insertion of an additional "vertex" $m_o^2 \phi_o^2$ with momentum zero into all internal lines of the original $\Gamma_o^{(N)}$. $\Gamma_{o\phi^2}^{(N)}$ as defined by (3.8) is renormalized not only by the factor $Z^{N/2}$ in (3.6) for the N external lines; we need also a renormalization for the local operator $\phi_o^2(x)$

$$\phi_o^2(x) = Z_{\phi^2} \phi^2 (x) \tag{3.9}$$

(3.6) with (3.8) and (3.9) becomes the full inhomogeneous Callan-Symanzik equation

$$\{m^2 \frac{\partial}{\partial m^2} + \beta \frac{\partial}{\partial g} - N\gamma\} \Gamma^{(N)} = i\alpha \Gamma^{(N)}_{\phi^2} \tag{3.10}$$

where

$$\alpha = - \frac{m^2}{2} Z_{\phi^2} \frac{\partial m_o^2}{\partial m^2} \qquad \beta = m^2 \frac{\partial m_o^2}{\partial m^2} \frac{\partial g}{\partial m_o^2}$$

$$\gamma = \frac{m^2}{2} \frac{\partial m_o^2}{\partial m^2} \frac{\partial \log Z_\phi}{\partial m_o^2} \quad . \tag{3.11}$$

Because $\Gamma^{(N)}$ and $\Gamma^{(N)}_{\phi^2}$ are finite, α, β, and γ must be finite too. This can be verified in perturbation theory, e.g.

$$\beta = \beta(g) = 3g^4/16\pi^2 + O(g^6) \qquad (3.12)$$

for the ϕ^4-theory.

If all momenta - in a Fourier transformed identical equation for $\tilde{\Gamma}$ - are submitted to scaling in the Euclidean region

$$p_i \to \lambda p_i, \qquad p_i^2 < 0, \qquad \lambda \to \infty$$

application of the Weinberg-theorem [21] tells us that to each order in perturbation theory $\tilde{\Gamma}^{(N)}_{\phi^2}$ is smaller by a factor λ^{-2} than $\tilde{\Gamma}^{(N)}$. The homogeneous equation obtained in this region

$$[m^2 \frac{\partial}{\partial m^2} + \beta \frac{\partial}{\partial g} - N\gamma]\Gamma^{(N)as} = 0 \qquad (3.13)$$

resembles the old renormalization group equation [22].

The generalization of (3.10) to s particles with k couplings for a Green's function with $N_1 \ldots N_s$ particles of the different types is rather obvious:

$$[\sum_{i=1}^{s} m_i^2 \frac{\partial}{\partial m_i^2} + \sum_{j=1}^{k} \beta_j(g_1 \ldots g_k)\frac{\partial}{\partial g_j} - \sum_{i=1}^{s} N_i\gamma_i]\ \Gamma^{(N_1 \ldots N_s)} =$$

$$= i \sum_{i=1}^{s} \alpha_i\ m_i^2\ \Gamma^{(N_1 \ldots N_s)}_{\phi_i^2} \qquad (3.14)$$

In the deep Euclidean region again the r.h.s. can be neglected and also all masses except one, say μ, as discussed in sect. 2c).

One notices the crucial dependence of Z_ϕ and g on the cutoff Λ at the intermediate stage of the calculation (3.11):

If Z_ϕ were independent of Λ, by simple dimension counting it had to be also independent of the single mass parameter μ_o. Hence γ and also β were zero. Introducing a dimensionless $\tilde{\Gamma}$ by $(q^2 = p_1^2,\ \omega_{ij} = (p_i p_j)/q^2)$

$$\tilde{\Gamma}^{as(N_1 \ldots N_s)} = (q^2)^D \ \hat{\tilde{\Gamma}}^{as(N_1 \ldots N_s)}(q^2/\mu^2, \omega_{ij}, g) \quad (3.15)$$

in the Fourier transformed $\tilde{\Gamma}$ one would arrive in this case at

$$\mu^2 \frac{\partial}{\partial \mu^2} \tilde{\Gamma}^{as} (q^2/\mu^2,\ \omega_{ij},\ g) = 0 \quad (3.16)$$

which is just the result of naive scaling. One sees that the regularization which is necessary because of the infinite renormalization is the deep reason for the fact that naive scaling is an untenable assumption in field-theory.

3b) C.-S.-Equation for Coefficient-Functions in the Light-Cone Expansion

For simplicity we derive this equation again for a field theory with only one scalar field. For a kinematical situation such that $(x-y)^2 \sim 0$ and hence for two operators

$$O'(x)O''(y) \underset{(x-y)^2\sim 0}{=} \sum_i C_{O(i)}[(x-y)^2]\, O_{(i)}(x-y;y) \tag{3.17}$$

a C.-S. equation can be written for "Green's functions" $\Gamma^{(N)}_{O',O''}$ where $O'(x)O''(y)$ is replaced by (3.17). We introduce renormalized operators O' and O'' in terms of unrenormalized ones O'_o and O''_o

$$\begin{aligned} O'_o &= Z'\,O' \\ O''_o &= Z''\,O'' \quad . \end{aligned} \tag{3.18}$$

The only change in the derivation of the C.-S. equation as compared to the one for $\Gamma^{(N)}$ in the last section is then the differentiation of Z' and Z'' in the form

$$\gamma'^{(\prime\prime)} = m^2 \frac{\partial m_o^2}{\partial m^2} \frac{\partial \log Z'^{(\prime\prime)}}{\partial m_o^2} \tag{3.19}$$

so that $\Gamma^{(N)}_{O';O''}$ - which is amputated now with respect to the N external ϕ-lines only - obeys

$$(D - N\gamma + \gamma' + \gamma'')\Gamma^{(N)}_{O'O''} = i\alpha\, \Gamma^{(N)}_{O';O'';\phi^2} \tag{3.20}$$

where

$$D = \mu^2 \frac{\partial}{\partial \mu^2} + \beta \frac{\partial}{\partial g} \quad . \tag{3.21}$$

After insertion of (3.17) we consider now for <u>each n</u> the local operators $O^{(n,i)}_{\alpha_1 \ldots \alpha_n}(x_n)$ in the series expansion of the bilocal operators $O_{(i)}$ in (3.17) (cf. (2.11)) separately, and we suppress the indices n and $\alpha_1 \ldots \alpha_n$ in what follows, (i) refers now to different operators at the same n:

$$O'O''\Big|_{\text{fixed } n} \to \sum_{(i)} C_{(i)} O_{(i)} .$$

This yields

$$\sum_i [(DC_{(i)})\Gamma^{(N)}_{O(i)} + C_{(i)}(D\Gamma^{(N)}_{O(i)}) - (N\gamma - \gamma' - \gamma'')C_{(i)}\Gamma^{(N)}_{O(i)}] =$$

$$= i\alpha \sum_i C_{(i)} \Gamma^{(N)}_{O(i)\phi^2} \qquad (3.22)$$

where

$$\Gamma_{O(i)} = \langle O|T(O_i(y)\phi(x_1)\ldots\phi(x_N)|O\rangle_{amp} . \qquad (3.23)$$

We have to postulate next a C.-S. equation similar to (3.20) for (3.23). If more than one operator contributes to $\sum_i$ i.e. has the same dimension (twist), one can expect only a linear combination of $O_{(i)}$ to be multiplicatively renormalizable

$$O_{(i)} = \sum_j (Z^{-1})_{ij} O^{o}_{(j)} .$$

With

$$\gamma_{ij} = \sum_k m^2 \frac{\partial m_o^2}{\partial m^2} (Z^{-1})_{ik} \frac{\partial Z_{kj}}{\partial m_o^2} \qquad (3.24)$$

the C.-S. equation for (3.23) becomes

$$\sum_j \{(D - N\gamma)\delta_{ij} + \gamma_{ij}\} \Gamma^{(N)}_{O(j)} = i\alpha \Gamma^{(N)}_{O(i)\phi^2} \qquad (3.25)$$

and we arrive from (3.22) and (3.25) finally at the homogeneous equation for the coefficient-functions in the light-cone expansion

$$\sum_j [(D + \gamma' + \gamma'')\delta_{ij} - \gamma^n_{ij}] C_{n,j} = 0 \qquad (3.26)$$

In this result the dependence on n has been indicated explicitely again, γ^n_{ij} is the quantity derived for the operator $O^{(n,i)}$ from its renormalization constant, equ. (3.24).

3c) Solution of the Homogeneous C.-S.-Equation

Fortunately for Green's functions in the asymptotic region and for the coefficients in a light cone expansion a solution of only the homogeneous C.-S., equation ((3.13) or (3.26)) is necessary. Let us consider one coupling constant but many masses, represented by a "scale" μ, and a general $\gamma(g)$ chosen appropriately according to the specific problem

$$(\mu^2 \frac{\partial}{\partial \mu^2} + \beta \frac{\partial}{\partial g} - \gamma)\, \tilde{\Gamma} = 0 \,. \tag{3.27}$$

For the dimensionless $\hat{\tilde{\Gamma}}$ (cf. (3.15)) with

$$t = \log q^2/\mu^2 \tag{3.28}$$

and

$$d\rho(g)/dg = \beta(g) \tag{3.29}$$

(3.27) can be rewritten as

$$[-\frac{\partial}{\partial t} + \frac{\partial}{\partial \rho} - \gamma(g(\rho))]\, \hat{\tilde{\Gamma}} = 0 \tag{3.30}$$

which is trivially solved in terms of new variables

$$u = \frac{\rho+t}{2}\,, \quad v = \frac{\rho-t}{2}$$

with the arbitrary function H

$$\hat{\tilde{\Gamma}}(g,t,\omega_{ij}) = e^{\int^{v} \gamma(\rho(v'+u))\,dv'}\, H(2u) =$$

$$= e^{\int^{\rho(g)} \gamma(g(\rho'))\,d\rho'}\, H(\rho(g) + t) \,.$$

We compare this solution with

$$\hat{\tilde{\Gamma}}(\bar{g}(t),0,\omega_{ij}) = e^{\int^{\rho(\bar{g}(t))} \gamma(g(\rho'))\,d\rho'}\, H(\rho(\bar{g}(t)))$$

where

$$\rho(g) + t = \rho(\bar{g}(t)) \tag{3.31}$$

so that the unknown H can be eliminated. From (3.31) follows

$$1 = \frac{d\rho}{d\bar{g}} \frac{d\bar{g}}{dt}$$

so that (cf. (3.29)) $\bar{g}(t)$ obeys the differential equation

$$\frac{d\bar{g}}{dt} = \beta(\bar{g}), \qquad \bar{g}(0) = g \tag{3.32}$$

and

$$\hat{\tilde{\Gamma}}(g,t,\omega_{ij}) = \hat{\tilde{\Gamma}}(\bar{g}(t),0,\omega_{ij})e^{-\int_0^t dt'\gamma(\bar{g}(t'))} \tag{3.33}$$

This equation provides an immensely important connection between $\hat{\tilde{\Gamma}}$ at arbitrary t and $\hat{\tilde{\Gamma}}$ at "medium" energies ($t \simeq 0$) with some effective coupling $\bar{g}(t)$. It may be used to investigate e.g. the ultraviolet (UV) limit $t = +\infty$ or the infrared (IR) limit $t = -\infty$.

If γ is represented by a matrix (3.30) reads

$$(\delta_{ij} \frac{\partial}{\partial v} - \gamma_{ij})\hat{\tilde{\Gamma}}_j = 0.$$

This is formally identical to the basic differential equation for time-dependent perturbation theory so that the solution, generalized from (3.33), becomes

$$\hat{\tilde{\Gamma}}_i = [\text{"t"}\ (e^{-\int_0^t dt'\gamma(\bar{g}(t'))})]_{ij}\Gamma_j(\bar{g}(t),0,\omega_{ij}) \qquad (3.34)$$

with "t-ordered" integrals in the expansion of the exponential.

3d) Asymptotic Regions, Approximate Scaling

Because the IR-case can be treated in a completely similar fashion we concentrate on $t \to +\infty$ (UV-limit) and consider again one coupling constant only. In this case everything depends on the zeros of $\beta(g)$. This is seen easily.

Near a zero g_{UV} (3.32) becomes

$$\frac{d\bar{g}}{dt} \simeq \beta'(g_{UV})\ (\bar{g} - g_{UV})$$

with the integral

$$|\bar{g} - g_{UV}| \propto e^{\beta'(g_{UV})t} . \qquad (3.35)$$

For $\beta'(g_{UV}) < 0$ this solution may be used only if $t \to +\infty$. It says that

$$\lim_{t\to+\infty} \bar{g}(t) = g_{UV}\ (\beta'(g_{UV}) < 0) \qquad (3.36)$$

and in a similar way

$$\lim_{t\to-\infty} \bar{g}(t) = g_{IR}\ (\beta'(g_{IR}) > 0).$$

The usual terminology for g_{UV} and g_{IR} is "UV-stable" or "IR-stable" fixed point. The limit of (3.33) at large t becomes then

$$\hat{\tilde{\Gamma}}(g,t,\omega_{ij}) \underset{t\to+\infty}{\to} e^{-t\gamma(g_{UV})} \Gamma(g_{UV},0;\omega_{ij}) \cdot \qquad (3.37)$$

$$\cdot(\text{undetermined const.})$$

The power $e^{-t\gamma} = (q^2/\mu^2)^{-\gamma(g_{UV})}$ is similar to the factor q^{2D} in $\tilde{\Gamma}$ (cf. (3.15)) which was determined by the dimension of $\tilde{\Gamma}$. γ is, therefore, called the "anomalous dimension".

The origin g = 0 is certainly a "fixed point" (cf. (3.12) for an IR-fixed point). It can be a UV-fixed point only if $b_o > 0$ in

$$\begin{aligned} \beta(g) &= - b_o\, g^3 + O(g^5) \\ \gamma(g) &= - a_o\, g^2 + O(g^4). \end{aligned} \qquad (3.38)$$

Using (3.32) near g = 0 one verifies easily that the power in (3.37) is changed here into a logarithmic factor

$$\Gamma(g,t;\omega_{ij}) \underset{t\to+\infty}{\to} \Gamma(0,0;\omega_{ij}) \cdot \qquad (3.39)$$

$$\cdot \exp\{- \frac{a_o}{2b_o} \log t + \text{undetermined const.}\}$$

We observe that we can never get rid fully of the exponential for all Green's functions in a certain field theory although the anomalous dimension may vanish for special operators [23]. Hence the most we can achieve is an approximate scaling behaviour at $t \to +\infty$. This is realized in (3.39) for $g_{UV} = 0$ ("asymptotically free case", AF) because the factor $(\log q^2/\mu^2)^{-a_o/2b_o}$ varies slowly, but also in (3.37) if $\gamma(g_{UV})$ is very small which can be true especially for $g_{UV} \ll 1$ ("asymptotically nearly free case", ANF).

These statements concerning scaling of asymptotic Green's functions can be translated immediately into identical statements for the moments $E_n(q^2)$ of structure functions in deep inelastic electron-proton scattering, because of (2.13), (2.16) and the equally homogeneous C.-S.-equation for C_n (3.26). The only difference is that a_o and the other coefficients in the expansion of must be replaced by a_o^n corresponding to γ^n of the operator $O^{(n)}$ which may also turn out to be a matrix (cf. (3.24), (3.26), (3.34)).

If g_{UV} is small or zero the general solution (3.37) or (3.39) for the asymptotic region is expressed in terms of a Γ at small or vanishing coupling. This implies that the structure of the asymptotic Γ (or E_n for deep inelastic scattering) is determined by the structure of the same quantity for a free field or a field with small coupling. This is true not only for internal symmetries, but also for the tensor-properties of Γ. <u>If</u> such a theory exists it explains, therefore, quite naturally the success of "parton" sum rules derived from internal symmetries and/or tensorial properties of free partons (quarks). E.g. the vanishing (or almost vanishing) of F_L in the experiments is related to the free field result $T_1 = 0$

(2.17) if the basic "matter" fields are fermions.

4. GAUGE FIELDS FOR STRONG INTERACTIONS

Consider first "ordinary" renormalizable field theories: massless and massive QED, fermions coupled to scalars or pseudoscalars. It can be verified by direct computation using (3.11) that in all these cases - as in the ϕ^4-theory, eq. (3.12) -

$$\beta'(0) > 0.$$

Therefore the AF situation is impossible [17,24]. Also the NAF is not very probable, because in

$$\beta = |b_o|\, g^3 - |b_1|\, g^5 + \ldots$$

even if $b_1 < 0$, it would require $|b_1| \gg |b_o|$ if g_{UV} is close $g \sim 0$. On the other hand there is no obvious reason why $|b_o|$ and $|b_1|$ should not be of the same order or magnitude in any "ordinary" field theory.

The situation is different for nonabelian gauge theories [6,7] which are also renormalizable [3].

The Lagrangian for massless vector fields A_μ^A coupled to massive fermions ψ

$$L = -\frac{1}{4} F_{\mu\nu}^A F^{\mu\nu A} + \bar{\psi}(i\not{D} - M)\psi$$

$$D_\mu = \partial_\mu + i\, g\, t^A A_\mu^A \qquad (4.1)$$

$$F^A_{\mu\nu} = \partial_\mu A^A_\nu - \partial_\nu A^A_\mu - g\, C^{ABC} A^B_\mu A^C_\nu$$

in which t^A are the generators of the gauge group with structure coefficients C^{ABC} in the representation of the fermions, can be quantized only by adding a gauge fixing L_g and a $L_{F.P.}$ with scalar anticommuting Faddeev-Popov ghosts ϕ^A [2,3]. In the Lorentz-gauge $\partial_\mu A^{\mu,A}$ = const.

$$L_g = -(2\alpha)^{-1} (\partial^\nu A^A_\nu)(\partial^\mu A^A_\mu)$$

$$L_{F.P.} = \partial^\mu \phi^{*A} \partial_\mu \phi^A - g(\partial_\mu \phi^{*A}) C^{ABC} A^B_\mu \phi^C \quad . \quad (4.2)$$

The advantage of a noncovariant gauge $n^\mu A_\mu$ = const. is that $L_{F.P.} = 0$ for $\beta = 0$ in

$$L_g = -(2\beta)^{-1}(n^\mu A^A_\mu)(n^\nu A^A_\nu) \quad (4.3)$$

but the renormalization is conceptually very difficult.

For the Lorentz-gauge one finds [6,7]

$$\gamma = -\frac{g^2}{16\pi^2}\left[\left(\frac{13}{2} - \alpha\right) C_2(G) - \frac{4}{3} T(R)\right] + \cdots$$

$$(4.4)$$

$$\beta = -b_o g^3 + \ldots = -\frac{g^3}{16\pi^2}[11C_2(G) - 8T(R)]/3$$

where C_2 is the quadratic Casimir operator of the gauge group G in the adjoint representation (e.g. $C_2(SU(n)) = n$) and T(R) which originates from the fermion loop is defined by $T_r(t^A t^B) = T(R)\delta_{AB}$. γ, the anomalous dimension of the vector field, is seen to be gauge dependent. The gauge invariant b_o in β remains positive as long as there are not too many multiplets of fermions. We can therefore have the AF case. But also the NAF case is possible here because $8T(R) - 11\,C_2(G) > 0$ (but small) does not mean that necessarily a similar "almost" cancellation must occur in the next term $O(g^5)$ in (4.4). From (3.37) or (3.39) we find a gauge dependent asymptotic behaviour of Green's functions.

For the <u>moments</u> E_n in (2.16) we obtain via (2.13) and (3.26) with $\gamma' + \gamma'' = 0$ [23] in general a solution of the type (3.34) because for each n there are <u>two</u> operators of lowest twist (i = F or V)

$$O^{(n,F)}_{\alpha_1 \ldots \alpha_n} = [(D_{\alpha_1} D_{\alpha_2} \ldots \bar{\psi})\gamma_{\alpha_k}(\ldots D_{\alpha_n}\psi) + \text{all possible permutations and combinations}]_{\text{traceless part}} \tag{4.5}$$

$$O^{(n,V)}_{\alpha_1 \ldots \alpha_n} = [(D_{\alpha_1} \ldots F^{A}_{\alpha_k \beta})(D_{\alpha_{k+1}} \ldots F_{\alpha_n}^{\beta,A}) + \ldots.]_{\text{traceless part}} .$$

If the gauge group G is related to colour in a system with coloured quarks [25] only the colour singlets must be taken because physical states and physical currents are colour singlets.

The symmetry group of hadrons is then, say, G $\times$ SU(3). No mixing of operators occurs only for the SU(3)-octet part, when only the first operator in (4.5) - with an additional SU(3)-matrix λ-contributes. In addition to (4.5) one must in the Lorentz-gauge, in principle, also include operators with ghost-fields. Therefore a result like [7]

$$a_o^n = C_2(R)[1 - \frac{2}{n(n+1)} + 4 \sum_{k=2}^{n} \frac{1}{k}] \tag{4.6}$$

in the SU(3)-octet part with the solution for C^n

$$C^n_{octet}(g,\log q^2/\mu^2) \to const \cdot C^n_{octet}(0,0)[\log q^2/\mu^2]^{-a_o^n/2b_o} \tag{4.7}$$

although it seems to be gauge invariant, must be checked to be independent of "ghost-mixing". As verified by Kainz and Schweda [26] in the ghostfree gauge (4.3) for an element of γ_{ij} which contributes to the operator mixing in the SU(3)-singlet case (for n = 2) this is indeed true [27]. The same may be expected for the other elements of γ_{ij} in the singlet case where mixing is important, but certainly not in other problems [28].

The logarithmic increase with n in (4.6)

$$a_o^n \underset{n\to\infty}{\simeq} \log n \tag{4.8}$$

points toward increasing violation of scaling in C^n of (4.7) and hence also in the leading terms in q^2 of E_n according to (2.16). It is amusing that this seems to be the case experimentally. Even some decrease with q^2 as implied by (4.7) or - for the NAF case - by (3.37) seems to be visible in a tentative analysis of the experiments with respect to the moments E_n [9].

5. CONCLUSIONS

Let us summarize first some positive features of gauge theories of strong interactions:

a) Although not compatible with full scaling, gauge theories are the only field theories to make a small violation of scaling understandable for small momenta of the structure functions.

b) If $\beta(g)$ has an ultraviolet stable g_{UV} at small values or at $g = 0$ the g_{IR} may lie at very large values. The "effective coupling" for low energies is then large, quarks etc. are in an "asymptotic prison" at low energies But at high energies their coupling is too small to make a production possible. This "explains" why they have not been observed so far.

c) The internal and angular-momentum symmetries of E_n are asymptotically the symmetries of a free field theory. This can be used as an argument why sum rules derived from free quarks or free partons continue to hold, as well as e.g. $F_L \simeq 0$.

d) The symmetry group of the gauge fields can be made to commute with the usual symmetries of strong interactions. This is in agreement with the idea of coloured quarks introduced for other reasons originally [25]. The gauge group is then coupled to "colour". Moreover such a scheme does not lead to obvious inconsistencies (like parity violating corrections to order α) if the hadrons are assumed to interact with gauge fields of unified weak and electromagnetic interactions [29].

The negative features include:

a') It seems almost impossible to find convincing

arguments for a specific model for hadronic interactions. Note the rather similar situation for gauge theories in weak and electromagnetic interactions!

One reason for this fact certainly is that all scaling and light cone ideas for current amplitudes are based so far on one single experiment, deep inelastic electron-nucleon scattering with data on even the so similar neutrino-nucleon scattering becoming available only very slowly.

b') So far we have assumed massless vector fields (cf. (4.1)). If one tries to implement massive fields via the Higgs-mechanism of spontaneous symmetry breaking [4] also here, it turns out that in order to have AF theories - and also NAF theories - it becomes impossible to make all vector fields massive [7]. Therefore one either accepts forces with infinite range and some unknown mechanism to straighten out the horrible IR-catastrophy of nonabelian massless gauge-theories or one relies on more or less "pen-waving" arguments using a dynamical mechanism for spontaneous breaking of the symmetry [30] to provide masses for the vector particles.

REFERENCES

1. C. N. Yang, R. L. Mills, Phys. Rev. 96, 191 (1956).
2. L. D. Faddeev, V. Popov, Phys. Letters 25B, 29 (1967).
3. G.'t Hooft, Nucl. Phys. B35, 167 (1971).
4. P. W. Higgs, Phys. Letters 12, 132 (1964); Phys. Rev. Lett. 13, 5o8 (1964); Phys. Rev. 145, 1156 (1966).

F. Englert, R. Brout, Phys. Rev. Letters 13, 321 (1964);
G. S. Guralnik, C. R. Hagen, T. W. B. Kibble,ibid.13, 585 (1964).

5. S. Weinberg, Phys. Rev. Letters 19, 1264 (1967), 27, 1688 (1971);
A. Salam, in "Elementary Particle Physics" (ed. N. Svartholm, Almquist and Wiksells, Stockholm, 1968), p. 367.

6. D. J. Gross, F. Wilczek, Phys. Rev. Letters 30, 1343 (1973);
N. D. Politzer, Phys. Rev. Letters 30, 1346 (1973).

7. D. J. Gross, F. Wilczek, Phys. Rev. D8, 3633 (1973) and "Asymptotically Free Gauge Theories II" prep. Princeton Univ.

8. S. Weinberg, Phys. Rev. Letters 31, 494 (1973).

9. cf. E. Bloom, SLAC-PUB 1319 (Report at Symposium on Electron and Photon Interactions at High Energies, Bonn, August 1973).

10. R. P. Feynman, unpublished and Phys. Rev. Letters 23, 1415 (1964).

11. J. D. Bjorken, Phys. Rev. 179, 1547 (1964).

12. Y. Frishman, Phys. Rev. Letters 25, 966 (1970);
R. A. Brandt, G. Preparata, Nucl. Phys. B27, 541 (1971).

13. For more general situations with light cone dominance cf. also W. Kummer, Nucl. Phys. B42, 141 (1972).

14. For electromagnetic currents we may disregard problems related to compatibility with equal time commutators, cf. W. Kummer, Phys. Rev. D6, 1670 (1972).

15. One usually excludes by means of perturbation theoretic arguments operators with odd twist like $\bar{\psi}\psi$, $\bar{\psi}\gamma_5\psi$, $\bar{\psi}\sigma_{\mu\nu}\psi$, whereas $\bar{\psi}\gamma_\mu\gamma_5\psi$ is absent in our parity conserving situation.

16. Inserting j_μ j_ν into a T-product, $z^2-i\varepsilon z_o$ must be replaced by $z^2-i\varepsilon$.

17. N. Christ, B. Hasslacher, A. H. Mueller, Phys. Rev. D6, 3543 (1972).

18. cf. e.g. N. N. Bogoliubov, D. V. Shirkov, "Introduction to the Theory of Quantized Fields", Interscience New York-London 1959, §50.

19. C. G. Callan, Phys. Rev. D2, 1541 (1970) and D5, 3202 (1972);
K. Symanzik, Commun. Math. Phys. 18, 227 (1970) and 23, 49 (1972).

20. "Amputation" means dropping the propagators at each external line.

21. S. Weinberg, Phys. Rev. 118, 838 (1960).

22. M. Gell-Mann, F. Low, Phys. Rev. 95, 1300 (1954).

23. E. g. the anomalous dimension $2\gamma_J = \gamma' + \gamma''$ for the two current operators needed in the application of (3.26) to deep-inelastic lepton-nucleon scattering vanishes. This can be seen as follows: Imagine j_μ to be coupled to an external field and inserted between one-particle states. The only renormalization effects originate then from the wave-function renormalization ("Z_1" in QED) and from the proper vertex renormalization ("Z_2^{-1}" in QED). Current conservation yields the Ward-identity ($Z_1/Z_2 = 1$) so that no renormalization is left over ($Z = 1$).

24. V. N. Gribov, L. N. Lipatov, Phys. Letters 37B, 78 (1971);
S. Coleman, D. J. Gross, Phys. Rev. Letters 31, 851 (1973).

25. M. Y. Han, Y. Nambu, Phys. Rev. 139, 1006 B(1965);
H. Fritzsch, M. Gell-Mann, Proc. of the Int. Conf. on High Energy Physics, NAL 1972.

26. W. Kainz, M. Schweda, unpublished.

27. A similar verification is claimed in ref. [7]. The relevant calculation has been performed, however, there for the inconsistent special case $n^2 = 0$, $(n.k) = 0$, where k is some external four-momentum. Nevertheless the result (4.6) as quoted from ref. [7] turns out to be correct for $n^2 \neq 0$ as well [26].

28. W. Kainz, W. Kummer, M. Schweda, in preparation.

29. S. Weinberg, Phys. Rev. Letters 31, 494 (1973).

30. S. Coleman, S. Weinberg, Phys. Rev. D7, 1888 (1973).

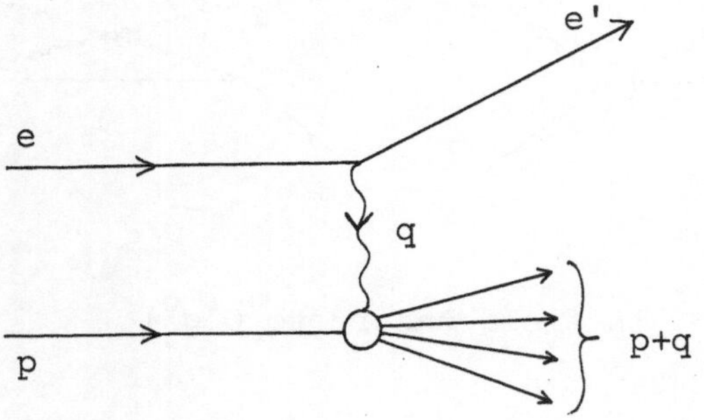

Fig. 1: Graph for deep inelastic e-p scattering

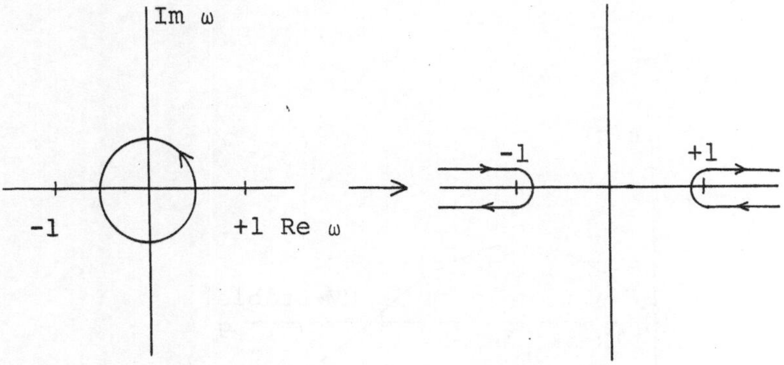

Fig. 2: Change of complex integration in eq. (2.15)

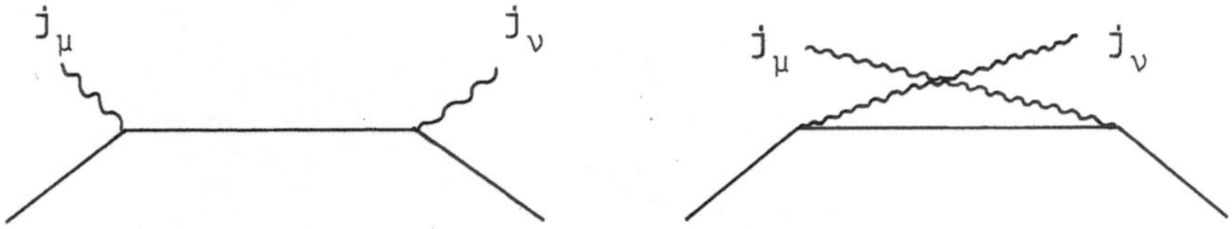

Fig. 3: $T_{\mu\nu}$ in field theory to $O(g^0)$

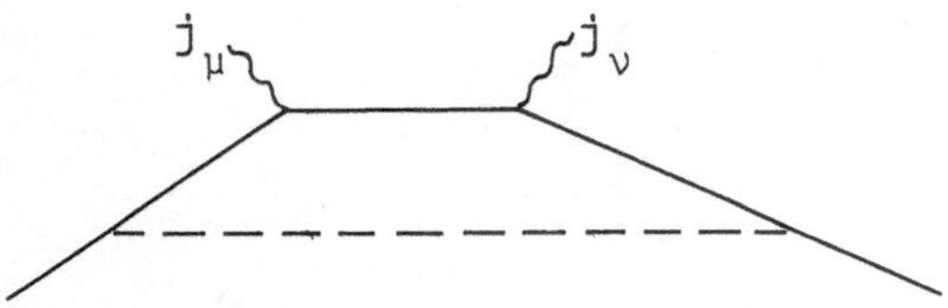

Fig. 4: Typical contribution to $T_{\mu\nu}$ to $O(g^2)$

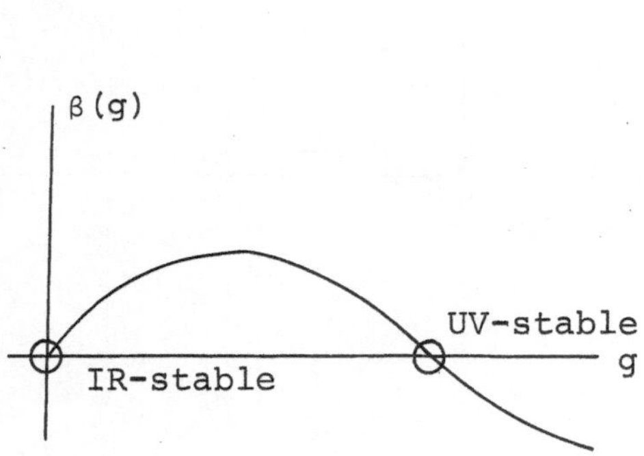

Fig. 5: UV and IR stable points in zeros of $T_{\mu\nu}$

Acta Physica Austriaca, Suppl. XIII, 395–445 (1974)

THE MELOSH TRANSFORMATION AND THE PRYCE-TANI-FOLDY-WOUTHUYSEN TRANSFORMATION

by

J. S. BELL

CERN, Geneva*)

1. INTRODUCTION

These lectures aim to give an account of "the Melosh transformation" and some of the main ideas leading up to it. The work of Melosh seemed to give theoretical support to suggestions, about the manner of breaking of relativistic SU(6) symmetry, which had already been advanced on empirical grounds. These phenomenological considerations will not be presented here. Attention will be focused on theoretical - in fact largely kinematical - considerations. The tale of relativistic SU(6) is a long, and often sad, one. To keep recent developments in perspective a particular aim here is to spell out the relation of the ideas of Melosh to older ideas based on the Foldy-Wouthuysen

*) Lecture given at XIII. Internationale Universitätswochen für Kernphysik, Schladming, Austria, February 4-15, 1974.

description of spin for relativistic particles. At the point of departure, the case of free quarks, the symmetry operators proposed by Melosh are actually identical with a subset of those of that earlier proposal.

For discussion of these matters I am indebted to many colleagues at Schladming and at CERN. In particular I have profited greatly from the expert guidance in this field of J. Weyers and A. Hey; Dr. Hey also read the manuscript and pointed out several obscurities and mistakes; I thank them both warmly.

1.1 Pryce [2]-Tani [3]-Foldy-Wouthuysen [4] transformation

Consider the Hamiltonian of the relativistic quark model

$$H = H_o + \ldots$$

where H_o is the free SU(3)-symmetric part

$$H_o = \int d^4x \; \delta(t) \; \psi^+(\beta m + \vec{\alpha}\cdot\vec{p})\psi \tag{1}$$

with $\vec{p} \equiv -i\partial/\partial\vec{x}$. In the no-interaction case ψ is a superposition

$$\psi = \sum_s \int \frac{d^3\vec{p}}{(2\pi)^3|2p_o|}\{a(p,s)u(p,s)e^{ipx}+b^+(p,s)u(-p,-s)e^{-ipx}\} \tag{2}$$

of solutions of the Dirac equation

$$(m + i\gamma p)u(p,s) = 0$$

Such solutions can be obtained by Lorentz boost from zero-momentum solutions w

$$u = e^{\frac{1}{2}\vec{\omega}\cdot\vec{\alpha}}\, w \qquad\qquad \beta w = \pm\, w$$

$$p_o = \pm\cosh\,\omega \qquad \vec{p} = \pm\,\vec{\omega}\;\;\omega^{-1}\,\sinh\omega \qquad \omega = |\vec{\omega}|$$

or

$$u = \{\cosh\tfrac{\omega}{2} + \omega^{-1}\;\vec{\omega}\cdot\vec{\alpha}\;\sinh\,\tfrac{\omega}{2}\}\;w$$

$$= \sqrt{\frac{|p_o|+m}{2m}}\{1 - \frac{i\vec{\gamma}\cdot\vec{p}}{|p_o|+m}\}\;\;w \quad .$$

The boost preserves the covariant norm, $\bar{w}w = \bar{u}u$. To obtain instead $w^+w = u^+u$, and so a unitary transformation, we renormalize w, so that

$$u = \frac{|p_o|+m\; - \;i\vec{\gamma}\cdot\vec{p}}{\sqrt{(|p_o|+m)^2 + \vec{p}^2}}\;\;w. \qquad (3)$$

This is then the unitary FW transformation:

$$u = e^{-iS(\vec{p})}w \qquad\qquad w = e^{iS(\vec{p})}u$$

$$S(\vec{p}) = \arctan\frac{\vec{\gamma}\cdot\vec{p}}{m+|p_o|} = \frac{1}{2}\,\arctan\frac{\vec{\gamma}\cdot\vec{p}}{m} \qquad (4)$$

$$|p_o| = \sqrt{m^2+\vec{p}^2} \quad .$$

Since for each of four linearly independent w's

$$(m\beta+\vec{\alpha}\cdot\vec{p})e^{-iS(\vec{p})}w = \pm\sqrt{m^2+\vec{p}^2}\; e^{-iS(\vec{p})}w = e^{-iS(\vec{p})}\sqrt{m^2+\vec{p}^2}\,\beta w$$

we have

$$e^{iS(\vec{p})}\;(m\beta+\vec{\alpha}\cdot\vec{p})e^{-iS(\vec{p})} = \sqrt{m^2+\vec{p}^2}\;\beta .$$

So in terms of the new field operator

$$\phi = e^{iS(\vec{p})}\psi \tag{5}$$

$$= \sum_s \int \frac{d^3\vec{p}}{(2\pi)^3|2p_o|}\{a(p,s)w(p,s)e^{ipx} + b^+(p,s)w(-p,-s)e^{-ipx}\} \tag{6}$$

we have

$$H_o = \int d^4x\;\delta(t)\;\phi^+\beta\sqrt{m^2+\vec{p}^2}\;\phi \tag{7}$$

and the non-vanishing equal-time anticommutator is again

$$\delta(t-t')\{\phi(x),\;\phi^+(x')\} = \delta(t-t')\{\psi(x),\psi^+(x')\} = \delta^4(x-x'). \tag{8}$$

1.2 U(6) x U(6) x O(3) symmetry

The free Hamiltonian

$$H_o = \int d^4x\;\delta(t)\;\phi^+\beta\sqrt{m^2+\vec{p}^2}\;\phi$$

is invariant under infinitesimal transformations

$$\delta\phi = -i\Lambda\ \phi \qquad (9)$$

where Λ is any constant Hermitian matrix (in Dirac- and unitary-spin indices) which commutes with β:

$$[\Lambda,\beta] = 0$$

i.e.

$$\Lambda = \frac{1}{2}(1 \pm \beta)(1,\vec{\sigma},\lambda,\lambda\vec{\sigma}) \qquad (10)$$

where β and $\vec{\sigma}$ are Dirac-matrices (conventions listed in Appendix) and the λ are Gell-Mann's 3 x 3 SU(3) matrices. This set of matrices forms a U(6) x U(6) algebra.

Of course we also have ordinary rotation invariance, under

$$\delta\phi = -i\vec{\omega}\cdot(\vec{r}\times\vec{p} + \frac{1}{2}\vec{\sigma})\ \phi\ . \qquad (11)$$

The spin part of this is already contained in U(6) x U(6), so we have a separate O(3) symmetry under purely orbital rotations:

$$\delta\phi = -i\ \vec{\omega}\cdot(\vec{r}\times\vec{p})\ \phi\ . \qquad (12)$$

So we have exhibited, for the relativistic quark model, a

U(6) x U(6) x O(3)

symmetry for the free part of the Hamiltonian. We might then suppose that the interactions are such that this symmetry remains fairly good, for the hypothesis of such an approximate symmetry [5] has had very considerable experimental success in classifying observed particles and resonances. We might even wish to consider, as a starting point, a model with interactions and bound states in which the above symmetry is exact. It seems that such notions, based on the FW description of relativistic particles, were among the very earliest conceptions of how to put SU(6) into the context of relativistic field theory [6,7].

1.3 Generators and charges

At time t = 0

$$\delta\phi = - i \Lambda \phi$$

is equivalent to

$$\delta\phi = [\phi, -iW(\Lambda)] \tag{13}$$

with

$$W(\Lambda) = \int d^4x\delta(t)\phi^+\Lambda\phi \tag{14}$$

(in virtue of the equal time anticommutation relations of the ϕ's); moreover the W's have the same algebra as the Λ's:

$$[W(\Lambda_1), W(\Lambda_2)] = W([\Lambda_1,\Lambda_2])$$

If we have a symmetry the W's commute with the Hamiltonian

$$[W(\Lambda)\ ,\ H] = 0\ .$$

Of more direct interest than the generators W are related "charges"

$$F(\Lambda) = \int d^4x\ \delta(t)\ \psi^+\ \Lambda\ \psi\ . \tag{15}$$

The particular interest of these is that some of the densities $\psi^+\Lambda\psi$ are, in the quark model, just those involved in weak and electromagnetic interactions. The quantities F, in virtue of the equal time anticommutators of the ψ's, have again the same algebra as the W's and Λ's:

$$[F(\Lambda_1),\ F(\Lambda_2)] = F([\Lambda_1,\Lambda_2])\ .$$

They are in fact related to the W's by unitary transformation:

$$\phi = e^{iS(\vec{p})}\ \psi = V\ \psi\ V^{-1} \tag{16}$$

where

$$V^{-1} = \exp\ i \int d^4x\ \delta(t)\ \psi^+\ S(\vec{p})\ \psi \tag{17}$$

so that

$$W(\Lambda) = V\ F(\Lambda)\ V^{-1} \tag{18}$$

1.4 Electromagnetic current

Suppose that the electromagnetic current is just

$$J_\mu = \bar{\psi}\, i\, \gamma_\mu\, Q\, \psi \tag{19}$$

where

$$Q = \begin{pmatrix} 2/3 & & \\ & -1/3 & \\ & & -1/3 \end{pmatrix}$$

is the quark charge. Write the FW transformation as

$$\begin{aligned} \psi &= (a - i\vec{\gamma}\cdot\vec{p}\ b)\ \phi \\ \psi^+ &= \phi^+(a' + i\vec{\gamma}\cdot\vec{p}'\ b') \end{aligned} \tag{20}$$

where

$$\vec{p}\ \phi = \frac{1}{i}\frac{\partial}{\partial\vec{x}}\ \phi \qquad\qquad \phi^+\ \vec{p}' = -\frac{1}{i}\frac{\partial}{\partial\vec{x}}\ \phi^+$$

and (a,b) are certain functions of $\vec{p}^2$, (a',b') the same functions of $\vec{p}'^2$. Then

$$\vec{J} = \phi^+\beta(a'b\vec{p} + ab'\vec{p}' + i\vec{\sigma}x[ab'\vec{p}' - a'b\vec{p}])Q\phi + \text{odd terms}\ . \tag{21}$$

These "odd terms" involve matrix elements that anti-commute with β; they transform therefore as $(\bar{6},6)$ and $(6,\bar{6})$ under U(6) x U(6).

1.5 Magnetic moments of (56,1) L = O baryons

Consider the lowest baryon multiplet, supposed to transform as (56,1) L = O under U(6) x U(6) x O(3). Denote such a particle with momentum $\vec{P}$ by

$$|\vec{P}, m, n>$$

where m is the eigenvalue of W $(\sigma_3/2)$, and n specifies the transformation character under ordinary SU(3). Consider matrix elements

$$<\vec{P}',m,n|\vec{J}(0)|\vec{P},m,n> .$$

The "odd" terms in $\vec{J}$ cannot contribute here, because under U(6) x U(6) they go as $(6,\bar{6})$ and $(\bar{6},6)$. In the other terms the spin-dependent part has precisely the same U(6) x U(6) character, $\vec{\sigma}Q$, as does the magnetic moment operator in the nonrelativistic quark model. So just as that model gives

$$\frac{\mu\ (\text{neutron})}{\mu\ (\text{proton})} = -\frac{2}{3} \qquad (22)$$

we have here, for example,

$$\frac{[<\vec{P}',m,\text{neutron}|J_2(0)|\vec{P},m,\text{neutron}>]^{m=+1/2}_{m=-1/2}}{[<\vec{P}',m,\text{proton}\ |J_2(0)|\vec{P},m,\text{proton}\ >]^{m=+1/2}_{m=-1/2}} = -\frac{2}{3} . \qquad (23)$$

But now, before drawing conclusions about magnetic moments, we have to relate the way in which spin states are labelled here, for moving particles, to more familiar ways.

There is no problem for particles at rest. Because we are concerned with O(3) singlets, W $(\vec{\sigma}/2)$ is equivalent to total angular moment $\vec{J}$, and m has the usual meaning. We can obtain moving particles, with momentum

$$\vec{P} = (\vec{\omega}/|\vec{\omega}|)\ M \sinh |\vec{\omega}|$$

by Lorentz boost

$$|0,m,n> \rightarrow e^{-i\vec{\omega}\cdot\vec{B}}|\ 0,m,n > \ .$$

Since the SU(3) character is supposed Lorentz invariant, there can at worst be a confusion of spin indices:

$$e^{-i\vec{\omega}\cdot\vec{B}}|0,m,n> \quad = \quad |\vec{P},m',n>\ C_{m'm}(\vec{P})$$

In the case of spin-1/2 particles, rotation invariance dictates the form

$$C_{m'm} = A(\vec{P}^2)\,\delta_{m'm} \quad + \quad B(\vec{P}^2)\ \vec{\sigma}_{m'm} \cdot \vec{P}\ .$$

The last term can be excluded by assuming inversion symmetry, and we can normalize so that A = 1. Then the states of given m are obtained by rotation-free boost of the corresponding rest states.

For a given spin-1/2 particle, Lorentz invariance and current conservation dictate the form

$$<\vec{P}',m'|J_\mu(0)|\vec{P},m> = \bar{u}'(f_1 i\gamma_\mu - f_2\sigma_{\mu\nu}iq_\nu/2M)u$$

$$= \bar{u}'([f_1+f_2]i\gamma_\mu - f_2\frac{P_\mu+P'_\mu}{2M})u \tag{24}$$

where f_1 and f_2 are functions of q^2; $f_1(0)$ gives the charge, and

$$\mu = [f_1(0) + f_2(0)]/2M$$

the total magnetic moment. Substitute

$$u = \begin{pmatrix} \sqrt{M+P_0}\,\phi \\ \dfrac{\vec{\sigma}\cdot\vec{P}}{\sqrt{M+P_0}}\,\phi \end{pmatrix} \qquad u' = \begin{pmatrix} \sqrt{M+P'_0}\,\phi' \\ \dfrac{\vec{\sigma}\cdot\vec{P}'}{\sqrt{M+P'_0}}\,\phi' \end{pmatrix}$$

obtained by boosting from rest

$$\binom{\phi}{0} \qquad \text{and} \qquad \binom{\phi'}{0}$$

where we use the old representation of the Appendix.

Then

$$<\vec{P}',m'|J_\mu(0)|\vec{P},m> =$$

$$\phi'^*[(AB'\vec{P}' + A'B\vec{P} + i\vec{\sigma}x(AB'\vec{P}' - A'B\vec{P}))(f_1 + f_2)$$

$$- \frac{\vec{P}+\vec{P}'}{2M}(AA'-BB'\vec{P}\cdot\vec{P}'-BB'\ i\vec{\sigma}\cdot\vec{P}X\vec{P}')f_2]\phi \tag{25}$$

where A(A'), B(B') are functions of $\vec{p}(\vec{p}')$ equal to $\sqrt{m}$ and $1/\sqrt{m}$ for zero momentum, and

$$[\langle\vec{P}';m|J_2(0)|\vec{P},m\rangle]\ _{m=-1/2}^{m=+1/2} =$$

$$2i(f_1+f_2)(AB'\ P_1' - A'BP_1) + 2if_2\ \frac{P_2+P_2'}{2M}(P_1P_2'-P_2P_1')BB'. \tag{26}$$

The last term here is relatively small for small momenta; if we ignore it, then comparison of (26) and (23) gives the desirable result (22). But if we take seriously also the last term in (26), then (22) requires, for all $q^2 \neq 0$

$$\frac{(f_1+f_2)_{\text{neutron}}}{(f_1+f_2)_{\text{proton}}} = \frac{f_{2\ \text{neutron}}}{f_{2\ \text{proton}}} = -\frac{2}{3}$$

which implies also

$$(f_{1\ \text{neutron}}/f_{1\ \text{proton}}) = -2/3$$

which is absurd, assuming just a little analyticity, since SU(3) alone gives

$$f_1(0)_{\text{neutron}} = 0 \qquad f_1(0)_{\text{proton}} = 1 .$$

Such trouble, with this kind of exact relativistic SU(6), was found at a very early stage. Thus Jordan [7] found that no scattering could occur. The root of the problem is that the FW transformation introduces a preferred frame of reference, that in which the particles are deboosted to rest. There is then a different U(6) x U(6) x O(3) symmetry for every choice of this frame, and Lorentz invariance requires that all are equally valid. While this is okay for free quarks, it is just too much symmetry to allow anything interesting to happen. One can reduce the symmetry by supposing that the U(6) x U(6) x O(3) defined above works only in the centre-of-mass system of the particles involved. But then there is trouble with locality. A remote particle of high energy can be important, because it contributes to fixing the centre-of-mass system, for local dynamics. Finally, however, one can restrict the symmetry to single (of course composite) particles at rest. This is enough to give the observed multiplet structure. If we insist strictly on particles at rest we cannot discuss magnetic moments, because they parametrize momentum transfer dependence of matrix elements. If we suppose that the symmetry remains good for slowly moving particles, we retain the nice result (23) for magnetic moments. If we admit in advance the possible failure of the symmetry for fast particles, we need not worry about the awkward last term in (26), which is of higher order in momenta. But the theoretical basis of such a position would be obscure.

2. CHARGES AND SYMMETRIES

Although the charges

$$F(\Lambda) = \int d^4x \ \delta(t) \ \psi^+ \ \Lambda \ \psi$$

do not commute with the free part of the Hamiltonian, there has been a sustained attempt, led by Gell-Mann, to use them somehow as symmetry operators.

2.1 Little leakage?

An early proposal was that the F's might be approximate symmetries of one-particle states in that for any one-particle state $|1\rangle$, $F|1\rangle$ is again mainly a one-particle state. This would be enough to impose an approximate multiplet structure on the one-particle states. However, Coleman [8] remarked that a necessary condition for little leakage from one-particle states would be that the F's do not excite the vacuum (for a simplified account of his reasoning see Bell [7]). The essential point is that one-particle states are mainly vacuum, and if the vacuum is transformed into something else, homogeneously, then because of its infinite extent the norm of the unwanted states will be infinite. Now in the case of non-interacting quarks one readily verifies that the F's do excite the vacuum, converting it into an infinite sea of particle-antiparticle pairs - except for those F's, to be referred to later as the "usual exceptions", which are related to the integrals of the time components of conserved vector currents, for which Λ is the unit matrix with respect to Dirac indices. Could this excitation of the vacuum be suppressed somehow by interactions? If $F|0\rangle$ were again vacuum (or zero), then since $|0\rangle$ is an eigenstate also of the Hamiltonian H,

$$[F, H]\ |0\rangle = 0 .$$

But again Coleman [8] remarked that in ordinary relativistic field theories

$$[F, H] = 0 \qquad (27)$$

would then follow, and the F's would have to be accurate symmetries in the conventional way; the essential point is that all matrix elements of [F,H] are related by crossing symmetry to those producing pairs from the vacuum.

Although any failure of (27) implies infinite leakage, this alone does not utterly condemn the F's as approximate symmetries. It means rather that leakage, in the sense of the norm of the unwanted many-body states which the F's generate from single-particle states, is a grossly exaggerated measure of symmetry breaking.

2.2 Easy saturation?

This particular embarrassment, with vacuum break-up, is avoided in the "easy saturation" philosophy. A commutator gives a sum rule:

$$\sum_n \langle a|F(\Lambda_1)|n\rangle\langle n|F(\Lambda_2)|b\rangle - (1 \leftrightarrow 2) = \langle a|F([\Lambda_1,\Lambda_2])|b\rangle. \qquad (28)$$

To each of the two terms on the left there will indeed be contributions, because of vacuum break-up, from states n with infinitely many particle-antiparticle pairs. But these contributions cancel when the difference of the two terms is formed, as is implied by the zero vacuum expectation value of $F([\Lambda_1,\Lambda_2])$. Then it can be conjectured that the sum rule is easily saturated in that it remains true in good approximation when the n-summation is restricted to a few states. In particular the hypothesis can be entertained that with one-particle states a and b, n can be restricted to one-particle states This dictates a multiplet structure for such states.

2.3 Infinite momentum frame

The operators $F(\Lambda)$ involve integration over all space at a particular time. This is not a Lorentz invariant construction. As a result the information contained in the above sum rules is different according to which Lorentz frame is used. Fubini and Furlan [9] argued that it is of particular interest to consider a frame in which the particles are moving very fast. In general the sum rule gives a non-linear relation between various form factors with different values of q^2. Suppose we have particles of momentum $\vec{P}$ (which has to be the same for a, b, n because the F's are uniform space integrals) then

$$q^2(n \leftarrow a) = - \left(\sqrt{\vec{P}^2+M_n^2} - \sqrt{\vec{P}^2+M_a^2}\right)^2 \quad .$$

For $P \rightarrow \infty$,

$$q^2(n \leftarrow a) \approx - \left(\frac{M_n^2-M_a^2}{2|\vec{P}|}\right)^2 \rightarrow 0 \quad .$$

Then we have relations between form factors all at $q^2 = 0$, which is simpler and of more immediate interest. Fubini and Furlan also gave arguments that the sum rules might converge most quickly, i.e. might be most readily saturated, in this limit. Combination of these ideas with PCAC led to the Adler-Weisberger relation. The $P \to \infty$ concept, in one form or another, has subsequently been of great importance.

2.4 Null-plane charges [10]

States of large P can be obtained by boosting states of fixed P:

$$|a\rangle \to e^{i\omega B_3} |a\rangle$$

boosting here in the 3-direction with rapidity ω. The boosts may be collected into the operators:

$$e^{i\omega B_3} \psi(x)\, e^{-i\omega B_3} = e^{\frac{1}{2}\omega\alpha_3} \psi(x')$$

$$\begin{array}{lll} x = x' & z = z' \cosh\omega + t' \sinh\omega & \qquad (29) \\ y = y' & t = t' \cosh\omega + z' \sinh\omega \,. & \end{array}$$

So instead of using states with large P we may use the modified charges, with large ω,

$$e^{i\omega B_3} F(\Lambda) e^{-i\omega B_3} = \int d^4x\delta(t)\psi^+(x')e^{\frac{1}{2}\omega\alpha_3} \Lambda e^{\frac{1}{2}\omega\alpha_3}\psi(x')$$

$$= 2\int d^4x'\delta(t'+z'\tanh\omega)\psi^+(x')\Lambda'\psi(x)$$

with

$$\Lambda' = \left(\frac{1+\alpha_3 \tanh\frac{\omega}{2}}{2}\right) \frac{\Lambda 2 \cosh^2\frac{\omega}{2}}{\cosh\omega} \left(\frac{1+\alpha_3 \tanh\frac{\omega}{2}}{2}\right) .$$

So naively we have the limit

$$\lim_{\omega\to\infty} e^{i\omega B_3} F(\Lambda)\, e^{-i\omega B_3} = \hat{F}(\Lambda)$$

$$\hat{F}(\Lambda) = 2\int d^4x\, \delta(t+z)\psi^+(x)\frac{1+\alpha_3}{2}\Lambda\frac{1+\alpha_3}{2}\psi(x) . \qquad (30)$$

Now many of the matrices (10) anticommute with α_3, and so give zero when multiplied from both sides by $(1+\alpha_3)/2$. The remaining matrices commute with α_3, and form a sub-algebra

$$\Lambda = (1,\lambda)(1,\beta\sigma_1,\beta\sigma_2,\sigma_3) \qquad (31)$$

denoted by

$$U(6)_W$$

whose particular interest was first noted by Barnes, Carruthers and von Hippel [11] and by Lipkin and Meshkov [11]. Their good parts (in the terminology of Gell-Mann and Fubini)

$$\frac{1+\alpha_3}{2} \quad \Lambda \quad \frac{1+\alpha_3}{2}$$

satisfy the same algebra. So also do the corresponding charges

$$[\hat{F}(\Lambda_1),\ \hat{F}(\Lambda_2)] = \hat{F}([\Lambda_1,\ \Lambda_2]) \tag{32}$$

- if we identify limits of commutators with commutators of limits. We tentatively do so for the set (31) which give non-zero $\hat{F}$. To do so more generally leads to contradictions [10].

2.5 Null-plane fields

Let us adopt for the moment a representation of Dirac matrices in which α_3 is diagonal:

$$\alpha_3 = \begin{pmatrix} +1 & 0 \\ 0 & -1 \end{pmatrix}$$

(see Appendix for details). Let then

$$\psi_g\ ,\quad \hat{\Lambda}$$

denote a 2-spinor and a 2x2 matrix obtained by restricting ψ and Λ to their "good" ($\alpha_3 = +1$) components. Then (30) can be rewritten

$$\hat{F}(\Lambda) = 2\int d^4x\ \delta(t+z)\psi_g^+(x)\ \hat{\Lambda}\ \psi_g(x)\ . \tag{33}$$

The assumption of simple null-plane anticommutation relations

$$\begin{aligned} 2\ \delta(t+z)\ \{\psi_{gi}(0),\ \psi_{gj}^+(x)\} &= \delta^4(x)\delta_{ij} \\ 2\ \delta(t+z)\ \{\psi_{gi}(0),\ \psi_{gj}(x)\} &= 0 \end{aligned} \tag{34}$$

(and others obtained by translation and/or Hermitian conjugation of these) gives again (32).

A "derivation" of (34) can be made as follows. Start, for example, from the canonical equal time anticommutator

$$\delta(t)\ \{\psi(0),\ \psi^{+}(x)\} = \delta^{4}(x)\ .$$

Multiplying from the left and right, respectively, by $e^{i\omega B_3}$ and $e^{-i\omega B_3}$, and using (29)

$$\delta(t) e^{\frac{1}{2}\omega\alpha_3}\{\psi(0),\psi^{+}(x')\}e^{\frac{1}{2}\omega\alpha_3} = \delta^{4}(x) = \delta^{4}(x')$$

or

$$\frac{\delta(t'+z'\ \tanh\ \omega)}{\cosh\ \omega}\ e^{\frac{1}{2}\omega\alpha_3}\{\psi(0),\psi^{+}(x')\}e^{\frac{1}{2}\omega\alpha_3} = \delta^{4}(x')\,.$$

In the case of good components, $\alpha_3 = +1$, naive passage to the limit $\omega \to \infty$ gives the first member of (34).

What happens if we attempt the same procedure with bad components? With one bad and one good component we obtain

$$\begin{aligned} \delta(t'+z'\ \tanh\omega)\{\psi(0),\psi^{+}(x')\} &= \delta^{4}(x')\cosh\ \omega \\ \delta(t'+z')\{\psi(0),\psi^{+}(x')\} &\overset{?}{=} \delta^{4}(x')\ \infty. \end{aligned} \tag{35}$$

In the particular case of free fields the result (34) can be confirmed explicitly. But in the case of bad components one does not get anything like (35), and can trace in detail how it comes about in this case that the null plane is not simply the limit $\omega \to \infty$. Since the limiting procedure is not always reliable, it would of course be very desirable to have a more serious derivation of (34) - say in the case of regularized field theory. In this connection the most relevant works that I know are the various demonstrations, usually formal but in special cases more serious, that starting from (34) as an axiom the usual perturbation expansion of S-matrix elements can be obtained [12].

Because of (34) the null-plane charges induce the transformation of fields

$$\delta\psi_g(x) = - i[\psi_g(x), \hat{F}(\Lambda)] = - i \hat{\Lambda} \psi_g(x) \qquad (36)$$

- for all x if $\hat{F}$ is a constant of the motion, and in any case for x on the plane $t + z = 0$ entering into the construction (33) of $\hat{F}$.

2.6 The null-plane charges are constants of the free motion

- unlike the original equal time charges (with the usual exceptions). To see this expand the non-interacting field ψ in creation and absorption operators

$$\psi = \sum a u e^{ik_\mu x_\mu}$$

where the u's are 4-spinors and a absorbs or creates according to the sign of

$$k_o = - i k_4 = \pm \sqrt{m^2 + \vec{k}^2} \quad .$$

Then the null-plane charges on the plane $t + z = \tau$ contain terms of the type

$$u^*\Lambda u' \; a^+ a' \int d^4x \; \delta(t+z-\tau) e^{i(\vec{k}'-\vec{k})\cdot\vec{x} - i(k_o'-k_o)t}$$

$$= u^*\Lambda u' \; a^+ a' \int d^3\vec{x} \; e^{i(\vec{k}_\perp'-\vec{k}_\perp)\cdot\vec{x} + iz(k_3'+k_o-k_3-k_o) - i\tau(k_o'-k_o)} .$$

Suppose first that k_o and k_o' are of opposite sign, so that we are considering pair creation. Then $(k_3' + k_o')$ and $-(k_3 + k_o)$, which have the same signs, respectively, as k_o' and $-k_o$, have the same sign. Such terms vanish on integration over z. When k_o and k_o' have the same sign, survival of the integration requires

$$k_3' + k_o' = k_3 + k_o \qquad \vec{k}_\perp' = \vec{k}_\perp$$

which, given the relation between k_o and $\vec{k}$, requires also $k_o = k_o'$, - in the absence always of quark-mass-splitting. The energy non-conserving terms do not survive.

2.7 The null-plane charges do not excite the vacuum

We saw above how the pair creation terms were eliminated in forming the null-plane charges, in the case of no interaction. It is quite generally true that the null-plane charges do not excite the vacuum. Because of the replacement of $\delta(t)$ by $\delta(t + z)$ the null-plane charges, unlike the equal time charges, are not space-displacement invariant, and do not conserve the component

P_3 of total momentum. But $\delta(t + z)$ is invariant under a combination of space and time displacement, so that the combination

$$P_3 \quad + \quad P_o$$

of momentum and energy is now conserved. Since for states other than vacuum (excluding the possibility of zero mass)

$$P_o > |P_3| \qquad\qquad P_3 + P_o > 0$$

the null-plane charges cannot excite the vacuum. Considering the $\hat{F}$'s, rather than the F's, as candidates for symmetry operators, the main objection to the little leakage hypothesis no longer applies. We are not obliged to go to easy saturation.

Before considering the F's as symmetries it is convenient to discuss their Lorentz transformation.

2.8 Lorentz transformation of null-plane charges

The null-plane charges were introduced as the limits of equal time charges under infinite boost in the 3-direction. They are then invariant under further boosting in that direction:

$$e^{i\omega B_3}\, \hat{F}(\Lambda)\, e^{-i\omega B_3} = \hat{F}(\Lambda) \ .$$

They also transform trivially under rotation about the 3-axis: those involving Dirac matrices 1 and σ_3 are invariant, and those involving $\beta\sigma_1$ and $\beta\sigma_2$ transform into one another as components of 2-vectors.

Other Lorentz transformations are, in general, not so simple, because they do not leave invariant the null plane $t + z = 0$. However, there are certain combinations which <u>do</u>; their generators are

$$E_1 = B_1 + J_2 \qquad\qquad E_2 = B_2 - J_1 \tag{37}$$

where $\vec{J}$ is the total angular momentum and $\vec{B}$ the corresponding generator of Lorentz boosts. The changes of coordinates consequent on infinitesimal transformations of magnitude ω are

$$\begin{array}{lllll} B_1: & \Delta x = \omega t & \Delta y = 0 & \Delta z = 0 & \Delta t = \omega x \\ J_2: & \Delta x = \omega z & \Delta y = 0 & \Delta z = -\omega x & \Delta t = 0 \end{array}$$

so that for the combination E_1

$$\Delta x = \omega(t+z) \qquad \Delta y = 0 \qquad \Delta z = -\omega x \qquad \Delta(z+t) = 0.$$

So the plane $(t + z) = 0$ just transforms into itself with a shear. Similarly with E_2.

The corresponding infinitesimal changes in $\psi(x_\mu)$ (apart from $x_\mu \to x_\mu - \delta x_\mu$) are

$$\begin{array}{ll} B_1: & \Delta\psi = \frac{1}{2}\omega\alpha_1\,\psi \\ J_2: & \Delta\psi = -\frac{i}{2}\omega\sigma_2\,\psi \\ E_1: & \Delta\psi = \frac{1}{2}\omega(\alpha_1 - i\sigma_2)\,\psi\,. \end{array} \tag{38}$$

Now it is a fact (see Appendix) that

$$(1 + \alpha_3)(\alpha_1 - i\sigma_2) = 0 \ .$$

So apart from the change of arguments discussed above, the good components of ψ are invariant under E_1 and E_2.

It follows that the good null-plane charges are also invariant

$$[E_1, \hat{F}] = [E_2, \hat{F}] = [B_3, \hat{F}] = 0 \ . \qquad (39)$$

2.9 $U(6)_W$ of null-plane charges as a particle symmetry

Suppose now that single particles with a given value of

$$\vec{h} = (P_x, P_y, h = P_z + P_o)$$

(all of which are unaltered by the $\hat{F}$'s) fall into multiplets which give irreducible representations of the algebra $U(6)_W$ of good null-plane charges

$$\hat{F}(\Lambda) = 2\int d^4x \ \delta(t+z)\psi^+(x) \ \frac{1+\alpha_3}{2} \Lambda \frac{1+\alpha_3}{2} \ \psi(x)$$

$$\Lambda = (1,\lambda)(1,\beta\sigma_x,\beta\sigma_y,\sigma_z) \ .$$

Of course the total angular momentum $\vec{J}$ must also be a symmetry, and we can define another symmetry generator, an "orbital" angular momentum, as

$$L_3 = J_3 - \hat{F}\left(\tfrac{1}{2}\sigma_3\right). \tag{40}$$

For particles moving in the 3-direction the multiplets should be invariant also under L_3, generating a group O(2). Altogether we are concerned with the symmetry

$$U(6)_W \times O(2)\ .$$

The octet containing the nucleons and the decuplet containing $N^*(1238)$ are supposed to form a

$$56\quad L_3 = 0$$

under this group. Let

$$|\vec{h},\, m,\, n\rangle$$

denote a state belonging to this multiplet, where m is the eigenvalue of $\hat{F}(\sigma_3/2)$, and n is collectively all the other quantum numbers (8 or 10, I, I_3, Y). Because the $\hat{F}$'s commute with the boosts B_3 and $\vec{E}_\perp$, we can choose phase and normalization conventions so that

$$|\vec{h},\, m,\, n\rangle = e^{-i\omega_1 E_1}\, e^{-i\omega_2 E_2}\, e^{-i\omega_3 B_3} |\vec{h}_o, m, n\rangle \tag{41}$$

where

$$\vec{h}_o = (0,\, 0,\, M)$$

M being nucleon mass, and

$$h(\equiv h_3) = e^{\omega_3 M} \qquad h_2 = \omega_2 h \qquad h_1 = \omega_1 h \ . \qquad (42)$$

2.9.1 Magnetic moments in the $56 L_3 = 0$

Consider, in the notation of the last section, the matrix element of the electromagnetic current

$$\langle \vec{h}', m', n' | J_\mu(0) | \vec{h}, m, n \rangle \ .$$

Consider, in particular, the good combination of components

$$J_3 + J_o = \psi^+(1 + \alpha_3)\ Q\ \psi \ . \qquad (43)$$

This is a W-spin singlet, i.e. commutes with the "spin" operators

$$\hat{F}(\beta\sigma_1) \qquad \hat{F}(\beta\sigma_2) \qquad \hat{F}(\sigma_3) \ .$$

The hypothesis that these are symmetries then requires the matrix element to be spin independent

$$\langle \vec{h}', m', n | J_3(0) + J_o(0) | \vec{h}, m, n \rangle = F(\vec{h}', \vec{h}, n)\,\delta_{m'm} \ . \qquad (44)$$

But we also have, for the nucleons of spin-1/2, the usual expression in terms of invariant form factors

$$\langle \vec{h}', m', n | J_\mu | \vec{h}, m, n \rangle = \bar{u}(\vec{h}', m')\left((f_1 + f_2)\, i\gamma_\mu - f_2 \frac{P_\mu + P'_\mu}{2M}\right) u(\vec{h}, m) \ . \qquad (45)$$

We have to be careful here to define the u's, not by the familiar rotation free Lorentz boosts, but by the sequence of special boosts in (41). The result is, in the good α_3-diagonal representation of the Appendix

$$u(\vec{h},m) = \frac{1}{\sqrt{h}} \begin{pmatrix} h\,\phi(m) \\ (M\sigma_3 - \vec{\sigma}_\perp \cdot \vec{P}_\perp)\,\phi(m) \end{pmatrix} \tag{46}$$

where ϕ is the ordinary Pauli 2-spinor for a particle at rest with $\sigma_3 = m$. Note that for particles at rest, in the case $L_3 = 0$, the index m is the eigenvalue both of J_3 and $\hat{F}$ $(\sigma_3/2)$. Then

$$\begin{aligned} \langle \vec{h}',m',n|J_3+J_0|\vec{h},m,n\rangle = {} & \delta_{m'm}\, 2\sqrt{h'h}(f_1+f_2) \\ & - f_2\frac{(h+h')^2}{2M\sqrt{hh'}}\delta_{m'm} - f_2\phi^*(m')\sigma_3\vec{\sigma}_\perp\cdot(\vec{P}_\perp' h-\vec{P}_\perp h')\phi(m)\,. \end{aligned} \tag{47}$$

Comparing (47) and (44)

$$f_2(q^2) = 0$$

except possibly at $q^2 = 0$, and there also assuming continuity. So in this scheme nucleons have zero anomalous magnetic moments [13].

Empirically, nucleons have large anomalous moments. So the conclusion is quite unwelcome. However, it should be noted that we have not here turned up any theoretical inconsistency, in the idea of exact $U(6)_W$ symmetry for bound states, of the kind found with the original relativistic SU(6) of Section 1.5. But it is very likely that such inconsistencies can be found.

3. MELOSH

The requirement of zero anomalous magnetic moments for nucleons, and other empirical failures, led to the conviction that the algebra in question, the $U(6)_W$ of good null-plane charges, was not a good symmetry. Various phenomenological mixing schemes were devised [14], and latterly the work of Melosh seemed to give some theoretical insight into the symmetry breaking.

3.1 First Melosh transformation

Melosh's first proposal [15] was directly inspired by the notion of null-plane charges as infinitely boosted limits of equal time charges

$$F(\Lambda) = \int d^4x \ \delta(t) \ \psi^+(x) \ \Lambda \ \psi(x) \ .$$

Now we noted that these F's were not symmetry operators even in the free field case, but that the related quantities

$$W(\Lambda) = \int d^4x \ \delta(t) \ \phi^+(x) \ \Lambda \ \phi(x)$$

did have at least this merit, where

$$\phi(x) = \frac{m + \sqrt{m^2 + \vec{p}^2} + i\vec{\gamma}\cdot\vec{p}}{\sqrt{(m + \sqrt{m^2+\vec{p}^2})^2 + \vec{p}^2}} \ \psi(x) \ .$$

What if we boosted the W's instead of the F's? The process is complicated by the appearance of γ_3, which anticommutes with α_3 and so mixes good and bad components, and of p_3, which is not invariant under boosts in the 3-direction. So Melosh considered instead

$$\phi'(x) = \frac{m+\sqrt{m^2+\vec{p}_\perp^2} + i\vec{\gamma}_\perp\cdot\vec{p}_\perp}{\sqrt{(m+\sqrt{m^2+\vec{p}_\perp^2})^2 + \vec{p}_\perp^2}} \ \psi(x) = e^{iS'}\psi(x) = V'\psi(x)V'^{-1}$$

$$S' = \arctan \frac{\vec{\gamma}_\perp\cdot\vec{p}_\perp}{m+\sqrt{m^2+\vec{p}_\perp^2}} = \frac{1}{2}\arctan\frac{\vec{\gamma}_\perp\cdot\vec{p}_\perp}{m} \tag{48}$$

$$V'^{-1} = \exp i \int d^4x \ \delta(t) \ \psi^+ \ S'(\vec{p}_\perp) \ \psi \ .$$

In terms of this the free Hamiltonian is

$$H_o = \int d^4x \ \delta(t) \ \phi'^+ \ \{\beta\sqrt{m^2+\vec{p}_\perp^2} + \alpha_3 p_3\} \ \phi \tag{49}$$

- i.e. α_1 and α_2 are eliminated, but not α_3. This Hamiltonian is not invariant under

$$\delta\phi' = - i \ \Lambda \ \phi'$$

except when

$$[\Lambda,\alpha_3] = [\Lambda,\beta] = 0 \ .$$

But this is just the restriction to the $U(6)_W$ subgroup of Λ's in which we are anyway now primarily interested. With Λ so restricted the quantities

$$W'(\Lambda) = \int d^4x \; \delta(t)\phi'^{+}(x)\Lambda \; \phi'(x) \tag{50}$$

are again constants of the free motion. Infinite boosting in the 3-direction then yields

$$\hat{W}'(\Lambda) = 2\int d^4x\delta(t+z)\phi'^{+}(x) \; \frac{1+\alpha_3}{2} \; \Lambda \; \frac{1+\alpha_3}{2} \; \phi'(x)$$

or in the good α_3-diagonal representation

$$\hat{W}'(\Lambda) = 2\int d^4x \; \delta(t+z) \; \phi_g'^{+} \; \hat{\Lambda} \; \phi_g'$$

$$= e^{iy'} \; \hat{F}(\Lambda) \; e^{-iy'} \tag{51}$$

$$y' = - 2\int d^4x \; \delta(t+z)\psi_g^{+} \; \arctan \frac{\vec{\gamma}_{\perp}\cdot\vec{p}_{\perp}}{m+\sqrt{m^2+\vec{p}_{\perp}^{\,2}}} \; \psi_g$$

Melosh's first proposal was that the $\hat{W}'$'s would be more plausible free-field prototype symmetry operators than the $\hat{F}$. But his proposal was severely criticized on grounds of lack of motivation [16]. Unlike the original equal time charges F, the null-plane charges $\hat{F}$ are already constants of the free motion - as are the Foldy-Wouthuysen transformed equal time charges W. By combining the FW transformation with the passage to the null plane we are redundantly combining two separately sufficient cures for the same disease, and the particular null-plane charges so generated are not necessarily healthier than others.

Melosh accepted this criticism [15], but felt that his transformation might prove to have other merits, concerned with rotation invariance [15,17]. Development of this idea led him in fact to a somewhat different transformation [17].

3.2 Rotation

The relation of ordinary rotation symmetry to the $U(6)_W$ algebra of null-plane charges is obscure, because of the definition of these quantities on a plane (t + z) = constant, which is symmetric under rotation about the 3-axis only, and because of the special role of the good $\alpha_3 = +1$ components of quark fields. This latter makes it useful to go to a representation of Dirac matrices in which α_3 is diagonal:

$$\alpha_3 = \begin{pmatrix} 1 & 0 \\ 0 & -1 \end{pmatrix} .$$

Such a "good" representation is set out in detail in the Appendix. Here we quote:

$$\alpha_1 = \begin{pmatrix} 0 & -\sigma_1 \\ -\sigma_1 & 0 \end{pmatrix} \qquad \alpha_2 = \begin{pmatrix} 0 & -\sigma_2 \\ -\sigma_2 & 0 \end{pmatrix} \qquad \beta = \begin{pmatrix} 0 & \sigma_3 \\ \sigma_3 & 0 \end{pmatrix}$$

$$\gamma_1 = i\begin{pmatrix} \sigma_3\sigma_1 & 0 \\ 0 & \sigma_3\sigma_1 \end{pmatrix} \quad \gamma_2 = i\begin{pmatrix} \sigma_3\sigma_2 & 0 \\ 0 & \sigma_3\sigma_2 \end{pmatrix} \quad \gamma_3 = i\begin{pmatrix} 0 & \sigma_3 \\ -\sigma_3 & 0 \end{pmatrix}$$

$$\sigma_1 = \begin{bmatrix} 0 & \sigma_1\sigma_3 \\ \sigma_3\sigma_1 & 0 \end{bmatrix} \qquad \sigma_2 = \begin{bmatrix} 0 & \sigma_2\sigma_3 \\ \sigma_3\sigma_2 & 0 \end{bmatrix} \qquad \sigma_3 = \begin{bmatrix} \sigma_3 & 0 \\ 0 & \sigma_3 \end{bmatrix} \tag{52}$$

In this representation the Dirac equation

$$(p_o - \vec{\alpha}\cdot\vec{p} - \beta m)\ \psi = 0$$

relates good and bad components:

$$\begin{aligned} (p_o - p_3)\psi_g &= (m\sigma_3 - \vec{\sigma}\cdot\vec{p}_\perp)\psi_b \\ (p_o + p_3)\psi_b &= (m\sigma_3 - \vec{\sigma}\cdot\vec{p}_\perp)\psi_g \quad . \end{aligned} \tag{53}$$

The FW transformed Dirac equation

$$(p_o - \beta\sqrt{m^2 + \vec{p}^2})\ \phi = 0 \tag{54}$$

yields

$$\begin{aligned} p_o\phi_g &= \sqrt{m^2 + \vec{p}^2}\ \sigma_3\ \phi_b \\ p_o\phi_b &= \sqrt{m^2 + \vec{p}^2}\ \sigma_3\ \phi_g \quad . \end{aligned} \tag{55}$$

Because the rotation matrices anticommute with α_3, the infinitesimal rotations

$$\begin{aligned} \delta\psi &= -\ i\vec{\omega}\cdot(\vec{r}\times\vec{p} + \tfrac{1}{2}\vec{\sigma})\ \psi \\ \delta\phi &= -\ i\vec{\omega}\cdot(\vec{r}\times\vec{p} + \tfrac{1}{2}\vec{\sigma})\ \phi \end{aligned} \tag{56}$$

mix good and bad components together. However, in the

particular case of free fields, regarding such fields as superpositions of plane waves each with definite $(p_o,\vec{p})$, the Dirac equation allows the elimination of the bad components:

$$\delta\psi_g = - i\vec{\omega} \cdot (\vec{r} \times \vec{p} + \frac{1}{2} \vec{\Sigma})\psi_g$$

$$\vec{\Sigma}_\perp = \vec{\sigma}_\perp \frac{m-\sigma_3\vec{\sigma}\cdot\vec{p}_\perp}{p_o+p_3}, \Sigma_3 = \sigma_3 \tag{57}$$

for the ordinary field ψ, and for the FW field

$$\delta\phi_g = - i\vec{\omega} \cdot (\vec{r} \times \vec{p} + \frac{1}{2} \vec{\Sigma}) \phi_g$$

$$\vec{\Sigma}_\perp = (p_o/|p_o|)\vec{\sigma}_\perp \qquad \Sigma_3 = \sigma_3 \quad . \tag{58}$$

Note that when interaction is introduced such formulae must in general acquire extra terms.

Because of the complication of (57) Melosh [18] argued that composite particles with definite spin could not at the same time belong to a definite representation of the $U(6)_W$ of null-plane charges.

We will illustrate his argument in terms of a particular particle π^+, and the "W-spin" subgroup of $U(6)_W$ generated by

$$\hat{F}(\tfrac{1}{2}\beta\sigma_1) \qquad \hat{F}(\tfrac{1}{2}\beta\sigma_2) \qquad \hat{F}(\tfrac{1}{2}\sigma_3)$$

- which are SU(3) scalars. Consider the matrix element

involving a π^+ at rest

$$F_{ij}(\vec{k}) = \int d^3\vec{x}\ e^{-i\vec{k}\cdot\vec{x}} \langle 0|p_i(\tfrac{\vec{x}}{2},0)\, n_j^+(-\tfrac{\vec{x}}{2},0)|\pi^+\rangle \qquad (59)$$

where p and n are p- and n-quark components of ψ and i,j are Dirac indices restricted to good values. In the good representation

$$\beta\sigma_1 = \begin{pmatrix} \sigma_1 & 0 \\ 0 & -\sigma_1 \end{pmatrix} \qquad \beta\sigma_2 = \begin{pmatrix} \sigma_2 & 0 \\ 0 & \sigma_2 \end{pmatrix} \qquad \sigma_3 = \begin{pmatrix} \sigma_3 & 0 \\ 0 & \sigma_3 \end{pmatrix} \qquad (60)$$

so that the W-spin operations are represented by (see Eq. (36))

$$\delta p_g = - i\vec{\omega}\cdot\vec{\sigma}p_g \qquad \delta n_g^+ = n_g^+\ i\vec{\omega}\cdot\vec{\sigma}$$

and

$$\delta F = - i[\ \vec{\omega}\cdot\vec{\sigma}, F\ ]$$

where F is a 2 x 2 matrix with elements F_{ij}. Now π^+ is supposed to be the third member of a W-spin triplet; it follows that

$$F_{ij} = f(\vec{k})\ (\sigma_3)_{ij} \qquad (61)$$

where σ_3 is as usual the 2 x 2 Pauli matrix and $f(\vec{k})$ is some function of $\vec{k}$.

Consider now the requirement that π^+ has zero ordinary spin, i.e. that $|\pi^+\rangle$ be invariant under ordinary rotations. Here we have the problem that interactions which can bind quark and antiquark into a pion will in general introduce extra terms into the rotation formula (57). However, following Melosh, we can consider in free field theory the possibility of constructing wave packets of free quark and antiquark with the desired transformation properties. So let $|\pi^+>$, in the above and in what follows, stand for such a wavepacket.

Rotational invariance about the 3-direction presents no problem, because Σ_3 in (57) is uncomplicated; it requires only cylindrical symmetry of $f(\vec{k})$. But consider rotation about the other axes, for example the x: from (57) (59) invariance requires

$$0 = \delta F = -i\{i \frac{\partial}{\partial k_2} k_3 - i\frac{\partial}{\partial k_3} k_2 + \frac{1}{2} \frac{m\sigma_1+\sigma_3 k_1+ik_2}{k_3+ \sqrt{m^2+\vec{k}^2}}\} F$$

$$+ iF\{\frac{1}{2} \frac{m\sigma_1+\sigma_3 k_1+ik_2}{k_3- \sqrt{m^2+\vec{k}^2}}\} \quad . \tag{62}$$

Substituting (61), invariance requires

$$k_1 f = k_2 f = k_3 f = 0 \; .$$

We would be obliged to suppose that the quarks also are at rest, which is not a very convincing model of a bound state.

If we play this game with ψ replaced everywhere by ϕ, the picture is simpler because of the simplicity of (58). Rotational invariance then requires

$$O = \delta F = -i\{i \frac{\partial}{\partial k_2} k_3 - i\frac{\partial}{\partial k_3} k_2 + \frac{1}{2}\sigma_1\} F$$
$$+ i F\{ - \frac{1}{2} \sigma_1\} \tag{63}$$

which is satisfied by

$$F = f(\vec{k}) \sigma_3 \tag{64}$$

with f spherically symmetrical.

3.3 Second Melosh transformation [1]

Melosh looked therefore for a transformation simplifying the rotation operation (57). In view of (58) we can guess that the result will be essentially the old FW transformation specialized to good components. In the FW transformation

$$\psi = \frac{m + |p_o| - i\vec{\gamma}\cdot\vec{p}}{\sqrt{(m + |p_o|)^2 + \vec{p}^2}} \phi \tag{65}$$

write

$$- i \gamma_3 p_3 = \alpha_3 p_3 \beta \tag{66}$$

and note that for free fields

$$\beta\phi = (p_o/|p_o|) \phi \tag{67}$$

Specializing then to good ($\alpha_3 = + 1$) components and noting that

$$|p_o| + p_3 p_o/|p_o| = |p_o + p_3| \tag{68}$$

we have

$$\psi_g = \frac{m+|p_o+p_3|-i\vec{\gamma}_\perp\cdot\vec{p}_\perp}{\sqrt{(m+|p_o|)^2+\vec{p}_\perp^2}}\,\phi_g \tag{69}$$

where the restriction of $\vec{\gamma}_\perp$ (γ_1 and γ_2, which commute with α_3) to good components is understood. Equivalently

$$\psi_g = \frac{m+|p_o+p_3|-i\vec{\gamma}_\perp\cdot\vec{p}_\perp}{\sqrt{(m+|p_o+p_3|)^2+\vec{p}_\perp^2}}\,\hat{\phi}_g \tag{70}$$

in terms of a trivially renormalized field

$$\hat{\phi}_g = \frac{\sqrt{(m+|p_o+p_3|)^2+\vec{p}_\perp^2}}{\sqrt{(m+|p_o|)^2+\vec{p}_\perp^2}}\,\phi_g = \sqrt{\frac{p_o+p_3}{p_o}}\,\phi_g\ . \tag{71}$$

We then have a unitary transformation, in the two-dimensional space of good indices, which may be rewritten

$$\begin{aligned}\hat{\phi}_g &= \frac{m+|p_o+p_3|+i\vec{\gamma}_\perp\cdot\vec{p}_\perp}{\sqrt{(m+|p_o+p_3|)^2+\vec{p}_\perp^2}}\,\psi_g \\ &= \exp\left(i\arctan\frac{\vec{\gamma}_\perp\cdot\vec{p}_\perp}{m+|p_o+p_3|}\right)\psi_g\end{aligned} \tag{72}$$

The modified null-plane charges

$$\begin{aligned}\hat{W}(\Lambda) &= 2\int d^4x\ \delta(t+z)\ \hat{\phi}_g^+\,\hat{\Lambda}\,\phi_g \\ &= e^{iy}\,\hat{F}(\Lambda)\,e^{-iy} \\ y &= -2\int d^4x\ \delta(t+z)\ \psi_g^+\arctan\frac{\vec{\gamma}_\perp\cdot\vec{p}_\perp}{m+|p_o+p_3|}\,\psi_g\end{aligned} \tag{73}$$

generate the infinitesimal transformation of Heisenberg field operators

$$\delta\hat{\phi}_g = - i \ \hat{\Lambda} \ \hat{\phi}_g \quad .$$

This is equivalent to

$$\delta\phi_g = - i \ \hat{\Lambda} \ \phi_g$$

which we have already noted to be free from the rotation difficulty of the last section.

This then is Melosh's proposed symmetry [18] - with one proviso. Comparing (73) with (51) we see that the modified charges have now lost not only transverse ($\vec{E}$) boost invariance (lost already in (51) because of the appearance of $\vec{p}_\perp$) but also longitudinal (B_3) boost invariance, because of the appearance of $|p_o + p_3|$. Melosh is more disturbed by this, and proposes that the new charges (73) should be symmetries only for systems at rest (as regards centre-of-mass motion) and that for moving systems one first deboosts to rest and afterwards reboosts.

3.4 Melosh-modified null-plane charges and FW equal time charges

In the case of no interaction the Melosh modified null-plane charges, like other null-plane charges (Section 2.6), are constants of the motion, and therefore symmetry generators. It follows that

$$\delta\phi = -i \ [\ \phi, \ \hat{W}(\Lambda) \ ]$$

is, like ϕ, a solution of the transformed Dirac equation (55). Then from

$$\delta\phi_g = - i \hat{\Lambda} \phi_g \tag{74}$$

it follows that

$$\delta\phi_b = - i\sigma_3 \hat{\Lambda} \sigma_3 \phi_b . \tag{75}$$

Now in the good representation the Λ's of the $U(6)_W$ set (31) take the form

$$\Lambda = \begin{bmatrix} \hat{\Lambda} & 0 \\ 0 & \sigma_3\hat{\Lambda}\sigma_3 \end{bmatrix}$$

- dictated by commutation with

$$\alpha_3 = \begin{bmatrix} 1 & 0 \\ 0 & -1 \end{bmatrix} \qquad \beta = \begin{bmatrix} 0 & \sigma_3 \\ \sigma_3 & 0 \end{bmatrix} .$$

Thus (74) and (75) together give

$$[\phi, \hat{W}(\Lambda)] = \Lambda\phi \quad . \tag{76}$$

This is precisely the same as for the appropriate subset of the original FW charges $W(\Lambda)$ of Section 1.2; see Eq. (13). Thus

$$[\phi, \hat{W}(\Lambda) - W(\Lambda)] = 0 .$$

It follows that $\hat{W}(\Lambda)$ and $W(\Lambda)$ differ at most by a c-number. Actually they do not differ at all - in the case of no interaction - as one finds by writing them out explicitly in terms of creation and destruction operators.

From (6)

$$\phi(x) = \sum_s \frac{d^3\vec{p}}{(2\pi)^3|2p_o|}(a(p,s)w(p,s)e^{ipx}+b^+(p,s)w(-p,-s)e^{-ipx})$$

one finds

$$W(\Lambda) = \int d^4x\ \delta(t)\ \phi^+\ \Lambda\ \phi \tag{77}$$

$$= \sum_{ss'} \int \{\frac{d^3\vec{p}}{(2\pi)^3}\ \frac{a^+(p,s')a(p,s)}{|2p_o|}\ \frac{w^*(p,s')\Lambda w(p,s)}{|2p_o|} + \frac{b(p,s')b^+(p,s)}{|2p_o|}\ \frac{w^*(-p,-s')\Lambda w(-p,-s)}{|2p_o|}\}. \tag{78}$$

Note that particle-antiparticle-pair-terms do not occur here because they would involve matrix elements like

$$w^*(-p,-s')\ \Lambda\ \ w(p,s)$$

which are zero because

$$[\Lambda,\beta] = 0 \qquad \beta w\ (p,s)\ =\ (p_o/|p_o|)w(p,s).$$

In computing instead

$$\hat{W}(\Lambda)\ =\ 2\int d^4x\ \delta(t+z)\ \hat{\phi}_g^+\ \ \Lambda\ \ \hat{\phi}_g \tag{79}$$

the following differences arise:

1) Integration over the null plane, $t + z = 0$, suppresses pair terms immediately (Sections 2.6, 2.7).

2) We obtain expressions

$$\int d^4x \; \delta(t+z) \; e^{i(\vec{p}_\perp - \vec{p}_\perp{}') \cdot \vec{x}_\perp + i(p_3 - p_o - p_3' + p_o')x_3}$$

$$= (2\pi)^3 \; \delta^2 \; (\vec{p}_\perp - \vec{p}_\perp{}')\, \delta(p_3 - p_o - p_3' + p_o') \qquad (80)$$

$$= (2\pi)^3 \; \delta^3 \; (\vec{p} - \vec{p}\,') \; \left| \frac{p_o}{p_3 + p_o} \right|$$

- where the last factor was absent in the equal time case.

3) We have to remember the renormalization factor in (71), which just cancels against the extra last factor in (80).

4) Instead of

$$w^*(p,s') \quad \Lambda \quad w(p,s)$$

we now have

$$2w_g^*(p,s') \; \Lambda \; w_g(p,s)$$

which is the same thing, because

$$\Lambda = \begin{pmatrix} \hat{\Lambda} & 0 \\ 0 & \sigma_3 \hat{\Lambda} \sigma_3 \end{pmatrix} \qquad w = \begin{pmatrix} w_g \\ \pm\sigma_3 w_g \end{pmatrix} .$$

Note that this demonstration of equality of $\hat{W}(\Lambda)$ and

$W(\Lambda)$ fails, as does the demonstration of compatibility of $\hat{W}$ invariance with rotational invariance, when interactions are present - because the free Dirac equation was explicitly invoked. Melosh insists strongly on the difference between his version of relativistic SU(6) and the old Foldy-Wouthuysen inspired version [19]. Whether he is aware of their identity in the case of no interaction, the case in which he explicitly developed his transformation, is not clear.

3.5 Concluding remarks

We have seen that the Melosh-modified null-plane charges are identical with some of the old Foldy-Wouthuysen modified equal time charges in the case of no interaction. Although this identity does not persist in the presence of interaction, the new and the old charges, considered as possible symmetry operations, continue to have much in common. In particular both involve in their construction a preferred frame of reference. Indeed the Melosh charges are particularly obscure in this respect; having lost the manifest E_1, E_2, and B_3 boost invariance of the unmodified null-plane charges, they have not acquired the manifestly simple rotational properties of the Foldy-Wouthuysen charges. It is very likely therefore that we would arrive at contradictions like those of Section 1.5 if we tried to use the charges $\hat{W}$ as symmetries of arbitrarily moving particles. To avoid such contradictions one can again suppose that only particles <u>at rest</u> transform simply, under the action of the $\hat{W}$'s. This is, effectively, what Melosh proposes.

There is then a puzzle in connection with his discussion of nucleon magnetic moments, giving again the famous and desirable $\mu_p/\mu_n = -3/2$. The discussion of magnetic moments requires consideration of small but non-zero momentum transfer, and therefore of states without a common rest system. Melosh seems to avoid this by using an expression in which a rather singular operator appears between particles at rest. But because of the singular nature of the operator these states cannot be momentum eigenstates. They have to be wavepackets, and their momentum spread although small is essential. Thus Melosh implicitly supposes that moving particles also transform simply under his charges. In this respect the discussion of de Alwis and Stern [16] seems to be more serious; considering explicitly the failure of $\hat{W}$ to commute with boosts they find that no definite value is obtained for μ_p/μ_n unless arbitrary additional assumptions are made.

In all this we have been supposing that, even in applications to composite systems, the Melosh or Foldy-Wouthuysen transformations are taken in the simple forms demonstrably effective for free quarks. It may well be that there are more complex forms, with explicitly interaction-dependent terms, which would be more appropriate. The possibility has been contemplated by Melosh [18], and explicitly developed in some cases by Foldy and Wouthuysen [4]. The investigation of such questions requires dynamics rather than kinematics. The outcome, in the shape of some kind of relativistic broken SU(6), will presumably depend on the strength and nature of the interaction. It may be that the primitive Melosh transformation will be found to contain some important part of the truth for some plausible interaction. Until then, as Melosh remarks, the empirical success of the free quark form remains unexplained.

APPENDIX

1. Notation (Central European Standard)

$\hbar = c = 1 \qquad x_4 = ix_o = it \qquad (x_1,x_2,x_3)=(x,y,z)\equiv\vec{x}$

$x_\mu = (x_1,x_2,x_3,x_4) \quad xp = x_\mu p_\mu = x_1p_1+x_2p_2+x_3p_3-x_op_o$

$\int d^4x = \int dx_1dx_2dx_3dx_o \qquad \delta^4(x) = \delta(x_1)\delta(x_2)\delta(x_3)\delta(x_o)$.

Dirac equation:

$(\vec{\alpha}\cdot\vec{p} + \beta m)u = p_ou \qquad \text{or} \qquad (m + i\gamma p)u = 0$

$\gamma_4 = \beta \quad \vec{\gamma} = -i\beta\vec{\alpha} \quad \gamma_5 = i\alpha_1\alpha_2\alpha_3 = \gamma_1\gamma_2\gamma_3\gamma_4$

$\gamma_\mu\gamma_\nu + \gamma_\nu\gamma_\mu = 2\,\delta_{\mu\nu} \qquad \gamma_\mu = \gamma_\mu^+ \qquad (\mu,\nu = 1\ldots5)$

$\sigma_{\mu\nu} = (\gamma_\mu\gamma_\nu - \gamma_\nu\gamma_\mu)/(2i) \qquad \vec{\sigma} = -\vec{\alpha}\gamma_5 = \vec{\gamma}x\vec{\gamma}/(2i)$

$\sigma_{ij} = \varepsilon_{ijk}\,\sigma_k \qquad \sigma_{i4} = \alpha_i \qquad i,j,k \neq 4$

Charge conjugation:

$$\psi_c = C\psi^+ \qquad C\vec{\alpha}^* C^{-1} = \vec{\alpha} \qquad C\vec{\sigma}^* C^{-1} = -\vec{\sigma}$$

Lorentz transformation:

$$\omega_{\mu\nu} = -\omega_{\nu\mu} \qquad \delta p_\mu = \omega_{\mu\nu}\, p_\nu \qquad \delta u = \frac{1}{4}\,\omega_{\mu\nu}\,\gamma_\mu\gamma_\nu\, u$$

rotation: $\delta\vec{p} = \vec{\omega}\times\vec{p} \qquad \delta p_o = 0 \qquad \delta u = -\frac{1}{2}\vec{\omega}\cdot\vec{\sigma}u$

boost: $\delta\vec{p} = \vec{\omega}p_o \qquad \delta p_o = \vec{\omega}\cdot\vec{p} \qquad \delta u = \frac{1}{2}\vec{\omega}\cdot\vec{\alpha}u$

2. Old Representation

$$\vec{\alpha} = \begin{pmatrix} 0 & \vec{\sigma} \\ \vec{\sigma} & 0 \end{pmatrix} \qquad \beta = \begin{pmatrix} 1 & 0 \\ 0 & -1 \end{pmatrix}$$

$$\vec{\gamma} = \begin{pmatrix} 0 & -i\vec{\sigma} \\ i\vec{\sigma} & 0 \end{pmatrix} \qquad \gamma_5 = \begin{pmatrix} 0 & -1 \\ -1 & 0 \end{pmatrix}$$

$$\vec{\sigma} = \begin{pmatrix} \vec{\sigma} & 0 \\ 0 & \vec{\sigma} \end{pmatrix} \qquad C = \begin{pmatrix} 0 & i\sigma_2 \\ -i\sigma_2 & 0 \end{pmatrix} = -\gamma_2$$

2 x 2 Pauli matrices (assumed in C)

$$\sigma_1 = \begin{pmatrix} 0 & 1 \\ 1 & 0 \end{pmatrix} \quad \sigma_2 = \begin{pmatrix} 0 & -i \\ i & 0 \end{pmatrix} \quad \sigma_3 = \begin{pmatrix} 1 & 0 \\ 0 & -1 \end{pmatrix}$$

$$\sigma_i \, \sigma_j = i \, \varepsilon_{ijk} \, \sigma_k$$

Rotation-free-boosted spinors

$$\begin{pmatrix} \sqrt{2m} & \phi \\ & 0 \end{pmatrix} \longrightarrow \begin{pmatrix} \sqrt{m + p_o} & \phi \\ \dfrac{\vec{\sigma}\cdot\vec{p}}{\sqrt{m+p_o}} & \phi \end{pmatrix}$$

3. Good Representation

$$(\vec{\alpha},\beta) \rightarrow U(\vec{\alpha},\beta)U^{-1} \qquad\qquad C \rightarrow U\, C\, \tilde{U}$$

$$U = \frac{1}{\sqrt{2}} (\alpha_3 + \beta) = U^{+} = \tilde{U} = U^{*} = U^{-1}$$

$$\vec{\alpha}_\perp = \begin{pmatrix} 0 & -\vec{\sigma}_\perp \\ -\vec{\sigma}_\perp & 0 \end{pmatrix} \qquad \alpha_3 = \begin{pmatrix} 1 & 0 \\ 0 & -1 \end{pmatrix} \qquad \beta = \begin{pmatrix} 0 & \sigma_3 \\ \sigma_3 & 0 \end{pmatrix}$$

$$\vec{\gamma}_\perp = \begin{pmatrix} i\sigma_3\vec{\sigma}_\perp & 0 \\ 0 & i\sigma_3\vec{\sigma}_\perp \end{pmatrix} \qquad \gamma_3 = \begin{pmatrix} 0 & i\sigma_3 \\ -i\sigma_3 & 0 \end{pmatrix} \qquad \gamma_5 = \begin{pmatrix} -\sigma_3 & 0 \\ 0 & \sigma_3 \end{pmatrix}$$

$$\sigma_1 = \begin{bmatrix} 0 & \sigma_1\sigma_3 \\ \sigma_3\sigma_1 & 0 \end{bmatrix} \qquad \sigma_2 = \begin{bmatrix} 0 & \sigma_2\sigma_3 \\ \sigma_3\sigma_2 & 0 \end{bmatrix} \qquad \sigma_3 = \begin{bmatrix} \sigma_3 & 0 \\ 0 & \sigma_3 \end{bmatrix}$$

Dirac equation:

$$(p_o - p_3)\, u_{1,2} = (m\sigma_3 - \vec{\sigma}\cdot\vec{p}\,)\, u_{3,4}$$

$$(p_o + p_3)\, u_{3,4} = (m\sigma_3 - \vec{\sigma}\cdot\vec{p}\,)\, u_{1,2}$$

Special boosts:

$$E_1 = B_1 + J_2 \qquad\qquad E_2 = B_2 - J_1$$

$$(\alpha_1 - i\sigma_2) = \begin{bmatrix} 0 & 0 \\ -2\sigma_1 & 0 \end{bmatrix} \qquad (\alpha_2 + i\sigma_1) = \begin{bmatrix} 0 & 0 \\ -2\sigma_2 & 0 \end{bmatrix}$$

Special-boosted spinors:

$$\text{rest spinor } \sqrt{2m} \begin{bmatrix} \phi/\sqrt{2} \\ \sigma_3\phi/\sqrt{2} \end{bmatrix} \xrightarrow{E_1E_2B_3} \begin{bmatrix} \sqrt{p_o+p_3} & \phi \\ \dfrac{m\sigma_3-\vec{\sigma}_\perp\cdot\vec{p}_\perp}{\sqrt{p_o+p_3}} & \phi \end{bmatrix}$$

REFERENCES

1. H. J. Melosh, preprint EFI 73/26, Chicago, August 1973.

2. M. H. L. Pryce, Proc. Roy. Soc. A195, 62 (1948). I am indebted to Prof. Bacry for drawing my attention to this reference in the paper of Foldy and Wouthuysen.

3. S. Tani, Soryushiron Kenku 1, 15 (1949); Progr. Theor. Phys. Kyoto 6, 267 (1951).

4. L. L. Foldy, S. A. Wouthuysen, Phys. Rev. 78, 29 (195o).

5. For a recent survey, and an account of the impact of the work of Melosh, see J. Weyers, Constituent quarks and current quarks, CERN TH. 1743.

6. F. Gürsey, Phys. Letters 14, 330 (1965).
K. T. Mahanthappa, E. C. G. Sudarshan, Phys. Rev. Letters 14, 458 (1965).
Riazuddin, L. K. Pandit, Phys. Rev. Letters 14, 462 (1965).

7. T. F. Jordan, Phys. Rev. 139, B148 (1965); 140, B766 (1965). A popular account of early difficulties with relativistic SU(6) is given by J. S. Bell in A. Zichichi (Ed.), Recent developments in particle symmetries, Erice Lectures, 1965 (Academic Press, N. Y., 1966).
More comprehensive early reviews of SU(6), non-relativistic and relativistic, are those of B. W. Lee in Particle Symmetries, Brandeis, 1965 (Gordon and Breach, N. Y., 1966); H. Ruegg, W. Rühl, Helv. Phys. Acta. 40, 9 (1967).

8. S. Coleman, Phys. Letters 19, 144 (1965); J. Math. Phys. 7, 787 (1966).

9. S. Fubini, G. Furlan, Physics 4, 229 (1965).

10. For a general account of null-plane charges, see H. Leutwyler, Springer Tracts in Modern Physics 50, 29 (1969).

11. K. J. Barnes, P. Carruthers, F. von Hippel, Phys. Rev. Letters 14, 82 (1965).
H. J. Lipkin, S. Meshkov, Phys. Rev. Letters 14, 670 (1965).

12. J. B. Kogut, D. E. Soper, Phys. Rev. D1, 2901 (1970).
F. Rohrlich, Acta Phys. Austriaca Suppl. VIII, 277 (1971).
C. Bouchiat, P. Fayet, N. Sourlas, Lettere Nuovo Cimento 4, 9 (1972).
For reviews of related developments, see J. Kogut and L. Susskind, Phys. Reports 8C (1973); Ph. Meyer, Cargèse Lectures 1972, Orsay preprint 72/40 (1972).

13. R. Dashen, M. Gell-Mann, Symmetry principles at high energy, Coral Gables Conference, 1966 (W. H. Freeman and Co., San Francisco, 1966).

14. R. Gatto, L. Maiani, G. Preparata, Phys. Rev. Letters 16, 377 (1966).
G. Altarelli, R. Gatto, L. Maiani, G. Preparata, Phys. Rev. Letters 16, 918 (1966).
H. Harari, Phys. Rev. Letters 16, 964 (1966).
I. S. Gerstein, B. W. Lee, Phys. Rev. Letters 16, 1060 (1966); Phys. Rev. 152, 1418 (1966).
D. Horn, Phys. Rev. Letters 17, 778 (1966).
N. Cabibbo, H. Ruegg, Phys. Letters 22, 85 (1966).
R. Gatto, L. Maiani, G. Preparata, Phys. Rev. Letters 18, 97 (1967); Physics 3, 7 (1967).
F. Buccella, H. Kleinert, C. A. Savoy, E. Celeghini,

and E. Sorace, Nuovo Cimento 69A, 133 (1970).
F. Buccella, E. Celeghini, C. A. Savoy, Nuovo Cimento 7A, 281 (1972).

15. H. Jay Melosh IV, Thesis, Cal. Inst. of Technology, Pasadena, 26 Feb. 1973.

16. S. P. de Alwis, Nuclear Phys. B55, 427 (1973).
E. Eichten, J. F. Willemsen, F. Feinberg, SLAC preprint (1973).
S. P. de Alwis, J. Stern, CERN preprint TH. 1679 (1973).

17. In discussions at CERN, J. Weyers insisted on the importance of spin considerations in motivating something like the Melosh transformation.

18. Ref. 1, Eq. (31). Our (73) can be identified with his (31) by making the following changes of notation in the former.

19. Ref. 1, page 31.

Acta Physica Austriaca, Suppl. XIII, 447–486 (1974)

ELECTROMAGNETIC INTERACTION OF HADRONS WITHOUT PARTONS?*)

by

P. STICHEL
Department of Theoretical Physics
University of Bielefeld
Germany

CONTENTS

---

*) Lecture given at XIII. Internationale Universitätswochen für Kernphysik, Schladming, Austria, February 4-15, 1974.

1. THE PARTON MODEL AND ITS DEFICIENCIES

1.1 Introduction

The parton model has been invented in 1969 by Feynman, Bjorken, Paschos and others [1] as a simple intuitive physical picture in order to explain the scaling phenomena in deep inelastic electron scattering observed at DESY and SLAC [2].

For the non-experts I will very quickly introduce some well-known notations and facts:

Deep inelastic eN scattering is defined as the process

$$e + N \to e' + X \qquad (1)$$

- where we have to integrate in the cross section formula over all degrees of freedom of the system X in the final state - in the kinematical region $\nu \to \infty$, $\omega: = -2\nu/q^2$ fixed (ν: energy loss of the scattered electron in the lab. frame; $q^2: =$ (invariant momentum transfer)^2).

In the one-photon exchange approximation this process looks diagramatically as follows:

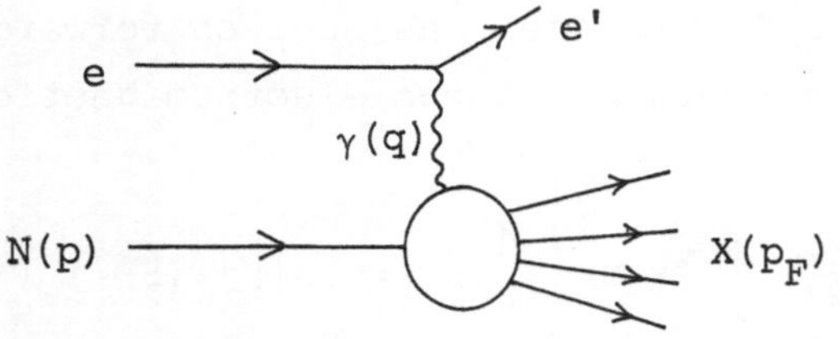

Fig. 1: Basic diagram for eN $\to$ e'X

where we put the corresponding momenta of the particles in parentheses. We have: $\nu = q.p$ (M: = 1).

The double differential cross section $\frac{d^2\sigma}{d\Omega dE'}$ in the one-photon exchange approximation is given by

$$\frac{d^2\sigma}{d\Omega dE'} = \frac{4\alpha^2 E'^2}{q^4} \left(\cos^2\frac{\theta}{2} W_2(\nu,q^2) + 2\sin^2\frac{\theta}{2} W_1(\nu,q^2)\right) \quad (2)$$

where E' and $\Omega(\theta,\Phi)$ are energy and solid angle of the scattered electron in the Lab. frame respectively. Instead of the structure functions $W_i(\nu,q^2)$ it is sometimes more useful to consider total cross sections for the absorption of longitudinal and transversal virtual photons in the Lab. frame

$$\sigma_T = \frac{4\pi^2\alpha}{\nu + q^2/2} W_1$$

$$\sigma_L = \frac{4\pi^2\alpha}{\nu + q^2/2} [W_2(1 - \frac{\nu^2}{q^2}) - W_1] \quad . \tag{3}$$

Due to the optical theorem these cross sections are proportional to the imaginary part of the virtual forward Compton scattering amplitude. Therefore we introduce the connected, spin averaged one-nucleon forward matrix element of the commutator of two electromagnetic currents

$$W_{\mu\nu}(q,p) := \frac{(2\pi)^2}{2} \int d^4x \, e^{iq.x} \, \frac{1}{2} \sum_s \langle p,s | [j_\mu(\tfrac{x}{2}), j_\nu(-\tfrac{x}{2})] | p,s \rangle. \tag{4}$$

By means of Lorentz covariance and current conservation we obtain the following relation between the W_i and $W_{\mu\nu}$

$$W_{\mu\nu} = (-g_{\mu\nu} + \frac{q_\mu q_\nu}{q^2}) W_1 + (p_\mu - q_\mu \frac{q.p}{q^2}).(p_\nu - q_\nu \frac{q.p}{q^2}) \, W_2 \tag{5}$$

Experimental results

Within experimental errors one obtains a constant value for the ratio

$$R := \sigma_L/\sigma_T \simeq 0.2 \tag{6}$$

- compare Fig. 2a (taken from Close [1]).

The most striking surprise of the data was the observation, that W_1 and νW_2 at large q^2 and ν but fixed ω become functions of the scaling variable ω only

$$W_1 \xrightarrow[\substack{\nu\to\infty \\ \omega \text{ fixed}}]{} F_1(\omega) \qquad (7)$$

$$\nu W_2 \xrightarrow[\substack{\nu\to\infty \\ \omega \text{ fixed}}]{} F_2(\omega) \qquad (8)$$

For the particular value $\omega = 4$ the scaling property (8) is shown in Fig. 2b (taken from Close [1]).

Due to $R \simeq$ const. scaling of W_1 follows from that of νW_2 and vice versa.

A scaling behaviour of the W_i as described in eqs. (7) and (8) has been suggested first by Bjorken [3]. Therefore we call the limit $\nu\to\infty$, ω fixed hereafter as Bjorken (Bj) - limit. If Bj-scaling holds with the particular powers of ν given in (7) and (8), we speak about "canonical" scaling (the reason for this notion will become clear later on).

The parton model may be defined in its most primitive form by means of the following two assumptions:

a) In the kinematical region of deep inelastic eN scattering the nucleon acts like a gas of free spin 1/2 point-like constituents, called partons.

b) The transverse momentum $\vec{p}_T$ of a parton inside the nucleon is small compared with $\vec{p}$ in the eN-CMS.

From assumption b) we conclude, that the scattering off partons is an incoherent one at high q^2. Therefore, the functions W_r ($r = 1,2$) have to be calculated as follows:

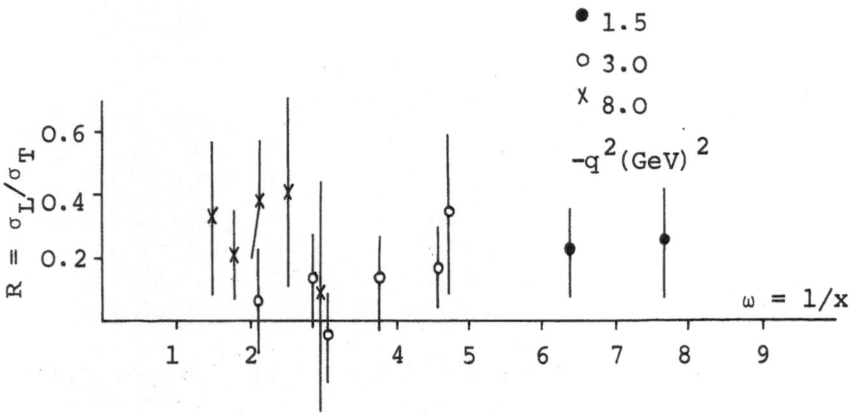

Fig. 2a $R = \sigma_L/\sigma_T$ data

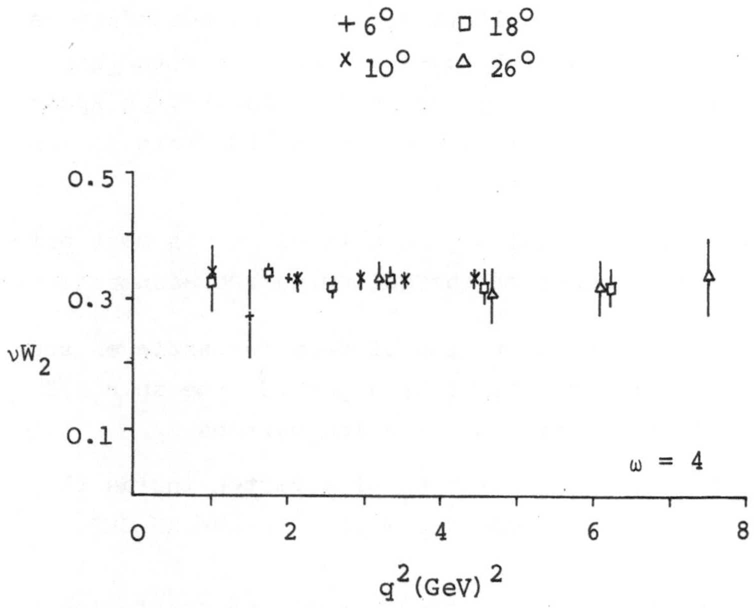

Fig. 2b Constancy of νW_2 at fixed ω

$$W_r = \sum_N P_N \sum_{i=1}^{N} \int_o^1 dx\ f_i^N(x)\ W_r^i(x,\nu,q^2) \tag{9}$$

where

W_r^i = contribution of the ith-parton;

P_N = probability, that there are N-partons within the nucleon;

$f_i^N(x)$ = probability of parton i having in a N-parton configuration a longitudinal momentum $x \cdot p$.

Because the electron is scattered elastically from a free parton we obtain for $W_{\mu\nu}^i$

$$W_{\mu\nu}^i = \frac{1}{2x} \sum_{\text{Spins}} \langle xp,i|j_\mu(0)|q+xp,i\rangle \langle q+xp,i| j_\nu(0)|xp,i\rangle \frac{1}{2}(2\pi)^6 \tag{10}$$

- or, after some algebra

$$W_1 = \frac{1}{2} \sum_N P_N \sum_{i=1}^{N} f_i^N(\tfrac{1}{\omega}) Q_i^2$$

$$\nu W_2 = \sum_N P_N \sum_{i=1}^{N} f_i^N(\tfrac{1}{\omega}) \frac{Q_i^2}{\omega} \tag{11}$$

where Q_i is the charge of the ith parton (in units of e). Canonical Bj-scaling is obviously fulfilled.

I will not discuss the above described form of the parton model any more - it serves only as an introduction to the subject [4].

1.2 The non-perturbative, covariant parton model (Cambridge parton model)

The primitive parton model as described above has been introduced as a simple intuitive physical picture without any reference to other models or principles in elementary particle theory. Furthermore it is thought to be valid only in a particular kinematical region or - in other words - in a distinguished class of Lorentz frames ($\vec{p}\to\infty$ frames).

The first connection of the parton idea to conventional field theory has been given by Drell, Levy and Yan [5] who calculated deep inelastic lepton scattering in cut-off field theories in perturbation theory. Scaling has been obtained in that way. Partons in a cut-off field theory are the bare particles, whose transverse momenta in the $\vec{p}\to\infty$ frame are restricted due to the cut-off. But one must note that a cut-off field theory violates unitarity.

The only no cut-off field theories which lead to canonical scaling in perturbation theory are the so-called superrenormalizable interactions (f.i. a Φ^3-theory - but no vacuum exists in such a field theory [6]). In all other renormalizable theories one observes a violation of scaling by logarithmic terms in any finite order of perturbation theory [7]. It is an open question up to now whether these logarithmic terms sum up to a power, because present techniques only allow the summation of the leading logarithmic terms (sometimes also the next to the leading terms may be summed up).

In considering this situation a non-perturbative treatment of the parton idea within the framework of field theory was necessary.

The non-perturbative covariant parton model of Landshoff, Polkinghorne and Short [8], may be characterized as follows:

(i) Hadrons are composite systems of partons.

(ii) Strong interaction amplitudes involving hadrons and virtual partons as external legs go rapidly to zero when the mass of any virtual parton becomes large.

(iii) The parton propagator is asymptotically proportional to the corresponding free propagator.

(iv) The e.m. formfactor of a parton with finite (virtual) parton masses goes asymptotically like a constant (point structure).

The requirements (iii) and (iv) are not completely independent of each other: From (iii) and the Ward-identity for the e.m. parton vertex we conclude (compare section 2.4), that the parton e.m. vertex behaves asymptotically at least like a constant if all external masses go to infinity. But present experience with perturbation theory tells us, that the asymptotic behaviour of a formfactor is independent of which external mass becomes large [9].

According to (i) photons couple directly to partons only and we have, due to (iv) in the Bj-limit for the virtual Compton amplitude

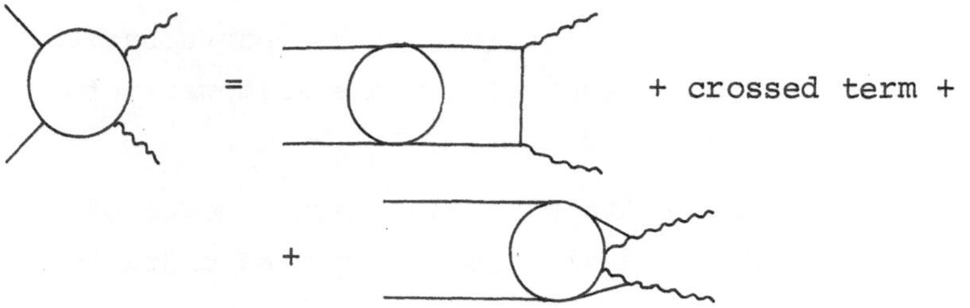

Fig. 3: Compton amplitude in the parton model

where the hadron-parton four- and six-point amplitudes are connected ones and internal lines are given by the full (renormalized) parton propagators. By introducing Sudakov variables for the internal loop momenta in Fig.3 and using assumptions (ii) and (iii) one can show, that the first diagram in Fig. 3 together with its crossed counterpart dominates in the Bj-limit leading to canonical Bj-scaling for the W_i [8]. Thereby the main contribution comes from finite masses of virtual partons. We will not go into technical details at this point, because we will use the Sudakov technique extensively in chapter 2.

The same parton picture has been successfully applied to the following inclusive processes:

a) one particle inclusive e^+e^- annihilation [10]: $e^+e^- \to h + X$

b) one-particle inclusive deep inelastic scattering

[10], i.e. the process

$$e + h \rightarrow h' + e' + X$$

and its crossed counterpart of inclusive heavy lepton pair production

$$h + h' \rightarrow \ell + \bar{\ell} + X$$

c) electron-positron annihilation [12]

$$e^+e^- \rightarrow X$$

and others [10].

For exclusive processes definite predictions have been obtained within this parton model only by means of additional assumptions:

α) Specific Regge assumptions in case of the e.m. form-factor [8] and single-particle electroproduction for small t [13] (t is here the momentum transfer squared between the baryons in the initial and final state).

β) Constituent interchange in case of single-particle photon- and electroproduction at large t [14].

1.3 Deficiencies of the Cambridge parton model

Despite of its great success the parton model of Landshoff and Polkinghorne shows a number of theoretical deficiencies. But recent storage ring experiments on the e^+e^- total annihilation cross section even raise serious doubts on its validity in nature. For the ratio

$$R: = \frac{\sigma(e^+e^- \to \text{hadrons})}{\sigma(e^+e^- \to \ell\bar{\ell})}$$

the parton model predicts assumptotically a constant [12], but recent experiments show [15]

$$R \sim c\sqrt{s} \; .$$

We will come back to this point in section 2.5. We now give a list of its theoretical deficiencies:

(i) The model does not answer the question about the physical nature of partons.

There exist three possibilities:

1) Partons are well known physical particles. This contradicts assumption (iv) of the model.

2) Partons are physical particles, perhaps quarks, which have not yet been seen experimentally either due to their high mass or due some unknown dynamical mechanism.

3) There exist no asymptotic parton fields, i.e. the parton propagator has no poles (compare [16]).

(ii) It has not been shown whether there exist interactions between partons such, that simultaneously hadrons appear as bound states and hadron-parton amplitudes show Regge behaviour and are fast decreasing for large external parton masses.

(iii) There are difficulties with gauge-invariance in case of additional photon couplings, because Green's functions involving hadrons, partons and e.m. currents, do not satisfy automatically the corresponding Ward-identities (WIs).

The last point (iii) demands for some discussion. Consider for example virtual Compton scattering described by the diagrams of Fig. 3 with the external hadron being off shell. If we denote by (p_1, q_1) the momenta of (h,γ) in the initial state and those in the final state by (p_2,q_2), then the WIs look as follows

$$q_1^\nu T_{\mu\nu}(p_2,q_2|p_1,q_1) = \Gamma_\mu(p_2-q_1|p_1)\Delta_F'^{-1}(p_2)\cdot \Delta_F'(p_2-q_1) - \Gamma_\mu(p_2|p_1+q_1)\Delta_F'^{-1}(p_1)\Delta_F'(p_1+q_1) \tag{12}$$

and a similar relation for $q_2^\mu T_{\mu\nu}$. Γ_μ is the e.m. vertex for the external hadron, which in the parton picture for large virtual photon mass has the structure given in Fig. 4,

$\Gamma_\mu(p'|p)$:

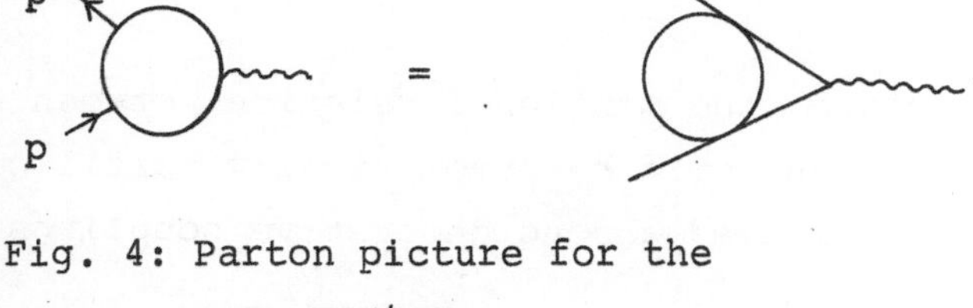

Fig. 4: Parton picture for the e.m. vertex

where Δ_F' describes the full renormalized hadron propagator, which may be calculated from Γ_μ by means of the corresponding WI

$$(p' - p)^\mu \Gamma_\mu(p'|p) = \Delta_F'^{-1}(p') - \Delta_F'^{-1}(p) . \qquad (13)$$

Now the trouble is obvious: The WI (12) requires a relation between hadron-parton four- and six-point amplitudes. It seems hard to understand this, because the principle of minimal e.m. interaction (i.e. the substitution $\partial_\mu \to \partial_\mu - ieA_\mu$ in the full strong interaction Lagrangian) tells us, how to introduce photon couplings for a given strong interaction. In case of the Compton amplitude (Fig. 3) as well as in case of the e.m. hadron vertex (Fig. 4) we coupled in agreement with assumption (iv) and the principle of minimal e.m. interaction the photon point-like to all charged partons. But this would be the full story if and only if

a) the couplings between either partons or partons and hadrons do not contain derivatives (in case of derivative couplings additional contact terms have to be introduced from the minimal substitution within the bare strong interaction vertex) and

b) hadron-parton amplitudes are constructed in agreement with the principle of locality, which is crucial for deriving WIs.

Therefore, the mentioned relation between hadron-parton amplitudes would be automatically fulfilled if locality were satisfied and no derivative couplings were present.

But the latter brings us back to point (ii):

For a introduction of e.m. interaction into any strong interaction model in agreement with the principle of minimal e.m. interaction we have to start with an expression for the strong interaction forces. This will be the subject of chapter 2.

One remark on the treatment of the gauge invariance problem in the recent literature. Brodsky, Close and Gunion [17], started with an expression for the hadronic self-energy $\Sigma(p)$ given by Fig. 5

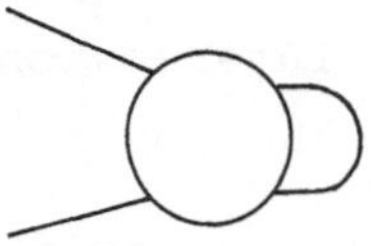

Fig. 5: Parton picture for the hadronic self-energy

put this by means of a double dispersion relation for the forward off-shell hadron-parton scattering amplitude (which is wrong in general, because anomalous complex singularities have been neglected) in the form of a linear operator applied to the 2nd order perturbation theory result and then introduced in this 2nd order expression e.m. interactions in accordance with the minimal substitution. No wonder, that the corresponding WIs are fulfilled. But their procedure has nothing to do (except in 2nd order) with a consequent application of the principle of minimal e.m. interaction, from which we obtain the only physical acceptable solution for WIs.

2. A CLASS OF BETHE-SALPETER (BS) MODELS

As pointed out in the last section, one has to start with the strong interaction force in order to treat the principle of minimal e.m. interaction consistently.

Because we are not yet able to solve strong interaction field equations we consider in the following phenomenological Bethe-Salpeter (BS) kernels for the hadron-hadron interaction as the basic strong interaction force. In this way we dispose of a theoretical laboratory for treating interactions of hadrons with each other and external photons at high energies.

We are not yet able to decide whether such a scheme leads to structures which are rich enough in order to explain at least qualitatively the most important experimental facts in high energy hadron physics. We are aware of the fact, that there might exist other alternatives to the parton picture which are more suitable for our purpose (compare f.i. the perturbative eikonal scheme of Orzalesi [18]).

2.1 Hadronic interactions in terms of effective BS-kernels

Consider the amputed off-shell $N\bar{N}$-scattering amplitude T given by the box of Fig. 6.

The BS-kernel K with the property of being two-particle irreducible in the t-channel (field theoretical analogue to the potential in the Schrödinger equat-

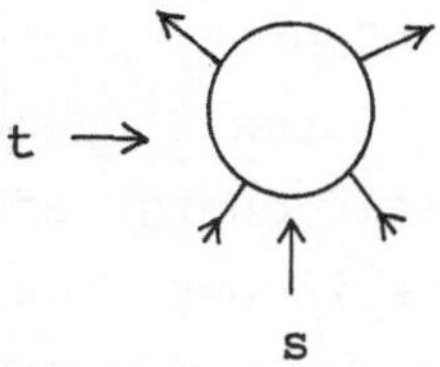

Fig. 6: $N\bar{N}$ off-shell scattering amplitude

ion) is related to T through the BS-equation [19].

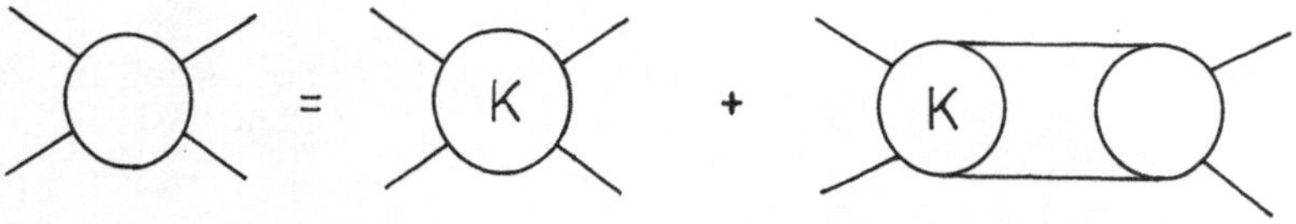

Fig. 7: BS-equation for the scattering amplitude

Here and in the following internal lines are given always by the full renormalized propagator Δ_F'.*)

What is the form of K in a given field theory? Consider as an example the usual PS-PS πN theory. K is given by the infinite sum of all t-channel $N\bar{N}$-irreducible sceleton graphs, i.e. we have

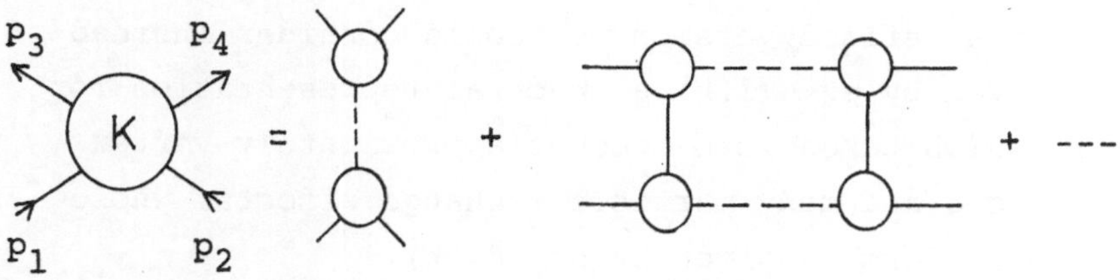

Fig. 8: Sceleton expansion of K

where  is the $\pi N\bar{N}$-vertex part. To get the exact kernel K is as hard as to solve our original field equation! Therefore, one has to introduce an approximat-

*) In order to avoid unnecessary complications due to the spin of hadrons we consider scalar hadrons only.

ion for K.

Best known is the ladder approximation of the BS-equation where one approximates K by the free meson propagator and neglects self-energy effects in internal nucleon lines (but this is inconsistent, as will be shown later on).

We make for K the following phenomenological ansatz

$$K(p_3,p_4,p_1,p_2) = \begin{cases} I((p_1+p_2)^2) & \text{for scalar coupling} \\ -(p_1-p_2)(p_3-p_4)I((p_1+p_2)^2) & \text{for vector coupling} \end{cases} \tag{12a}$$

$$\text{with} \quad I(k^2) := \frac{-ig^2}{(k^2-m^2)^\delta} \tag{12b}$$

where g, m and δ ($\delta > 0$) are free parameters and K has to be understood as an exchange of a neutral system.

In a refined version we should consider instead of $I(k^2)$ given by eq. (12b) a spectral representation for $I(k^2)$, which takes into account approximately vertex corrections and many particle exchange effects, whose asymptotic form is given by eq. (12b).

This effective kernel K we represent graphically in the following by a double line.

In order to obtain a consistent approximation scheme we must approximate Dyson's equation for the nucleon self-energy in an analogous manner. We obtain

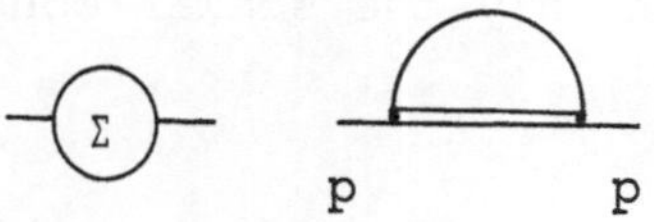

$+ \delta M^2 - (p^2 - M^2)(Z_2 - 1)$

Fig. 9: Approximate Dyson equation for Σ

where the renormalization constants δM^2 and Z_2 have to be adjusted as usual, i.e. such, that (M = nucleon mass)

$$\Sigma(M^2) = 0, \qquad \Sigma'(M^2) = 0 \; . \tag{13}$$

The Dyson equ. in Fig. 9 is a nonlinear integral equation for $\Sigma(p^2)$, which has in the cases of scalar resp. vector coupling the form [20]

$$\Sigma(p^2) = - \int \frac{d^4k}{(2\pi)^4} I((k+p)^2) \frac{\begin{pmatrix} 1 \\ -(p-k)^2 \end{pmatrix}}{k^2 - M^2 - \Sigma(k^2)} + \delta M^2 - (p^2 - M^2)(Z_2 - 1) \, . \tag{14}$$

We call the approximation of the BS-equation by means of the effective kernel (12) and nucleon propagators given by (14) dressed ladder approximation [21].

The additional coupling of pions to the $N\bar{N}$ systems will be treated in a zeroth approximation in connection with pion electroproduction in section 2.7.

2.2 Asymptotics for propagators and the elastic scattering amplitude

In this lecture we consider only the scalar coupling case systematically.

Propagator

We have for $\Sigma(p^2)$ the integral equation (14) (scalar coupling case) with the constraints (13).

Z_2 is a function of the four parameters g, δ, m and M depending on them for dimensional reasons in the following form only

$$Z_2 = Z_2(\frac{m}{M}, \frac{g}{M^\delta}, \delta) .$$

We have to distinguish two cases.

(i) $Z_2 > 0$ - canonical case.

The only asymptotic behaviour of $\Delta_F'(p^2)$ consistent with the integral equation (14) is

$$\Delta_F'(p^2) \xrightarrow[p^2\to\infty]{} Z_2^{-1}.\frac{1}{p^2} \tag{15}$$

as the integral in (14) gives a contribution to $\Sigma(p^2)$ which behaves asymptotically like $(p^2)^{-\delta+1}$ (for $\delta \leq 1$ the integral diverges, one has first to perform the mass renormalization). We have called this case the canonical one, because we obtain for $p^2\to\infty$, up to a finite constant, the free propagator.

(ii) $Z_2 = 0$ - non-canonical case

If we define

$$\Sigma_o(p^2) := \Sigma(p^2) - p^2 + M^2 \tag{16}$$

then (14) takes the form

$$\Sigma_o(p^2) = \int \frac{d^4k}{(2\pi)^4} I((k+p)^2) \frac{1}{\Sigma_o(k^2)} + \delta M^2. \tag{17}$$

As for the determination of the asymptotic behaviour of $\Sigma_o(p^2)$ the ultraviolet region in the integral in (17) dominates, we may neglect the "mass" term in I.

Then we get immediately from the scaling property of I

$$\Sigma_o(p^2) \underset{p^2\to\infty}{\sim} (p^2)^{-\delta/2+1}, \qquad \delta < 2 \tag{18}$$

and therefore

$$\Delta_F'(p^2) \underset{p^2\to\infty}{\sim} (p^2)^{\delta/2-1}, \qquad \delta < 2\ . \tag{19}$$

We note, that each solution of equ. (17) is a non-perturbative solution which in case of fixed δ, m and M is valid for those values of the coupling constant g only which satisfy $Z_2 = 0$. Whether such solutions exist has not yet been shown.

BS-amplitude

Regge behaviour can be analyzed essentially in the same manner as done by Nakanishi in the case of the ladder approximation [22]. We don't think, that new features will appear.

We now treat that limit which will be important for the analysis of deep inelastic electron scattering: the Bj-limit for the forward $N\bar{N}$-scattering amplitude with the nucleon in the initial and final state on its mass shell. Our BS-amplitude (expressed as a function of the total energy squared s and the $\bar{N}$ mass^2) satisfies the BS-equ.

$$T((p_1 + p_2)^2, p_2^2) = I((p_1 + p_2)^2) - \tag{20}$$

$$- \int \frac{d^4k}{(2\pi)^4} T((p_1+p_2-k)^2, (p_2-k)^2) \Delta_F'^2((p_2-k)^2) I(k^2),$$

where $p_1^2 = M^2$.

We are interested in the limit of T for $p_1 \cdot p_2 := \nu \to \infty$, but $\omega := -2\nu/p_2^2$ fixed. The arguments in the following go parallel to the analysis of the first diagram in Fig. 3 in the Bj-limit by Polkinghorne et al. [8]. Introduce Sudakov variables

$$k = xp_1 + yp_2 + \kappa \quad \text{with} \quad \kappa \perp p_1, p_2 . \tag{21}$$

After the transformation $y = y' - \frac{xM^2}{2\nu}$ we obtain for the invariants under the integral in (20) at large ν

$$k^2 \simeq \kappa^2 - \frac{2\nu}{\omega} y'^2 + 2xy'\nu - 2y'x \frac{M^2}{\omega}$$

$$(p_1+p_2-K)^2 \simeq (x-1)^2 M^2 + \kappa^2 +$$

$$+ [2\nu(y'-1) - xM^2].[\frac{1-y'}{\omega} + x - 1] + \frac{M^2}{\omega} (y'-1) \tag{22}$$

$$(p_2-K)^2 \simeq \kappa^2 + (y'-1)\ 2\nu[- \frac{y'-1}{\omega} + x] + xM^2 \frac{y'-1}{\omega}$$

Let us first consider the contribution from $y' \simeq 1$, i.e. from finite arguments of the BS-amplitude in the integral. With the substitution $y' = 1 + \bar{y}/2\nu$ and the fact, that the Jacobian J of the transformation (21) behaves for large ν as

$$J \sim \nu, \tag{23}$$

we obtain from the integral in (20) a contribution proportional to $\nu^{-\delta}$ arising from $I(k^2)$. All other non-ultraviolet contributions to (20) can be analyzed in the same way as Polkinghorne did, with the result that they are negligible. For the ultraviolet contribution to the integral in (20) we must distinguish the cases of canonical resp. non-canonical asymptotic behaviour of the propagator. With (22), (23) and the substitution $\kappa \to \kappa\sqrt{\nu}$ we obtain at large ν for $T \xrightarrow[Bj]{} \nu^{-\gamma}$ as input a contribution proportional to $\nu^{-\gamma-\delta}$ in the canonical case resp. $\nu^{-\gamma}$ in the non-canonical case to the integral in equ. (20). Supposed the solution of equ. (20) in the non-canonical case may be obtained by means of an iterative procedure, we get finally in both cases

$$T \xrightarrow[Bj]{} \nu^{-\delta} . \tag{24}$$

2.3. Minimal electromagnetic interaction and Ward-identities

We are interested in the rules, which lead in case of hadronic Green's functions resulting from a given interaction force to the corresponding Green's functions with additional n-external photon lines without any e.m. radiative corrections.

In the Lagrangian framework this rule is given by the principle of minimal e.m. interaction, i.e., the substitution $\partial_\mu \to \partial_\mu - ieA_\mu$ in the strong interaction Lagrangian.

This we have to translate into the language of graphs for the construction of proper photon-hadron Green's functions.

Rule 1 for the insertion of n-photon lines into strong interaction proper Feynman graphs:

Start with any proper renormalized strong interaction Feynman graph and couple n photon lines successively to all internal charged lines, add the corresponding seagull graphs in case of $n \geq 2$ for scalar lines and the appropriate contact graphs in case of derivative couplings (if derivatives operate on charged lines).

We don't intend to formalize this rule, which certainly can be done. I only remind you that for each particular photon-hadron Green's function constructed according to this rule the corresponding WIs are automatically fulfilled.

Our object are Green's functions defined by means of integral equations, i.e. we sum up in closed form an infinite set of Feynman graphs. Therefore, it would be to cumbersome to apply rule 1 directly.

It is more appropriate in our case to start with a sceleton expansion of proper Green's functions in terms of the effective BS-kernel K and the renormalized hadron propagators Δ_F'.

The following rule 2 is not a general one, it can be applied to our BS-model only provided the BS-kernel transforms like an isoscalar.

Rule 2 for the insertion of n photon lines into a strong interaction proper sceleton graph:

n = 1: Insert into each Δ_F' of a charged line the e.m. vertex part Γ_μ and add the appropriate contact graphs in the case of vector couplings.

n = 2: If the two photons have to be coupled to the same charged line insert into the corresponding Δ_F' the full amputated off-shell Compton amplitude. In all other cases go ahead successively with the rule for n = 1.

n $\geq$ 3: Generalize the case n = 2.

In particular we have to note that no e.m. form factor multiplies the contact terms!

For n = 1 the rule 2 follows from rule 1 by partial summation. For n = 2 we obtain rule 2 for the simultaneous coupling of both photons to the same charged line by means of rule 1 for n = 2 and again partial summation.

Rule 2 will be illustrated by the examples treated in the following sections.

2.4 The e.m. vertex and its asymptotic behaviour

The proper hadronic amplitude we have to start with is the self-energy $\Sigma(p^2)$ which according to Fig. 9 is given by only one sceleton graph. According to rule 2 we obtain for the e.m. vertex part Γ_μ in the scalar coupling case

Fig. 10: BS-equation for Γ_μ-scalar coupling case

The first diagram on the r.h. side of Fig. 10 is the improper contribution, the 2nd one the proper contribution to Γ_μ.

In the vector coupling case we obtain from rule 2

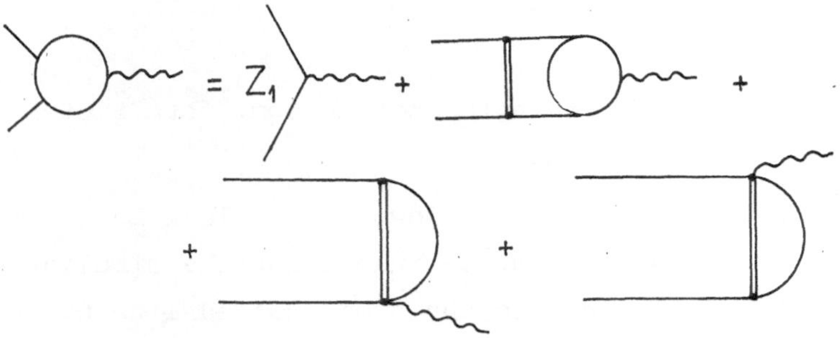

Fig. 11: BS-equation for Γ_μ-vector coupling case

It is easy to convince oneself that the Γ_μ satisfying the BS-equ. Fig. 10 or 11 respectively fulfill the WI

$$(p_2-p_1)^\mu \, \Gamma_\mu(p_2|p_1) = \Delta_F'^{-1}(p_2) - \Delta_F'^{-1}(p_1) \qquad (25)$$

if and only if the integral equ. (14) for $\Sigma(p^2)$ holds and we have $Z_1 = Z_2$.

Asymptotic behaviour

We will discuss the case of scalar BS-coupling only. We introduce as usual two e.m. form factors F and G through the decomposition ($q := p_2 - p_1$)

$$\Gamma_\mu(p_2|p_1) = (p_1+p_2)_\mu \, F(p_2^2,p_1^2,q^2) + (p_2-p_1)_\mu \, G(p_2^2,p_1^2,q^2). \qquad (26)$$

By means of the WI (25) we may express G in terms of F and Σ. We are interested therefore in the asymptotic behaviour of $F(p_2^2,p_1^2,q^2)$ for any or several of its arguments going to infinity.

(i) $Z_2 = Z_1 > 0$.

In that case Z_1 is fixed by the normalization condition $F(M^2,M^2,0) = 1$. From the BS-equ. Fig. 10 it is at least plausible, that if any of its arguments goes to infinity,

$$F \to Z_1 \qquad (27)$$

(ii) $Z_2 = Z_1 = 0$.

We have now a homogeneous integral equation for F. By using Sudakov's technique and eq. (19) one discovers again a universal asymptotic limit of the kind*)

$$F \sim (p^2)^{-\gamma} \quad \text{with} \quad p^2 = \max(p_2^2, p_1^2, q^2) \ . \tag{28}$$

In order to identify the power γ we use the WI (25) in the particular form

$$F(p^2, p^2, 0) = (\Delta_F'^{-1}(p^2))' \tag{29}$$

and obtain, due to eq. (19)

$$\gamma = \frac{\delta}{2} \ . \tag{30}$$

Note that the non-canonical case for scalar BS-coupling leads to an asymptotic decreasing e.m. form factor!

2.5 e^+e^--annihilation: contact terms and asymptotic increasing R

e^+e^--annihilation is the crucial test for each parton alternative, if the present experimental tendency [15] of

$$R \simeq c\sqrt{s}$$

---

*) Such a form independence of the vertex asymptotic behaviour has been observed by Appelquist and Primack for the summation of leading terms in perturbation theory [9].

will be confirmed by the data to be taken in the next future.

We have to consider the absorptive part of the hadronic contribution to the photon self-energy

$$\Pi_{\mu\nu}(q^2) := \int d^4x e^{iq\cdot x} \langle 0|T^* (j_\mu(\tfrac{x}{2}), j_\nu(-\tfrac{x}{2}))|0\rangle .$$

This is a vacuum amplitude, i.e. no hadronic amplitude to start with exists. Therefore we have to consider the Dyson-equation for $\Pi_{\mu\nu}$ [19], which is given for our BS-model by the diagrams of Fig. 12 and Fig. 13 respectively.

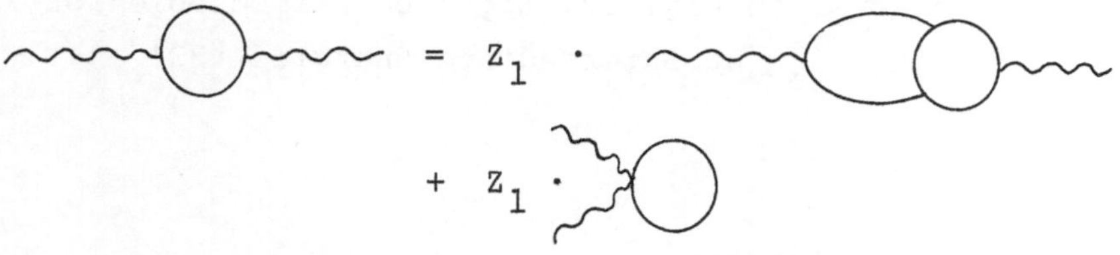

Fig. 12: Dyson equation for $\Pi_{\mu\nu}$-scalar coupling case

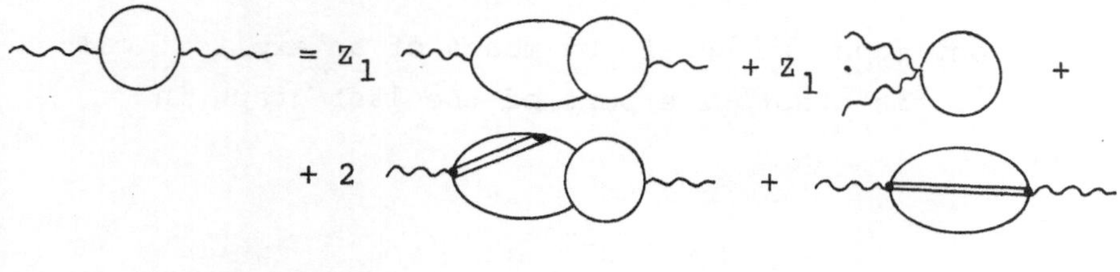

Fig. 13: Dyson equation for $\Pi_{\mu\nu}$-vector coupling case

A full analysis of the eq.'s given by Fig. 12 and 13 has not yet been performed. We will only briefly sketch what are the main features we expect.

Scalar coupling case with $Z_1 > 0$

According to the foregoing results hadrons show a partonlike behaviour in that case, i.e. we expect in the asymptotic region the dominance of the contribution $Z_1^2 \cdot$
$\cdot$  leading to R = const. [12].

As an alternative one should consider the possibility, that present energies are not yet asymptotic. A strong hadron-hadron interaction, contributing to the first diagram in Fig. 12, might be responsible for an increasing σ_{tot} at intermediate energies [23].

Vector coupling case

It has been quite recently suggested by Rubinstein that contact graphs give the dominating contribution [24].

In our model we obtain by means of an explicit calculation of the absorptive part of the last graph in Fig. 13

$$R \sim s^{1-\delta} .$$

In particular we have $R \sim \sqrt{s}$ for $\delta = 1/2$.

It is remarkable in this context that contact graphs find a natural place in a scheme starting with an effective vector coupling BS-kernel. Very important is the remark made after stating rule 2: no e.m. form factors multiply the contact terms as an immediate consequence of the principle of minimal e.m. interaction.

2.6 Deep inelastic eN-scattering: breakdown of canonical scaling?

Let us first construct the off shell Compton amplitude $T_{\mu\nu}$. We have

Fig. 14: General structure of $T_{\mu\nu}$

with

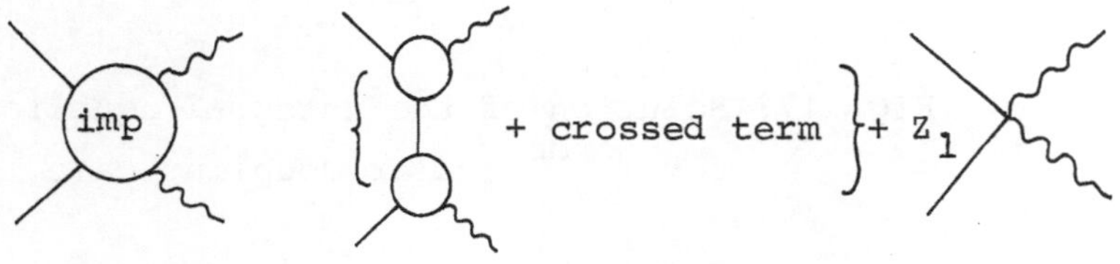

Fig. 15: Improper part of $T_{\mu\nu}$

and due to our rule 2, starting with the hadronic amplitude $\Sigma(p^2)$, for the proper part of $T_{\mu\nu}$ in the scalar coupling case.

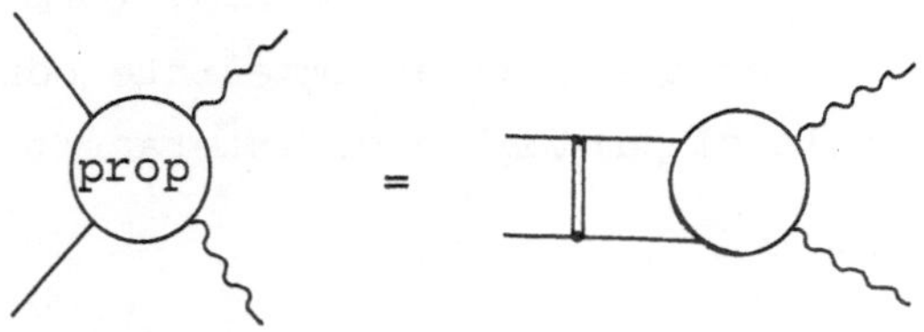

Fig. 16: $T_{\mu\nu}^{prop}$-scalar coupling case

The equations given by Fig. 14, 15 and 16 lead to an integral equation for $T_{\mu\nu}^{prop}$, which due to the BS-equation for T possesses the solution given in Fig. 17.

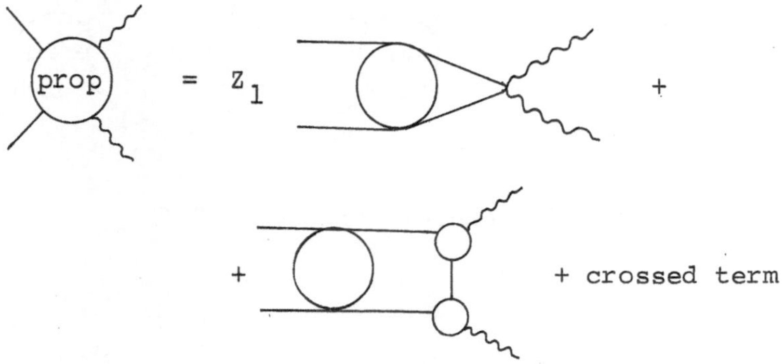

Fig. 17: Solution of the integral equation for $T_{\mu\nu}^{prop}$-scalar coupling case

In the vector coupling case we obtain for $T_{\mu\nu}^{prop}$ by means of rule 2 the diagrams of Fig. 18.

In the <u>case of scalar coupling and $Z_1 > 0$</u> the analysis of the behaviour in the Bj-limit is completely the same as for the Cambridge parton model, because e.m. forı factors behave asymptotically like a constant and the fall-off property of T for virtual external masses is

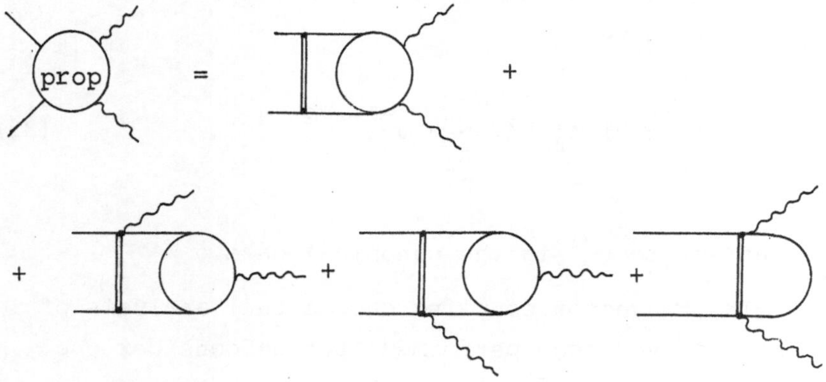

Fig. 18: $T^{prop}_{\mu\nu}$-vector coupling case

guaranteed by the Bj-limit result for T (eq. (24)). We say, that nucleons show a parton-like behaviour in that case.

In the case of scalar coupling and $Z_1 = 0$ we perform an analysis of the graphs in Fig. 17 by means of the Sudakov technique.*) Seagull graphs need not to be considered, as they contribute to the real part of $T_{\mu\nu}$ only. For the handbag diagram of Fig. 17 and its crossed counterpart we find from finite internal nucleon masses as well as from the ultraviolet region in the loop integration

$$T_2 \underset{Bj}{\sim} \nu^{-1-\delta/2} \qquad (31)$$

i.e. a scaling law corresponding to an anomal dimension.

*) Compare section 2.2

The reason for that can best be seen from the contribution of finite internal masses to the second graph in Fig. 17. We have

$$F^2(-q^2)\ \Delta_F'(k^2) \underset{Bj}{\sim} \nu^{-1-\delta/2} \qquad (32)$$

in contrast to ν^{-1} in the canonical case.

In the vector coupling case a full analysis of $T_{\mu\nu}$ has not yet been performed. Let us consider one particular contribution - the last contact graph in Fig. 18. It contributes to W_1 only with the result

$$W_1 \underset{Bj}{\sim} \nu^{-\delta+1} \qquad (33)$$

i.e. we get canonical scaling only if $\delta = 1$.

From the work of Fishbane and Sullivan [25] we know that in the vector-gluon model scaling is violated by logarithmic terms in perturbation theory. In which form these logarithms sum up in the solution of our integral equation Fig. 18 we don't know yet.

The results of this section have to be compared with the recent work of Gutbrod and Schröder [26].

2.7 Exclusive pion electroproduction

The simplest mechanisms for an additional $\pi N\bar{N}$ coupling within our scheme is a weak (i.e. 1st order) $\pi N\bar{N}$ coupling such, that the $\pi N\bar{N}$-vertex part Γ_5 satisfies the following inhomogeneous BS-equation

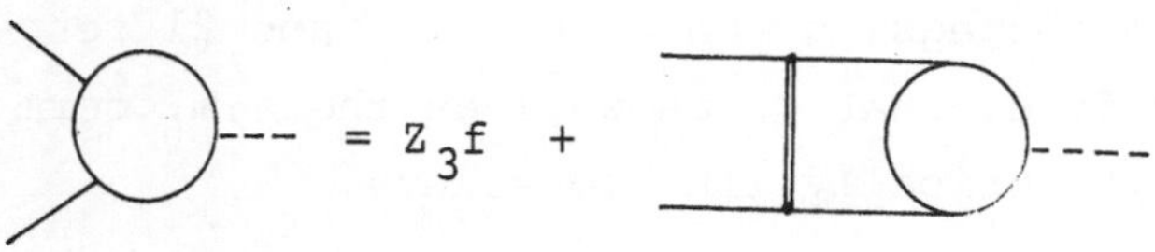

Fig. 19: BS-equation for the $\pi N\bar{N}$-vertex part Γ_5.

where Z_3 is defined by the normalization condition $\Gamma_5(M^2, M^2, \mu^2) = f$. The physical picture for such a mechanism is given by $\vec{\Phi}_\pi(x) := \text{const}\ \partial^\mu \vec{A}_\mu(x)$ with $\vec{A}_\mu$ being the hadronic part of the weak interaction axialvector current.

The corresponding self-energy Σ_π we find from the Dyson equation

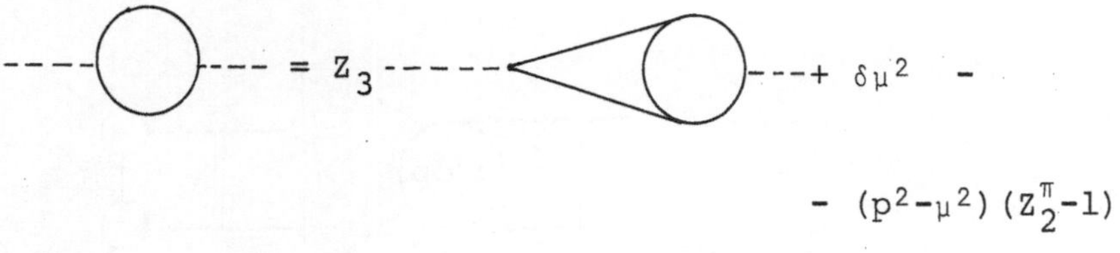

Fig. 20: Dyson equation for Σ_π

By means of the equations given in Figs. 19 and 20 it is easy to construct the pion e.m. vertex part with the aid of rule 2. We obtain

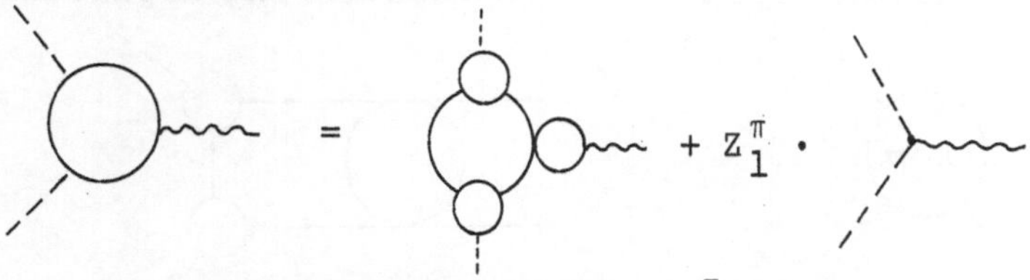

Fig. 21: Equation for Γ_μ^π

The WI for Γ_μ^π requires as usual $Z_1^\pi = Z_2^\pi$.

From the eqs. given in Fig. 19 and 21 respectively we easily find, that Γ_5 as well as the e.m. form factor behave asymptotically like constants.

The amplitude for single π^+ electroproduction T_μ is now again constructed by means of our rule 2, starting with $\Gamma_5 - Z_3 f$ as the proper hadronic vertex part. In this way we obtain in the scalar coupling case

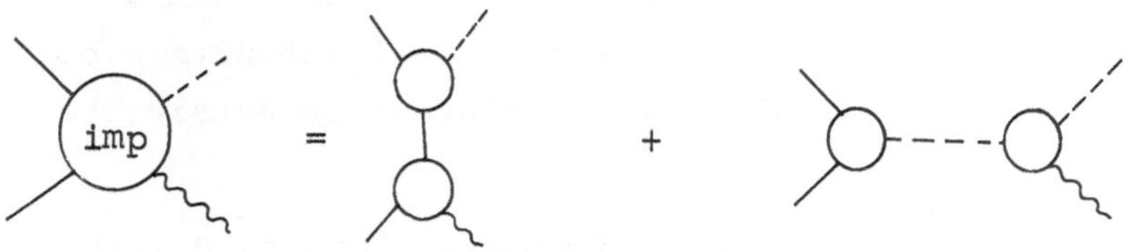

Fig. 22: Electroproduction amplitude - improper part

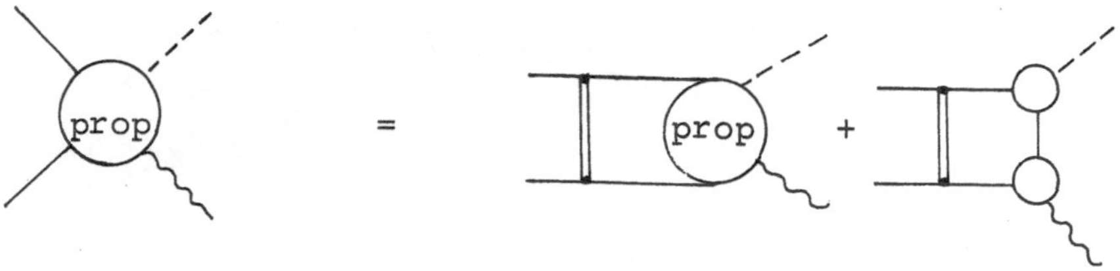

Fig. 23: Electroproduction amplitude - proper part

The solution of the integral equ. for the proper part may be given in terms of the BS-scattering amplitude T

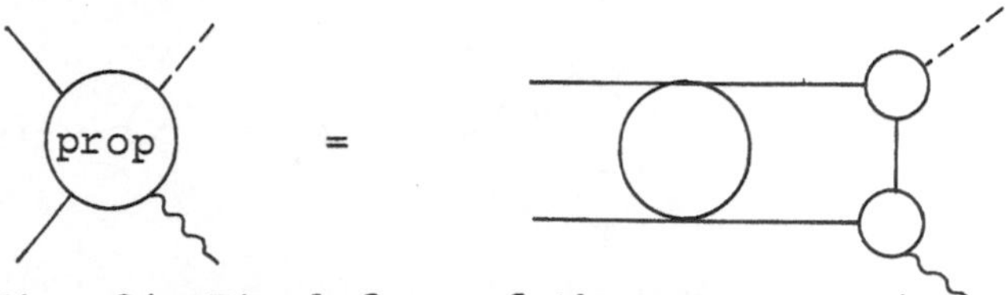

Fig. 24: Final form of the proper part

<u>Regge behaviour</u> (scalar coupling, $Z_1 > 0$)

If we analyse the Regge-behaviour of the proper part (Fig. 24) by means of the Sudakov technique we rediscover the fixed pole known from the ladder model [27].

$$T_\mu^{proper} \underset{R\text{-lim}}{\simeq} \frac{(p_1+p_2)_\mu}{2\nu} [\Gamma_5(p_2|p_1) - Z_3 f] \qquad (34)$$

$$+ \text{ moving Regge-pole contributions}$$

<u>Bj-limit</u> (scalar coupling, $Z_1 > 0$)

The analysis of the behaviour of the proper part (Fig.24) in the Bj-limit goes again completely parallel to the parton model analysis in deep inelastic scattering (parton like behaviour of N in that case) with the result

$$A_i \underset{Bj}{\sim} \nu^{-1} \qquad (35)$$

where the A_i are the invariant functions in the usual decomposition

$$T_\mu^{proper} = A_1 \ (p_1 + p_2)_\mu + A_2 \ (p_2 - p_1)_\mu + ck_\mu .$$

The result (35) means again canonical scaling.

I remind you that the Cambridge parton model needs additional asumptions (dominance of certain Regge graphs) in order to deal with such an exclusive process.

3. CONCLUSIONS

In order to introduce into a given strong interaction scheme additional couplings of hadrons with external photons in agreement with the principle of minimal electromagnetic interaction we have to start with the basic hadronic force and not with amplitudes. This is the main lesson we learned from a critical discussion of the parton model in chapter 1.

We realized this idea in chapter 2 by means of phenomenological BS-kernels as the basic hadronic force. The strong interaction scheme has been made selfconsistent by using the same BS-kernel in Dyson's equation for propagators.

A discussion of electromagnetic processes within our framework led to a parton-like behaviour of hadrons in the scalar coupling case for $Z_2 > 0$. Only a partial discussion of the interesting new features showing up for $Z_2 = 0$ and in the vector coupling case have been given in the present paper. The first results encourage us to follow up this line.

REFERENCES

1. Compare some recent books and reviews and the literature cited therein:
R. P. Feynman, Photon-Hadron Interaction, Addison Wesley, New York, 1972.
F. E. Close, Partons and Quarks, Daresbury: Lecture Notes Series, No. 12, DNPL IR31 (1973).

P. V. Landshoff, J. L. Polkinghorne, Physics Report 5c, 1 (1972).

2. Proc. Intern. Symp. on Electron and Photon Interaction at High Energies, Bonn, 1973.

3. J. D. Bjorken, Phys. Rev. 179, 1547 (1969).

4. A more refined form of the original parton model has been discussed by Kuti, Weißkopf: Phys. Rev. D4, 3418 (1971).

5. D. S. Drell, D. J. Levy, T. M. Yan, Phys. Rev. D1, 1035, 1617, 2402 (1970).

6. K. Osterwalder, Fortschr. Physik 19, 43 (1971).

7. S. J. Chang, P. M. Fishbane, Phys. Rev. D2, 1084 (1970).
V. N. Gribov, L. N. Lipatov, Sov. J. Nucl. Phys. 15, 438 (1972).
P. M. Fishbane, J. D. Sullivan, Phys. Rev. D4, 2516 (1971).
A. L. Mason, J. Math. Phys. 14, 1601 (1973).

8. P. V. Landshoff, J. C. Polkinghorne, R. D. Short, Nucl. Phys. B28, 225 (1971).

9. T. Appelquist, J. R. Primack, Phys. Rev. D1, 1144 (1970).

10. P. V. Landshoff, J. C. Polkinghorne, Phys. Reports 5c, 1 (1972).

11. P. V. Landshoff, J. C. Polkinghorne, Nucl. Phys. B33, 221 (1971).
G. Altarelli, L. Maiani, Nucl. Phys. B56, 477 (1973).

12. N. Cabibbo, G. Parisi, M. Testa, Nuovo Cim. Lett. 4, 35 (1970).

13. P. V. Landshoff, J. C. Polkinghorne, Nucl. Phys. 43B, 279 (1971).

14. D. M. Scott, Nuovo Cim. 18A, 271 (1973).

15. Talk given by G. Goldhaber at the "Frühjahrstagung der Deutschen Physikalischen Gesellschaft", Hamburg, Febr. 20 - 22, 1974.

16. C. A. Orzalesi, Nuovo Cim. Lett. 5, 581 (1972).

17. S. J. Brodsky, F. E. Close, J. F. Gunion, Phys. Rev. D8, 3678 (1973).

18. C. A. Orzalesi, Trieste preprint IC/73/148.

19. Compare:
J. D. Bjorken, S. D. Drell, Relativistic Quantum Field Theory, McGraw Hill Book Co, New York, 1965.

20. For spinor QED such an integral equation for $\Sigma(p^2)$ has been introduced by
R. Haag, Th. A. J. Maris, Phys. Rev. 132, 2325 (1963).

21. For the particular case $\delta = 1$ this dressed ladder approximation has been discussed by
R. Saenger, Phys. Rev. 159, 1433 (1967).

22. N. Nakanishi, Phys. Rev. 133, B214 (1964).

23. T. Muta, Preprint, Kyoto University, Nov. 1973.

24. C. Ferro Fontan, H. R. Rubinstein, CERN-preprint, Ref. TH. 1810.

25. P. M. Fishbane, J. D. Sullivan, Phys. Rev. D4, 2516 (1971).

26. F. Gutbrod, U. E. Schröder, Nucl. Phys. B62, 381 (1973).

27. J. C. Polkinghorne, Nucl. Phys. B3, 395 (1967).

Acta Physica Austriaca, Suppl. XIII, 487–533 (1974)

MASS SHELL AND NON-LOCALITY IN ELECTRODYNAMICS

by

F. ROHRLICH
Syracuse University, Syracuse, New York

1. MOTIVATION

Much criticism can be brought against our present theory of electromagnetic interactions. Most of it can be summarized as follows.

(A) Conceptually, classical and quantum electrodynamics are not very close; for example, the photon picture of the latter is extended to virtual photons to describe what are essentially classical interactions such as the Coulomb interaction. A larger common ground is not only desirable on its own accord but also for reasons of logical coherence: the classical limit of quantum electrodynamics has never been carried through satisfactorily. Not unrelated to this problem is the lack of intuitive understanding which was emphasized so well in Feynman's Nobel Lecture [1]. The underlying physical model is not capable of permitting simple

+) Lectures given at XIII. Internationale Universitätswochen für Kernphysik, Schladming, February 4 - 15, 1974.

qualitative, order of magnitude estimates of basic phenomena without falling back on computational recipes; examples are the "anomalous" magnetic moment of the electron, radiative level shifts, etc.

(B) Mathematically, it is well known that quantum electrodynamics, because of its infinite range poses even more difficulties than other types of field theories.

This criticism is to be contrasted to the excellent predictive power of the theory. There never was in the history of physics a theory which agrees so well with experiment and which is mathematically so unsatisfactory. It poses a tremendous challenge to the theorist.

Now one can take the point of view, as we shall do, that good physics breeds good mathematics and that the above cited ills are due to a physically poorly formulated theory. One can argue that in the past, in the effort to extract information from the theory, one gave into what seemed to be the <u>mathematically</u> most convenient formulation: manifest covariance, locality of interacting fields, easy computational schemes, an approximation of free charged particles by plane waves, etc. If one emphasizes the physics, however, keeping as close as possible to the observable situation, one lets the mathematical structure develop into whatever it may, possibly into a rather difficult but hopefully well defined problem. In this endeavor one tries to minimize the theoretical constructs which do not involve observables. This will hopefully counter some of the criticism expressed in (A), and in its wake perhaps provide at least a partial answer to (B).

The following lectures report on the first attempts

in this direction. Naturally, the existence of a seemingly quite satisfactory classical formulation [2,3] of the theory of point charges has influenced this development. The classical theory referred to is free of all divergences and is "already renormalized." It involves no cut-offs of any sort and - remarkably - accomplishes its success pretty much within the conventional framework using only certain conceptual modifications.

The present point of view is based on the separation of electromagnetic interactions into local and non-local (action-at-a-distance) interactions, in the quantum case as in the classical case. This separation corresponds closely to what is being observed; and this separation leads directly to the free field as the only independent freedom of the electromagnetic field. We are thus led to undo in part what Maxwell has done: the beauty and simplicity of his equations rely heavily on treating all electromagnetic interactions as local. This works fine for phenomenological electrodynamics. But on the fundamental level the price is too high. That price includes self-energy problems, gauge problems, quantized Coulomb fields, an indefinite metric Hilbert space, and a physical picture which mixes real and virtual electromagnetic fields. A separation of these two will be shown to be of great value.

We shall therefore first review the classical theory very briefly and discuss its reasons for success as compared to the conventional theory. We shall then construct a quantum electrodynamics in analogy. But we shall not "quantize the classical theory". Such a notion had its place in the historical development of quantum mechanics, but is entirely outdated. It is in no way better than the notion of "deriving special relativity from Newtonian mechanics". Obviously, it's the other way around: classical electrodynamics (CED) is a limit of quantum electrodynamics (QED).

The only electromagnetic field in the new form of QED is the free field A_{in} (or equivalently A_{out}). We shall then show that, as a consequence, there is no longer a gauge problem and that one is led to a theory with a positive definite Hilbert space. Furthermore, causality can no longer be introduced into the theory in the usual fashion by imposing locality on the interacting fields. Finally, we shall cast the theory in a form in which there are no infrared divergences. The mass shell approach, i.e. the separation into local interactions (with real electromagnetic fields) and nonlocal interactions will also be valuable in this task.

2. RECALL OF CLASSICAL POINT ELECTRODYNAMICS

(2a) General Theory

Consider a system of n point particles of masses m_k and charges e_k, and with worldlines $z_k^\mu(\lambda_k)$ (k=1,2, ...,n), which depend on the monotonically increasing parameters λ_k. This system is in interaction with two electromagnetic fields $\bar{F}^{\mu\nu}$ and $F_P^{\mu\nu}$ describing all free and all "bound" fields, respectively. The system is completely specified by the following assumptions [3]:

(A) The action integral is

$$I = -\sum_{k=1}^{n} m_k \int_{-\infty}^{\infty} \sqrt{-\frac{dz_k^\alpha}{d\lambda_k}\frac{dz_{\alpha k}}{d\lambda_k}}\, d\lambda_k + \sum_{k=1}^{n} \int_{-\infty}^{\infty} \bar{A}_\alpha(x) j_k^\alpha(x) d^4x$$

(2.1)

$$-\frac{1}{2}\int \bar{F}_{\alpha\beta}(x) F_P^{\alpha\beta}(x) d^4x + \frac{1}{2}\sum_{\substack{k\ell \\ (k\neq\ell)}}\int j_{\alpha k}(x) D_P(x-x') j_\ell^\alpha(x') d^4x d^4x'$$

with

$$j_k^\mu(x) \equiv e_k \int_{-\infty}^{\infty} \delta_4(x-z_k) \frac{dz_k^\mu}{d\lambda_k} d\lambda_k \qquad (2.2)$$

$$D_P(x) \equiv \frac{1}{(2\pi)^4} P\!\!\!\int \frac{e^{ik\cdot x} d^4k}{k^2} \; ; \qquad (2.3)$$

Here $\bar{F}^{\mu\nu} \equiv \partial^\mu \bar{A}^\nu - \partial^\nu \bar{A}^\mu$, and similarly for $F_P^{\mu\nu}$.

(B) This action integral is not only Poincaré invariant under the connected group, but also time reversal invariant, and each field is also time reversal covariant. The latter means that under a time reversal transfromation a Lorentz tensor is mapped into the contragredient space, apart from an overall sign factor. For example, for $x^\mu \to x'^\mu = x_\mu$

$$j'^\mu(x') = -j_\mu(x); \quad \bar{F}'^{\mu\nu}(x') = -\bar{F}_{\mu\nu}(x); \quad F_P'^{\mu\nu}(x') = -F_{P\mu\nu}(x).$$

(2.4)

(C) The total electromagnetic field $F^{\mu\nu} \equiv \bar{F}^{\mu\nu} + F_P^{\mu\nu}$

satisfies the asymptotic conditions, conveniently expressed in terms of the potential:

$$A^{\mu}_{in}(x) = \lim_{t'\to-\infty} \int D(x-x') \overleftrightarrow{\partial}'_{o} A^{\mu}(x')d^3x'$$
$$A^{\mu}_{out}(x) = \lim_{t'\to\infty} \int D(x-x') \overleftrightarrow{\partial}'_{o} A^{\mu}(x')d^3x' \tag{2.5}$$

with

$$D = 2 \text{ Re } D_+, \quad D_+ \equiv \frac{i}{(2\pi)^3} \int e^{ik\cdot x} \frac{d^3k}{2\omega} \tag{2.6}$$

where $\omega = |\vec{k}| = k^o$.

From part (A) of these three postulates one obtains first, via the Euler-Lagrange equations, in the standard way

$$\partial_{\alpha} \bar{F}^{\alpha\mu} = 0 \tag{2.7}$$

$$\partial_{\alpha} F^{\alpha\mu}_{P} = - \sum_{k=1}^{n} j^{\mu}_{k}(x) \equiv - j^{\mu}(x) \tag{2.8}$$

$$m_k \frac{dr^{\mu}_k}{d\tau_k} \equiv e_k \bar{F}^{\mu\alpha}(z_k)v_{k\alpha} + e_k v_{k\nu} \sum_{\substack{\ell=1 \\ (\ell\neq k)}}^{n} e_{\ell} \cdot$$
$$\cdot \int (v^{\mu}_{\ell}\partial^{\nu}_{k} - v^{\nu}_{\ell}\partial^{\mu}_{k}) D_P(z_k - z_{\ell}) \quad . \tag{2.9}$$

The proper time of particle k, t_k, is defined by

$$d\tau_k = d\lambda_k \sqrt{- \frac{dz^{\alpha}_k}{d\lambda_k} \frac{dz_{k\alpha}}{d\lambda_k}} \quad . \tag{2.10}$$

Next, from part (B) one finds using (2.4)

$$\bar{A}'^{\mu}(x') = -\bar{A}_{\mu}(x) \tag{2.11}$$

$$A'^{\mu}_{P}(x') = -A_{P\mu}(x) . \tag{2.12}$$

This implies that the most general time covariant solution of (2.8) is

$$A^{\mu}_{P}(x) = D_P * j^{\mu}(x) . \tag{2.13}$$

The star denotes the convolution product. An additional solution of the homogeneous equation proportional to D_1*j ($D_1 \equiv 2 \text{ Im } D_+$, (2.6)) would not violate time reversal covariance but would be inconsistent with the assumption that <u>all</u> the (time covariant) free fields are described by $\bar{A}$.

Finally, from part (C)

$$A_{in} = \bar{A} - \tfrac{1}{2}D*j$$
$$A_{out} = \bar{A} + \tfrac{1}{2}D*j \tag{2.14}$$

or, equivalently,

$$\bar{A} = \tfrac{1}{2}(A_{in} + A_{out}) \tag{2.15}$$

$$A_{out} = A_{in} + D*j \equiv A_{in} + 2A_- . \tag{2.16}$$

This identifies $\bar{A}$ as the arithmetic mean of the free <u>in</u> and <u>out</u> fields. From (2.13) follows

$$A^\mu_P(x) = \sum_{k=1}^{n} A^\mu_{Pk}(x)$$
$$A^\mu_{P_k}(x) = e_k \int D_P(x-z_k) v^\mu_k \, d\tau_k \tag{2.17}$$

The charge e_k thus produces the bound field $A^\mu_{Pk}(x)$ at the point x. This auxiliary construct (a measurement at x cannot determine A^μ_{Pk} but only the total field) can be used conveniently in the equation of motion (2.9) to yield the conventional form of these equations,

$$m_k \frac{dv^\mu_k}{d\tau_k} = e_k \bar{F}^{\mu\nu}(z_k) v_{k\nu} + e_k v_{k\nu} \sum_{\substack{\ell=1 \\ (\ell\neq k)}}^{n} F^{\mu\nu}_{P\ell}(z_k)$$
$$= e_k [F^{\mu\nu}_{in}(z_k) + F^{\mu\nu}_{k,-}(z_k) + \sum_{\substack{\ell=1 \\ (\ell\neq k)}}^{n} F^{\mu\nu}_{\ell\, ret}(z_k)] v_{k\nu} . \tag{2.18}$$

Here we used (2.15), (2.16) and the definition

$$A^\mu_{ret} = D_R * j^\mu \tag{2.19}$$

where

$$D_R \equiv D_P + \tfrac{1}{2} D .$$

The conventional form of the relativistic equation of motion for a system of point charges has thus been obtained in (2.18). Note that the conventional fields

A_{ret} are linear combinations of free and bound fields.

The following remarks are relevant:

(a) There is no electromagnetic self-energy, the masses m_k which enter the Lagrangian are the experimental masses and no mass renormalization needs to be carried out. In conventional language: the theory is already mass renormalized. The masses are therefore entirely of non-electromagnetic nature. If the charged particles were not points but extended due to the presence of some other interaction, as is the case e. g. for hyperons, a finite electromagnetic self-energy would of course be present.

(b) The absence of electromagnetic self-energy is due to the exclusion of $k = \ell$ in the current-current interaction of (2.1). This exclusion which also occurs in the Wheeler-Feynman theory [4], occurs here in a theory with radiation fields. Action-at-a-distance occurs only for part of the interaction, the rest, the one involving free fields is a field theoretic local interaction. If one assumes that there is no free field, $\bar{A} = 0$, one recovers the Wheeler-Feynman theory. $\bar{A} = 0$ is just the absorber condition of that theory [5].

(c) Not all the self-interaction is eliminated by the exclusion: $k \neq \ell$. The second force-term in the equation of motion is a self-interaction. It emerges from the free field. In terms of the particle kinematics it can be written as follows

$$e_k F^{\mu\nu}_{k-}(z_k) v_{k\nu} = \frac{e_k^2}{6\pi}(\ddot{v}^\mu - \dot{v}_k^2 v_k^\mu) \equiv \Gamma_k^\mu . \qquad (2.20)$$

It describes radiation reaction and is known as the

Abraham four-vector. Insertion into (2.18) gives the Lorentz-Dirac equation for a many particle system [6].

(d) This radiation reaction force is a fourvector and is spacelike. It cannot be equal to the negative of the timelike fourvector of radiation rate

$$\left(\frac{dP^{\mu}_{rad}}{d\tau}\right)_k = \frac{e_k^2}{6\pi}\,\dot{v}_k^2\, v_k^{\mu} \equiv R v_k^{\mu} \quad . \tag{2.21}$$

In fact the relation is according to (2.20)

$$-\left(\frac{dP^{\mu}_{rad}}{d\tau}\right)_k = \Gamma_k^{\mu} - \frac{e_k^2}{6\pi}\,\ddot{v}_k^{\mu} \quad . \qquad (2.22)\ \text{local}$$

They differ by the Schott term. However, while such a local equality obviously cannot exist, there does exist a global equality between these two quantities:

$$-\int_{-\infty}^{\infty}\left(\frac{dP^{\mu}_{rad}}{d\tau}\right)_k d\tau_k = \int_{-\infty}^{\infty}\Gamma_k^{\mu}(\tau_k)\,d\tau_k \quad . \qquad (2.22)\ \text{global}$$

The Schott term does not contribute here because the accelerations vanish asymptotically. Eq. $(2.22)_{global}$ can be regarded as a nonlocal conservation law of energy and momentum.

(4) Current conservation is ensured by (2.8). Therefore, A^{μ}_{ret} of (2.19) satisfies the Lorentz condition

$$\partial\cdot A_{ret} = \partial\cdot(D_R * j) = D_R * (\partial\cdot j) = 0.$$

Similarly, $\partial \cdot A_P = 0$ which ensures that (2.13) is the solution of (2.8). The free field $\bar{A}$ of (2.15) involves only one indpendent field which can be chosen either as A_{in} or as A_{out} because of (2.16). We choose A_{in}; then this free field can be taken in the Coulomb gauge $A^o = 0$ so that $\partial \cdot A_{in} = \nabla \cdot \vec{A}_{in} = 0$. It follows that the total field

$$A \equiv \bar{A} + A_P = A_{in} + A_{ret} \tag{2.23}$$

satisfies the Lorentz condition

$$\partial \cdot A = 0 \, . \tag{2.24}$$

(f) The nonlocal conservation law (2.22) also appears in a somewhat different form in the equations of motion for the point charges. The case of a single charge will suffice to illustrate the point. As is well known the Lorentz-Dirac equation must be combined with the asymptotic condition to turn it into an equation of motion (i.e. into a second order equation). The asymptotic condition is

$$\lim_{\tau \to \infty} e^{-\tau/\tau_o} \dot{v}(\tau) = 0 \qquad (\tau_o \equiv \frac{e^2}{6\pi m}) \, , \tag{2.25}$$

which is a very weak constraint on the acceleration, since the critical time $\tau_o \sim 10^{-23}$ sec is very small. When combined with the Lorentz-Dirac equation, (2.9), it yields, for a single charge,

$$m\dot{v}^\mu(\tau) = \int_0^\infty e^{-\alpha} K^\mu(\tau + \alpha\tau_o) d\alpha \, , \tag{2.26}$$

K^{μ} being the effective force, i.e. the force exerted by the incident field $F^{\mu\nu}_{in}$ less the rate of momentum loss due to radiation, (2.21),

$$K^{\mu} \equiv e\, F^{\mu\alpha}_{in}\, v_{\alpha} - Rv^{\mu}. \tag{2.27}$$

For $\mu = 0$ (2.26) expresses a nonlocal energy conservation law.

(2b) The two-field point of view

The novel feature of the above theory lies in the introduction of two electromagnetic fields, one free $\bar{F}$ and one "bound field" F_P, or equivalently $\bar{A}$ and A_P. Of these, the free field plays a dominant role, while the "bound field" is eliminated in favor of a nonlocal current-current interaction.

The best way to introduce the separation of the usual single electromagnetic field into two fields is perhaps as follows. Consider any linear inhomogeneous wave equation, of which

$$\Box A = - j \tag{2.28}$$

may serve as an example. The absence of a mass term is here irrelevant. The general solution of (2.28) is

$$A = A_H + A_P \tag{2.29}$$

where A_H is any solution of the homogeneous equation $\Box A_H = 0$ and A_P is a particular solution. Now the homogeneous and the inhomogeneous wave equations describe two different situations and have entirely different physical interpretation. The latter depends crucially on the source j, the former does not. Of course the separation (2.29) is not unique since one can always include a solution of the homogeneous solution in A_P. The initial conditions do not help, because they only fix the total A. What physical property is it then which provides for a unique separation of any A in the manner (2.29). This becomes obvious when we look at the corresponding Green functions. The intersection of all Green functions D_I of the inhomogeneous equation, $\Box D_I = -\delta$ is unique. It is D_P as given by (2.3). Its Fourier transform is entirely off mass shell. All Green functions of the homogeneous equation, D_H are entirely on mass shell.

The physical distinction between A_H and A_P is now obvious and makes the separation (2.29) unique: A_H corresponds to free particles, because it is entirely on mass shell; A_P is entirely off mass shell corresponding to "virtual particles" only. The field A_P thus does not correspond to observable particles and can be called a virtual field. It can be eliminated entirely in favor of an action-at-a-distance nonlocal interaction between the particles. This interaction is already present in the action integral (2.1). It reappears in the equations of motion (2.9). These equations involve $\bar{F}$ whereas F_P does not appear; it is completely superfluous once the equation of motion (2.9) has been derived; it is reintroduced via (2.17) only to give the equations of motion the conventional appearance (2.18).

A separation similar to (2.29) already appears in the elementary study of electrodynamics. One learns that the electromagnetic field contains a $\frac{1}{r^2}$ part and a $\frac{1}{r}$ part, describing Coulomb interactions and radiation fields, respectively. Actually it is both the Coulomb interactions and the "Biot-Savart interactions" which involve a $\frac{1}{r^2}$ field. This separation into $\frac{1}{r^2}$ and $\frac{1}{r}$ parts can be carried out for both the retarded and the advanced fields. It is Lorentz invariant [7], and one can write $F = F_2 + F_1$. However, this separation is not very useful for two reasons. First, it cannot be carried out for the potentials, and secondly, the fields F_2 and F_1 do not satisfy local differential equations. This can be seen as follows.

With $R^\mu = x^\mu - z^\mu$ and $R^0 \gtrless 0$ for the retarded and advanced case, respectively, we have for the field at a point x associated to the source point z along the light cone ($R^\alpha R_\alpha = 0$) and for $x \neq z$,

$$\partial_\alpha F_2^{\alpha\mu}(x)\Big|_{\substack{\text{ret}\\ \text{adv}}} = \pm \frac{2e}{\rho^4_{\substack{\text{ret}\\ \text{adv}}}} R\cdot\dot{v}_{\substack{\text{ret}\\ \text{adv}}} R^\mu = -\partial_\alpha F_1^{\alpha\mu}(x)\Big|_{\substack{\text{ret}\\ \text{adv}}}. \quad (2.30)$$

Here ρ is the invariant spatial distance $\rho = u\cdot R$ with $u^2 = 1$, $u\cdot v = 0$, so that $R = \rho(u \pm v)$ for the two cases.

While the separation $F = F_2 + F_1$ is thus not suitable for either the retarded or the advanced case, another covariant separation presents itself, viz. the separation

$$F_{\text{ret}} = \tfrac{1}{2}(F_{\text{ret}} + F_{\text{adv}}) + \tfrac{1}{2}(F_{\text{ret}} - F_{\text{adv}}) \equiv F_+ + F_- \,, \quad (2.31)$$

so successfully employed by Dirac [6]. Looking at the corresponding Green functions we note

$$\frac{1}{2}(D_R + D_A) = D_P$$
$$\frac{1}{2}(D_R - D_A) = \frac{1}{2}D. \qquad (2.32)$$

Thus, F_+ is to be identified with our F_P while F_- is an F_H. This is exactly the separation (2.29) for the retarded field.

Now the separation into F_2 and F_1 is physically related to the separation (2.31). Asymptotically, F and F_1 are equal in the following sense: with the limits taken along a null ray,

$$\lim_{\substack{\rho\to\infty \\ \tau\to+\infty}} \rho(2F_- - F_{1\,\mathrm{ret}}) = 0; \qquad \lim_{\substack{\rho\to\infty \\ \tau\to-\infty}} \rho(2F_- + F_{1\,\mathrm{adv}}) = 0. \qquad (2.33)$$

Similarly, near the origin

$$\lim_{\rho\to 0} \rho^2 \, (F-F_2)_{\substack{\mathrm{ret}\\ \mathrm{adv}}} = \lim_{\rho\to 0} \rho^2 \, (F_+-F_{2\,\substack{\mathrm{ret}\\ \mathrm{adv}}}) = 0. \qquad (2.34)$$

This concludes our discussion of the classical two-field point of view.

3. QUANTUM ELECTRODYNAMICS

(a) Lagrangian formulation

A quantum field theory cannot be cast into a Lagrangian formulation in the classical sense because that formulation depends on the calculus of variations to yield - via Noether's theorem - the desired field equations and conservation laws. This calculus is not applicable to operators. Nevertheless, such a formulation is used widely in a heuristic way in order to "justify" the field equations. We, too, shall use it here for heuristic reasons, but primarily we use it in order to ensure that the quantized theory is analogous to the classical one.

Thus, we start with the action integral modeled after (2.1)

$$I = \int d^4x[-\bar{\Psi}(\gamma\cdot\partial+m)\Psi + J_\mu\bar{A}^\mu - \tfrac{1}{2}\bar{F}_{\alpha\beta}F_P^{\alpha\beta}] + \tfrac{1}{2}\int d^4x\int d^4x' J_\alpha(x)D_P(x-x')J^\alpha(x') \tag{3.1}$$

where

$$J^\mu(x) = -\frac{ie}{2}[\bar{\Psi}(x), \gamma^\mu\Psi(x)]. \tag{3.2}$$

The Euler-Lagrange equations are

$$\partial_\alpha \bar{F}^{\alpha\mu} = 0 \tag{3.3}$$

$$\partial_\alpha F_P^{\alpha\mu} = -J^\mu \tag{3.4}$$

$$(\gamma\cdot\partial+m)\Psi = -ie[\bar{A} + D_P*J]\cdot\gamma\Psi . \tag{3.5}$$

The solution of (3.4) is again

$$A_P^\mu = D_P*J^\mu . \tag{3.6}$$

It is unique by the requirement of time reversal covariance and the understanding that all free fields are in $\bar{F}^{\mu\nu}$. The latter is equivalent to the requirement that $F_P^{\mu\nu}$ be a virtual field, i.e. have no support on the photon mass shell.

The standard equation for Ψ follows from (3.5) and (3.6),

$$(\gamma\cdot\partial+m)\Psi = -ie\gamma\cdot A\ \Psi \tag{3.7}$$

$$A \equiv \bar{A} + A_P\ . \tag{3.8}$$

It must be emphasized that (3.8) is a <u>definition</u> for A. In order to identify A one needs again the asymptotic conditions which can now be written in terms of positive mass shell test functions f_+,

$$\begin{aligned} A_{in}(f_+) &= \lim_{t\to\infty} \int f_+^*(x')\overleftrightarrow{\partial_o'} A(x')d^3x' \\ A_{out}(f_+) &= \lim_{t'\to\infty} \int f_+^*(x')\overleftrightarrow{\partial_o'} A(x')d^3x'\ . \end{aligned} \tag{3.9}$$

As in (2.14) - (2.16) one finds

$$\bar{A} = \frac{1}{2}(A_{in} + A_{out}) \tag{3.10}$$

$$A_{out} - A_{in} = D*J\ . \tag{3.11}$$

The identities (2.39) now permit one to express A in terms of the independent fields, A_{in} and Ψ or, alternatively, A_{out} and Ψ,

$$A = A_{in} + D_R * J = A_{out} + D_A * J \; ; \qquad (3.12)$$

these are the Källén-Yang-Feldman equations.

The identification of $\bar{A}$, (3.10) and (3.11) shows that A_{in} and A_{out} are not independent, so that if A_{in} and Ψ, say, are taken as the independent fields, the field equation (3.5) becomes

$$(\gamma \cdot \partial + m)\Psi = -ie[A_{in} + \tfrac{1}{2}D * J + D_P * J] \cdot \gamma\Psi \; . \qquad (3.13)$$

We now discard the Lagrangian formulation having exploited its heuristic value; we can take this field equation at the basis of the theory. It must be supplemented by the information that A_{in} is a free field, i.e. it satisfies the appropriate commutation relations and the field equation

$$\Box A^{\mu}_{in} = 0 \; . \qquad (3.14)$$

We see that A_P has completely disappeared from the set of necessary quantities.

At this point it may seem that we have accomplished nothing: the "result" (3.13) is a trivial combination of the conventional field equation (3.7) and the Maxwell equations in the Källén-Yang-Feldman form (3.12). But what this derivation shows is that the conventional starting point with Ψ and A as fundamental fields is here replaced by a physical argument which leads to Ψ and A_{in} as fundamental fields. Whatever independent degree of freedom may be contained in the conventional interacting field A, it is already contained in the free field A_{in}. <u>We have</u>

no freedom to impose conditions on the total field A (such as commutation relations) as is conventionally done, because the difference, $A-A_{in}$ is completely determined by Ψ and the interaction.

We shall therefore now turn to the commutation relations.

(b) Commutation relations

The field equations (3.13) and (3.14), when taken as the starting point of QED must be supplemented by commutation relations. The free field A^{μ}_{in} must be restricted to its two independent degrees of freedom. The radiation gauge (or Coulomb gauge) is here the natural choice. Thus, we assume

$$A^{o}_{in} = 0 \qquad (3.15)$$

$$\partial \cdot A_{in} = \nabla \cdot \vec{A}_{in} = 0 \qquad (3.16)$$

and the equal time commutation relations

$$[A^{k}_{in}(x), A^{\ell}_{in}(x')] = 0 \qquad (t=t') \qquad (3.17)$$

$$[\dot{A}^{k}_{in}(x), A^{\ell}_{in}(x')] = -i(\delta^{kl} - \frac{\partial^{k}\partial^{\ell}}{\nabla^{2}})\delta_{3}(\vec{x}-\vec{x}') \; . \qquad (3.18)$$

The operator ∇^{-2} is defined by

$$\nabla^{-2} f(\vec{x}) = - \frac{1}{4\pi} \int \frac{d^{3}x'}{|\vec{x}-\vec{x}'|} f(\vec{x}') \; . \qquad (3.19)$$

The mathematical advantage of this gauge is of course

that it permits one to keep a positive definite Hilbert space [8]. The main reason for the choice of this gauge is however the physical interpretation: we do not want longitudinal or Coulomb photons introduced into the theory.

At this point the following question arises: how is it possible that one particular gauge is physically preferred over all other gauges, if the observable quantities, the field strengths, are completely insensitive to the choice of gauge?

The answer can be given on the classical level: the equations for A^μ which are used above and which are those of the Coulomb gauge can be regarded as a particular choice of variables dictated by physical considerations but <u>independent</u> of the choice of gauge. Let A' be the potentials in any arbitrary gauge satisfying the Maxwell-Lorentz equations:

$$\partial_\alpha (\partial^\alpha A'^\mu - \partial^\mu A'^\alpha) = - J^\mu \; . \tag{3.20}$$

Define the new variables

$$A^\mu \equiv A'^\mu - \partial^\mu \nabla^{-2} \nabla \cdot \vec{A}' \; . \tag{3.21}$$

Then it is easily verified that the A^μ satisfy the equations

$$\begin{aligned} \nabla^2 A^0 &= - J^0 \\ \Box \vec{A} &= -\vec{J}_\perp \equiv - (\vec{J} - \nabla\nabla^{-2} \nabla \cdot \vec{J}) \end{aligned} \tag{3.22}$$

and the <u>identity</u>

$$\nabla \cdot \vec{A} \equiv 0 \; . \tag{3.23}$$

These equations differ from the field equations in the Coulomb gauge only by the fact that (3.23) is an identity and not a gauge condition. In terms of the original A'^{μ} which are in an arbitrary gauge, (3.22) and (3.23) are indentical to the field equations (3.20). If one chooses $J^{\mu} = 0$ one obtains the free field equations.

Thus, (3.21) should not be regarded as a gauge transformation but as the transformation to the physical choice of variables, given any A'^{μ}. The transformation (3.21) does not depend on the gauge of A'. It was pointed out to me by Professor Gordon Feldman.

It is evident that (3.15) and (3.16) are not manifestly Lorentz invariant, i.e. that A^{μ} cannot transform as a four-vector. In fact, it transforms like a four-vector transformation together with a gauge transformation. The latter depends on the Lorentz transformation in question and is such that (3.15) and (3.16) are preserved.

Actually, this transformation law is of much deeper significance within the Poincaré group than a four-vector law would be. The free field is to describe particles of mass zero and spin 1. Correspondingly, the transformation must be that of a representation of such particles. In fact, it is the representation $(1,1) \oplus (1,-1)$ in the notation of Gel'-fand, Minlos, and Shapiro [9]. It is a non-unitary infinite dimensional representation. I cannot enter into a discussion of this matter here but I refer to the paper by Bender [10].

The commutation relations for A^k_{in} show clearly that the fields commute at two different space points at equal time. However, this does not ensure that $\vec{A}_{in}(x)$

is a local field. In fact, $\vec{A}_{in}(x)$ is <u>not a local field.</u> To see this we solve (3.14) with the initial conditions (3.17) and (3.18); we find the well-known commutation relations

$$[A^k_{in}(x), A^\ell_{in}(x')] = -i(\delta^{kl} - \frac{\partial^k \partial^\ell}{\nabla^2}) D(x-x'). \qquad (3.24)$$

Now ∇^{-2} is defined by (3.19) and a simple computation leads to the result

$$-4\pi\nabla^{-2} D(x) = \varepsilon(t)\theta(|t| - r) + \frac{t}{r}\,\theta(r - |t|) \qquad (3.25)$$

which is indicated in the following figure:

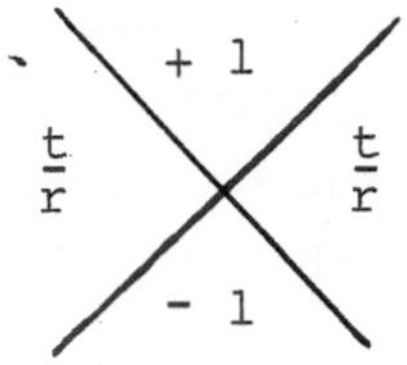

The proof of (3.25) is as follows. Starting with

$$D(x) = \frac{1}{2\pi}\varepsilon(t)\delta(x^2)$$

we have from (3.19) using $|\vec{x}| = r$

$$4\pi\nabla^{-2} D(x) = -\int \frac{d^3x'}{|\vec{x}-\vec{x}'|} \frac{1}{4\pi r'} [\delta(t-r') - \delta(t+r')]$$

$$= -\int_0^\infty \frac{r'^2 dr'}{r_>} \frac{1}{r'} [\ldots] = -(\int_0^r \frac{r'}{r} dr' + \int_r^\infty dr')[\ldots]$$

$$= -\frac{t}{r}\,\theta(r-|t|) - \varepsilon(t)\theta(|t|-r).$$

Thus, when the field $A^k_{in}(f)$ averaged over a test function f on R^4 is compared with $A^\ell_{in}(g)$ where f and g have supports which are spacelike with respect to one another, one finds that they do <u>not</u> commute. Therefore, A^k_{in} is not a local field. The field strengths F_{in} of course do commute on spacelike separated supports.

It should also be noted that this nonlocality does not show up in the Schroedinger picture where the A^k_{in} are explicitly time independent and are averaged only over f on R^3, corresponding to (3.17) and (3.18), but not to (3.24). The latter equation is only applicable to the Dirac or Heisenberg picture.

In addition to the above commutation relations we have for the independent field Ψ,

$$[A_{in}(x), \Psi(x')] = 0 \qquad (t=t') \qquad (3.26)$$

which is assumed to satisfy the conventional equal time commutation relations. [As will become obvious in the following, there are reasons to suspect that this assumption may have to be modified in the future].

$$\{\Psi(x), \bar{\Psi}(x')\} = i\gamma^o\delta_3(\vec{x}-\vec{x}') \qquad (t=t') \qquad (3.27)$$

$$\{\Psi(x), \Psi(x')\} = 0 \qquad (t=t'). \qquad (3.28)$$

(Since we are paying no attention here to renormalization, Z-factors are being ignored here.)

Now, we chose A_{in} arbitrarily as the independent field; had we chosen A_{out}, this field would have to satisfy the same commutation relations as A_{in}, i. e. (3.24) since it, too, is to describe free particles. But the relation (3.11) will permit this only with suitable restrictions: we must have

$$[A^{\mu}_{in}(x),(D*J^{\nu})(x')] + [(D*J^{\mu})(x),A^{\nu}_{in}(x')] +$$

$$+ [(D*J^{\mu})(x\ ),(D*J^{\nu})(x')] = 0. \ (\forall x,x') \ . \tag{3.29}$$

Since this relation involves commutators between A_{in} and J and between J and J for timelike separations, it is far from obvious that it is valid.

This situation is of course independent of the choice of gauge. In order to explore its consequences one can make use of the technique of operator derivatives [11] and one finds from (3.27)

$$-i\int D(x-\xi)d^4\xi \ \frac{\delta(D*J^{\nu})(x')}{\delta A_{in\ \mu}(\xi)} + i\int D(x'-\xi')d^4\xi' \ \frac{\delta(D*J^{\mu})(x)}{\delta A_{in\ \nu}(\xi')}$$

$$+ \ D(x-\xi)d^4\xi D(x'-\xi')d^4\xi'[J^{\mu}(\xi),J^{\nu}(\xi')] = 0$$

or

$$\int D(x-\xi)D(x'-\xi')d^4\xi d^4\xi' \ [i \ \frac{\delta J^{\nu}(\xi')}{\delta A_{in\ \mu}(\xi)} - i \ \frac{\delta J^{\mu}(\xi)}{\delta A_{in\ \nu}(\xi')} -$$

$$- [J^{\mu}(\xi),J^{\nu}(\xi')]] = 0 \ . \tag{3.30}$$

This means that

$$i \ \frac{\delta J^{\nu}(\xi')}{\delta A_{in\ \mu}(\xi)} - i \ \frac{\delta J^{\mu}(\xi)}{A_{in\ \nu}(\xi')} = [J^{\mu}(\xi),J^{\nu}(\xi')] \tag{3.31}$$

is to hold at least on the mass shell. One can show that this is indeed a correct equation, if one accepts off-mass shell unitarity and the following relation between the current and the S-operator, [12]

$$J^{\mu}(x) = - iS^{\dagger} \frac{\delta S}{\delta A_{in\ \mu}(x)} . \qquad (3.32)$$

When one now tries to recover the conventional theory, one sees that in that theory the field A^{μ} defined in (3.8), or equivalently the square bracket in (3.13) is required to be <u>local.</u> An argument similar to the above then shows that this can be achieved if one assumes not only (3.32) but also a suitable <u>causality condition.</u> This condition is here best expressed in the way suggested by Bogoliubov, [13]

$$\text{supp} \frac{\delta J(x)}{\delta A_{in}(x')} \subset \bar{V}_{+}(x-x') \qquad (3.33)$$

where

$$\bar{V}_{+}(x-x') = \{(x-x') \mid (x-x')^2 \leq 0, x^0 \geq x'^0\} .$$

This is one of the basic assumptions of asymptotic quantum field theory [14]. But in one form or another this causality requirement is made in any form of the conventional theory. And it is a highly nontrivial assumption. In the formulation we have developed here this requirement does not appear and, in fact, there seems to be no room for it. We are thus led to question the locality condition of the conventional theory for interacting fields.

Since locality, causality, and analyticity are well known to be closely linked to one another, and have been very successfully employed in the past, we see that we face here a very difficult and very fundamental problem. From our present point of view the usual approach is simply unacceptable: one cannot linearize the field equation for Ψ, (3.13), and put it in the form (3.7) by simply <u>postulating</u> A to be local. This would require a difficult consistancy proof and in any case is likely to be incorrect. If it were valid, the Haag-Ruelle theory could be generalized to include zero mass fields and nonlocal fields of the type (3.17), (3.18).

Rather, from the present point of view, the causality and analyticity properties must be derived from the assumptions already made. This involves such considerations as the observation that D and D_P in (3.13) vanish outside the light cone, and that $D_P + \frac{1}{2}D \equiv D_R$ has very special support properties and that its Fourier transform has very special analyticity properties. Various recent papers may be relevant for such a study [15]. It seems doubtful that the conventional commutation relations for the interacting field A^{μ} will be validated. But in any case, we are led here to the need for a <u>proof</u> of locality and causality properties of the interaction field A^{μ} starting with the assumptions already made.

Finally, we note that the physical basis of the commutation relations of fields lies in the symmetry property with which we wish to endow the associated particles: Bose or Fermi statistics. This means that such commutation relations are imposed on <u>free</u> fields since only such fields are related in a unique way to physical particles. There is no such physical justification for the commutation relations of interacting fields.

4. THE S-MATRIX WITH NONLOCAL INTERACTION AND WITHOUT INFRARED DIVERGENCES

(a) Physical Considerations

In the conventional formulation of quantum electrodynamics the S-matrix contains infrared divergences. These arise from two quite different origins; some arise in the phase factor of Coulomb scattering, while others arise in low frequency bremsstrahlung. We shall show that the former arise from off-mass shell fields, the latter from on-mass shell fields.

The usual perturbation expansion [Born expansion or Dirac picture (interaction representation) calculation] yields the exact (nonrelativistic) cross section already in second order, i.e. the magnitude of the scattering amplitude is exact. The higher order corrections contribute only to the phase of the amplitude. But these corrections are divergent. The divergence is closely related to the infinite range of the Coulomb potential, since scattering in a Yukawa potential $e^{-\lambda r}/r$ yields a finite phase. In the limit $\lambda \to 0$ the phase diverges [16].

This infrared divergence arose because the conventional perturbation expansion approximates a charged particle in a Coulomb field asymptotically by a plane wave; [17] this is incorrect because a Coulomb wave does not become plane even asymptotically. The phase factor $\exp [i \frac{\alpha Z}{v} \ln (2pr \sin^2 \frac{\theta}{2}]$ does not have a vanishing phase in the limit $\theta \to 0$ or $r \to \infty$; in fact, it diverges. The zero order approximation of the conventional theory is therefore incorrectly chosen.

It is evident from the physical origin of this

divergence that it is related entirely to off-mass shell fields.

Having established that a charged particle can never completely escape from electromagnetic interaction with other charges, it follows that it will be ever so slightly accelerated, even at very large distances. Consequently, it will always radiate a little: there is no finite time interval for an electromagnetic scattering process, outside of which there is no more radiation emitted or absorbed. Of course, this radiation will involve very small energies. But since it takes place over an infinite time interval, an infinite number of photons will be emitted.

The on-mass shell origin of the infrared divergence is thus connected with the fact that every charged particle is at all times accompanied by a "cloud" of soft photons. This cloud is treated in perturbation expansion as one photon at a time. The divergence can be made to cancel in each order only in the transition probability and not in the S-matrix itself [18,17]. In order to accomplish the latter the cloud as a whole must be described [19].

We conclude that the conventional theory must be modified in two respects: one must abandon Fock space for photons and one must use the correct zero-order approximations, i.e. take into account that a free charge is asymptotically a "distorted plane wave."

It is instructive to introduce the latter modification first and remain in Fock space, which is perfectly allright as long as one remains within finite time intervals. This was done by Eriksson [20]. It leads naturally to a space of coherent states which is unitarily equivalent to Fock space, containing a vacuum state and a finite average number of emitted photons.

For infinite time intervals, however, one must abandon Fock space in favor of the space of coherent states which corresponds to a representation of the commutation relations which is unitarily inequivalent to the Fock representation. It has no vacuum. This was discussed by various investigators, including Chung [19], Kibble [21], Kulish and Faddeyev [22], and Zwanziger [23].

(b) Asymptotics

The interaction Lagrangian L follows from the classical action I, (2.1), as

$$L(t) \equiv \frac{\delta I^{int}}{\delta t} = \sum_{k=1}^{n} \int j_k \bar{A} d^3x + \sum_{k=1}^{n} \sum_{\substack{\ell=1 \\ (k \neq \ell)}}^{n} \iint j_k(x) \cdot j_\ell(x') \cdot D_P(x-x') d^3x d^4x' . \tag{4.1}$$

Since this is a nonlocal Lagrangian we shall not attempt a canonical formalism leading to a Hamiltonian. Rather, we shall later use the Lagrangian as the generator of the quantum mechanical transition amplitude, describing the development in time in an integral rather than a differential way, following Feynman's ideas.

It is obvious that L(t) does not vanish for $|t| \to \infty$. Let us in particular compute the asymptotic form of the double sum in (4.1). We have, aymbolically.

$$\lim_{t\to\pm\infty} D_P(x) = \frac{1}{2}D_{\substack{R\\A}}(x) = \pm\,\frac{1}{2}\theta(\pm t)D(x) = \frac{1}{8\pi r}\delta(t\mp r). \qquad (4.2)$$

Furthermore, the currents are asymptotically those of uniformly moving particles, i.e.

$$j_k^\mu(x) = e_k\delta_3(\vec{x}-\vec{z}_k)\,\frac{dz_k^\mu}{dt}\,(z_k^o = t) \qquad (4.3)$$

with $dz_k^\mu/dt = v_k^\mu d\tau_k/dt$ a constant. Therefore, the double sum $L_P(t)$ in (4.1) becomes for large t

$$L_P^{as}(t) = \sum_{\substack{k=1\\(k\neq\ell)}}^{n}\sum_{\ell=1}^{n}\int\frac{j_k(\vec{x},t)\cdot j_\ell(\vec{x},t-|\vec{x}-\vec{x}'|)}{8\pi\,|\vec{x}-\vec{x}'|}\,d^3x d^3x' =$$

$$= \sum_{\substack{k=1\\(k\neq\ell)}}^{n}\sum_{\ell=1}^{n}\frac{e_k e_\ell}{8\pi}\,\frac{dz_k}{dt}\cdot\frac{dz_\ell}{dt}\,\frac{1}{|\vec{z}_k-\vec{z}_\ell|}\;. \qquad (4.4)$$

The retardation term was dropped here because t is assumed so large that the retarded time is already on the straight line portion of the particle world line, $t' > t_o$.

Now $L_P(t)dt$ is a Lorentz invariant and, in fact, the double sum can be written accordingly,

$$\frac{dz_k}{dt}\,\frac{dz_\ell}{dt}\,\frac{dt}{|\vec{z}_k-\vec{z}_\ell|} = \frac{v_k\cdot v_\ell}{\sqrt{(v_k\cdot v_\ell)^2-1}}\,\frac{dt}{|t|}\;. \qquad (4.5)$$

The quantity

$$v_{kl} = \frac{\sqrt{(v_k \cdot v_\ell)^2 - 1}}{v_k \cdot v_\ell} = \sqrt{1 - \frac{1}{(v_k \cdot v_\ell)^2}} \qquad (4.6)$$

is the relativistically invariant relative velocity of two uniformly moving particles k and ℓ. Thus,

$$\int_{t_o}^{T} L_P^{as}(t)dt = \sum_{\substack{k=1 \\ (k \neq \ell)}}^{n} \sum_{\ell=1}^{n} \frac{e_k e_\ell}{8\pi v_{kl}} \ln \frac{T}{t_o} \quad (T > t_o > 0). \qquad (4.7)$$

A completely analogous derivation can be carried out for the integral over times in the distant past yielding

$$\int_{-T}^{-t_o} L_P^{as}(t)dt = \sum_{\substack{k=1 \\ (k \neq \ell)}}^{n} \sum_{\ell=1}^{n} \frac{e_k e_\ell}{8\pi v_{kl}} \ln \frac{T}{t_o} \quad (T > t_o > 0) \qquad (4.8)$$

Of course, v_{kl} in (4.7) refers to the relative velocity of the outgoing particles, while in (4.8) it refers to the ingoing particles.

For the time T we can take the proper time of particle k; if we take $t_o = 1/m_k$, (our units are $\hbar = c = 1$) then with $z_k = v_k \tau_k + a_k$ we have

$$\ln \frac{T}{t_o} = \ln|p_k \cdot (z_k - a_k)|.$$

If we are dealing with only two particles, $e_\ell = Ze$, m_ℓ

large, $e_k = e$, $m_k = m$, $v_k = v$, then the coefficient of the logarithm in (4.7) is

$$\frac{e \cdot Ze}{4\pi v} = \frac{\alpha Z}{v} \tag{4.9}$$

which agrees with the coefficient of the well known Coulomb phase of non-relativistic scattering theory. It is obvious that the limit $T \to \infty$ is meaningless.

In the quantized case the situation is very similar. The main formal difference between (3.1) and (2.1) lies in the expression for the current. Since we are dealing with free or almost free particles asymptotically, it is sufficient to approximate the asymptotic current by

$$j^\mu(x) = -\frac{ie}{2}[\bar{\Psi}(x), \gamma^\mu \Psi(x)] \tag{4.10}$$

where $\Psi(x)$ is a solution of the free Dirac equation. Taking the limit to large $|t|$ one finds [22],

$$j^\mu_{as}(x) = \int \frac{d^3p}{\varepsilon} p^\mu \rho(\vec{p}) \; \delta_3(\vec{x} - \frac{\vec{p}}{\varepsilon} t) \tag{4.11}$$

where the only remanent of the quantized fields is the operator charge density

$$\rho(\vec{p}) = -\frac{e}{\varepsilon} \sum_r [a_r^\dagger(\vec{p}) a_r(\vec{p}) - b_r^\dagger(\vec{p}) b_r(\vec{p})]. \tag{4.12}$$

The creation and annihilation operators are here invariantly normalized according to

$$\{a_r(\vec{p}), a_s^\dagger(\vec{p}\,')\} = \delta_{rs}\varepsilon\delta_3(\vec{p}-\vec{p}\,') \ , \qquad (4.13)$$

and similarly for b_r. Thus, the asymptotic QM current (4.11) differs from the classical one, (4.3) essentially by the replacement of e by $\int\rho(\vec{p})d^3p$, since dz^μ/dz^o corresponds exactly to p^μ/ε.

The previous (classical) calculation can thus be repeated with only minor changes. In particular, we use the momenta corresponding to the classical $p_k = m_k v_k$, since the current (4.11) is so expressed. Then one finds in analogy to (4.7) and (4.8)

$$\Phi(t_2,t_1) \equiv \int_{t_1}^{t_2} L_P^{as}dt = \int_{t_1}^{t_2} \varepsilon(t)\frac{dt}{t} \int \frac{d^3p}{\varepsilon}\frac{d^3p'}{\varepsilon'}\frac{\rho(\vec{p})\rho(\vec{p}\,')}{8\pi} \cdot \frac{p\cdot p'}{\sqrt{(p\cdot p')^2-m^4}} \ . \qquad (4.14)$$

The last factor is the reciprocal relative velocity (4.6) expressed in terms of the momenta. In order to avoid the divergence at the origin, an interval $|t| < 1/m$ can again be excluded from the integration.

The expression (4.14) is identical with the result obtained by very different methods using local interactions and a Hamiltonian approach [22].

We now turn to the discussion of the mass shell term of the interaction Lagrangian in (4.1) or of its analogue in (3.1)

$$L_1(t) = \int J\bar{A}d^3x = \int JA_{in}d^3x + \tfrac{1}{2}\int J(x)\cdot J(x')D(x-x')d^3xd^3x' \ . \qquad (4.15)$$

In the asymptotic limit, J becomes again the free current (4.11) and the second term does not contribute by symmetry ($D = D_R - D_A$ and the contributions which added in the case of D_P now cancel). Thus,

$$R(t_2,t_1) \equiv \int_{t_1}^{t_2} L_1^{as} dt = \int_{t_1}^{t_2} j_{as} \cdot A_{in} d^4x. \qquad (4.16)$$

Before proceeding further we observe that the j_{as} commute with one another for all space-time points. Thus, Φ and R also commute.

Let us now consider the asymptotic system, i. e. the idealized system which is characterized by the free Lagrangian and the asymptotic interaction Lagrangian only. Follwing the approach indicated at the beginning of this section (b), we start with the free Lagrangian in zeroth approximation and assume as known and solved the associated field equations and commutation relations. Thus, we have a Fock space of free Dirac particles and photons.

We are now encountering the basic difficulty in generalizing the above classical theory into quantum electrodynamics: it is not known how to relate the dynamics of the quantized system to the Lagrangian functions which we have discussed.

Let us consider the asymptotic dynamics. The usual procedure would be to integrate

$$|t + \delta t>^{as} = (1 - i\, H^{as}(t)\,\delta t)\ |t>^{as} \qquad (4.17)$$

with $H^{as}(t) = -L_1^{as}(t)$ in the interaction picture. This leads to the time development [22]

$$|t_2\rangle^{as} = e^{iR(t_2,t_1)+i\Phi(t_2,t_1)}|t_1\rangle^{as} \equiv e^{iC(t_2,t_1)}|t_1\rangle^{as}. \tag{4.18}$$

On the other hand, if one starts with

$$|t + \delta t\rangle^{as} = (1 + iL^{as}(t)dt)|t\rangle^{as} \tag{4.17)'$$

where $L^{as} = L_1^{as} + L_p^{as}$ one has because of the commutativity of the j_{as}

$$|t_2\rangle^{as} = T_+(e^{\int_{t_i}^{t_2} L^{as}(t)dt})|t_1\rangle^{as}$$

$$= T_+^{(A_{in})}(e^{iR(t_2,t_1)})\, e^{i\Phi(t_2,t_1)}|t_1\rangle^{as}$$

$$= e^{iR(t_2,t_1)}\, e^{\frac{i}{2}\int_{t_1}^{t_2} j_{as}(x)D_A(x-x')j_{as}(x')d^4x d^4x'}$$

$$e^{i\Phi(t_2,t_1)}|t_1\rangle^{as}\; e^{iR(t_2,t_1)}\, e^{2i\Phi(t_2,t_1)}|t_1\rangle^{as}.$$

Thus, an extra factor $\exp(i\Phi)$ appears. Eq. (4.17)' therefore does not seem to be acceptable.

For finite times t the $|t\rangle^{as}$ are vectors in Fock spac

Mathematically, this space is a direct product space of the Fock space of Dirac particles and the space describing the photons. The latter needs careful attention. While Φ gives only a relatively trivial factor, the exponent R is the interesting operator,

$$R(t_2,t_1) = \frac{i}{(2\pi)^{3/2}}\int\frac{d^3k}{2\omega}d^3p\rho(\vec{p})\sum_j[c_j^\dagger(\vec{k})(e^{-ik\cdot\frac{p}{\varepsilon}t_2} - e^{-ik\frac{p}{\varepsilon}t_1})\;\frac{e_j^*\cdot p}{k\cdot p} + h.\,c.] \tag{4.19}$$

where both p and k are on their positive mass shells $p^o = \varepsilon$ and $k^o = \omega$; the c_j are the photon annihilation operators. It is convenient to express R in the following form:

$$R(t_2,t_1)=i(c^\dagger(f)-c(f)); \quad c^\dagger(f)=\int\frac{d^3k}{2\omega}\sum_j c_j^\dagger(\vec{k})f_j(\vec{k};t_2,t_1) \tag{4.2o}$$

with

$$f_j(\vec{k};t_2,t_1)\equiv\frac{1}{(2\pi)^{3/2}}\int\rho(\vec{p})d^3p\frac{e_j^*(\vec{k})\cdot p}{k\cdot p}(e^{-ik\cdot\frac{p}{\varepsilon}t_2}-e^{-ik\cdot\frac{p}{\varepsilon}t_1}). \tag{4.21}$$

It is important to note that the f_j are effectively Abelian operators. The operators c(f) satisfy the commutation relations

$$[c(f),c^\dagger(f)] = \sum_j \int\frac{d^3k}{2\omega}|f_j(\vec{k};t_2,t_1)|^2 \equiv \|f\|^2. \tag{4.22}$$

Now the operator exp(iR) with R of the form (4.2O) is well known in the theory of coherent states [24]. The Campbell-Baker-Hausdorff formula permits one to cast it into normal ordered form,

$$e^{iR} = e^{-c^\dagger(f)+c(f)} = e^{-\frac{1}{2}\|f\|^2} e^{-c^\dagger(f)} e^{c(f)} . \tag{4.23}$$

As long as $\|f\| < \infty$ this operator is unitary on Fock space. In fact, the completeness relation

$$k^2 \sum_j e_j^\mu(\vec{k})e_j^{*\nu}(\vec{k}) = k^2 g^{\mu\nu} - k^\mu k^\nu \tag{4.24}$$

yields

$$||f||^2 = \frac{1}{(2\pi)^3}\int \rho(\vec{p})\rho(\vec{p}')d^3pd^3p'\int \frac{d^3k}{2\omega}\left(\frac{p'\cdot p}{k\cdot p'k\cdot p} - \frac{1}{k^2}\right)$$

$$\left(e^{-ik\cdot\frac{p'}{\varepsilon'}t_2} - e^{-ik\cdot\frac{p'}{\varepsilon'}t_1}\right)\left(e^{ik\cdot\frac{p}{\varepsilon}t_2} - e^{ik\cdot\frac{p}{\varepsilon}t_1}\right) \tag{4.25}$$

and this integral converges at the lower limit ($\omega \to 0$) as long as t_2 and t_1 are finite.

The integration over d^3k is an integration over all photons which are emitted and absorbed and which belong to the photon cloud surrounding the charges asymptotically. Thus, these are not observed and must therefore have an energy below threshold of observability. The finite energy resolution of any experiment enters here as the natural cut-off at the upper limit of the k-integration. The norm (4.25) is thus finite for all $|t_2 - t_1| < \infty$.

However, if one takes the reference time $t_1 \to -\infty$ in the usual way [25] so that the exponential containing t_1 reduces to zero by adiabatic switching, the norm (4.25) will diverge at the lower limit and (4.23) will no longer be unitary on Fock space.

But while the operator (4.22) acting on Fock space does not map back into Fock space, the asymptotic space is a meaningful space. This space can be expressed as an infinite tensor product [26]. It corresponds to a representation of the c-operator algebra which is not unitarily equivalent to the Fock representation. It is in this space that the asymptotic states (4.18) are defined when the limit $t_1 \to -\infty$ is taken.

(c) The S-operator

Once the asymptotic states are defined, one can turn to the definition of the S-operator. As in the asymptotic case we construct an "L-function." This is the Lagrangian of the action principle with the solutions of the field equations substituted. After that substitution this Lagrangian function will of course no longer yield the field equations according to Lagrange's equations. Therefore we call the resultant quantity an "L-function."

It was first observed by Dollard [27] that the infrared divergences of nonrelativistic quantum mechanics can be traced to the asymptotically too slow fall-off of the electromagnetic interaction. His suggestion to use not the free but the asymptotic states amounts to using the interaction L-function.

$$L' = L - L^{as} \tag{4.26}$$

instead of L as a perturbation.

The L'-function plays the role of an infinitesimal propagator of interaction analagous to L^{as} in (4.17). But we are now dealing with the full interaction so that the currents are no longer semiclassical and are no longer associated with constant velocity. Therefore the distinction between retarded and advanced propagation is no longer trivial. Now, the semiclassical form of the L-function is

$$L = \int J.A_{in} d^3x + \frac{1}{2}\int J\cdot D{*}Jd^3x + \int J\cdot D_P{*}Jd^3x \ , \tag{4.27}$$

from (3.1), (3.10), and (3.11). While D_P is symmetric,

the D term does not describe the relativistic, second-quantized, on-mass shell field propagation correctly. The difference between the classical and the quantized on-mass shell field lies in the physical meaning of the Fourier transform. The quantized free field always involves both particles and antiparticles and a suitable reinterpretation is necessary in order to deal with antiparticles rather than with particles of negative energy. This point is of course well known: it means that retarded interaction goes hand-in-hand with positive frequencies, corresponding to emission at the retarded time point, while advanced interaction goes with negative frequencies corresponding to absorption at the advanced point. Thus,

$$\frac{1}{2}D+D_P = D_R = D_R\Big|_{k^o>0} + D_R\Big|_{k^o<0} \rightarrow D_R\Big|_{k^o>0} + D_A\Big|_{k^o<0} =$$

$$= D_c = \frac{i}{2}D_1 + D_P \tag{4.28}$$

where D_c is the causal function of Stückelberg [28], (also called the "Feynman propagator"). Thus, the quantization of the radiation field has the effect of changing the homogeneous Green function $D = D_+ + D_-$ to $i\, D_1 = D_+ - D_-$. This sign change is an alternative and equivalent way of expressing (4.28).

Thus, we find for the L-function (4.26) in the quantum electrodynamic case, using the results of section (4b),

$$L'(t) = \int (J-j_{as})\cdot A_{in} d^3x + \int J\cdot D_c * J d^3x - \int j_{as}\cdot D_P * j_{as} d^3x. \tag{4.29}$$

At this point we are again faced with the same problem as in the case of the asymptotic dynamics: how can one determine the S-matrix from the knowledge of the L-function? In (4.17)' the L-function, L^{as}, was at least local; in (4.29) the L-function, L'(t), is not even local in time because of the $D_c * J$ factor.

The problem is of course one of operator ordering. To find the answer to this question of ordering we begin with the conventional theory. In the Dirac (interaction) picture

$$S_{fi}(t_2,t_1) = \langle f\ t_2|\ S_D\ (t_2,t_1)|t_1 i\rangle \tag{4.30}$$

describes the time development from the initial state at t_1 to the final state at t_2. The operator S_D is given by

$$S_D(t_2,t_1) = T_+(\exp\ [i\int_{t_1}^{t_2} J.A_{in}\ d^4x]) \tag{4.31}$$

where J and A_{in} are in the Dirac picture. For large $|t_1|$ and $|t_2|$ the states in (4.30) can be replaced by the asymptotic states yielding

$$S_{fi}(t_2,t_1) = \langle f|e^{-iC(t_2,t_o)}\ T_+(e^{i\int_{t_i}^{t_2} J\cdot A_{in} d^4x)}\ e^{iC(t_1,t_o)}|i\rangle . \tag{4.32}$$

The operators C are given in (4.18). The time t_o is a reference time, i.e. the time at which the state vectors of the Dirac and Heisenberg pictures coincide. It can be chosen to be $t_o = 0$ and will then yield a symmetric S-operator. One then goes from an initial coherent state to a final coherent state. Alternatively, can choose

$t_o = -\infty$, as is usually done. In that case one goes from an initial Fockstate to a final coherent state. We shall presently adopt the latter.

The T_+ product in (4.31) can be converted into a normal ordered product $N^{(A_{in})}$ for the electromagnetic fields only. That relation is known e.g. from the functional formulation of quantum electrodynamics:

$$T_+(e^{i\int J\cdot A_{in}}) = T_+^{(\psi)}\, N^{(A_{in})}(e^{i\int J\cdot A_{in} + \frac{i}{2}\int J\, D_c^* J}) \,. \qquad (4.33)$$

In the limits $t_2 \to \infty$, $t_1 \to -\infty$, which must of course be carefully defined, one obtains from (4.32) and (4.33)

$$S_{op} = e^{-\frac{i}{2}\int j_{as}(x)d^4x\, D_P(x-x')d^4x' j_{as}(x') - i\int j_{as}A_{in}d^4x}$$

$$\cdot T_+^{(\psi)} N^{(A_{in})}(e^{i\int J\cdot A_{in}d^4x + \frac{i}{2}\int J(x)d^4x D_c(x-x')d^4x' J(x')}) . \qquad (4.34)$$

All integrals here extend over all space-time. And now one sees that, except for the operator ordering, this is exactly the expression

$$e^{i\int L'(t)dt} \qquad (4.35)$$

with L' given by (4.29). No derivation of (4.34) is known which goes directly from L'(t) to the S-operator. Nor is this ordering in (4.34) obvious intuitively.

The S-operator (4.34) is free of infra-red divergences as was shown by several authors [22]. In particular, it describes all electrons as surrounded by a cloud of soft photons.

(d) Action-at-a-distance QED

Let us recall that we started out in classical electrodynamics with a time symmetric action. This time-symmetry is still preserved in the action (3.1). But in view of the interdependence of A_{in} and A_{out}, (2.16), we had to pick one of these as independent variable, thus introducing an apparent time asymmetry.

In classical electrodynamics it is possible to preserve the time symmetry by casting the theory into the action-at-a-distance form [4]. For this theory Wheeler and Feynman introduced the "complete absorber condition" which states that there are no free photons in the system: all radiation emitted is reabsorbed before the charges reach $t = +\infty$, and there are no free incident photons. The system is consequently characterized by an action (2.1) in which the total free field $\bar{A}$ is absent from the start. This theory leads to physical predictions which are identical to those of the conventional theory, as was shown by Wheeler and Feynman [4]. This is also evident from the formulation of classical electrodynamics which I proposed ten years ago [3], and in which the Wheeler-Feynman theory emerges as a special case.

The formulation of QED presented here suggests one to obtain the generalization of the Wheeler-Feynman theory to the QED level [29]. We can start with (3.1), but without the $\bar{A}$ field. Then (4.1) will also contain only the off-shell term leading to the asymptotic states analogous to (4.17)', but without the R term

$$|t_2\rangle^{as} = e^{i\Phi(t_2,t_1)}|t_1\rangle^{as} . \tag{4.36}$$

Thus, there are no free photons and no photon cloud, but only the Coulomb phase. Consequently, one can stay in Fock space, because the complete absorber condition ensures the absence of soft photons irrespective of the energy resolution of the measurement. The full interaction (4.29) then leads to the S-operator

$$S_{op} = e^{-\frac{i}{2}\int j_{as} \cdot D_{P*} j_{as} d^4x} T_+^{(\psi)} [\exp(\tfrac{i}{2}\int J \cdot D_{P*} J d^4x)] \qquad (4.37)$$

as is evident from (4.34) if one takes the photon vacuum expectation value (complete absorber condition) and then, consistently, excludes all on-mass-shell photon propagators.

As can now be easily verified, this operator is not unitary. Since it differs only by the (unitary) factor $\exp(-i\Phi)$ from the result of Davies [29] we see that also his S-operator is not unitary. No mention of this fact is made in that reference, [30].

We conclude that it does not seem possible to generalize the Wheeler-Feynman action-at-a-distance electrodynamics from the classical to the quantum level.

5. CONCLUSIONS

The starting point of these lectures was a formulation of classical electrodynamics in which the interaction is partly via fields (radiation fields) and partly via action-at-a-distance. In the quantized form this distinction is seen to be equivalent to a separation of on-mass shell and off-mass shell electromagnetic fields. Only the former are expressed in terms of field operators.

Casting the theory into such a form eliminated the unpleasant gauge problems which beset the conventional theory. It also permits a positive definite Hilbert space. The cost of this very physical formulation is a non-local interaction and the absence of local interacting electromagnetic fields. The field equations are consequently nonlocal.

The difficult problem of solving these equation suggests of course a perturbation expansion. But in view of the already existing knowledge and experience in QED, an S-matrix would be desirable. But the L-function which characterizes the physical system in the formulation presented here does not seem to permit one to obtain such a matrix. The conventional theory however leads to an S-operators which is just the expected exponentiated L-function, but operator ordered in a very specific way, (4.34). This derivation can be done taking full account of the correct asymptotic behavior of charged particles (which is not of the plane wave type), so that this S-operator contains no infra-red divergences.

The generalization of the classical action-at-a-distance theory of Wheeler and Feynman to the quantum level fails because it does not lead to a unitary S-operator. But the partially action-at-a-distance description which is proposed here does lead to satisfactory results.

REFERENCES

1. R. P. Feynman, Science 153, 699 (1966).

2. F. Rohrlich, Classical Charged Particles, Addison-Wesley Publishing Co., Reading, Mass. 1965.

3. F. Rohrlich, Phys. Rev. Lett. 12, 375 (1964); see also ref. 2.

4. H. Tetrode, 2. Phys. 10, 317 (1922); A. D. Fokker, ibid. 58, 386 (1929); J. A. Wheeler, R. P. Feynman, Rev. Mod. Phys. 17, 157 (1945) and 21, 425 (1949).

5. Ref. 2, Section 7-2.

6. P. A. M. Dirac, Proc. Roy. Soc. A 167, 148 (1938).

7. Ref. 2, Section 4-8.

8. F. Strocchi, Phys. Rev. D2, 2334 (1970).

9. I. M. Gel'fand, R. A. Minlos, Z. Ya. Shapiro, Representations of the Rotation and Lorentz Groups and Their Applications, the Macmillan Co., New York, 1963.

10. C. M. Bender, Phys. Rev. 168, 1809 (1968).

11. F. Rohrlich, J. Math. Phys. 5, 324 (1964) and also in Analytic Methods in Mathematical Physics, edited by R. P. Gilbert and R. G. Newton, Gordon and Breach, New York 1970.

12. F. Rohrlich, J. C. Stoddart, J. Math. Phys. 6, 495 (1965); T. W. Chen, F. Rohrlich, M. Wilner, ibid. 7, 1365 (1966).

13. N. N. Bogoliubov, D. V. Shirkov, Introduction to the Theory of Quantized Fields, Interscience Publishers, Inc. New York, 1959. See also ref. 12.

14. For a review of asymptotic quantum field theory see my Schladming lectures of 1967 (Acta Phys. Austriaca Suppl. IV, 228 (1967); see also the ensuring Theory of the S-operator: F. Rohrlich, Phys. Rev. 183, 1359 (1969), and A. Pagnamenta and F. Rohrlich, Phys. Rev. D1, 1640 (197o).

15. D. Iagolnitzer, H. P. Stapp, Commun. Math. Phys. 14, 15 (1969); J. Bros, D. Iagolnitzer, Ann. Inst. H. Poincaré A XVIII No. 2 (1973).

16. R. H. Dalitz, Proc. Roy. Soc. (London) A 2o6, 5o9 (1951).

17. See for example J. M. Jauch and F. Rohrlich Theory of Photons and Electrons, Addison-Wesley Publishing Company, Reading, Mass. 1955, Eq. (16-8).

18. J. M. Jauch, F. Rohrlich, Helv. Phys. Acta 27, 613 (1954).

19. V. Chung, Phys. Rev. 14oB, 111o (1965).

2o. K.-E. Eriksson, Physica Scripta 1, 3 (1970).

21. T. W. B. Kibble, J. Math. Phys. 9, 315 (1968); Phys. Rev. 173, 1527 (1968); 174, 1882 (1968); 175, 1624 (1968).

22. P. Kulish, L. D. Faddeyev, Theor. and Math. Phys. 4, 745 (1971).

23. D. Zwanziger, Phys. Rev. D7, 1o82 (1973).

24. The literature on this subject is very extensive. An excellent introduction is given in John R. Klauder and E. C. G. Sudarshan, Fundamentals of Quantum Optics. Extensions to quantum field theory are given in the second reference quoted in (11) and (19)-(22). These references also lead to the earlier literature.

25. See reference (22) and also S. Schweber, 1972 Cargèse Summer School Lecture.

26. J. von Neumann, Comp. Math. 6, 1 (1938).

27. J. Dollard, J. Math. Phys. 5, 729 (1964).

28. E. C. G. Stückelberg, Helv. Phys. Acta 19, 241 (1946); D. Rivier, Helv. Phys. Acta 22, 265 (1949); M. Fierz, Helv. Phys. Acta 23, 731 (1950).

29. Attempts to generalize the Wheeler-Feynman theory to QED are not new. We mentioned in particular F. Hoyle and J. V. Narlikar, Ann. Phys. 54, 2o7 (1969) and 62, 44 (1971); Nature 228, 544 (197o); P. C. W. Davies, J. Phys. A4,836 (1971) and 5, 1o25 (1972); Proc. Camb. Phil. Soc. 68, 751 (197o). Although these papers have necessarily certain features in common with our work, the approach taken here is different even though the results are essentially the same. We also note that these authors do not pay attention to the infrared problem.

30. I am indebted to Prof. I. Bialynicki-Birula for a valuable discussion on this point.

Acta Physica Austriaca, Suppl. XIII, 535–568 (1974)

UNIFIED MODEL OF WEAK AND ELECTROMAGNETIC INTERACTIONS+)*)

by

H. PIETSCHMANN
Institut für Theoretische Physik
der Universität Wien
and
Institut für Hochenergiephysik der
Österreichischen Akademie der Wissenschaften

1. INTRODUCTION

If one measures the success of a physical theory by the accuracy of its theoretical predictions as compared with experiments, Quantum Electrodynamics is the most successful theory in science. No wonder, that it was then chosen as a model after which other physical theories - in particular Weak Interactions - where designed. Naturally, there are some distinct differences such as violation of parity or hypercharge in

---

+) Lectures given at XIII. Internationale Universitätswochen für Kernphysik, Schladming, February 4 - 15, 1974.

*) Supported by Fonds zur Förderung der wissenschaftlichen Forschung.

Weak Interactions. The stumbling block which prevented the "classical" V-A theory of Weak Interactions to become as elegant and as powerfully predictive as Quantum Electrodynamics, is the fact that no quanta - let alone low mass quanta - of Weak Interactions have been found so far. The celebrated intermediate boson shall be assumed to exist throughout our lectures but it is its high mass (low mass values being ruled out experimentally) which renders the classical theory of Weak Interactions non-renormalizable.

In what follows, we shall interest ourselves in the fascinating development of the last few years which has led to a possibility of marrying Quantum Electrodynamics with Weak Interactions. The result is a unified, renormalizable model which comprises both interactions. Like any good theory it makes definite predictions.

Like always in our days, there are many alternative models with different predictions. Each of them, however, can definitely be supported or ruled out by crucial experiments. The fog was lifting a bit just now before our eyes when neutral weak currents were reported by some experimental groups. This justifies the emphasize on the one model which predicts these currents and only these currents.

In order to get an easy start into the subject, let us begin by contemplating the common features as well as some of the distinct differences in Quantum Electrodynamics and standard Weak Interactions.

2. COMMON AND DIFFERENT FEATURES OF QUANTUM ELECTRODYNAMICS AND STANDARD WEAK INTERACTIONS

The main similarity which springs to our mind is the fact that both weak and electromagnetic currents are of vector nature. The first step in the direction of a unification was actually taken very early in the concept of the Conserved Vector Current (CVC).

Let us write the total weak current as

$$J_\lambda^+(x) = \ell_\lambda^+(x) + \cos\theta\, j_\lambda^{\pi+}(x) + \sin\theta\, j_\lambda^{K+}(x) \tag{1}$$

$$\ell_\lambda^+(x) = \sum_\ell \bar{\psi}_{\nu_\ell}(x)\gamma_\lambda(1+\gamma_5)\psi_\ell(x) \qquad \ell = e,\mu \tag{2}$$

where θ is the Cabibbo angle and the superscripts on the hadronic currents denote the $\Delta Y = 0$ and $\Delta Y = \Delta Q = 1$ parts. (In our metric, the square of time-like vectors is positive).

The weak current with opposite change of charge is

$$J_\lambda^-(x) = (J_\lambda^+(x))^\dagger \quad . \tag{3}$$

Let us write the total electromagnetic current as

$$J_\lambda(x) = \ell_\lambda(x) + j_\lambda^3(x) + j_\lambda^0(x) \tag{4}$$

where the superscripts denote isovector and isoscalar parts.

$$\ell_\lambda(x) = \sum_\ell \bar{\psi}_\ell(x)\gamma_\lambda \psi_\ell(x) \qquad \ell = e,\mu \quad . \tag{5}$$

Then CVC tells us, that the vector part of the hadronic weak currents j^π_λ and the isovector part of the hadronic electromagnetic current form an isotriplet. This is more precisely expressed by means of the commutation relation

$$[I^+,I^-] = 2I^3 \tag{6}$$

where

$$I^\pm = \int d^3x \; j_o^{\pi\pm,V}(x) \tag{7}$$

$$I^3 = \int d^3x \; j_o^3(x) \quad . \tag{8}$$

After the success of current algebra, one is justified in assuming also the following commutation relation for the axial currents

$$[Q_a^+,Q_a^-] = 2I^3 \tag{9}$$

$$Q_a^\pm = \int d^3x \; j_o^{\pi^\pm,A}(x) \quad . \tag{10}$$

This scheme is very elegant and it also closely relates electromagnetic and weak currents. Let us therefore see what happens if we try the same game with the leptonic currents. Define

$$i^\pm = \int d^3x \; \ell_o^{\pm V}(x) \tag{11}$$

$$i^3 = \int d^3x \; \ell_o(x) \quad . \tag{12}$$

Using the canonical commutation relations we find

$$[i^+,i^-] = 2i^3 + \int d^3x \sum_\ell \psi^\dagger_{\nu_\ell}(x)\ \psi_{\nu_\ell}(x) \ . \qquad (13)$$

It is seen, that the commutation relation does not "close", i.e. a new piece occurs on the right hand side of eq.(13): a neutral current which is composed of neutrinos only and must therefore be a neutral weak current.

Before we go on to elaborate on neutral weak currents, let us first see whether they cannot be avoided. What, in fact, have we done? In defining eqs. (11) and (12) we have effectively introduced some sort of weak isospin space spanned by the doublet

$$\ell = \begin{bmatrix} \nu_e \\ e \end{bmatrix} \qquad (14)$$

and a similar one for muon type leptons. (We shall, for simplicity, not always write out the doubling of leptons because it only amounts to a trivial generalization!). In this space, we can write the lepton currents of eqs. (2) as

$$\ell^{\pm}_\lambda(x) = \bar{\ell}(x)\gamma_\lambda(1+\gamma_5)\ \tau^{\pm}\ \ell(x) \ . \qquad (15)$$

However, the leptonic part of the electromagnetic current given in eq. (5) cannot be written as

$$\ell_\lambda(x) \neq \bar{\ell}(x)\ \gamma_\lambda\ \tau^3\ \ell(x) \ . \qquad (16)$$

A way out would be to choose a triplet representation for the leptons instead of eq. (14).

$$E = \begin{pmatrix} e' \\ \nu \\ e \end{pmatrix} \tag{17}$$

with the weak lepton current

$$\ell_\lambda^{\pm}(x) = \bar{E}(x)\gamma_\lambda(1+\gamma_5)\ t^{\pm}\ E(x) \tag{18}$$

t being the 3 by 3 isospin matrices.

Now indeed the electromagnetic lepton current can be written as

$$\ell_\lambda(x) = \bar{E}(x)\ \gamma_\lambda\ t^3\ E(x) = \bar{e}'\gamma_\lambda\ e' - \bar{e}\ \gamma_\lambda\ e \tag{19}$$

showing that the commutation relation closes as well as that the (as yet unknown) "heavy electron" e' has opposite charge as the electron, i.e. positive particle and negative antiparticle.

The lesson we have learned is that weak neutral currents can indeed be avoided. The price is the "prediction" of heavy leptons. (Of course, one can always manufacture a more complicated scheme which predicts both!).

The first alternative - weak neutral currents - has originally been proposed by Salam, Ward and Weinberg [1] whereas the second one belongs to Georgi and Glashow [2]. From now on, we shall only pursue the first alternative.

3. THE SALAM-WEINBERG MODEL

In section 2, we have qualitatively discussed possible unification schemes without attempting at a complete description. Now, we have to remember that both vector and axial vector currents have to be incorporated. To achieve this, we are helped by the fact that only left-handed combinations occur in weak currents. Thus we define a left-handed lepton doublet

$$L(x) = \frac{1+\gamma_5}{2} \begin{pmatrix} \nu_e \\ e \end{pmatrix} \quad . \tag{20}$$

As in section 2, we skip the trivial complication due to the doubling of leptons (electronic and muonic). Necessary generalizations to include both present no difficulty.

The leptonic weak currents can now be written as

$$\ell_\lambda^{\pm}(x) = 2\bar{L}(x)\ \gamma_\lambda\ \tau^{\pm}\ L(x) \quad . \tag{21}$$

In order to be able to write the electromagnetic current also, we still have to define the right-handed electron, remembering that the neutrino never enters with right-handed projections.

$$R(x) = \frac{1-\gamma_5}{2}\ e(x)\ . \tag{22}$$

The electromagnetic current becomes

$$\ell_\lambda(x) = \bar{e}\ \gamma_\lambda\ e = \frac{1}{2}(\bar{L}\ \gamma_\lambda\ L - \bar{L}\ \gamma_\lambda\ \tau^3\ L) + \bar{R}\ \gamma_\lambda\ R. \tag{23}$$

Notice that both a triplet component and a singlet enters eq. (23). The weak currents (21) provide exactly the remaining triplet components so that the interaction Lagrangian can be neatly written in an invariant way:

$$L_1(x) = \frac{g}{2}\bar{L}\,\gamma_\lambda\,\vec{\tau}\,L\,\vec{A}^\lambda + g'[\tfrac{1}{2}\bar{L}\,\gamma_\lambda\,L + \bar{R}\,\gamma_\lambda\,R]\,B^\lambda \quad . \qquad (24)$$

Notice the different coupling constants g for the triplet interaction and g' for the singlet interaction. It shall turn out that the ratio of these two coupling constants is the only undetermined parameter in the model.

However, leaving this freedom of different coupling, neither the neutral triplet boson A_λ^3 nor the singlet boson B_λ can be identified with the photon. Rather, the electromagnetic current (23) is coupled to a certain mixture of them. To bear this out, let us explicitly write down this mixture

$$\begin{aligned} A_\mu^3 &= \cos\phi\; Z_\mu - \sin\phi\; A_\mu \\ B_\mu &= \sin\phi\; Z_\mu + \cos\phi\; A_\mu \end{aligned} \qquad (25)$$

ϕ is called Salam-Weinberg mixing angle.

Inserting the mixture (25) into the interaction Lagrangian (24) yields

$$L_1(x) = \text{charged part} + [g'\cos\phi\,(\tfrac{1}{2}\bar{L}\gamma_\mu L+\bar{R}\gamma_\mu R) -$$

$$- \frac{g}{2}\sin\phi\,\bar{L}\gamma_\mu\tau^3 L]A^\mu + [g'\sin\phi\,(\tfrac{1}{2}\bar{L}\gamma_\mu L+\bar{R}\gamma_\mu R) +$$

$$+ \frac{g}{2} \cos\phi \, \bar{L} \gamma_\mu \tau^3 L] Z^\mu \quad . \tag{26}$$

Needless to say that we identify A_μ with the photon field. A comparison with eq. (23) then gives

$$e = g \, . \, \sin\phi = g' \, . \, \cos\phi \tag{27}$$

so that

$$\frac{g'}{g} = \mathrm{tg}\ \phi \ . \tag{28}$$

Much more exciting is the question as to the coupling of the new neutral boson Z_μ to the leptons. The coefficient of Z_μ in eq. (26) can be written in the following way

$$\begin{aligned} & g'\sin\phi[\tfrac{1}{2}\bar{L}\gamma_\mu L+\bar{R}\gamma_\mu R] + \frac{g}{2} \cos\phi \, \bar{L}\gamma_\mu \tau^3 L = \\ & = \frac{g}{4\cos\phi}[\bar{\nu}\gamma_\mu (1+\gamma_5)\nu - \bar{e}\gamma_\mu (1-4\sin^2\phi+\gamma_5)e] \ . \end{aligned} \tag{29}$$

It is seen, that the neutral boson Z_μ couples to an ordinary V-A neutral neutrino current. The coupling to the charged leptons is - apart from a different sign - modified in its vector part. The effective vector coupling constant is

$$C_V = 1 - 4 \sin^2\phi \quad . \tag{30}$$

Let us speculate in passing that there exists an interesting possibility: Whereas the photon A_μ couples only to the vector current of the electron, the boson Z_μ would only couple to the electrons axial vector provided $\sin\phi$ is just 1/2.

4. THE PROBLEM OF MASSES AND THE HIGGS PHENOMENON

In our attempts so far we had in mind only the unification of weak and electromagnetic interactions. But in order to overcome the non-renormalizability of weak interactions, we have to get rid of the mass term of intermediate bosons in the free Lagrangian.

Let us recall the problem! The propagator of a vector boson is

$$\Delta^F_{\mu\nu}(p) = \frac{g_{\mu\nu} - \frac{p_\mu p_\nu}{M^2}}{p^2 - M^2} \tag{31}$$

which shows the bad high energy behaviour causing non-renormalizability. In passing, we observe that for very large masses of the intermediate boson we have

$$g^2_{SW}\,\Delta^F_{\mu\nu}(p) \to -\,g_{\mu\nu}\frac{g^2_{SW}}{M^2} + O\left(\frac{p^2}{M^2}\right) \tag{32}$$

where g_{SW} is some "semi-weak" coupling constant. Eq. (32) gives us the relation between the semi-weak coupling constant and the ordinary four-fermi coupling constant $G/\sqrt{2}$:

$$\frac{G}{\sqrt{2}} = \frac{g^2_{SW}}{M^2} \quad . \tag{33}$$

The only way to avoid troubles due to the propagator is to skip the mass term in the free Lagrangian. Where, then, do we get the physical mass of the intermediate bosons from? The answer has been provided by Higgs and others [3]. To indicate the idea before going into details, let us suppose a vector boson interacts with a scalar field in the following way

$$L_{int} = g\, A^{\mu} A_{\mu}\, \Phi \,. \qquad (34)$$

Suppose now, that due to some mechanism the scalar field has a non-vanishing vacuum-expectation value

$$\langle 0|\Phi|0\rangle = \lambda \neq 0 \quad . \qquad (35)$$

We can then split off the (real) constant λ from the field

$$\Phi = \lambda + \Phi' \qquad (36)$$

where the residual field Φ' is supposed to have vanishing vacuum-expectation value.

Inserting eq. (36) into eq. (34) yields

$$L_{int} = g\, \lambda A^{\mu} A_{\mu} + g\, A^{\mu} A_{\mu}\, \Phi' \,. \qquad (37)$$

The first term of eq. (37) is nothing but a perfect mass term for the vector field which we have now obtained out of the interaction Lagrangian.

What is the mechanism that induces a non-vanishing

vacuum expectation value in the scalar field? Let us be guided by a classical example. Suppose we have a harmonic potential with a fourth order term

$$V(x) = \mu^2 x^2 + hx^4 \qquad h > 0 \quad . \tag{38}$$

This potential is drawn in Fig. 1a. Clearly, the groundstate is centered around x = 0. The same potential, but for negative μ^2, is drawn in Fig. 1b. It has a local maximum at x = 0 and minima at

$$\frac{\lambda}{\sqrt{2}} = \sqrt{\frac{-\mu^2}{2h}} \quad . \tag{39}$$

The ground state is now centered around λ.

With these preliminaries at hand, let us now learn about the full Higgs-model.

The Higgs Lagrangian is

$$L_H = \frac{1}{4}(F_{\mu\nu})^2 + |(\partial_\mu - ieA_\mu)\Phi|^2 - \mu^2|\Phi|^2 - h|\Phi|^4 \tag{40}$$

$$F_{\mu\nu} = \partial_\mu A_\nu - \partial_\nu A_\mu \quad . \tag{41}$$

For positive μ^2, this is nothing but the ordinary Lagrangian of a scalar field interacting with the electromagnetic field. Clearly, it is renormalizable and invariant under the gauge transformations

$$\begin{aligned} A_\mu &\to A_\mu + \partial_\mu \Lambda \\ \Phi &\to \Phi\, e^{ie\Lambda} \quad . \end{aligned} \tag{42}$$

Let us now see what happens if μ^2 becomes negative. As we infer from our previous example, Φ will develop a non-vanishing vacuum expectation value

$$<0|\Phi|0> = \frac{\lambda}{\sqrt{2}} \neq 0 \tag{43}$$

and we split the residual (complex) field in two real fields χ and θ in the following way

$$\Phi(x) = \frac{1}{\sqrt{2}} \{\lambda + \chi(x)\} e^{\frac{i}{\lambda}\theta(x)} . \tag{44}$$

Because of the gauge freedom, the field $\theta(x)$ can be eliminated by the following choice of gauge

$$\Lambda(x) = - \frac{1}{e\lambda} \theta(x) . \tag{45}$$

Inserting these substitutions in eq. (40) the Higgs Lagrangian becomes

$$L_H = \frac{1}{4}(F_{\mu\nu})^2 + \frac{1}{2}(\partial_\mu \chi)^2 - \lambda \chi(\mu^2 + h \lambda^2) +$$

$$+ \frac{1}{2} e^2 \lambda^2 (A_\mu)^2 - h \lambda^2 \chi^2 + \text{coupling terms.} \tag{46}$$

Neglecting higher order corrections, the linear term in χ of eq. (46) has to vanish if we insist in a vanishing vacuum-expectation value for this field. Since λ should be different from zero, the only alternative is to fix it at its "classical" value (39).

Moreover, we can infer the following mass values from eq. (46):

$$m_A = e\,\lambda$$
$$m_\chi^2 = 2\,h\,\lambda^2\,. \tag{47}$$

If we want to use the same technique for generating masses in our realistic model of section 3 we have to realize that eq. (24) is invariant under a larger gauge group than (42). In fact, we have a triplet and a singlet of vector mesons and the invariance group is the (non-abelian) group $SU_2 \otimes U_1$. Thus we need a doublet of Higgs-mesons

$$\Phi = \begin{pmatrix} \Phi^+ \\ \Phi^0 \end{pmatrix} \tag{48}$$

with a self interaction similar to eq. (40) to ensure the non-vanishing vacuum expectation value

$$L_2 = -\mu^2|\Phi|^2 - h|\Phi|^4 \quad , \tag{49}$$

where, of course,

$$|\Phi|^2 = |\Phi^+|^2 + |\Phi^0|^2. \tag{50}$$

The gauge invariant coupling of the Higgs meson Φ to the intermediate vector bosons is

$$L_3 = |(\partial_\mu - i \frac{g}{2} \vec{\tau} \vec{A}_\mu + i \frac{g'}{2} B_\mu) \Phi|^2 . \tag{51}$$

Consider, for the moment, the neutral part of this Lagrangian. By means of eqs. (25) and (27) it can be written in the following form:

$$(-i \frac{g}{2} \tau_3 A_\mu^3 + i \frac{g'}{2} B_\mu) \Phi =$$

$$= \frac{ig}{2\cos\phi} \begin{pmatrix} (A_\mu \sin 2\phi - Z_\mu \cos 2\phi) \Phi^+ \\ Z_\mu \quad \Phi^0 \end{pmatrix} . \tag{52}$$

Eq. (52) shows that the neutral Higgs-meson does not couple to the photon. This is important to render the photon massless.

Let us now turn the crank and invoke the Higgs-mechanism; in analogy to eq. (44) we write

$$\Phi = e^{\frac{i}{\lambda} \vec{\tau} \vec{\theta}(x)} \begin{pmatrix} 0 \\ \frac{1}{\sqrt{2}}[\lambda + \chi(x)] \end{pmatrix} . \tag{53}$$

A special choice of gauge allows for the elimination of the 3 fields $\vec{\theta}(x)$ so that

$$\Phi(x) \to \begin{pmatrix} 0 \\ \frac{1}{\sqrt{2}}[\lambda + \chi(x)] \end{pmatrix} . \tag{54}$$

Before we can add up all the pieces of Lagrangians we have yet to add the free Lagrangian

$$L_o = \frac{1}{4}(\vec{A}_{\mu\nu})^2 + \frac{1}{4}(B_{\mu\nu})^2 + i\,\bar{R}\,\not{\partial}\,R + i\,\bar{L}\,\not{\partial}\,L \tag{55}$$

with

$$\vec{A}_{\mu\nu} = \partial_\mu \vec{A}_\nu - \partial_\nu \vec{A}_\mu + g\,\vec{A}_\mu \times \vec{A}_\nu \tag{56}$$

$$B_{\mu\nu} = \partial_\mu B_\nu - \partial_\nu B_\mu \quad . \tag{57}$$

L_o does not contain any mass terms. The masses of the intermediate bosons is induced by the Higgs mechanism. For the leptons we still have to add a mass term which is invariant under the gauge group and leaves the neutrino massless:

$$L_4 = \sqrt{2}\,\frac{m_e}{\lambda}\,(\bar{R}\,\Phi^+\,L + \bar{L}\,\Phi\,R) \ . \tag{58}$$

Eq. (58) automatically induces a coupling of the Higgs meson to electrons!

Adding up all pieces, i.e. eqs. (24), (49), (51), (55) and (58), the total Salam-Weinberg Lagrangian becomes

$$L_{SW} = L_o + L_1 + L_2 + L_3 + L_4 \ . \tag{59}$$

To obtain a Lagrangian directly comparable with experiments we use eqs. (25), (27) and (54). This gives us the Lagrangian in the so-called "U-form" from which all unphysical particles have been removed. It thus gives an explicitly unitary S-operator. Adding the suppressed muonic leptons, the Salam-Weinberg Lagrangian becomes

$$
\begin{aligned}
L = L_o &+ \left(\frac{g}{2\sqrt{2}}\, \ell^{\dagger}_{\lambda}\, W^{\lambda} + \text{h.c.}\right) + e\, \ell_{\lambda}\, A^{\lambda} + \\
&+ \sum_{\ell} \frac{g}{4\cos\phi}\, \{\bar{\nu}_{\ell}\gamma_{\lambda}(1+\gamma_5)\nu_{\ell} - \bar{\ell}\,\gamma_{\lambda}(C_V+\gamma_5)\ell\} Z^{\lambda} - \\
&- h\,\lambda^2\,\chi^2 - h\,\lambda\,\chi^3 - \frac{h}{4}\,\chi^4 + \frac{1}{2}(\partial_{\mu}\,\chi)^2 + \qquad (60) \\
&+ \frac{g^2}{4}\left(W^{\dagger}_{\mu}\, W^{\mu} + \frac{1}{2\cos^2\phi}\, Z_{\mu}\, Z^{\mu}\right)(\lambda^2 + 2\lambda\chi + \chi^2) + \\
&+ \sum_{\ell} m_{\ell}\left(1 + \frac{1}{\lambda}\,\chi\right) \bar{\ell}\ell \quad ,
\end{aligned}
$$

C_V is given in eq. (30).

We can immediately read off the following relations from eq. (60):

$$\frac{G}{\sqrt{2}} = \frac{g^2}{8M_W^2} \qquad (61)$$

$$M_W = \frac{g\lambda}{2} \qquad (62)$$

$$M_Z = \frac{M_W}{\cos\phi} \qquad (63)$$

$$m_{\chi} = \lambda\sqrt{2h}\ . \qquad (64)$$

Combining these relations we can write

$$\frac{G}{\sqrt{2}} = \frac{1}{2\lambda^2} = \frac{e^2}{8M_W^2 \sin^2\phi} \tag{65}$$

which gives a relation between M_W and $\sin\phi$. Eq. (63) is particularly noticable since it gives a new meaning to the Salam-Weinberg mixing angle. Its cosine now appears as the mass ratio of charged versus neutral intermediate bosons.

The relation between M_Z, M_W and $\sin\phi$ is drawn in Fig. 2.

The renormalizability of the Lagrangian (59) can better be shown in its "R-form" where all the Higgs fields remain. We have no space to go into this interesting but involved question and refer the diligent reader to the literature [4].

5. THE PROBLEM OF HADRONS

Weak Interactions are more versatile than Electromagnetism also because they allow for $\Delta Y \neq 0$ transitions. It is precisely due to this fact that we run into another problem when we try to incorporate hadrons: There are no neutral currents in $|\Delta Y| = 1$ transitions. This fact is rather well established as can be seen from table 1.

To specify the problem, let us see what happens if we try to naively couple also hadrons in a gauge invariant way. We shall write the hadronic currents in terms of quarks and define

$$q = \begin{bmatrix} p \\ n \\ \lambda \end{bmatrix} . \tag{66}$$

The ordinary Cabibbo current can then be written in the following way:

$$\bar{p}(n \cos\theta + \lambda \sin\theta) = \bar{q}\, C_+ q \tag{67}$$

where θ is the Cabibbo angle and

$$C_+ = \begin{bmatrix} 0 & \cos\theta & \sin\theta \\ 0 & 0 & 0 \\ 0 & 0 & 0 \end{bmatrix} . \tag{68}$$

Also, we define

$$C_- = C_+^T . \tag{69}$$

For an invariant coupling, the neutral triplet boson has to be coupled to the current with

$$C_3 = [C_+, C_-] = \begin{bmatrix} 1 & 0 & 0 \\ 0 & -\cos^2\theta & -\frac{1}{2}\sin 2\theta \\ 0 & -\frac{1}{2}\sin 2\theta & -\sin^2\theta \end{bmatrix} . \tag{70}$$

This current is

$$\bar{q}C_3 q = \bar{p}p - \cos^2\theta\,\bar{n}n - \sin^2\theta\,\bar{\lambda}\lambda - \frac{1}{2}\sin 2\theta(\bar{n}\lambda + \bar{\lambda}n) . \tag{71}$$

Because of the off-diagonal elements in the matrix (70) there are neutral strangeness changing currents in (71). They are forbidden by experiments.

A way out has been shown by Glashow, Iliopoulos and Maiani [5]. The prize, they are willing to pay, is the introduction of a fourth quark which carries a new quantum number called charm. Charmed particles have not been observed so far and thus must be heavy enough to have escaped detection.

With the charmed quark p' we can define the quadruplet

$$q' = \begin{bmatrix} p' \\ p \\ n \\ \lambda \end{bmatrix} \qquad (72)$$

and the 4 by 4 matrix

$$C'_+ = \begin{bmatrix} 0 & A \\ 0 & 0 \end{bmatrix} \qquad (73)$$

$$A = \begin{bmatrix} -\sin\theta & \cos\theta \\ \cos\theta & \sin\theta \end{bmatrix} . \qquad (74)$$

The Cabibbo current then becomes

$$\bar{q}'C'_+q' = \bar{p}'(-\sin\theta\ n + \cos\theta\ \lambda) + \bar{p}(\cos\theta\ n + \sin\theta\ \lambda) . \qquad (75)$$

In its uncharmed part it is identical to eq. (67). However, the commutator between C'_+ and its transpose is now diagonal

$$[C'_+, C'_-] = \begin{bmatrix} I & 0 \\ 0 & -I \end{bmatrix} . \qquad (76)$$

Therefore, there are no neutral strangeness changing currents.

A very physical way to see these connections has been indicated by Llewellyn-Smith [6]. He starts from the observation that in a renormalizable theory, cross-sections cannot rise proportional to the laboratory energy, because such a rise stems precisely from the bad high-energy behaviour leading to unacceptable divergences in higher order graphs. Therefore, there must be cancellations. It is known, that these cancellations do occur in gauge theories. (What Llewellyn-Smith actually shows is that the inverse is also true!).

In a "Gedankenexperiment", let us consider the process

$$\nu + \bar{\nu} \to W^+ + W^- . \qquad (77)$$

In lowest order, the corresponding graphs are shown in Fig. 3. Fig. 3a gives the graph of the standard theory. Cancellations are either due to Fig. 3b (Salam-Weinberg) or to Fig. 3c (Georgi-Glashow). In a very similar way, we can now describe the problem of hadrons. To this end, let us consider the ficticious "quark process"

$$n + \bar{\lambda} \to W^+ + W^- . \qquad (78)$$

Again, this process shows unacceptable high energy behaviour in lowest order and cancellations must therefore occur. (The possibility that process (78) does not occur in the first place because strangeness changing transitions might be V + A instead of V - A has been ruled out by Λ decay experiments!).

In complete analogy to Fig. 3, we have 2 alternatives: Either neutral currents or new "heavy" fermions (quarks). Neutral currents are ruled out experimentally (see table 1) so that only the possibility of new "charmed" quarks remains. This is precisely what we have learned from our more quantitative analysis above.

We should mention that there are models which do not rely on quarks at all and which also achieve agreement with data [7].

6. SOME TESTS OF THE MODEL

The cleanest tests of theoretical predictions are always achieved with purely leptonic processes. To sort out weak contributions, one can either look for parity violating effects or for processes involving neutrinos. We shall describe an example for both possibilities here.

First, let us turn to elastic scattering of a neutrino on a charged lepton. Lowest order graphs are shown in Fig. 4. The graph of Fig. 4a is the one of the classical V-A theory. It is complemented in the Salam-Weinberg theory by Fig. 4b.

By means of a Fierz re-ordering identity we can write the matrix-element for graph Fig. 4a in the following way:

$$M^{(a)} = \frac{G}{\sqrt{2}} \bar{\nu}_e \gamma_\lambda (1+\gamma_5) \nu_e \; \bar{e} \gamma^\lambda (1+\gamma_5) e \; . \tag{79}$$

In this way of writing it is particularly suited for a comparison with the neutral current contribution. Notice that we have already approximated the W-propagator by its low energy limit (32).

The graph of Fig. 4b leads to the matrix element (compare eq. 60)

$$\begin{aligned} M^{(b)} &= \frac{1}{M_Z^2} \frac{-g^2}{16\cos^2\phi} \bar{\nu}_e \gamma_\lambda (1+\gamma_5) \nu_e \; \bar{e} \gamma^\lambda (C_V+\gamma_5) e = \\ &= \frac{-G}{2\sqrt{2}} \bar{\nu}_e \gamma_\lambda (1+\gamma_5) \nu_e \; \bar{e} \gamma^\lambda (C_V + \gamma_5) e \end{aligned} \tag{80}$$

where we have made use of eq. (63). C_V is given by eq. (30).

Adding the 2 matrix elements (79) and (80) leads to a total transition matrix element for neutrino electron scattering with effective couplings in the electron current given by

$$\begin{aligned} C_V^{e\nu_e} &= \frac{1}{2} + 2 \sin^2\phi \\ C_A^{e\nu_e} &= \frac{1}{2} \; . \end{aligned} \tag{81}$$

For electron scattering on a neutretto (muonic neutrino) graph Fig. 4a does not contribute and we arrive at the effective couplings of eq. (80), namely

$$C_V^{e\nu_\mu} = -\frac{1}{2} + 2\sin^2\phi$$

$$C_A^{e\nu_\mu} = -\frac{1}{2} \quad . \tag{82}$$

It is clear, that these experiments provide a clear cut test for the Salam-Weinberg model together with a determination of $\sin\phi$. The single, celebrated candidate for a $\bar{\nu}_\mu e$ elastic event [8] leads to a first rough estimate

$$(\bar{\nu}_\mu e): \qquad \sin^2\phi = 0{,}35 \pm 0{,}25 \ . \tag{83}$$

The experiments on elastic scattering of antineutrinos from a reactor on electrons [9] yields

$$(\bar{\nu}_e e): \qquad \sin^2\phi < 0{,}3 \quad . \tag{84}$$

Much better determinations are in principle possible from semi-leptonic events [10]. The CERN experiment leads to [11]

$$(\text{semi-leptonic}): \ \sin^2\phi = 0{,}4 \pm 0{,}1 \quad . \tag{85}$$

Let us now turn to the process

$$e^+ + e^- \rightarrow \mu^+ + \mu^- \quad . \tag{86}$$

Due to the possible exchange of a Z-boson and its interference with the photon exchange, the cross-section becomes

$$\frac{d\sigma}{d\cos\theta} = A(1+\cos^2\theta) - B\cos\theta + O(m_\mu^2) \qquad (87)$$

with

$$A = \frac{\pi}{2}\frac{\alpha^2}{s} - \frac{G\alpha C_V^2}{128\sqrt{2}}$$

$$(88)$$

$$B = \frac{G\alpha}{64\sqrt{s}} \quad .$$

C_V is again given by eq. (3o). Higher order photon exchanges have not been included in eqs. (87) - (89). The process is discussed in detail in reference 12.

At higher energy storage rings for electrons and positrons, weak interaction effects should become dominant by a process similar to the 2 photon exchange [13]. So far, this has only been explored in the standard V-A theory [14]. In the Salam-Weinberg model, the effects should come in even earlier due to interference between Z and γ contributions. The rather exciting phenomenon is graphically shown in Fig. 5.

REFERENCES

1. A. Salam, J. C. Ward, Phys. Lett. 13, 168 (1964). A. Salam in Elementary Particle Theory, ed. N. Svartholm. (Almquist and Wiksell, Stockholm 1968). S. Weinberg, Phys. Rev. Lett. 19, 1264 (1967).
2. H. Georgi, S. L. Glashow, Phys. Rev. Lett. 28, 1494 (1972).

3. P. W. Higgs, Phys. Rev. Lett. 13, 5o8 (1964).
Phys. Rev. 145, 1156 (1966).
T. W. B. Kibble, Phys. Rev. 155, 1554 (1967).

4. G.'t Hooft, Nucl. Phys. B33, 173 (1971) and B35, 167 (1971).
E. S. Abers, B. W. Lee, Phys. Rep. 9c/1 (1973).
G.'Hooft, M. Veltman "Diagrammar" CERN73-9.

5. S. L. Glashow, J. Iliopoulos, L. Maiani, Phys. Rev. D2, 1285 (1970).

6. C. H. Llewellyn-Smith, Phys. Lett. 46B, 233 (1973), see also J. Bell, Nucl. Phys. B60, 427, (1973).
J. Schechter, Y. Ueda, Lett. Nuovo Cim. 8, 991 (1973).

7. I. Bars, M. S. Halpern, M. Yoshimura, Phys. Rev. D7, 1233 (1973).

8. F. J. Hasert et al., Phys. Lett. 46B, 121 (1973).

9. F. Reines, Neutrino Symposium Trieste (1973).

10. R. B. Palmer, Phys. Lett. 46B, 240 (1973).
A. de Rujula, S. L. Glashow, Phys. Lett. 46B, 377 (1973).
L. M. Sehgal, Aachen preprint (1973).

11. F. J. Hasert et al., Phys. Lett. 46B, 138 (1973).

12. G. V. Dass, DESY 73/40 (1973).
R. Gatto, G. Preparata, Lett. Nuovo Cim. 7, 89 (1973).
V. K. Cung, A. K. Mann, E. A. Paschos, Phys. Lett. 41B, 355 (1972).

13. S. J. Brodsky, T. Kinoshita, H. Terazawa, Phys. Rev. D4, 1532 (1971).
N. A. Romero, A. Jaccarini, P. Kessler, J. Parisi,

Compt. Rend. B298, 153, 1129 (1969).
Phys. Rev. D4, 2927 (1971).
R. W. Brown, I. J. Muzinich, Phys. Rev. D4, 1496 (1971).
M. Greco, Nuovo Cim. 4A, 689 (1971).
C. E. Carlson, Wu-Ki Tung, Phys. Rev. D4, 2873 (1971).

14. R. Kögerler, H. Pietschmann, Lett. Nuovo Cim. 18, 1013 (1973).

FIGURE CAPTIONS

1. Harmonic potential with 4th order term

   a. $\mu^2 > 0$
   b. $\mu^2 < 0$

2. Relation between M_Z, M_W and $\sin\phi$

3. Lowest order graphs for the process $\nu + \bar{\nu} \rightarrow W^+ + W^-$

4. Lowest order graphs for neutrino-electron scattering

5. Various contributions to electron-positron processes

TABLE 1: Absence of Neutral Currents with $\Delta Y \neq 0$

$K^\pm \rightarrow \pi^\pm e^+ e^-$	$< 4.10^{-7}$
$K^\pm \rightarrow \pi^\pm \mu^+ \mu^-$	$< 2,4.10^{-6}$
$K^\pm \rightarrow \pi^\pm \nu\bar{\nu}$	$< 1,4.10^{-6}$
$\Sigma^+ \rightarrow p e^+ e^-$	$< 7.10^{-6}$
$K_L^o \rightarrow \mu^+ \mu^-$	$= (1,2 \begin{smallmatrix} +\ 0,8 \\ -\ 0,4 \end{smallmatrix}).10^{-8}$

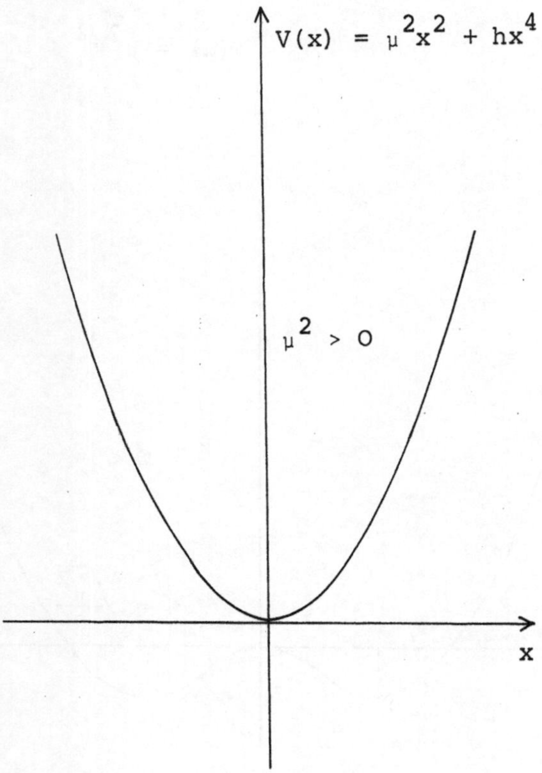

Fig. 1 (a)

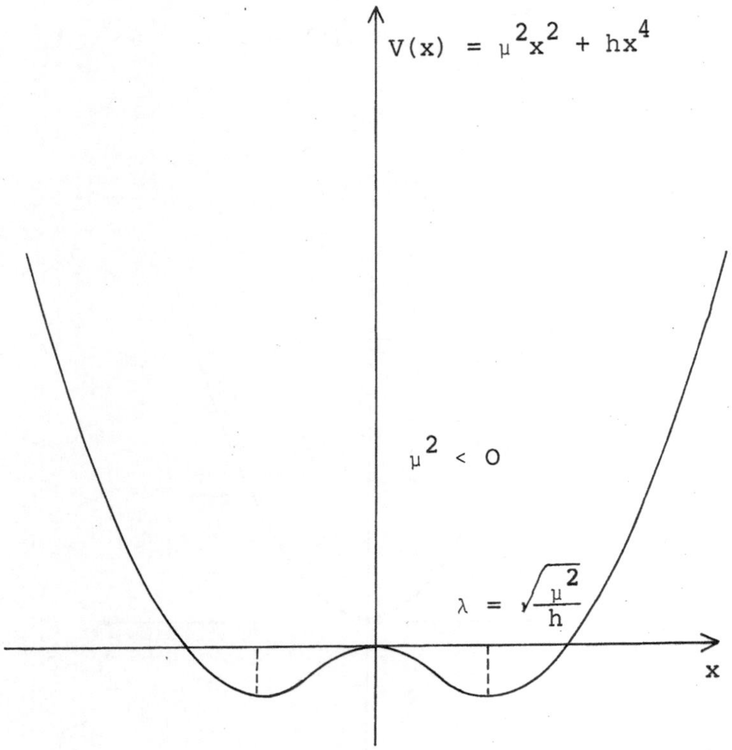

Fig. 1 (b)

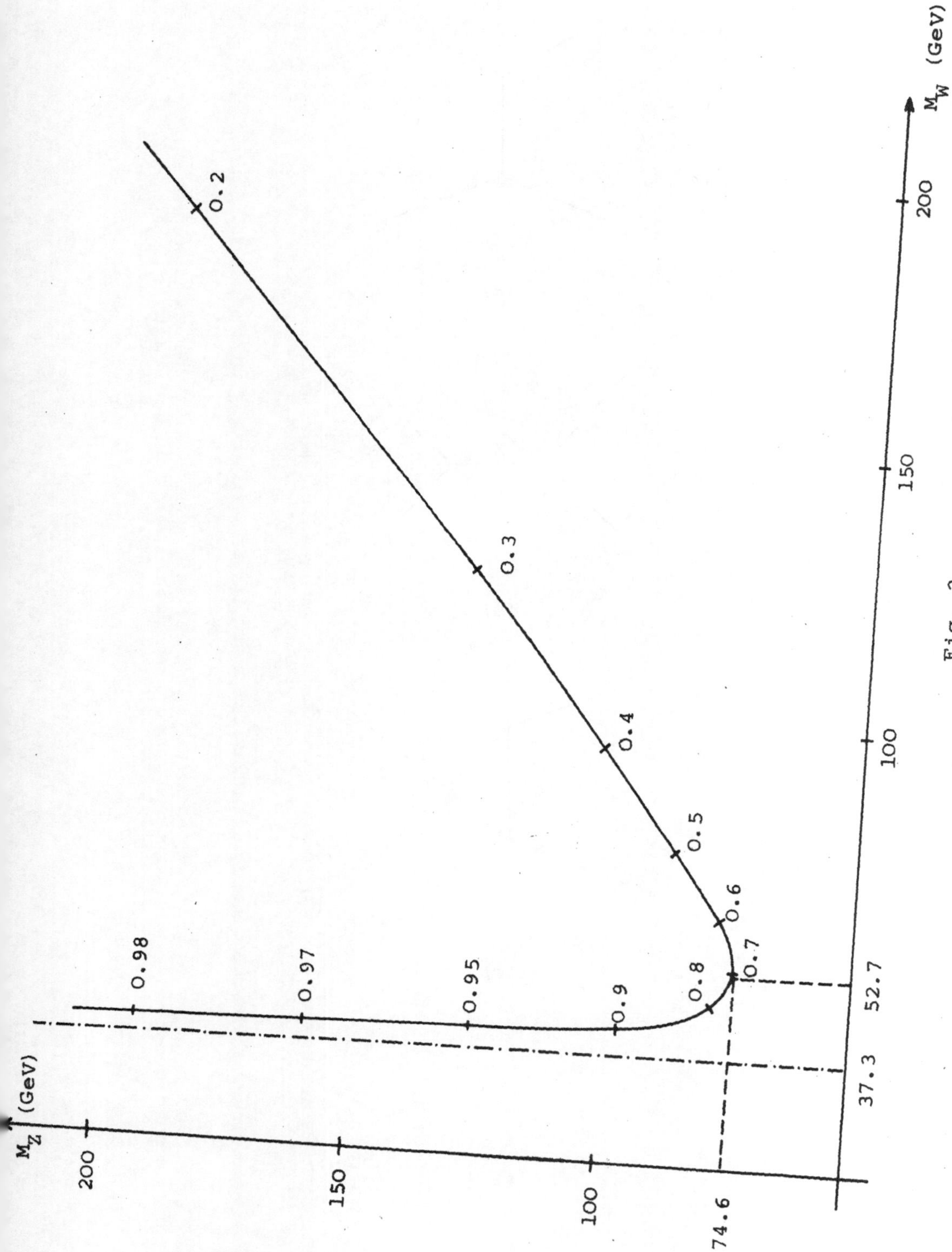

M_W (GeV)
200
150
100
52.7
37.3
M_Z (GeV)
200
150
100
74.6
0.2
0.3
0.4
0.5
0.6
0.7
0.8
0.9
0.95
0.97
0.98

Fig. 2

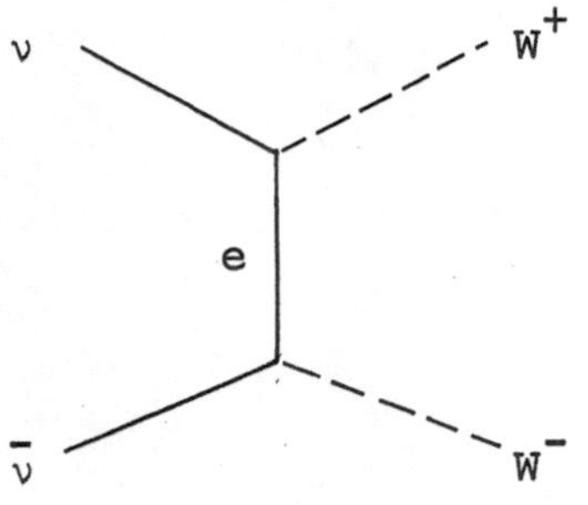

(a)

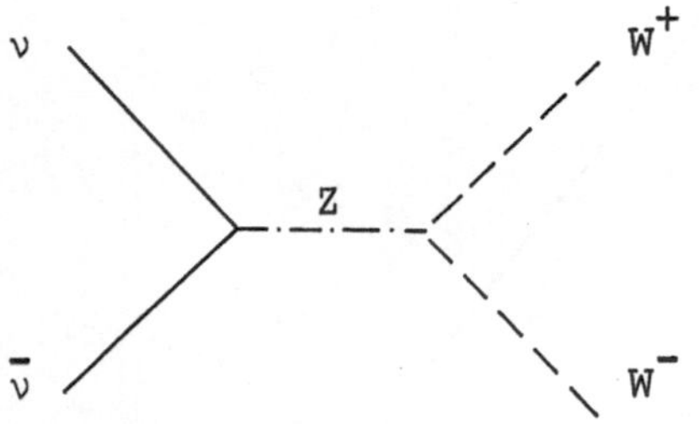

(b)

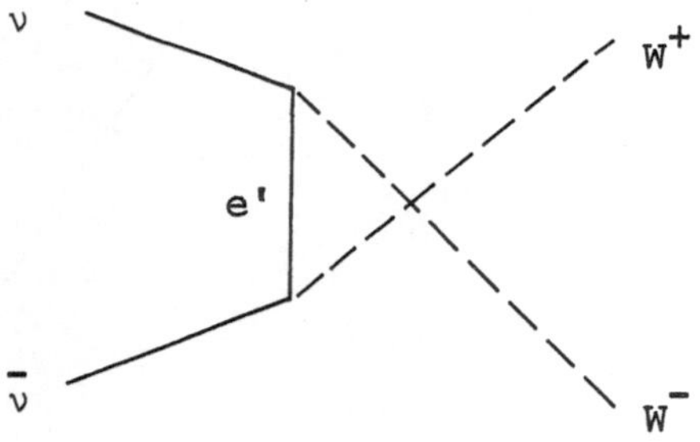

(c)

Fig. 3

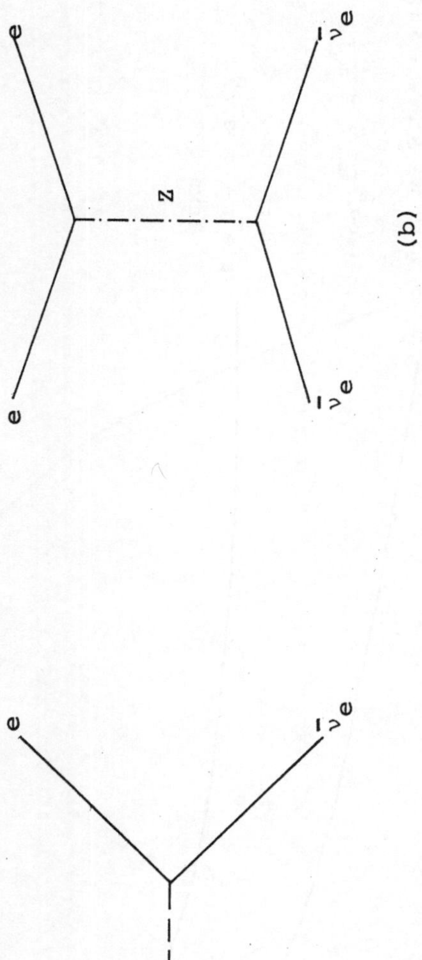

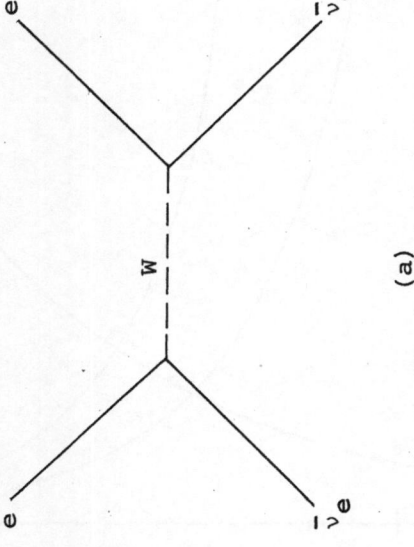

Fig. 4

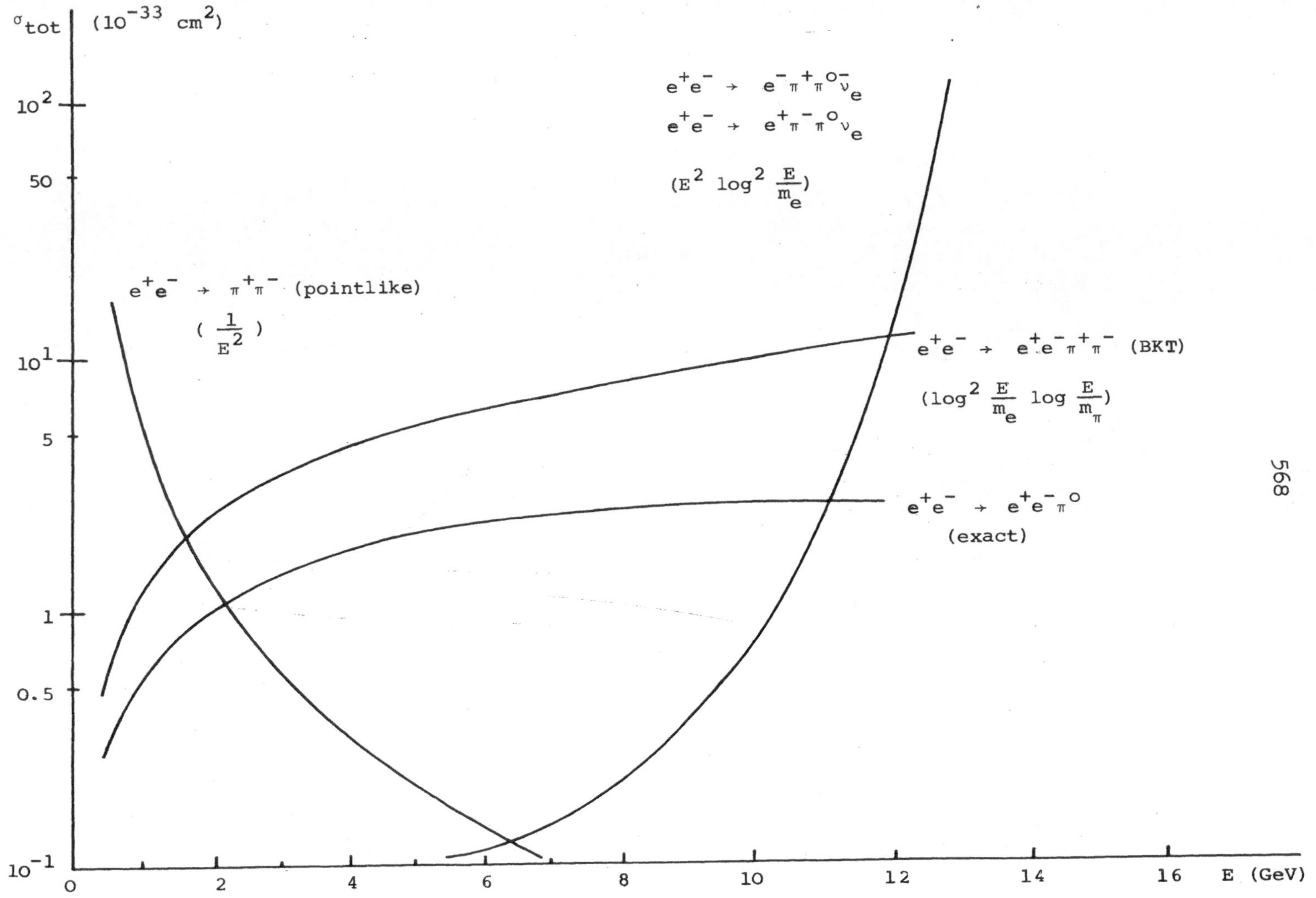
σtot (10^-33 cm^2)
10^2
50
10^1
5
1
0.5
10^-1
0
2
4
6
8
10
12
14
16
E (GeV)
e+e- → π+π- (pointlike)
(1/E^2)
e+e- → e-π+π0ν̄e
e+e- → e+π-π0νe
(E^2 log^2 E/me)
e+e- → e+e-π+π- (BKT)
(log^2 E/me log E/mπ)
e+e- → e+e-π0
(exact)

Fig. 5

Acta Physica Austriaca, Suppl. XIII, 569–594 (1974)

WEAK INTERACTIONS OUTSIDE THE LABORATORY*)

by

G. MARX
Department of Atomic Physics
Eötvös University in Budapest

1. ARE WEAK FORCES ALWAYS REALLY WEAK?

Nature shows us four faces.
The world of "strong" nuclear interactions is to be seen within the nucleus and in high energy collisions, where this strongest force of nature is a real autocrat. The coupling constant is big, $g^2/\hbar c \gtrsim 1$, it produces large masses and as a consequence of the Wick relation $r = \hbar/mc$ the massiv hadrons transmit momentum over short distances: $r \lesssim 2.10^{-13}$ cm. The strong interaction vanishes beyond the nuclear surface because it is strong.

Outside the nucleus, in the atoms and molecules the ruling force is the electromagnetic one, $e^2/\hbar c = 1/137$. We are able to lengthen the range of Coulomb force to macroscopic distances, because the light quanta have a vanishing rest mass, m = 0. This makes electrical engeneering possi-

*) Lecture given at XIII. Internationale Universitätswochen für Kernphysik, Schladming, Austria, February 4-15, 1974.

ble. On the other hand, the electromagnetic field is a vector field, coupled to a vector current. The fourth component of this current is indefinit: one has + and - charges. This produces a screening effect; the electric force of + charges is screened by the presence of - charges in condensed matter, so electricity does not manifest itself outside macroscopic bodies.

On astronomical scale the only force to be observed is gravity. It is a very faint phenomenon, its coupling constant is

$$\gamma = 6.67 \times 10^{-8} \ \mathrm{cm}^3 \ \mathrm{g}^{-1} \ \mathrm{s}^{-2}$$

$$l_g = (\hbar \gamma c^{-3})^{\frac{1}{2}} = 1.6 \times 10^{-33} \ \mathrm{cm} \ .$$

This gives a very small pure number in combination with any physical length. But the quanta of gravity have a vanishing rest mass, they form a tensor field with a positive definite source density (the mass density), so gravity is never screened: its lines of force are of infinite length, reaching from one astronomical body to the other. It is really a wonderfull luck of the searching man, that the worlds of gravitational, electric and nuclear forces have been separated by rather sharp boundaries: this enabled us to understand nature step by step.

The fourth force of nature is beta radioactivity. It is called weak interaction, because its coupling constant is

$$G = 10^{-5} \frac{\hbar^3}{M_p^2 c} \quad \text{or} \quad l_w = (G/\hbar c)^{\frac{1}{2}} = 6.65 \times 10^{-17} \ \mathrm{cm} \ .$$

Its range is much smaller than the nuclear range, it is practically zero. The beta coupling is almost everywhere negligible. The only exceptions are the cases, where the nuclear or electric interactions are supressed by some selection rules. The radioactive beta decays are always slow. In nuclear decays, where $E \lesssim 1$ MeV,

$$t_\beta \simeq \lambda^5/cl_w^4 \gg 1 \text{ sec} ,$$

if $\lambda = h/p$ stands for the Broglie wave length. (In particle physics $E \approx 100$ MeV, $t_\beta \approx 10^{-10}$s, this life time is long enough for an unstable particle to produce an observable track.) Now we may ask the question: where to look to see the weakface of nature? What is the role of weak interactions in our world?

The answer of high energy physicists known. The strength of the weak coupling becomes comparable to the strong and electric interactions, if the Broglie wave length of interacting particles becomes comparable to the weak length l_w. This happens at $E \approx 100$ GeV in center of mass system, which means $E \approx 10^4$ GeV in laboratory frame. These energies are beyond the reach of our present machines and they are at the end of the cosmic ray spectrum. Does it mean, that the weak interactions are completely unimportant for our world? Would a man of the street notice, if the value of G changed by a factor of ten?

The answer is rather sophisticated. There exists a particle, which is subjected only to weak and gravitational forces: the neutrino. Its cross section is $\sigma = l_w^4/\lambda^2$, so its mean free path comes out to be

$L = d^3/\sigma = d^3\lambda^2/l_w^4$. By writing $\lambda = 10^{-11}$ cm (MeV region) and $d = 10^{-8}$ cm (separation of atoms in condensed matter) we arrive at L = several light years. The planets and stars are impermeable for photons, but they are transparent for neutrinos: $L_\nu \gg \phi$ of $* \gg L_\gamma$. This has the consequence, that all the astronomical bodies are surface sources of light but volume sources of neutrinos. To find any importance of weak interactions, we must leave our laboratories and we must look at the sky.

2. EARTH

The heavy elements of Earth have been formed in the central parts of massive stars some 5 - 10 billion years ago. The $10^9 - 10^{10}$ K^o temperature produced excited nuclear states which cool down slowly by radioactive decays. This radioactivity heated up the cold Earth (vulcanos, thermal sources etc.) Each gram of the continental crust radiates about one antineutrino per second. (Half of them comes from the β decay of K40.) In the case of a homogeneous Earth this would give a surface flux density $I = 3.10^9$ $\bar{\nu}/cm^2$ sec. The two possibilities of $\bar{\nu}$ detection are induced β^+ decay and induced K capture, e.g. $\bar{\nu} + e^- + {}^{112}Sn \rightarrow {}^{112}I \rightarrow {}^{112}Sn + e^- + \bar{\nu}$. The capture probability turns out to be $10^{-4} - 10^{-6}$ events per mol per year, depending on the target nucleus (I.Lux-G.Marx).

The result is highly dependent on the distribution of chemical elements within the Earth. Such a measurement would be of high value for understanding the birth of our home planet.

3. SUN

Earth is an old and dead piece of matter from the point of view of nuclear physics. The Sun is alive. Its energy comes from the work done by nuclear forces during the fusion $4H \rightarrow He$. But the Sun is not an H bomb. The solar activity is regulated by the weak interactions. In the first step of the fusion chain one has $^{1}H + {}^{1}H \rightarrow {}^{2}He$, but the strong interaction is not strong enough to produce a stable di-proton. $^{2}He \rightarrow {}^{1}H + {}^{1}H$. It needs the help of the weak interaction: $^{2}He \rightarrow {}^{2}H + e^{+} + \nu$. This makes the life time of the solar H finite, but very long: several billion years. The brightness of the Sun is proportional to G^2.

$$^{1}H + {}^{1}H \rightarrow {}^{2}He + e^{+} + \nu_1 \quad (100\ \%)$$

$$^{1}H + e^{-} + {}^{1}H \rightarrow {}^{2}He + \nu_2 \quad (0.25\ \%)$$

$$^{1}H + {}^{2}H \rightarrow {}^{3}He + \gamma \quad (100\ \%)$$

$$^{3}He + {}^{3}He \rightarrow {}^{4}He + {}^{1}H + {}^{1}H \quad (91\ \%)$$

$$^{3}He + {}^{4}He \rightarrow {}^{7}Be + \gamma \quad (9\ \%)$$

$$e^{-} + {}^{7}Be \rightarrow {}^{7}Li + \nu_3$$
$$^{1}H + {}^{7}Li \rightarrow {}^{8}Be + \gamma$$
$$^{8}Be \rightarrow {}^{4}He + {}^{4}He$$
$$(9\ \%)$$

$$^{1}H + {}^{7}Be \rightarrow {}^{8}B + \gamma$$
$$^{8}B \rightarrow {}^{8}Be + e^{+} + \nu_4$$
$$^{8}Be \rightarrow {}^{4}He + {}^{4}He$$
$$(0.01\ \%)$$

The branching ratios have been calculated from the standard solar model. (The initial composition has been assumed to agree with the present surface composition. The age of the Sun is given by the time, when its luminosity reaches the today observed value.)

It is a very atttractive possibility to test the theory by observing the solar neutrinos, coming directly from the active centre (N. Menyhard-G.Marx). Practically the most convenient capture reaction is the following:

$$\nu + {}^{37}Cl \rightarrow {}^{37}A + e^{-}; \quad {}^{37}A \rightarrow {}^{37}Cl + e^{+} + \nu$$

(Pontecorvo-Alvarez). Its energy threshold is 0.8 MeV. The energies of the solar neutrinos and the corresponding capture cross sections have been given by E.Bahcall:

$E_1 = 0.42$ MeV	$\sigma_1 = 0$
$E_2 = 1.44$ MeV	$\sigma_2 = 1.7 \times 10^{-45}$ cm^2
$E_3 = 0.861$ and 3.385 MeV	$\sigma_3 = 0.3 \times 10^{-45}$ cm^2
$E_4 = 14.06$ MeV	$\sigma_4 = 1400 \times 10^{-45}$ cm^2

Each $4H \rightarrow He$ fusion produces 28 MeV and two neutrinos. Knowing the present solar luminosity, one gets a flux density $I = 7.10^{10} \nu/cm^2$ sec at the Earth. The corresponding capture rate in predicted to be $R = 14 \times 10^{-5}$ capture per mol per year. 80% of the events expected to be produced by the energetic B neutrinos (ν_4).

The neutrino detector has been realized by R.Davis

Jr. (USA). 400 000 liters of tetrachloroethylene (C_2Cl_4) was located deep underground. The produced argon was a noble gas, it did not stick to anything. The A atoms were washed out by the He bubbles, collected and counted by their radioactive decays with nearly loo % efficiency. No signal stronger than 2.10^{-5} capture per mol per year has been found. The factor of seven inconsistency made the solar neutrino experiment to one of the great scientific puzzles of our time.

a) Are the solar neutrinos unstable and do not reach the Earth? (This would need a new interaction and new sort of particles as decay products.)

b) Is our knowledge about the nuclear reactions poor? (A $^3He + {}^3He \rightarrow {}^6Be$ resonance would enhance this branch and suppress the B branch, but this assumption has not been confirmed by subsequent experiments.)

c) Does the Davis experiment contain a systematic error? (Nobody has found any. Alvarez suggested to test the detector with a giant β^+ active source. Davis is going to prepare a second experiment with Li instead of Cl.)

d) Or is our knowledge about the origin of solar energy wrong? (Are no H → He fusion reactions going at the centre of the Sun?) As a matter of fact, the last possibility cannot be excluded. The solar model has been constructed to explain the present optical luminosity and it has not been tested up to now. Its first test (measuring the neutrino luminosity) has gone wrong. The B neutrino test is a very sensitive one. The B branching ratio depends on the value of the Gamov factor (strong Coulomb repulsion between H and Be) and it is very sensitive for temperature ($\sim T^{25}$). A slight drop of T would suppress the B neutrinos, but it would be inconsistent with the accurately known present optical luminosity L_o.

From completely positivistic point of view, there is no contradiction between the observed optical luminosity L_o and the low neutrino luminosity. The neutrinos produced in the Sun reach our Earth in 8 minutes without any interaction. The heat produced by the nuclear fusion diffuses slowly toward the solar surface. The diffusion time is of the order $t = (\gamma M^2/R) : L_o \approx 10^7$ years. The sunshine proves that the solar reactor worked 10^7 years ago. The Davis experiment proves that it was turned off 8 minutes ago. Is our Sun a variable star? What would this mean for the mankind?

A sudden drop in the central temperature might be caused by a partial mixing of the solar matter. The outer layers are richer in H. A mixing would decrease the average molecular weight μ at the centre and this would produce a drop of temperature at given equilibrium pressure and density, according to the gas law $p/\rho = RT/\mu$ (W.Fowler) Such a mixing might be explained by an increase of the temperature gradient above a critical value, where the steady state solution is already unstable against perturbations (Dilke, Gough). A slight drop of central temperature would cause suddenly a strong drop in L_ν, but a very delayed smooth drop in L_o. It is not yet estimated accurately, what is the building-up time of a critical temperature gradient. It is regulated anyway by the weak ventil $^1H + {}^1H \rightarrow {}^2H + e^+ + \nu$ and it is proportional to G^{-2}. The estimated value is about $10^8 - 10^9$ years.

At first sight we may argue, that the continuity of life on Earth does not tolerate such changes. But there is a geological evidence, that the warm steady state of the terrestrial climate was interrupted in every 250 million years by cooler unstable periods of a few million years duration. This has been found to have happend at least three times (Precambrium, Perm, Pleistocen). Just now we live in such a cooler period with frequent ice ages. The long time period years has been a puzzle of geology since decades: 250 million years cannot be explained by any planetary constant (Lande et al.). It may be a manifestation of the decreased solar temperature suggested by the Davis result.

Life needs long quiet periods for spreading and it needs changes for the genetic evolution. Climatic changes present always a great challenge for the living beings. The inert life forms die out, the elastically adapting life forms change their habits and survive. This is the moral of the ice ages. The last "bad weather" created the man. May be, these waves of conservation and revolution are related to the smallness of the weak coupling constant. May be, the whole story of biological evolution would have produced a different world, if G had been a bit different from its real laboratory value.

4. UNIVERSE

The biggest object of the world is the Universe itself. The red shift of distant galaxies shows, that the Universe is expanding. The relict electromagnetic radiation (being now of 2.7°K temperature) proves, that the Universe was hot in the past. An extrapolation with the help of the equations of theoretical physics indicates a singularity on the time axis.

Let R(t) be the universal scale function ("radius of curvature"). Its time dependence is influenced by the ρ mass and p pressure distribution, according to the Einstein equations. The two consequences of the Einstein equations are the energy equation

$$\frac{1}{2}\dot{R}^2 - \frac{\gamma}{R}\left(\frac{4\pi}{3}R^3\right)\rho = E = -\frac{1}{2}\varepsilon c^2$$

and the adiabatic equation

$$\frac{d}{dt}\left(\rho c^2 \frac{4\pi}{3} R^3\right) + p\frac{d}{dt}\left(\frac{4\pi}{3}R^3\right) = 0 \ .$$

Here $\varepsilon^2 = \varepsilon$ by the scale convention. ($\varepsilon = +1, 0, -1$ in closed spherical, flat, open hyperbolical space.)

According to these equations the matter was very hot ($T > 10^{12}\ {}^{\circ}K$) and very dense ($\rho >$ nuclear density) in the first $t = 10^{-4}$ seconds. Different unstable forms of particles were created and destroyed continuously (hadron era). When the density was larger, than that of the nucleus ($d < 10^{-13}$ cm), a neutrino with an energy $E > 1$ MeV ($\lambda < 10^{-11}$ cm) had a mean free path $L = d^3\lambda^2 l_w^{-4}$ few meters. That meant, that the neutrinos had been thermalised within $t = L/c < 10^{-6}$ sec. The matter consisted of different components:

$$\rho = \rho_\gamma + \rho_e + \rho_\mu + \rho_\nu + \rho_{hadron} \ .$$

Here $\rho_\gamma c^2 = aT^4$ according to the Stefan-Boltzmann law. $e^\pm, \mu^\pm$ and ν were extrem relativistic Fermion gases. Because of their different statistics (factor $\frac{7}{8}$) and different degrees of freedom (2)

$$\rho_e = \rho_\mu = \rho_\nu = \frac{7}{4}\rho_\gamma \; .$$

When the temperature dropped below $kT = m_\pi c^2$, the hadrons annihilated and the leptons made the main constituent of matter (lepton era). From the adiabatic equation one can get T(R):

$$\frac{dT}{dR} = 3\frac{\rho c^2 + p}{R d\rho/dT} \simeq \text{const}\,\frac{T}{R}\text{ , consequently } TR = \text{const.}$$

That means, that the relativistic forms of matter obey the ρR^4 = const law. (There are only exceptional points: $\mu^\pm$ annihilation at $kT = m_\mu c^2$, $e^\pm$ annihilation at $kT = m_e c^2$). After the positron annihilation ($T = 6.10^9$ K^o, t = 3 sec) the photons became the most aboundant particles, giving the largest contribution to the mass and pressure (radiation era). Finally the temperature dropped below T = 4000°K. The protons and electrons combined to H atoms, the photons lost their overweight because of the Hubble shift ($t = 3.10^5$ years). From this point on the density function was determined by the baryon conservation: ρR^3 = const (Atomic era).

The R(t) function can be computed from the energy equation. Let us start at the end of hadron era ($t_i = 10^{-4}$ sec $\simeq$ 0). The initial value R_i is a free input parameter. By knowing $\rho(R)$, as indicated above, we can compute R(t) and T(t). The computation steps, when we reach the present temperature T_o = 2.7^{o}K. The computer gives the time t_o, the present radius R_o, the present rate of expansion $H_o = \dot{R}_o/R_o$ and the present deceleration $q_o = - \ddot{R}_o R_o \dot{R}_o^{-2}$. The present value of Hubble shift is known: $H_o = 55 \pm 5$ km/sec/Mpc. We may choose R_i so, that the integration gives this measured H_o value. On this way the dynamical history of the universe is given uniquely.

According to astronomical observations, the present atomic density of the universe is $\rho_{hadron}(t_o) = 3 \times 10^{-31}$ g cm^{-3}. This means, that the number of photons (corresponding to $T = 2.7^oK$) is about 10^9 times larger than the number of nucleons, but their small energy ($kT_o = 10^{-4}$ eV) makes their present contribution to gravity negligible: $\rho_\gamma = 4.10^{-34}$ g cm^{-3}. The same is true for the neutrinos, if they are massless.

The rest mass of neutrinos is rather poorly known. Rosenfeld's Table of Particle Properties has given $m(\nu_\ell) < 60$ eV, $m(\nu_\mu) < 800$ keV. One has got 10^{-9} as much relict neutrinos as nucleons in the Universe. If their rest mass is tiny but nonvanishing ($m_\nu \gtrsim 1$ eV), they shall be the decisive component of the gravitating mass. Let us repeat the calculation, by taking a possible $m_\nu \neq 0$ into account. Here $m_\nu = \max \{m(\nu_\ell), m(\nu_\mu)\}$.

In the early lepton era the neutrinos are in thermal equilibrium with the other particles, consequently their occupation number is

$$N = \left[1 + \exp \frac{1}{kT} \sqrt{m^2 + p^2}\right]^{-1} ,$$

where T is the overall temperature. During the expansion T and ρ decrease, so does the kinetic energy, the mean free path becomes larger and larger. When the average time between two collisions comes out to be larger than the cosmic time variable t, the interaction of neutrinos ceases with other particles. Their number does not change any longer, their momentum $p = h/\lambda$ decreases according to the Hubble law $\lambda \sim R$. On this way $\rho_\nu (m_\nu, R)$ can be calculated. Let us use this new function in the integration of the Einstein equation (Sralay-Marx).

We start at $t_i = 0$, $T_i = 10^{12}\ K^o$, R_i = arbitrary. We compute R(t), T(t). We stop the computation at $t = t_o$, when $T (t_o) = 2.7^oK$. We obtain t_o, R_o, H_o, q_o in terms of the input parameters m_ν, R_i. We can eliminate R_i with the help of H_o. On this way for each value of m_ν and H_o one obtains a definit value for the age of the universe (t_o) and for the present deceleration (q_o); the latter one is related to the present total mass density ρ_o (Figure 1). The main conclusions are the following: The inequality $t_o > 4.5 \times 10^9$ years (oldest Moon rocks) gives $m_\nu < 200$ eV. The inequality $t_o > 10^{10}$ years (nuclear chronology) or $q_o < 1$ (Hubble shift of distant objects) gives $m_\nu < 30$ eV. <u>On this way we succeeded to improve the empirical mass limit of neutrinos considerably!</u>

The Universe is open, if motion dominates over masses ($q_o < \frac{1}{2}$, $\rho_o < \rho_{crit}$). The Universe is closed, if gravity dominates over the kinetic energies ($q_o > \frac{1}{2}$, $\rho > \rho_{crit}$). The critical density (corresponding to a flat universe) is given by the present value of the Hubble constant H_o:

$$\rho_{crit} = \frac{3H_o^2}{8\pi\gamma} = 2.5 \times 10^{-29}\ g/cm^3 .$$

Very light neutrinos ($m_\nu < 1$ eV) would give an open Universe. Massive ones ($m_\nu > 15$ eV) would give a closed one.

If $\rho \ll \rho_{crit}$, the expansion of the universe is very fast, the gravity is very weak. There is no time enough for galaxy formation. If $\rho \gg \rho_{crit}$, the whole expansion period of the closed universe (before contraction) is too short for the evolution of stars and life. Consequently the actual mass density cannot differ too much from ρ_{crit}. But ρ_{crit} is apparently much larger than the astronomically observed density ρ_{hadron}. About 90-99% of the required mass is missing. A possible explanation is, that this unvisible missing mass is present in form of massive neutrinos: $\rho_{crit} - \rho_{hadron} = \rho_\nu$. The rest mass required for this explanation is a few eV.

Is it possible, that in our present Universe the most common form of matter are neutrinos ($N_\nu \approx 10^9\ N_p$) nearly at rest ($E_{kin} = 10^{-4}$ eV, $m_\nu\ c^2 \approx$ few eV)?

5. GALAXY

The galaxies have been concentrated in huge clusters. By counting the galaxies one can calculate the potential energy of a cluster. By measuring the Doppler shifts one can calculate the kinetic energy of a cluster. As a rule, the kinetic energy comes out to be larger, than the absolute value of the potential energy. This observation is in sharp contradiction with the other empirical fact, that most of the clusters look very old and very stable. This is a well-known puzzle of the extragalactic astronomy. One has to assume an unknown component, which helps to stabilize the cluster with its additional mass attraction. The missing mass of the Coma Cluster is at least 80 % of the required total mass, in other galaxies it is even larger.

Let us try to explain this phenomenon with the help of the massive relict neutrinos! (Szalay-Marx). The galaxies attract the slow neutrinos, a neutrino halo has been formed around the cluster and its gravitational field helps to stabilize the cluster. We shall try to make a self-consistent calculation of the potential:

$$\nabla^2\phi = 4\pi\gamma\ (\rho_{gal} + \rho_\nu)$$

with $\phi(\infty) = 0$. The density of galaxies is given by the barometric formula

$$\rho_g(r) = \rho_g(\infty)\ \exp\left(-\frac{M\phi}{kT}\right)$$

where the "cluster temperature" T can be obtained from the Doppler shifts:

$$\frac{1}{2}\ kT_c = \frac{1}{2}M\ \overline{v^2_{rad}}\ .$$

The neutrino density is given by the Thomas Fermi formula:

$$\rho_\nu(r) = \frac{2m_\nu}{3\pi^2\hbar^3} p^3_{Fermi}(r) = \frac{4\sqrt{2}m_\nu^4}{3\pi^2\hbar^3} [-\phi(r)]^{3/2} .$$

It follows from this formulas, that neutrino matter will dominate over galactic matter, if

$$m_\nu \left(\frac{m_\nu v_g}{\hbar}\right)^3 > \rho_g, \text{ i.e. } m_\nu > 1 \text{ eV}.$$

The strong dependence of the formulas on m_ν is evident. Bludman assumes the dominance of neutrino matter (formation of neutrino stars as first step of the gravitationa condensation). According to the detailed calculations one can explain the size and shape of the clusters by assuming $\frac{1}{4}$ eV < m_ν < 1 eV (Figure 2, Szalay.) This means that galactic matter density dominates at the center, neutrino density dominates in the outer region.

The conclusion from the point of view of particle physics is the following: with the help of clusters of galaxies one has been able to push down the limit of neutrino masses into the region of 1 eV. The shape of different clusters may give a possibility to test the assumption $m_\nu \neq 0$ quantitatively.

The conclusion from the point of view of cosmogony: the formation of the first condensations (clusters, galaxies) from the original H gas may be correlated with the formation of neutrino clouds, if the neutrinos have a tiny rest mass. Otherwise the galaxy formation remains an open problem.

Is the birth of galaxies a work of neutrinos? The question is worth for a more detailed investigation.

6. LIFE

Many living beings show a definit "handedness": left and right are not equivalent for us. The expressions "helicity" and "chirality" have biological origin. The root of this reflexion asymmetry is the form of the enzyme molecules, which catalyse the biochemical reactions. Deeper one finds the α-helix structure of the proteins and the double helix structure of the DNA. These helices are righthanded in every terrestrial species. The origin of this complete asymmetry has been asked by biologists since a century, as first by Pasteur, who observed the optical activity of organic compounds. The theoretical alternatives are the following:

a) Terrestrial life started with the spontaneous formation of one single helical molecule. All the living beings are blueprints (i.e. children) of this first living molecule (with some "errors" caused by mutations). It is a pure chance, that this first molecule happened to be a right helix. If there is life on the Mars, independent from that on Earth, it will show right or left structure with 50 - 50 % probability.

b) The differential equations describing the reaction kinetics of biochemistry, i.e. the multiplication of living matter have unstable solutions. If the concentration of the righthanded molecules shows a very tiny overweight at the beginning, they will be the dominant and only forms very soon. (Stronger even as the weak ones. This mechanism amplifies any small asymmetry, caused by random, statistical fluctuations.)

c) There is an intrinsic preference of Nature for one version (for righthanded helix). The original asymmetry of the laws of Nature may be a very little one, it will be enlarged by the biological amplification. (If this is true, the Mars-life will be based also on right helices.)

If two molecules, being mirror images of each other, are not completely equivalent, their Hamiltonian must not commute with parity. The only known term of the world Hamiltonian, which does not commute with parity, is the weak interaction. Let us trace the possibility, that the fundamental weak asymmetry of matter may be related to the actual maximum asymmetry of terrestrial living matter.

The biochemistry is a manifestation of electromagnetic interactions. The weak interactions are so weak at low energies, ($G_{eff} = l_w^2/\lambda^2 < 10^{-20}$) that any connection between the two asymmetries must be a rather indirect one.

As first approach let us consider the experiment of A. Garay. He started with a racemic (optically inactive) mixture of right and left amino acids. After a long irradiation with β^--rays (electrons, showing an overweight of the negative helicity) he found, that the mixture became optically active in the same sense, as the amino acids of biological origin. Later L. Keszthelyi made an other experiment: irradiation of racemic mixture of amino acid with β^+-rays (positrons with positive helicity) caused again optical activity in the correct sense. If these experiments are right, the small driving effect (amplified to dominance by life) was the selective destruction of enantimorphic carbon compounds by beta radioactivity of the early Earth produced considerable temperature differences.)

A possible theoretical explanation of the Garay-Keszthelyi effect is given by P. Hraskö. Let us consider a right helical molecule. An electron is able to move along the vertical helix like in a screw-driver-shaped wire. The electron moving upwards produces a circular current, it works like a tiny electromagnet. The outside magnetic field shows upwards and creates a magnetic polarisation in the neighbouring matter. The spin of the electron interacts with this magnetisation, in ground state the spin being directed upwards, i.e. the electron moving freely along a right helix will show a preference for positive helicity. (The same is true for the electron moving down.) This means that a matter consisting of right-helical structures can be modelized with a gas of positive helical electrons. (A matter consisting of left-helical molecules can be modelized with a gas of negative helicity electrons.) Now a positron coming with positive helicity from radioactive decay and slowing down gradually in condensed matter will pick up a polarized electron moving in the same direction within the right helix to have a small relative velocity and will form orthopositronium. The positron (with positive helicity) with the polarized electron in a left helix will form parapositronium. The formation and annihilation probabilities of the ortho- and parapositronium are different, consequently the β^+-rays will destroy the left-helicies more effectively than the right-helices. Evidently, before accepting this explanation, one needs experimental checks, e.g. irradiation of amino acid by electrons (positrons) showing "artificial" positive (negative) helicity.

These experiments are under preparation in Budapest.

A second approach to the origin of molecular asymmetry is offered by the possible existence of weak currents. CERN experiments indicate the existence of a coupling between muon neutrinos and charged particles (electrons and protons), as predicted by a wide class of gauge theories of weak interactions. There is a possibility for a parity-asymmetric weak force between electron and nucleon. (According to the Lee-Prentki-Zumino model this may be true even in the case, when neutrinos do not form neutral current.) Evidently such an interaction would give a small parity-asymmetric correction to the Coulomb-force, so it could create parity-impurities in atoms and molecules.

An asymmetric μ-p interaction may be observed by looking for the helicity-dependence of scattering of high energy muons on protons (Serpukov). Probably a more sensitive experiment is the observation of parity impurities in atomic spectroscopy (C. and M. A. Bouchiat). In low energy approximation the vertex of Fig. 3 gives the following odd e-p potential:

$$H'(r) = i \frac{G}{\sqrt{2}} \left(\frac{\vec{\sigma}_e \cdot \vec{p}_e}{m_e c}\right) \delta (\vec{x}_p - \vec{x}_n) \begin{cases} 2\sin^2\theta_W + 1 \\ \cos^2 \theta_W \\ 10^{-4} \end{cases}$$

in the Weinberg or Lee-Prentki-Zumino or Giorgi-Glashow model of weak interaction. Its expectation value is very small compared to the molecular energies:

$$\langle H' \rangle \sim Ga^{-4} m_e^{-1} \sim (Gm_e^2) \cdot \alpha^4 \cdot m_e c^2 \sim 10^{-15} \text{ eV} .$$

(Here a is the "radius" of the corresponding atomic orbit, coming from $\psi^{\dagger}(0)$ grad $\psi(0)$. It is still possible to find situations, where this leads to observable effects. Let us consider the $2S_{1/2}$ state of the H atom (E. Gajzágö). H' mixes a $2p_{1/2}$ impurity to this state. According to perturbation theory

$$\psi(2S_{1/2}) =$$

$$= \phi(2S_{1/2}) + \frac{<2p|H'|2S>}{E_{2S} - E_{2p}} \phi(2p_{1/2}) .$$

The energy denominator is just the Lamb shift:

$$E_{Lamb} = E_{2S} - E_{2p} \sim \alpha^5 m_e c^2 \sim 10^{-5} \text{ eV} ,$$

consequently

$$\psi(2S) = \phi(2S) + \varepsilon\phi(2p) \text{ with } \varepsilon = \frac{<H'>}{E_{Lamb}} = 10^{-10} .$$

Let us consider now the decay of this state into ground state. The emitted radiation will be a superposition of the forbidden 2s → 1s magnetic dipol transition and the allowed 2p → 1s electric dipol transition.

$$\tau_s = \tau(2S_{1/2} \rightarrow 1S_{1/2}) = 10^7 \text{sec}, \ \tau_p = \tau(2p_{1/2} \rightarrow 1S_{1/2}) = 10^{-8} \text{sec}.$$

The interference of the electric and magnetic dipole transitions will give circular polarisation:

$$C = \frac{2\mathrm{Re}\ MD\cdot ED^{*}}{(MD)^2+(ED)^2} \sim 2\,\varepsilon\,\frac{\tau_s^{-\frac{1}{2}}\tau_p^{-\frac{1}{2}}}{\tau_s^{-1}+\tau_p^{-1}} \sim 2\,\varepsilon\,\left(\frac{\tau_s}{\tau_p}\right)^{\frac{1}{2}} \sim 10^{-3}\ ,$$

which is an encouragingly large value. (Smaller circular polarisation has been measured already in nuclear transitions.)

The effect may be improved, if we go from electronic atom to muonic atom (the orbit radius becomes smaller) and if we go from H to heavier elements (more interacting nucleons, smaller orbits). An optical experiment with Cs atoms ($Z = 55$) is under preparation in the Ecole Normale (M. A. Bouchiat).

Let us now assume, that the electronic neutral weak current does exist and it produces an asymmetric potential in the atomic and molecular physics. It will have the consequence, that two asymmetric molecules, which are mirror images of each other, will not have identical binding energies. The difference of the energies will be of the order of $\langle H'\rangle$ or even larger, due to the many interacting e-p, e-n, e-e pairs in one molecule. The qualitative situation is easy to be understood in the case of the helical molecules. If the right helix molecule prefers positive helicity for some electrons and the left helix molecule prefers negative helicity for some electrons, $\langle H'\rangle$ will have opposite sign in the two cases. A difference in the spectra of the two molecules may become observable with the use of nonlinear frequency mixing (Bay-Marx).

It is still questionable, how such a tiny effect

in molecular binding energy can give rise to considerable chemical asymmetries, required to understand the handedness of life. A remarkable observation has been published by W. Thiemann (Jülich). A racemic (optically inactive) solution of Na-NH_4-tartarat was cooled down below the solubility limit. When about 2% of the tartarat formed a precipitate, it was separated from the solution and was dissolved again in water. Its optical activity was measured with respect to the solution as standard. Repeated experiments gave the same negative optical rotation of the precipitate: 10^{-3} degree of arc. The possibilities of an optically active impurity or a bacterial contamination had been eliminated by carefull handling and extra experimental checks. So the chemists concluded, that in the lattice energies of the left- and right-oriented tartarat crystals the experiments show a relative difference $\sim 10^{-5}$. Evidently, this surprising large molecular effect needs confirmation by new experiments and by theory before being accepted.

The organic molecules, observed in the outer space by radio astronomy may have time enough to reach their absolute ground states. (E.g. their right modifications.) If the planetary system has been formed from an interstellar matter contaminated by righthanded compounds, this may explain the little asymmetry of the beginning, which has been amplified by the biological multiplication. (Even if there is a very small difference in the speed of chemical reaction in one step, biochemistry works with millions or billions of steps, and the final difference may be considerable.)

Are our biological asymmetries related to the weak asymmetries in the fundamental laws of matter? We do not know enough to give an answer now. The parity impurity in the atomic structure is awfully small, if it exists at all. There are too many gaps in our knowledge. But gaps of knowledge are, what particle physicists like the best: they call for further work.

FOR REFERENCES see the proceedings of the following meetings:

Seminar on Weak Interactions at Low Energies, Debrecen (Hungary), 29-30 March 1974; (Dept. of Atomic Physics, Eötvös University Budapest).

Conference on Neutrino Science, Philadelphia (USA), 26-29 April 1974; (Dept. of Theor. Physics, Pennsylvania University, Philadelphia).

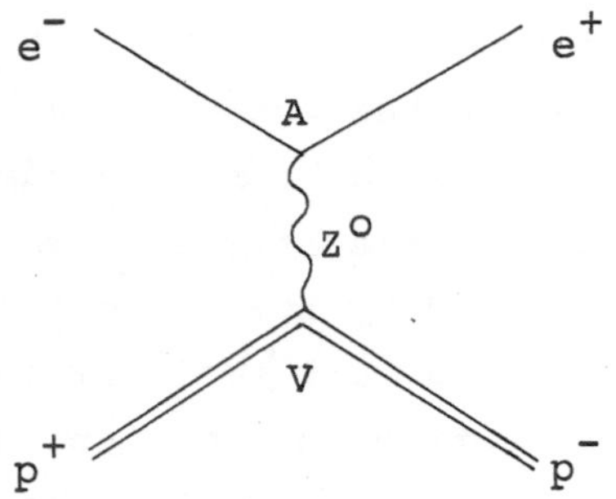

Fig. 3

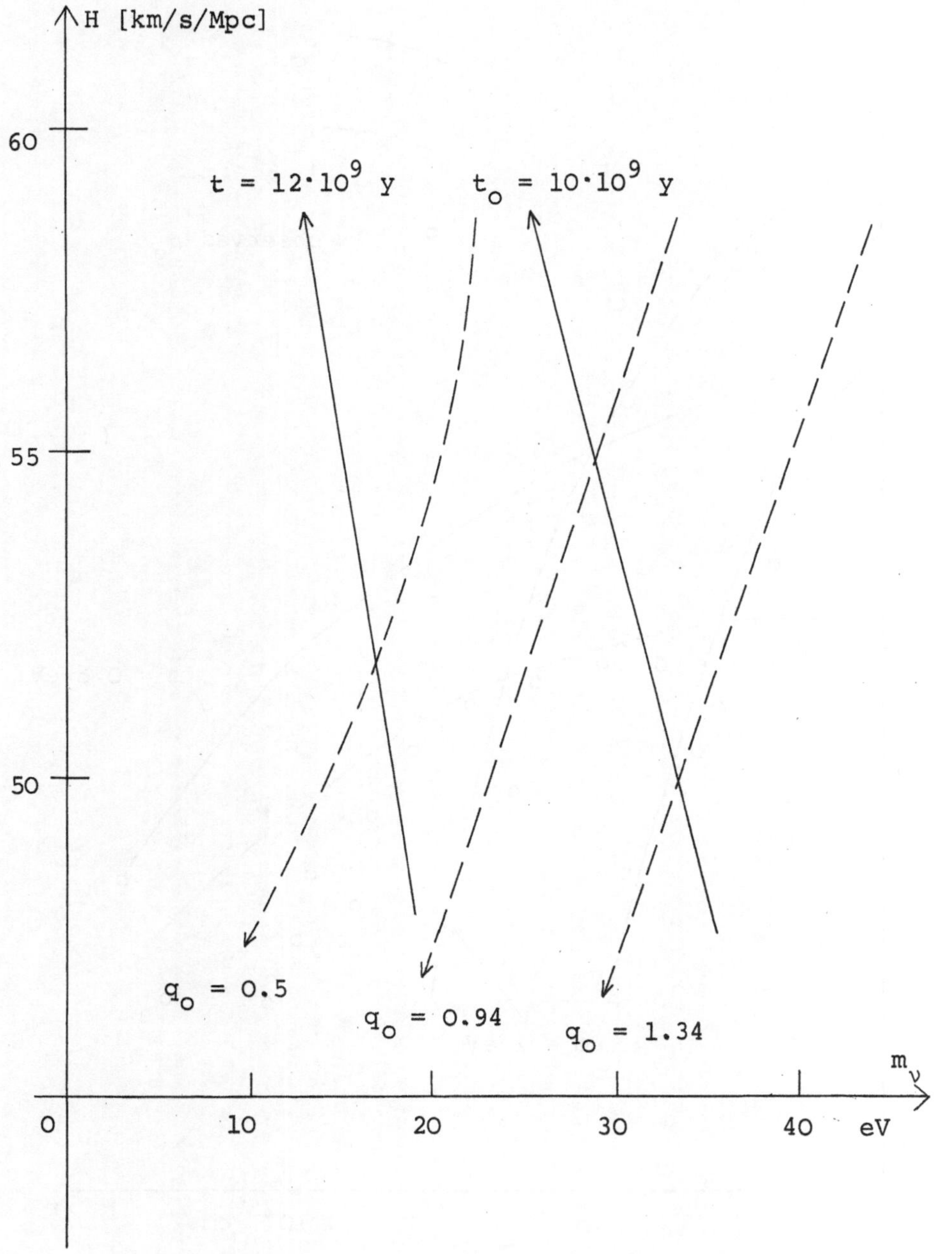
H [km/s/Mpc]
60
55
50
t = 12·10⁹ y
t₀ = 10·10⁹ y
q₀ = 0.5
q₀ = 0.94
q₀ = 1.34
m_ν
0
10
20
30
40
eV

Fig. 1

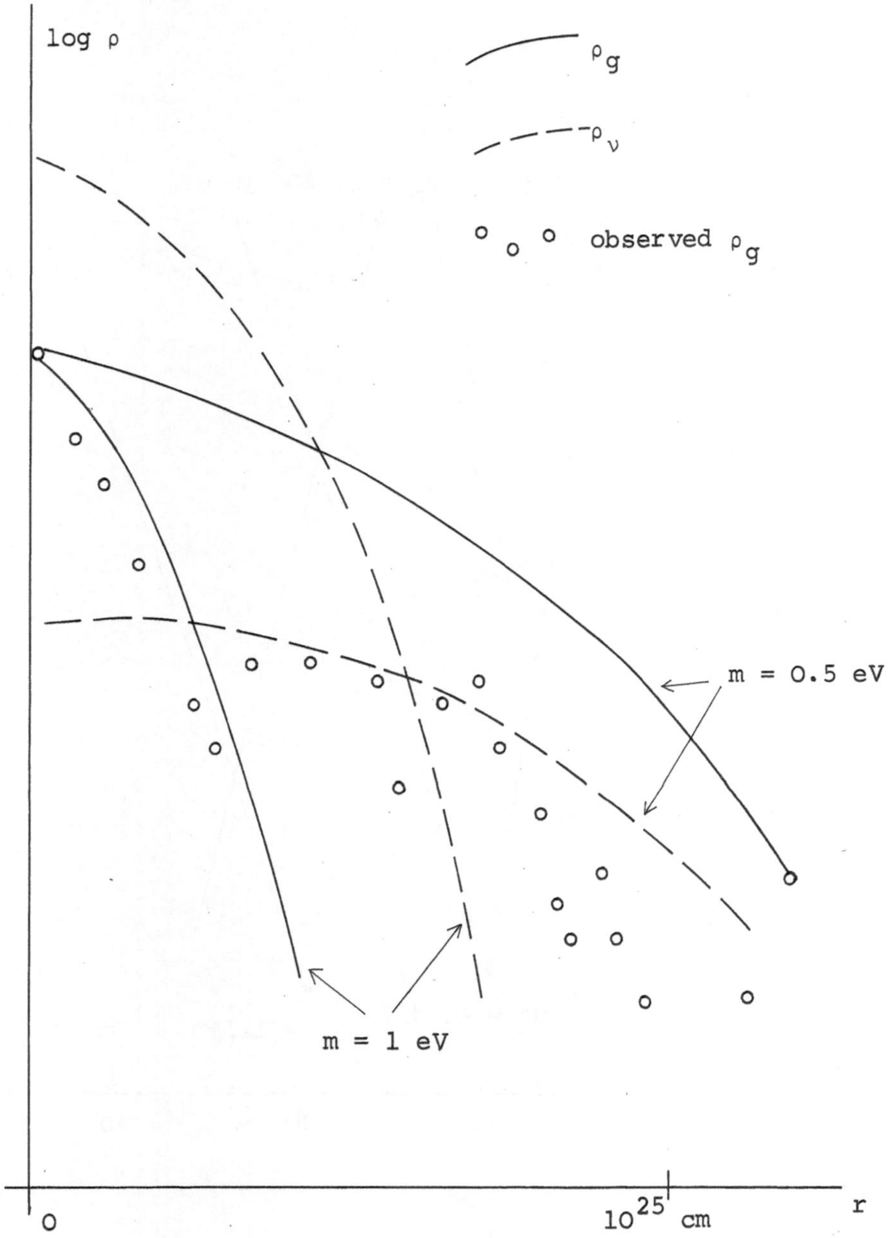

Fig. 2

Acta Physica Austriaca, Suppl. XIII, 595–647 (1974)

SYMBOLIC COMPUTING AND ITS RELATIONSHIP TO PARTICLE PHYSICS*)

by

J. A. CAMPBELL
King's College Research Centre, King's College
Cambridge, England

WHAT IS SYMBOLIC COMPUTING?

Regrettably, it is still true that "computing" means "FORTRAN" to a majority of physicists M, and that in Europe it means "ALGOL" to a majority of those physicists not included in M. The various alternative names for the type of computing which I shall describe have evolved as defences against the FORTRAN-ALGOL ideology. Firstly, as FORTRAN in particular is used for numerical computation, there is the name "non-numerical computing" for computations in systems which allow answers to be given in terms of symbols. This name is good, but not perfect, because it does not do justice to the possibility of adjoining to symbolic results integer or rational or floating-point coefficients, or even to the use of symbolic means to calculate purely numerical rational

*) Lecture given at the XIII. Internationale Universitätswochen für Kernphysik, Schladming, February 4 - 15, 1974

quantities (e.g. Bernoulli numbers, Clebsch-Gordan coefficients) with programs that are much more compact than programs in FORTRAN. "Algebraic computing" is a name which gives the same flavour without the limitations of the previous suggestion, but it suffers from the rather more serious defect that most "algebraic" computations in the past (e.g. (3) below) have actually dealt with problems in analysis and not algebra. "Symbolic computing" (abbreviated from now onwards as SC) is probably the best compromise as a choice: it conveys to a mathematical physicist the idea of a program whose inputs and outputs are in terms of symbols, while carrying only the slight risk that a student of the arts may misunderstand it to mean pseudo-computing or a form of computing which is not really there.

The standard example which contrasts numerical with symbolic computing at a simple level is the computation of Legendre polynomials. Consider the follwoing piece of a FORTRAN (?) program:

```
      P(O)=1
      P(1)=X
      DO 2J=2,20
      P(J)=((2*J - 1)*X*P(J - 1) - (J - 1)*P(J - 2))/J
    2 CONTINUE
```
(2)

which expresses the first 20 Legendre polynomials of some variable X. It is reasonable to suppose that the preceding two lines of the program can easily be fitted in - as a DIMENSION statement for P followed by a binding for the required numerical value of X - and also that we can arrange for the floating-point numerical values of $P_n(x)$

to be printed out as they are computed if "CONTINUE" is replaced by a suitable output command with P(J) as argument. But what if the veil over this preceding part is lifted to reveal, instead, the two lines

```
SERIES P(20)
POLY X
```
(1)

Then the program cannot be in FORTRAN, and the unexpected lines are strong clues that each P_n is to be calculated as a series in the polynomial or symbolic variable X. In fact this program makes good sense in the SC system TRIGMAN [1], whose programs are almost identical in form with FORTRAN except that the basic non-numerical symbols and symbolic ("series") variables must be declared at the beginning, as in (1). Further, there is no trace of floating-point arithmetic in (2), so that coefficients are accumulated as rational numbers. Again, if the statement numbered 2 in (2) is replaced by a suitable output statement, the results are printed on calculation. Now, however, they are

```
-1/2 + 3/2*X**2

-3/2*X + 5/2*X**3

        .........

46189/262144 - 4849845/131072*X**2 + 334639305/262144*X**4
 - 557732175/32768*X**6 + 15058768725/131072*X**8
 - 29113619535/65536*X**10 + 136745788725/131072*X**12
 - 49589132175/32768*X**14 + 347123925225/262144*X**16
 - 83945001525/131072*X**18 + 34461632205/262144*X**20
```
(3)

The result (3) is one striking answer to the question "What is symbolic computing?".

The basic point that I have made so far is that SC can exist. To answer the next question "When is SC useful?", an enlightening course is to read the various reviews and histories of the subject and its internal problems. Computations in particle physics make up a large number of the applications of SC, as these reviews demonstrate. I shall return to the most interesting applications later. At this stage it is enough to mentions reviews of relevant work: in theoretical physics [2] with the help of the now-famous symbolic language LISP [3], and in general theoretical physics [4,5] and celestial mechanics [6]. These surveys cover more than just the territory of particle physics, but each type of application ultimately bears on all other types through its contributions to improvements in the use of languages, systems, programming techniques and the analysis of algorithms. The most recent and comprehensive summary of physical applications of SC, by Barton and Fitch [7], deserves close study.

HOW IS SYMBOLIC COMPUTING POSSIBLE?

With some overlap, there are four ways to express a symbolic computation. All methods rely on the fact it is equally easy to represent symbols (through their character codes) and numbers inside the storage of a computer. Alternatively, where the representation is free from possible ambiguities, a representation which omits the explicit symbols may be a good thing (e.g. $\frac{54}{7}x^3yz^2$

in a polynomial in the four ordered variables x, y, z, w is described unambiguously by a sequence 54 7 3 1 2 0). If we decide to work upwards from such a grass-roots level, we may design a very compact occupancy of storage and a fast system, but we then have a long way to go even before we can program a simple symbolic operation like differentiation. Alternatively, we may ask for a medium in which the programming of high-level concepts is immediately available. There are advantages in both approaches. In the "fourfold way" mentioned above, the first two ways are related to the former approach, and the remaining two to the latter.

The first method of SC is to choose a compressed representation for the basic items of data in storage, and then write the necessary programs in assembly code for a given computer. The earliest experiments with celestial mechanics and Dirac-matrix algebra were of this type, probably because the limited storage space on early computers had to be used as economically as possible. Kaiser's report [8] of such a program for a small computer was historically the first to be published. One general-purpose SC system in present use, M.Veltman's system SCHOONSCHIP [9], has the same background, but it can now carry out a wide range of SC thanks to the large amount of work which has been put into its development. Because of the efficient use of storage which a compressed representation of data makes, and because of the efficiency of direct programming in assembly code, such a system runs very quickly and economically for problems which it is specifically designed to solve, but requires extensive and laborious rebuilding (a task for designers and not for casual users!) for other problems. For those who wish to see how a compact assembly-code representation for Dirac algebra is managed, Kaiser [8] gives some hints, and Strubbe [25] describes the internal conventions of SCHOONSCHIP in great detail. Veltman [10] declares himself strongly in favour of this point of view in a survey article which is an interesting expression of a minority opinion. The majority opinion is that it is better for a truly general system to sacrifice efficiency in order to allow the casual user some facilities for extending its range by reprogramming. (The other three methods below accommodate this opinion). To put the matter into perspective, there are several typical exercises in symbolic computing where the most efficient system (and the most efficient system for any one exercise is certainly not guaranteed to be the most efficient for the next problem) may need a few minutes for a computation, while the least efficient system needs a small number of tens of minutes. Note that a symbolic computation is usually "one-off", in the sense that it must be done only to produce the required general answer, rather than being run many times with different initial data as in the cases of the standard big numerical computations in nuclear physics and in chemistry. Thus the question of total computing time for SC is usually not important. In principle, the question of space is different: I shall come back to that point in a later section.

The second approach is to make use of a well-known high-level language ("high-level" means in practice anything higher than assembly code) as far as possible, to permit quick reprogramming, but to maintain the idea of compressed representations in storage for the basic symbolic quantities. If FORTRAN is the high-level language, for example, it may be possible to write the symbolic extensions into a small amount of assembly code. The best-known SC systems of this type are TRIGMAN [1] and SAC [11]. TRIGMAN, which allows all the obvious symbolic operations on expressions which are made up of terms of the form

$$KM(y_1,y_2,\ldots)\begin{matrix}\sin\\\cos\end{matrix}(a_1x_1 + a_2x_2 + \ldots) , \qquad (4)$$

where K is a rational number, the a_i are integers, x_i and y_i are symbols, and M is a multinomial in the y_i with rational coefficients, is built for celestial mechanics. It is a system mostly written in FORTRAN, and is very portable. Its assembly-code section, consisting of subroutines to handle a compressed representation of (4), typically consist of less than 100 instructions which are devided into small groups and which are easy even for a relative amateur to rewrite for a new computer. The disadvantage of the simplicity here is that it is not safe to mix positive and negative powers of y_i in (4), and that no symbolic denominator which is not simply a product of powers of the y_i is allowed. SAC is a more complex system intended initially to facilitate experiments with the fastest algorithms for the basic operations on symbolic terms; in particular, it rejoices in computing greatest common divisors of polynomials, which is <u>not</u> a simple computing problem, so that it can therefore deal happily with symbolic denominators.

If we use a FORTRAN-based system, without further help, to make a computation which is primarily symbolic, the average line of the program will be a call to one of the symbolic subroutines. TRIGMAN as seen in (2) is not "pure" TRIGMAN, but a super-TRIGMAN dialect which is first expanded into the pure form by a translator [12] written in SNOBOL [13]. The stage of translation is needed because we are asking for the specification of a symbolic computation to be given through a series of references to special cases (i.e. subroutines) in a numerical language. A genuinely symbolic language permits a third approach to SC - our choice of representation for the basic data must fit the syntax of the language, and therefore only exceptionally can it be as compact as the choices in the first two approaches, but we gain freedom from all assembly-code programming and the maximum flexibility both in the writing of our program and in changing our minds later about how it should be written. Although there are other symbolic languages of probable interest [14] to particle physicists, I shall take only the example of LISP [3] here. LISP stands for "List Processing Language", as its basic items of data (apart from the "atom" like X or AB or NIL or 2) from the viewpoint of the causal user (not of the professional analyser of syntax), a structure like (AB CD 2.6 D) or (AB (CD 2) ((D))) or (TIMES 2 3 4 7) or (TIMES (PLUS AB CD) (G L1 MU)). We select the meanings of the elements of list data for ourselves, and write programs accordingly, e.g. particle physicists may like to regard the last element of the last list as the Dirac matrix γ_μ on a fermion line labelled L1. It is clear that in a good list processing language there must be at least three new primitive functions or operations: one to pick up the head or first element of the list which is its argument, one to pick up the tail or remainder of the list, and one to combine its two arguments into a single list structure. The respective names for these primitives in LISP are CAR, CDR and CONS. Thus CAR applied to a list (A B) gives A, and CDR gives (B). CONS applied to arguments A and (B) gives back (A B), while for (A) and (B) it returns the different result ((A) B). Further, while CAR applied to (B) gives B, CDR must logically give (), which is equivalent to the special atom NIL. With this background, it is instructive to consider the appearance of some LISP functions as the user may choose to write them. For ease of comparison with a known numerical operation, first, there is the LISP factorial function, here called FAC:

```
(FAC (LAMBDA (N) (COND ((ZEROP N) 1)
       (T (TIMES N (FAC (SUB1 N)) )) )))
```

Next, for a purely symbolic example, take the case of a function DELETE:

```
(DELETE (LAMBDA (X Y) (COND ((NULL Y) NIL)
       ((EQUAL X (CAR Y)) (CDR Y))
       (T (CONS (CAR Y) (DELETE X (CDR Y)) )) ))) ,
```

(5)

which gives back a copy of the value of Y less the first occurrence (if any) of the value of X. These examples illustrate several distinctive properties of LISP. The non-specialist is likely to notice immediately the property of bracket-proliferation and the prefix notation for functions and arguments. Both of these distract attention from the main issue of easy computation for particle physicists. The final approach, mentioned below, is intended to keep the advantages of LISP while removing these undesirable properties from view. However, one helpful characteristic of (5) is recursion, or the ability to define a function in terms of itself. More conventional FORTRAN-like organization of functions is also possible, e.g. in the definition of a function which reverses a list:

```
(REVERSE (LAMBDA (U) (PROG (V)
       G (COND ((NULL U) (RETURN V)))
         (SETQ V (CONS (CAR U) V))
         (SETQ U (CDR U))    (GO G)   )))          (6)
```

A LISP program consists of the definitions of many relatively short functions, which may be built to call themselves and each other freely. To understand the LISP language fully, Weissmann's book [15] is required reading, and Lurié's article [16] for physicists also contains useful information. When these two have been assimilated, the LISP 1.5 Manual [3] stands ready to deliver up the rest of the story.

Fourthly, there is the promised method which removes the drawbacks of the LISP approach, and which has a bonus the availability of all the basic functional operations which the particle physicist may want. Thus, for example, the differentiation function and a function to compute the trace of an arbitrary product of Dirac matrices are already parts of the foundation. This method is the use of a symbolic programming <u>system</u> which has been constructed in a symbolic <u>language</u> L (e.g. LISP), which contains a large number of special functions that the designer has previously written in L, and which allows the user all the flexibility of L in rewriting or extending the system. The best-known system with special facilities for particle physicists is unquestionably REDUCE [4,17], designed by A.C. Hearn. To demonstrate that the bracket and prefix conventions of LISP may be laid to rest, it is sufficient to show how (5) and (6) can be expressed within REDUCE:

```
PROCEDURE DELETE (X,Y);
   IF Y = NIL THEN NIL
      ELSE IF X = CAR Y THEN CDR Y
      ELSE CAR Y . DELETE (X,CDR Y) ;
```

and

```
PROCEDURE REVERSE U ;
   BEGIN SCALAR V ;
      G: IF U = NIL THEN RETURN V ;
         V := CAR U . V ;
         U := CDR U ;    GO TO G
   END ;                                    (7)
```

For users with applications in mind, the fourth method (or the use of any advanced system (e.g. SCHOONSCHIP)) is the most highly recommended, because all of the basic programming and concepts have been taken care of in advance. Suppose, for illustration, that we wish to compute the trace part (i.e. excluding scalar multipliers) of the differential cross-section for Compton scattering of an electron of initial 4-momentum p_i and mass m to a final 4-momentum p_f by a photon with 4-momentum k_i, energy k and polarization vector e_i. The corresponding final-state quantities for the photon are k_f, k' and e_f. Then a suitable REDUCE program, whose meaning (and power) should be evident, is:

```
MASS  KI = O, KF = O, PI = M, PF = M ;
MSHELL KI,KF,PI,PF ;
COMMENT  G(L,P,Q, ... ) MEANS
   (GAMMA.P)*(GAMMA.Q)* ... ON LINE L ;
FOR ALL P LET GP(P) = G(L,P) + M ;
COMMENT P.Q IS A SCALAR PRODUCT OF 4-VECTORS ;
LET PI.EI = O, PI.EF = O, PI.PF = M**2 + KI.KF,
   PI.KI = M*K, PI.KF = M*KPR, PF.EI = - KF.EI,
   PF.EF = KI.EF, PF.KI = M*KPR, PF.KF = M*K, KI.EI = O,
   KI.KF = M*(K - KPR), KF.EF = O, EI.EI = -1, EF.EF = -1
COMMENT THE NEXT INSTRUCTION EVALUATES THE TRACE ;
   GP(PF) * (G(L,EF,EI,KI)/(2*KI.PI) +
   G(L,EI,EF,KF)/(2*KF.PI)) *
   GP(PI) * (G(L,KI,EI,EF)/(2*KI.PI) +
   G(L,KF,EF,EI)/(2*KF.PI)) $
WRITE "THE CROSS-SECTION IS ", !*ANS ;                 (8)
```

If a REDUCE flag is suitably preset to produce picturesque 2-dimensional output, the printed answer will be

THE CROSS-SECTION IS $1/2*K*KPR^{(-1)}$ +

$$1/2{*}K^{(-1)}{*}KPR + 2{*}EI.EF^{2} - 1 \tag{9}$$

Thus REDUCE and its partners may be regarded as utilities to help particle physicists with calculations in field theory. For people who are happy with this category of computing, later sections will cover the past history of SC in more detail. For those who have an interest in less orthodox uses, or who are still curious about some of the first three approaches, it is now necessary to consider a few of the hidden subtleties of SC.

DISTINCTIVE PROPERTIES OF SYMBOLIC COMPUTATIONS

In numerical computing, the demands on the allocation of storage are usually fixed and known at the time that a program begins to run. In SC, either the demands are probably irrelevant to assess (as in an estimate of the maximum number of terms present at any stage of the computation of the trace in (8)) or impossible to guess accurately. This is a most significant difference between the two computations. If a piece of symbolic information (I_1 I_2 I_3 ... I_N) will grow to an unpredictable size, storage for that information cannot be reserved economically at compile time. Therefore the organization of storage for each extensible item will have the form of Fig. 1,

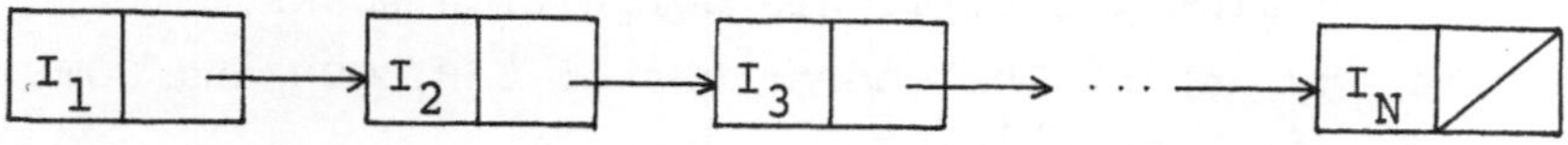

(Fig. 1: Division of storage (schematic) for symbolic information)

where a box indicates some continuous run of words (including 1 word) in storage, and an arrow marks some type of pointer, or possibly multiple pointers, to the next block(s) in the item. The concept of the pointer is central to SC. It allows programs to build structures of arbitrary length by establishing links (e.g. in LISP, each use of CONS creates in effect one more pointer) between blocks without needing to carry data analogous to the arguments of DIMENSION statements in FORTRAN. The structure of Fig. 1, for data and often for programs too, underlies all SC languages and systems.

There needs to be no special positional relationship in storage between adjacent blocks of Fig. 1. If the address of the first word of the block containing I_k is n_k, all that is necessary is that the pointer from the block which holds I_{k-1} should contain some representation of n_k. Often this is just n_k itself, in binary notation. A frequent convention for the marker (diagonal bar in Fig. 1) to indicate the end of a compound item is a pointer to the word with address 0. In LISP, this is the job of the atom NIL.

While there are languages where the blocks of Fig.1 have complicated subdivisions, LISP chooses the simplest convention: each block is one word long, and for the basic list structure it is partitioned into two equal pieces. Fig. 1 is exactly the representation in storage of the LISP list (I_1 I_2 I_3 ... I_N). If a list has a sublist at position k, I_k is itself a pointer to that sublist (which is of course also a list). This definition works recursively to any depth. The operations of CAR, CDR and CONS now become clear by reference to Fig. 1.

Obviously a program cannot go on establishing

pointers indefinitely. Unless a computation is short, all of the available storage will eventually be filled with lists. The nature of SC almost always guarantees that some of the lists in a filled storage will no longer be needed. The process of putting all the words from useless lists onto one new list, the list of renewed free storage, is called garbage collection. Most symbolic languages and systems perform garbage collection automatically, and require a number b_g of bits in each word in the region of storage in which lists are accommodated to be set aside to carry indicators for the garbage-collecting subroutines. This number can be taken as 2. Garbage collection, a rewarding subject in itself, is treated by Knuth [18], in an excellent book which touches on many other questions of importance for SC. I mention automatic garbage collection here because most systems adopt this method, instead of requiring the user to exert partial or total control over when a given list should be returned to free storage. It is true nevertheless that systems of the latter type, of which CAMAL [19] is probably the best example, are remarkably fast. Recently, improved facilities for the measurement of performance in SC systems have shown that an automatic garbage collector in a big computation can steal large amounts of the total computing time if not watched carefully. Thus the debate between the two schools of thought on garbage collection may be reopened soon. In any case, intending users of symbolic languages or systems should be aware of both the virtues and dangers of the garbage collector, an assistant which is hardly ever needed in numerical programming.

We have now reached the stage of being able to understand a simple relationship between word length and size of storage in symbolic work. The LISP interpretation of Fig. 1 provides the best case for some threatening inequalities. Suppose that the total number of words of storage to be accessible to pointers is $W = 2^N$. To provide unique pointers for all words, we therefore need N bits for every pointer. The partitioning in Fig. 1, where a pointer occupies half of each box, therefore demands 2N bits for each box. A "box" is not quite a word, because we must add a further b bits for miscellaneous uses (where certainly $b \geq b_g$) like the labelling of types of pointers and data in a multi-type system. Hence, the word length must be at least 2N + b bits for unique addressing, so that

$$L \geq 2 \log_2 W + b \qquad \text{or} \qquad W \leq 2^{(L - b)/2} \tag{10}$$

We may gain factors of 2 here and there by regarding W as the size of the store used for lists only, but occasionally a pointer in a list may point to a part of the remaining storage (e.g. for the system itself), and in any case the storage for lists becomes the dominant region for sufficiently long and interesting problems. Therefore it is not unreasonable to take W as the total size of fast storage. Manufacturers in these days fix L but are prepared to offer a W which is limited only by the size of our budgets. While this is fine for numerical work, (10) tells us that their generosity has a saturation point for SC. (Of course there are ways around this difficulty, but strongly at the expense of computing speed and efficiency). Over a wide range of problems, it seems to be a fact of life that 32K of storage (1K = 1024 words) allows non-trivial applications of SC, while 16K is not enough. If $b_g = 2$, (10) then shows that $L \geq 32$.

Further, if $b > b_g$, then $L > 32$. Thus the IBM 36o and 37o computers are marginally inefficient for SC, and machines with smaller L are not impressive. The useful advances all appear to have come first in connection with machines with $L \geq 36$ (e.g. 36 for the IBM 7o9o, 48 for the D.E.C. PDP-1O and Ferranti Atlas, and 6o for the CDC 6OOO series).

Numerical programs are usually compiled before use. In most symbolic systems, this process is not compulsory. The alternative is that a system contains (or even "is") a program which interprets the user's program P without any transformation, and carries out the work prescribed by P. Although this is not very helpful in batch processing, unless we want P to change its own definition during a run, it is perfectly adapted to interactive computing. Many of the early SC works in particle physics, which used nothing more than traces of products of Dirac matrices, had no need of interactive computing, but now there is an increasing need in SC for interactive facilities, even if they are used only to play with the results of simplified traces. Playing is an essential part of all but the simplest work, because results of new computations tend to be long (another point of difference from numerical computations, where the length can always be controlled if the program is free from errors). Therefore the user may engage in transformations, substitutions etc. at an interactive console, on the basis of what he sees in the intermediate results, to make the final output as readable as possible.

There is one more significant difference between numerical computing and SC which is not hard to understand. In many computations there is some analogue of an order in a perturbation theory which may serve as a measuring parameter for the demands on time or storage space. In FORTRAN it may be (for the sake of illustration) that each iteration fills up one column of an M x N array by performing a computation which inspects each non-zero element. Suppose that all elements are initially zero except for one column of starting values. In that case, the variation of total computing time T with iteration n is proportional to Mn, so that T is quadratic in n. If inspection covers only a fixed number of elements each time, which is another well-defined type of FORTRAN exercise, T depends linearly on n. If inspection of all pairs is required, T is cubic in n, and so on. However, because of the fixed bounds on the number of elements, this increase is polynomial in n. In symbolic work, on the other hand, suppose that the only interesting action is to use CONS to create a new list element, following inspection of all (E) list elements which have been created up to that point. Then dT/dn is proportional to E (and remains so if "all" is replaced by "a fixed fraction of", which is characteristic of much of SC), and E is very nearly T/k, where k is the time taken to inspect one element. Hence T is exponential in n. Although this is an idealized "computation", it is close enough to the truth to display an observed behaviour of many exercises in SC, both in time and in demands on storage space. The freedom from the constricting DIMENSION statement of FORTRAN which is expressed in pointers has its price! It is essential for users to keep this fact in mind, and to avoid bad programming practices. FORTRAN is friendly towards inefficiencies (indeed, it tries to hide them at times), but SC demands that its practitioners stay wide awake.

FACILITIES FOR SYMBOLIC MATHEMATICS

I have presented an outline of all approaches to symbolic computing above. The more rudimentary methods are described as an aid to "do-it-yourself" projects which require a system that must be built from the ground upwards. Additionally, those methods are of some historical value, because even advanced systems for physicists were presumably once under construction at ground level. For example, REDUCE [17] was in that condition in 1964 [2o], and required roughly six years to evolve into its present state of sophistication. From here onwards, I shall assume that the interest is confined to ready-made SC systems which are available for immediate use by the physicist. What facilities can he expect to find in a good general system?

Firstly, (almost) all types of polynomial or multinomial manipulation should be available. The word "almost" makes a concession in the case of symbolic division, which is very difficult to do well in a confined region of storage. Fortunately there is now less excuse than previously for the omission of division, because extensive work on better methods than symbolic versions of the well-known Euclidean greatest-common-divisor algorithm have now led to short and fast symbolic subroutines for this job. It is now reasonable, therefore, to expect the handling of rational functions of polynomials in a symbolic system. For those who are interested to see in detail how what is an easy operation in numerical mathematics becomes hard in SC, Hearn's practical description [5] of polynomial division is highly recommended. Knuth also surveys the whole subject [21].

Secondly, a standard function for differentiation should be available. Construction of such a function in any symbolic language is not difficult: in fact, there is some evidence that the first-ever symbolic computation was Kahrimanian's [80] 1953 computation of derivatives. All good systems can differentiate polynomials without any help. Opinions vary on what other forms they should recognize. Particle physicists will probably require the sine, cosine and natural logarithm to be included. However, as no system can be expected to satisfy every user, there should be a facility for adding new forms and new rules for differentiation to what is already present. This is a particular case of the more general question of substitution.

Substitution covers the whole range of transformations (as distinct from simplifications) on user-generated expressions. Historically it tends to be the first general-purpose problem to be tackled after a physicist has built a successful program for Dirac-matrix algebra which produces its results in terms of scalar products of 4-vectors. (Users prefer expressions like $E_1E_2(1 - \cos\theta_{12})$ to $k_1 \cdot k_2$ in their results!). The two systems which began life as Dirac-algebra programs but which are now widely used for other problems are SCHOONSCHIP [9] and REDUCE [5,17]. It may be said that they survived because they could handle general substitutions, while many other Dirac-algebra programs (including one of my own [22] faded away because they were never developed to that point. A substitution may have the form

```
LET K1.K2 = E1*E2*(1 - COS(TH12))
```
(11)

or

FOR ALL X LET DF(TAN(X),X) = 1/COS(X)**2 . (12)

Both (11) and (12) have analogues in (8), but (12) also shows how argumentation of a system's function for differentiation should be possible. The distinction between particular symbols in (11) and the general dummy variable X in (12) should also be possible. I have used REDUCE notation here, but equivalent notations exist in other systems.

It is an important point that we make particular substitutions like (11) to allow further simplification of an expression (where "simplification" means any operation which improves our understanding of the expression). Thus systems normally attempt resimplification after each substitution. Therefore the ideal system is one in which a command such as LET is generalized to accept every piece of information U that describes a potential simplification. In many cases U can be a pattern, e.g. "replace by 0 any term with at least 3 factors of f(x), for any x except 0 or 1, and a coefficient greater than 133/9", but simplification by detection of patterns is still very much a subject for research. Unconditional patterns of products are pratical keys for substitution, e.g.

FOR ALL A, B LET F(A)*F(B) = SIGMA(J)*(2*J + 1)*F(A-B), (13)

but the other side of the present boundary between achievement and future research is represented by an unconditional pattern of a well-known sum, i.e.

FOR ALL A LET COS(A)**2 + SIN(A)**2 = 1 , (14)

which is impracticable at present.

In any system for particle physics where all these basic facilities are present, the first special feature which is necessary is a section to do all of the usual manipulations of Dirac γ-matrices. Luckily there is no need to discuss the requirement further; systems written by particle physicists have normally started from that point, and have added substitution, differentiation and polynomial-handling later.

Quite apart from Dirac matrices, which are processed in terms of their properties only, without the employment of any explicit 4 x 4 representations, a general system is rendered more useful if it can carry out general matrix operations. For example, it can then be called on to solve simultaneous equations. Once again I refer to REDUCE notation for clarity. A matrix quantity is declared as a special type of array, e.g. RZ(3,3) for the matrix of rotations through θ about the Z axis in 3 dimensions. A value is attached to RZ through the binding

```
RZ := MAT((COS(TH),SIN(TH),O),(- SIN(TH),COS(TH),O),
      (O,O,1))    ,
```

and RZ is then available for all the obvious matrix operations, e.g. RZ**(-1) causes the inverse to be computed, and DET(RZ), TRACE(RZ) and TP(RZ)(transpose) may also be called. If a general system does not possess matrix operations explicitly, it is always possible to write them into the system. In practice the best systems can be expected to invert 8 x 8 matrices with no particular symmetries without getting into difficulties (with space), and larger matrices with special symmetries or zero elements are amenable to treatment.

Since we have differentiation, can we also ask for integration? Unfortunately the answer must be given, cautiously, as "no". Differentiation is algorithmic, as taught even in secondary schools, but there is little that is algorithmic in university courses in integral calculus beyond the rule for integrating x with respect to itself. It is only recently that a truly algorithmic theory of integration in finite terms has been developed (mainly by Risch [23]) which covers the majority of the functions of one variable that are of interest to particle physicists. Programming of the complete Risch algorithm is not trivial. Parts of the algorithm have been combined with heuristic methods, particularly by Moses [81], to make symbolic systems devoted entirely to integration, but these are among the longest symbolic systems in existence. In principle Moses' system SIN can be added to REDUCE (both being written in LISP), to give a capability for integration, and this has actually been done in the SCRATCHPAD [24] and MACSYMA [26] systems, but the resulting combinations run slowly in huge allocations of storage that are only available in a small number of privileged places, e.g. MIT and the IBM Yorktown Heights laboratory. In practice, these combined systems have had no effect on research in particle physics. On the question of integration in finite terms, Moses [81] gives a survey which does not minimise the difficulties, and Barton and Fitch [7] present the best outline of Risch's algorithm for pedestrians, with references.

A second class of integrals involves integrations over internal 4-momenta in Feynman diagrams with closed internal loops. This case is peculiar enough to be considered by itself, with no help from Risch's algorithm. No system includes functions for symbolic closed-loop integrations directly, but it is possible to add relevant functions on demand. The first computation of this type which did not rely exclusively on table look-up was reported in 1967 [27]. Lately Fox and Hearn [28] have formalized the approach, using integrals in fourth-order quantum electrodynamics (henceforth abbreviated as QED) as examples, and have provided explicit definitions of the relevant functions in REDUCE in an appendix.

To go one stage further than Dirac algebra, it is reasonable to expect that general tensor algebra should be possible in symbolic systems. Much of the information about any tensor behaviour can be inserted into a system at the "substitution" level, if the tensor is regarded as a function of its indices and this function is described to the system through rules having the form of (13). Moreover, if the bias is general-relativistic to the extent that the tensors under examination are compounded from the metric tensor $g_{\mu\nu}$, particle physicists can make profitable use of the two fast systems CAMAL [19] and ALAM or CLAM [29], or the program of Payne [30] which its owner claims to be faster than anything else in computing the Einstein tensor from the metric (although this claim largely ignores the fact that systems run slower on slower computers!). Problems arise, however, when elements of the tensor algebra are non-commuting objects. We cannot use substitution successfully to say LET A*B - B*A = C, for the same reason that (14) fails. Unless tricks special to the calculation in hand can be programmed (and this requires ingenuity or inside knowledge of how a system behaves), we cannot make much progress in general. There exists a small system [31] whose internal state is adaptable to descriptions of non-scalar objects, but its influence so far has been confined to celestial mechanics and the symbolic structure of numerical analysis. More research on systems of this type is needed.

A SURVEY OF SYMBOLIC COMPUTING IN PARTICLE PHYSICS

Historically, Dirac-matrix algebra was the first subject in SC to engage the attention of particle physicists. It is embarrassing to have to list the 12 independent rediscoveries of the concept of the Dirac-algebra program. Therefore I shall not do so. Historians of science can reassemble them all by reading the papers of Kaiser [8], Campbell and Hearn [32] and Barton and Fitch [7]. As with most new subjects falling between established subjects S_1 and S_2 and combining their features, journals in S_1 tend to reject papers on the grounds that they are really papers about S_2, and journals in S_2 advise the authors that the proper place for the material is a journal in S_1. Thus the subject remains new in effect, for want of publicity, long after it has ceased to be new in fact. The first Dirac-algebra programs were written no later than 196o, yet in 1968 at least one student with many trace calculations to complete wrote his own program because of the lack of information about the existence of other programs. News of future developments in SC will probably not take so long to spread, because among particle physicists there is a large informed public with some experience of SCHOONSCHIP and REDUCE.

The most extensive use of SC systems in particle

physics has been on calculations which have contained some traces over products of Dirac matrices. The earliest complete example which I can describe from first-hand knowledge is the Compton scattering case considered in (8) and (9), but programmed (without substitutions) in PDP-1 LISP at the end of 1963. The calculation was set as an exercise in a graduate-level course. Although correct, the answer was awarded 0 marks, because computation with anything other than numbers was well known in that time and place to be impossible and hence illegal. The first long computed result to be published was that of Tsai and Hearn [33], who gave cross-sections for

$$e^- + e^+ \rightarrow W^- + W^+ \rightarrow e^- + \mu^+ + \bar{\nu}_e + \nu_\mu \tag{15}$$

evaluated in 1964 and 1965 in an early version of REDUCE. Computations of this type are still the mainstay of applications in physics. For example, Brown and Smith [34] in 1971 have exercised the same facilities in SCHOONSCHIP to determine production cross-sections for the intermediate vector boson W in nuclear fields Z via

$$\mu^- + Z \rightarrow Z + W^- + \nu_\mu \quad , \qquad \bar{\nu}_\mu + Z \rightarrow Z + W^- + \mu^+ .$$

SCHOONSCHIP is popular with physicists associated with CERN and Brookhaven, while users elsewhere tend to turn to REDUCE. Nevertheless, other systems are still in use for these basic computations. M. J. Levine's system ASHMEDAI [35] has performed [63] the traces occurring in the astrophysical reactions

$$\gamma + \gamma \rightarrow \nu + \bar{\nu} \ , \qquad \gamma + \gamma \rightarrow \gamma + \nu + \bar{\nu}, \tag{16}$$

where W bosons are included in the intermediate states, and has been put to use in the much more taxing computation of the sixth-order part of the magnetic moment of the electron by Levine and Wright [36]. Here, symbolic integration must enter too, but it is subsidiary to trace calculations.

Are there any topics which owe all their significant information to the help of SC? One obvious example is W-boson physics, where SC has derived all the interesting information about cross-sections (e.g. in (15) and (16)). Without the programs, the information would have been confined to order-of-magnitude estimates and simple production cross-sections in reactions of no experimental interest. Unfortunately the fact that nobody has yet seen a W boson makes this example less valuable. The most substantial example is certainly in different category: the sixth-order parts of the magnetic moments of the electron and muon have now been determined completely, and every stage of the work except the very first approximate estimate needed the help of SC systems. The history of these computations is interesting enough to be reported in detail. In what follows, the fine-structure constant will be written as πp.

The first correct results for the fourth-order part of the electron's moment were achieved [37] in 1957. Sommerfield's value of $-0.328p^2$ was obtained with a decomposition of the electron mass operator in a method developed by Schwinger [78]. After that, there was no further reported progress until 1965. During that period Feynman [79] issued a challenge to particle physicists to obtain a plausible estimate of the sixth order part without doing an exact calculation by the laborious methods of perturbation theory in QED. As an answer to the challenge, Drell and Pagels [38] used a dispersive approach into which exact information about the low-energy Compton scattering amplitude was substituted, to obtain an estimate of $0.15\rho^3$. A typical sub-diagram is shown in Fig. 2.

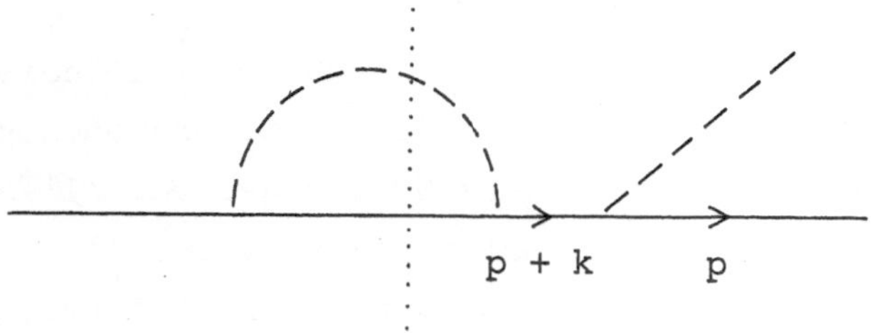

(Fig. 2: A contribution to the magnetic moment of the electron).

The broken lines indicate photons, and the diagram is partitioned along the dotted line to calculate the substitution for Im F_2 in the dispersion relation (17) below. The right-hand partition thus contains the Compton scattering information. In 1968 Parsons [39] repeated the calculation with the help of REDUCE, and was therefore able to retain many more symbolic terms. The method common to the two calculations is to write the electromagnetic current as an expression in which one lepton is off its mass shell. In terms of Fig. 2, $(p + k)^2 = W^2$. The most general form for the current through a vertex is

$$\frac{e}{2m}\,\bar{u}(p)\{[\gamma_\mu + F_2^+(W^2)(-i\sigma_{\mu\nu}k^\nu/2m)]\;[\gamma\cdot(p+k)+m]$$

$$+[\gamma_\mu + F_2^-(W^2)(-i\sigma_{\mu\nu}k^\nu/2m)][-\gamma(p+k)+m]\}\;u(p+k)$$

after current conservation and the Ward-Takahashi identities have been imposed. Here, $\sigma_{\mu\nu} = i(\gamma_\mu\gamma_\nu - \gamma_\nu\gamma_\mu)/2$, and $F_2^+(m^2)$ is just the anomalous magnetic moment of the electron. The dispersion relation used is

$$F_2^\pm(W^2) = \frac{1}{\pi}\int_{m^2}^{\infty} \frac{dW'^2 \, \mathrm{Im}\, F_2^\pm(W'^2)}{W'^2 - W^2 - i\varepsilon}, \qquad (17)$$

where only 2-particle intermediate states are retained in Im $F_2^\pm$ to any order. It is assumed in one case [38] that the low-energy part of the integral in (17) is dominant, so that the infinite limit is replaced by $6m^2$, and in the other case [39] that the upper limit is a free parameter chosen so that the calculation reproduces the fourth-order result $-0.328\rho^2$. By this latter method, Parsons found an estimate of $0.13\rho^3$. It is amusing that this work agrees in spirit with Feynman's challenge, but eventually turns out to be wrong by almost exactly one order of magnitude.

The 72 Feynman diagrams contributing to the sixth-order part of the magnetic moment may be classified into 4 groups: 1) the 6 diagrams containing a subdiagram for scattering of photons by photons (Delbrück scattering), 2) the 4 diagrams containing a fourth-order insertion of vacuum polarization as a subdiagram, 3) the 12 diagrams containing at least one second-order insertion of vacuum polarization, and 4) the remaining 50 diagrams. Group 1 has the property that its contribution is independent of cut-offs, but also that each diagram is so independent. (Each of the others may require cut-offs in individual diagrams, but the sums are independent of cut-offs). The first published report of an exact sixth-order computation, by Aldins, Kinoshita, Brodsky and Dufner [4o], gave the entire contribution from Group 1: for the electron's moment, it was (o.36 ± o.o4)p^3. One of the 6 diagrams is shown in Fig. 3.

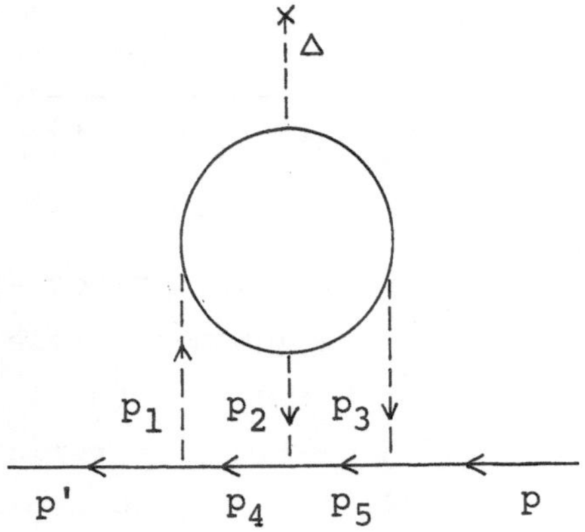

(Fig. 3: A diagram with photon-photon scattering as a subdiagram).

The contribution to the magnetic moment in the case of Fig. 3, where the lepton in the loop has mass m and the other lepton has mass m_μ, may be found as

$$\frac{1}{48m_\mu} \lim_{p'\to p} \mathrm{Tr}(\gamma{\cdot}p+m_\mu)(\gamma^\mu\gamma^\nu-\gamma^\nu\gamma^\mu)(\gamma{\cdot}p'+m_\mu)M_{\mu\nu}(\Delta=0) \qquad (18)$$

by standard methods of QED. The cross in Fig. 3 marks the photon of the external field. In (19), Π is the polarization tensor for light-by-light scattering, or in effect a Delbrück scattering amplitude, and

$$M_{\mu\nu} = -\frac{e^2}{(2\pi)^8}\int d^4p_1\int d^4p_3 \frac{1}{p_1^2p_2^2p_3^2}\left[\frac{\partial}{\partial\Delta^\mu}\Pi_{\kappa\rho\sigma\nu}(-p_1,p_2,p_3,-\Delta)\right]$$

$$\cdot\gamma^\kappa(\gamma{\cdot}p_4-n_\mu)^{-1}\gamma^\rho(\gamma{\cdot}p_5-n_\mu)^{-1}\gamma^\sigma \quad . \qquad (19)$$

The integrands in (19) are reduced to integrals over one-dimensional Feynman parameters with ranges between 0 and 1 by steps which are generalizations of the simple integral identity

$$\frac{1}{ab} = \int_0^1 \frac{dx}{[ax+b(1-x)]^2} \quad .$$

For (19), this gives an integral over a 7-dimensional unit hypercube. (Cut-offs may be introduced into the integrals analogous to (19) for other diagrams if needed (e.g. $p_i^2 \to p_i^2 - \lambda^2$, $i = 1,2,3$)). As the authors can find no further analytic simplification of (19) by integration over Feynman parameters, the 7-dimensional integrals in ref. 4o are performed by an adaptive Riemann-summation program (RIWIAD) in FORTRAN, developed by G. Sheppey and subsequently by B. Lautrup. By regarding m_μ/m as a variable numerical parameter, one can complete the muon and electron calculations at the same time. The adaptive section of RIWIAD varies the size of the mesh on any axis A in the hypercube at each iteration according to the rate of variation of the integrand near A in the previous iteration. The number of small hypercubes in the volume of integration is held roughly constant at a prescribed value - 500,000 to 1,000,000 in integrals of interest in this problem. The numerical work therefore requires much more time than the SC, and has been practicable only on the fastest computers of the period: the CDC 6600 and the IBM 360/91. Along with its estimate of an integral, numerical programs of this type must be able to calculate an error on the estimate, which may be reduced somewhat as the number of iterations increases. Hence all theoretical values obtained by adaptive

Riemann summation must come with a quoted error.

At about the same time that ref. 4o appeared, Mignaco and Remiddi [41] reported a computation of the Group 2 diagrams, with the result of $0.055p^3$ for the electron's moment. This result was supported by Aldins, Brodsky, Dufner and Kinoshita [42], who later gave a value of $(-0.154 \pm 0.009)p^3$ for the total contribution of Groups 2 + 3. Their paper contains a full account of the method of computation via (18) and (19), the optimum Feynman parametrization for any diagram, and the use of RIWIAD. It is the only detailed published account of how to set about such a calculation.

In the following year (1971), Calmet and Perrottet [43] presented a formally different account of the computation for Groups 2 + 3 (although the same numerical methods as in ref. 42 are used in the final stage), which arrived at a value of $(-0.1507 \pm 0.0566)p^3$.

It is somewhat easier to calculate the overall difference between the sixth-order magnetic moment of the muon and that of the electron than to complete the computation in Group 4 for the electron. The next achievement was the evaluation of this difference by Brodsky and Kinoshita [44], who gave a result of $(20.3 \pm 1.3)p^3$. Levine and Wright [36] later remarked that the computed error could be reduced to $1.1p^3$. However, the highlight of the paper of Levine and Wright was the abbreviated presentation of a complete result for all four groups: $(1.49 \pm 0.25)p^3$. Meanwhile, in a continuation of the methods of refs. 4o and 42 to Group 4, Kinoshita and Cvitanovic [47] found an overall result of $(1.29 \pm 0.06)p^3$, which is now regarded as the standard value for the electron. Lautrup, Petermann and de Rafael [46] and Remiddi [82] review the state of this calculation, omitting only the final touch added by the last-named reference. Levine and Wright used ASHMEDAI[35] for the SC, both Mignaco and Remiddi and Lautrup and his collaborators used SCHOONSCHIP, while the SC behind the other papers was done in REDUCE.

Most recently, Levine and Wright [83] have given a detailed account of their work, together with a restated value of $(1.21 \pm 0.07)p^3$ in the case of the electron. The error is decreased from their previous estimate by more iterations in numerical integration, with up to 10^7 points per unit hypercube, while the changed central number follows from a correction of one isolated programming fault.

For the record, I should add that I began a parallel program of work in 1966, using REDUCE and another small symbolic program [22]. The computation was completed in 1969, but I was unable to satisfy myself about the convergent behaviour of the accompanying adaptive numerical program for the integrations, and I was initially unable to tell whether or not the Group 1 diagrams were included in the answer because I started from the mass-operator formalism [78] instead of from diagrams. The overall result of $1.6p^3$ which was quoted by Gilleland and Rich [48] was thus presented explicitly as a sum of my own computations and the contribution $0.36p^3$ of ref. 4o, at a seminar in the Physics Department of the University of Michigan in September 1969. However, with the help of Mr. R. Carroll of that Department, I was later able to determine that the Group 1 diagrams were indeed present. With further tests of the program for adaptive integration, and the removal of one mistake in axis labelling in one case, the first result has stabilized at $1.27p^3$, with a generous computed error of $\pm\ 0.19p^3$. As both Dr. J. Calmet and I have used the numerical program (NAZDAR: = "good luck" in Czech) and RIWIAD to obtain consistent results for specific examples, and as RIWIAD has provided the numerical results of ref. 47, this is strong independent support for the value $(1.29 \pm 0.06)p^3$ of Kinoshita and Cvitanovic.

The present state of the magnetic-moment calculation through sixth order is unsatisfactory to the extent that the computed theoretical error (from the integrals) prevents the best possible comparison with accurate experimental results for the muon. This is not a shortcoming of SC! Nevertheless, it may have indirect influence on SC through a stimulus to further efforts to find analytic results from general techniques. In this connection there is the successful study of two Group 3 diagrams by Billi, Caffo and Remiddi [84], and a remarkable new prescription of Levine and Roskies [85] that works algorithmically for all planar diagrams of Group 4 and is obviously good for some non-planar diagrams as well. It consists of a Wick rotation of four-momenta to Euclidean space ($p,q \rightarrow$ "$\underline{p}$", "$\underline{q}$"), followed by replacement of propagator denominators (which can be rewritten in terms of no more than two momenta in these cases) by a series expansion in Gegenbauer polynomials:

$$\frac{1}{(\underline{p}-\underline{q})^2+m^2} = \frac{Z}{M} \sum_{n=0} Z^n C_n(\hat{\underline{p}}\cdot\hat{\underline{q}})$$

where $\quad 2MZ = \underline{p}^2 + \underline{q}^2 + m^2 - \{[\underline{p}^2 + \underline{q}^2 + m^2]^2 - 4M^2\}^{1/2}$

and $\quad M = |\underline{p}||\underline{q}|$.

There is some hope for SC of integrals with the help of such expansions, because C_n (the Gegenbauer polynomial $C_n^{(0)}$) is most simply expressed by $nC_n(u) = 2\cos(n\cdot\arccos u)$, and a conversion of the arccos function to its logarithmic representation leaves the resulting QED integrands in a form which is suitable for processing with Risch's original algorithm [23], at least for integration over one variable. How much further this procedure can be carried in general is a good question for further exploration.

Less spectacular than the evaluation of the lepton magnetic moments, the various SC systems have rendered quiet but steady service over the past 10 years in the evaluation of traces of Dirac-matrix products and resolution of such products into sums in the 16 basis quantities 1, γ_μ, γ_5, $\gamma_5\gamma_\mu$ and $\gamma_\mu\gamma_\nu$, in many problems. Barton and Fitch [7] list a good selection of papers which would never have appeared, or which would have been greatly delayed at best, without the help of SC systems. The heavy demands for Dirac algebra have prompted careful mathematical study of the properties of the products, with a view to the improvement of the simple but inefficient algorithms found in standard texts on QED. I shall quote two examples of the theoretical work which has been provoked by the existence of SC systems, and which in turn has improved the performance of some systems. Firstly, there is Kahane's [49] algorithm for the summation over μ in products $\gamma_\mu F\gamma^\mu$, where F is an expression of arbitrary length in Dirac matrices. Barton and Fitch [7] describe this algorithm in simplified form. Secondly, Chisholm [50] has presented a number of rules for computations involving electromagnetic vertices γmm, where the photon of 4-momentum k may be off the mass shell. Displaying some of the systematic behaviour which any graduate student who has calculated Compton scattering or any longer cross-section will believe to be present (although mostly hidden), the answers which follow from Chisholm's rules are expressed as Chebyshev polynomials of scalar products of 4-vectors, and of $2m^2 - k^2$ in particular. There is no reason to believe that we have yet found the optimal algorithms in this subject, i.e. it is an interesting field for further mathematical work.

As the analysis of reactions involving the production of high-spin particles becomes more popular, the need for sets of independent amplitudes to describe the processes will increase. (This is not to say that the field is unpopular even now). On most occasions helicity amplitudes are required, to make the applications and predictions of Reggeology more transparent, but frequently it is helpful to have a set of covariant and gauge-invariant amplitudes A(s,t) to remove kinematical singularities and thus make the investigation of analyticity rather easier. The two sets are connected by a matrix transformation T. For example, for photoproduction of a 0^- meson and a baryon ("3") of spin no higher than 3/2 from a nucleon ("1"), the connection has the form

$$M_{\lambda_3\lambda_1\lambda_4\lambda_2}(s,t) = \bar{u}_\nu(p_3,\lambda_3)\sum_j A_j(s,t)T_j^{\nu\mu}\varepsilon_\mu(p_2,\lambda_2)u(p_1,\lambda_1), \tag{2o}$$

with $s = (p_1 + p_2)^2$ and $t = (p_1 - p_3)^2$, and where the photon has a polarization vector e_μ, and λ_i are the helicities. One must first find all $T_j^{\nu\mu}$ for a given spin, and then it is a relatively simple matter to go via the $A_j(s,t)$, whose determination by the methods of Hearn [51] is most practical, to the helicity amplitudes for a particular spin and parity with the aid of (2o). A tabulation of M in any given case is an ideal SC problem for any system which recognizes Dirac matrices. REDUCE is so far the leader in this industry, as the paper of Clark, Manny and Parsons [52] (with $J^P = \frac{1}{2}^{\pm}, \frac{3}{2}^{\pm}$ for the final-state baryon) shows after only a moment's inspection.

Up to this point I have concentrated on the symbolic simplification of the absolute square of an amplitude for a reaction, or on computations which have used the same facilities as the simplification. On the other hand, every prior stage of a calculation, from the point that a Hamiltonian of interaction is written down, can be tedious on occasions. The symbolic squaring of a sum of amplitudes is easy to compute. This leaves 3 stages at which SC may cut down the amount of hard labour: 1) derivation of all the Feynman diagrams of a specified order, from a statement of the Hamiltonian of interaction, 2) derivation of an amplitude from a given diagram by the application of Feynman rules, and 3) renormalization of amplitudes, wherever this procedure is non-trivial.

Hearn and I [32] have written a system of programs in LISP to handle the first two cases under quite general conditions. Our paper gives a full account of how this writing is done. The second case obviously requires SC, but the first case is largely a question in graph theory which can be answered faster by the manipulation of matrices whose elements, integers, describe the properties of the graphs. As Veltman [10] has said, it is always wise to examine any potential SC problem to see if it may actually be represented numerically, even if this representation is possibly only in a small percentage of examples. The most effective graph-generating programs are in FORTRAN: details are available in papers by Calmet and Perrottet [53], James and North [54] and Perrottet [55]. Although no graph-processing languages have yet been used to produce published work in theoretical physics, I recommend GRASPE [56] to anyone who is interested in research in this area. It has recently helped in a check of the assertion of Samuel [45] that there are 1330 Feynman diagrams which contribute to the eighth-order part of the magnetic moment of the muon.

Renormalization is in a less satisfactory state in general. It is evident from advanced textbooks in quantum field theory that hardly any renormalizations are suited to algorithmic treatment at present. Here is another research problem that combines physics and computer science in an interesting way. However, there is one successful algorithmic scheme for renormalization, which may suffice here as an example of what is needed in general. This has been developed for scalar electrodynamics by A. Visconti and collaborators [57,58,59] but is now extended to regular QED. The quantity f, which is simply related to the S-matrix, is developed in a perturbative series whose general term is $e^n f^{(n)}/n!$, where e is the electronic charge. A starting value f_0 is given by

$$f_0 = \exp(-\tfrac{1}{2}\int \bar{\eta}_\alpha(x) S_{\alpha\beta}(x-x';m)\eta_\beta(x')d^4x d^4x')\cdot$$

$$\cdot\exp(-\tfrac{1}{4}\int J_\mu(x)\Delta(x-x')J_\mu(x')d^4x d^4x') \quad , \qquad (21)$$

where $(\eta, \bar{\eta})$ and J are sources of the lepton and photon fields respectively. The mass of the lepton is m. Standard definitions apply for the free photon and lepton Green's functions:

$$\Delta(x) = -\frac{2i}{(2\pi)^4}\lim_{\varepsilon\downarrow 0}\int d^4k\,\frac{e^{ik\cdot x}}{k^2-i\varepsilon}\quad ,$$

$$S(x) = -\frac{2i}{(2\pi)^4}\lim_{\varepsilon\downarrow 0}\int d^4k\,\frac{e^{ik\cdot x}}{k^2+m^2-i\varepsilon}\,(i\gamma{\cdot}k - m)\quad .$$

The general function f is related to f_0 by

$$f = \exp\left(ie\int \Gamma(\xi)\,d\xi\right) f_0\ ,$$

where

$$\Gamma(\xi) = \frac{\delta}{\delta\eta_\alpha(\xi)}\,\gamma^\nu_{\alpha\beta}\,\frac{\delta}{\delta\bar{\eta}_\beta(\xi)}\,\frac{\delta}{\delta J^\nu(\xi)}$$

and δ indicates a functional derivative at the point ξ. The finite (renormalized) part $Pf^{(n)}$ of the nth-order term in the perturbative series is then given algorithmically by

$$Pf^{(0)} = f_0\quad , \tag{22}$$

$$Pf^{(n+1)} = i\int :\Gamma(\xi):Pf^{(n)}d\xi + \sum_{k=1}^{n}\binom{n}{k}D^{(k)}Pf^{(n-k)}\ , \tag{23}$$

where the differential operators $D^{(k)}$ are prescribed for a given field theory in terms of its renormalization constants [59], and :: denotes Wick ordering of the factors in the product Γ. The structure of this algorithmic method, which gives a clear prescription of initial conditions (21) and (22) followed by a recurrence (23), is in essence no more complicated than the structure of the Legendre-polynomial algorithm in (2). Calmet and Perrottet [43] have used this renormalization procedure to assist in their calculation of parts of the sixth-order magnetic moment of the electron. What is now needed is a reduction of the process of renormalization to an equally uncomplicated algorithmic form for arbitrary quantum field theories!

There remains one aspect of the determination of amplitudes where there can be no general solutions. I refer to the SC of integrals over unknown 4-momenta of intermediate-state particles in any Feynman diagram with closed internal loops. The standard transformation to integrals over one-dimensional Feynman parameters brings the problem into line with the ordinary exercises of integral calculus by steps which I have mentioned already, but I have also mentioned that those exercises are difficult. In QED, many of the interesting results are given in terms of Spence functions [60]:

$$\mathrm{Li}_2(x) = -\int_0^x \frac{\log(1-y)}{y}\,dy, \quad \mathrm{Li}_{n+j}(x) = \int_0^x \frac{\mathrm{Li}_n(y)\,dy}{y},$$

or more formally in terms of Nielsen's generalized polylogarithms [61]

$$S_{n,p}(x) = \frac{(-1)^{n+p+1}}{(n-1)!p!}\int_0^1 \frac{[\log y]^{n-1}\,\log[1-xy]^p}{y}\,dy$$

and the associated subtracted functions, but these are

not completely covered by Risch's algorithm. Moses [62] has outlined a procedure for integration in finite terms where Spence functions of general order are concerned, but further work of value to particle physicists may not be forthcoming for some time. Nevertheless, if the integrands involved are sums of rational functions of symbolic variables times Spence functions of the same variables, and if integrals thus obtained still have the same form (as is the case, at least up to fourth order in QED, for multiple integrals over Feynman parameters), Moses' decision procedure is applicable. The most advanced paper in physics which uses integration in finite terms together with SC is by Fox and Hearn [28], who evaluate integrals

$$I(k,m,n,r,s)=\int_0^1 dy\int_0^1 dx\int_0^1 dw\int_0^1 dz\ \frac{z^r w^s x^{2k+n-m-3} y^{k+n-2} d^n}{b^n[w^2xb+yd^2]^k}\ , \tag{24}$$

where

$$b = 1 - xy(1-wz(1-wz))$$

and

$$d = 1 - x\ (1-wz(1-w))\quad .$$

The other end of the spectrum in integration is represented by the program which basically looks up tables of integrals, and composes new results by substitution for general constant symbols, differentiation, integration by parts and so on, but not by a Risch-type decision procedure. This form of SC was very popular at an early stage of the use of SC in physical chemistry. Levine appears to have made extensive use of the method in calculating astrophysical rates of neutrino production

[63]. Presumably it is also used in the computation of the magnetic moment of the electron [36]. A program which combines table-searching with rudimentary decision procedures [32] has calculated radiative corrections to lepton-proton scattering [27] with much greater accuracy than was available in manual calculations. Most recently, as a sign that for physicists' problems table searches can still keep pace with decision procdures, Petermann [64] has written a SCHOONSCHIP program to evaluate integrals of the type (24).

This is necessarily an incomplete survey of SC in particle physics, but it is intended only to show the type of work which is possible. Having now had the main course, so to speak, I shall add only two small items as desserts. Firstly, at the same time as the work on the lepton magnetic moments, SC has been responsible for the completion of the calculation of the fourth-order Lamb shift [65] - an effort which is almost equally heroic. Secondly, as Gammel and Menzel [66] indicaté, there is scope for the adaptation of existing symbolic programs to the examination of properties of the Bethe-Salpeter equation.

A VIEW OF APPLICATIONS OUTSIDE PARTICLE PHYSICS

SC techniques for particle physics have centred on Dirac algebra and its accompanying simplifications. Other subjects have encompassed other central techniques. A "complete computational physicist" should be acquainted with the computing problems of all subjects. Then, if the emphasis on new problems in particle physics shifts to something which is already well understood computationally in (say) celestial mechanics or physical chemistry, the relevant techniques or systems can be borrowed immediately to good effect. Outside physical sciences, it is advisable to keep abreast of applications of SC to "artificial intelligence" research. It is not appropriate to give a long survey here, but two papers which may be read profitably are Hewitt's [67] remarks about his system PLANNER and Winograd's [68] account of an application of PLANNER to a restricted natural-language problem.

Barton and Fitch [7] and Jefferys [6] describe the nature of SC in celestial mechanics. Most of the computations with trigonometric (Poisson) series have no relationship to particle physics, but it is worth studying both the mechanization of Lie-transform theory in Hamiltonian dynamics by Deprit [69] and collaborators and the methods of compact representation of symbolic expressions in storage which are characteristic of all good Poisson-series processors. Celestial mechanics and particle physics meet in a paper [31] which quotes some symbolically-computed connections between the non-commuting differential operators of two different versions of the Lie-transform theory, because the program used to generate equations for the operators in one version in terms of nested commutators of operators in the other was first built to solve a problem [70] of current algebra posed within a set of commutation and anti-commutation relations.

Physical chemistry offers the computation of expectation values of energy operators in problems having a large number of linearly independent wave-functions as a basis. The bulk of the computing in present cases deals with the SC of integrals, where often an integral carries several discrete quantum indices or is evaluated in one of the 8 unfamiliar coordinate systems [71] in which wave equations have separable solutions. In the former situation, integrands with the highest indices are usually defined through recurrence relations involving lower members of the set. Techniques of optimal programming for these integrals, where there may be several paths through a collection of recurrence relations to any one member, will probably be useful in applications outside physical chemistry in the future. Ref. 72 describes a typical SC exercise in physical chemistry.

Probably the most interesting field for study of SC outside particle physics is general relativity. Again, Barton and Fitch [7] provide the best survey. The central problem is the computation of various tensors and the curvature scalar R from given elements of a metric tensor $g_{\mu\nu}$. The usual sequence is as follows:

$$[\mu\nu,\rho] = \frac{1}{2}\left(-\frac{\partial g_{\mu\nu}}{\partial x^{\rho}} + \frac{\partial g_{\mu\rho}}{\partial x^{\nu}} + \frac{\partial g_{\nu\rho}}{\partial x^{\mu}}\right) \quad \text{(Riemann-Christoffel symbols)}$$

$$\left\{{\rho \atop \mu\nu}\right\} = g^{\rho\sigma}\,[\mu\nu,\sigma] \quad \text{(Christoffel symbols of the second kind)}$$

$$R_{\mu\nu\rho\sigma} = \frac{\partial}{\partial x^{\rho}}[\nu\sigma,\mu] - \frac{\partial}{\partial x^{\sigma}}[\nu\rho,\mu] + \left\{{\alpha \atop \nu\rho}\right\}[\mu\sigma,\alpha] - \left\{{\alpha \atop \nu\sigma}\right\}[\mu\rho,\alpha]$$

(Riemann tensor)

$$R_{\mu\nu} = g^{\rho\sigma}\,R_{\sigma\mu\rho\nu} \quad \text{(Ricci tensor)}$$

$$R = g^{\mu\nu}\,R_{\mu\nu} ,$$

and

$$T_{\mu\nu} = R_{\mu\nu} - \tfrac{1}{2}Rg_{\mu\nu} \qquad (25)$$

where $T_{\mu\nu}$ is the energy-momentum tensor and (25) represents the Einstein field equations of general relativity, modulo factors of π and small powers of 2. Looking ahead, we may hope that various models in group theory and current algebras will eventually specifiy a direct computation of a $T_{\mu\nu}$, or at least a T_{00}. If this ever happens, the specification is almost certainly likely to be algorithmic, and so difficult to follow up by hand that SC will be essential. In that event, there will be something to learn from the existing fast systems for the derivation of (25). Similarly, where properties of physical systems (e.g. metric tensors, Hamiltonians of interaction, sets of commutation relations) are expressed or classified through algebraic or analytic constraints, SC is invaluable for the determination of concrete results. An example is the Petrov classification of a family of metrics by d'Inverno and Russell-Clark [73]. Not only was this a computation which would have been impracticable by hand, but it uncovered several errors in previous assertions about the properties of these metrics: in particular, that four proposed members of the family were not genuine members at all. More recently, in connection with a solution of Einstein's vacuum field equations which was put forward as a possible description of a black hole, Gibbons and Russell-Clark [74], using CAMAL [19], showed that the solution lacked several of the necessary properties. This is a model computation of a type which I hope will become more common in the future: a quick use of an existing SC system to settle a pressing question of physics which would otherwise have been difficult or impossible to resolve within any reasonable time.

NEW DIRECTIONS

I have suggested questions for continued research, in passing, at several places above. I would like to end by collecting together a few suggestions in one place.

Although SC seemed a short time ago to have exhausted all the interesting problems in QED, a new problem may have emerged recently. The sixth-order part of the magnetic moment of the muon is about $20p^3$. In the spirit of Feynman's old challenge [79], Lautrup [75] has made an estimate of the eighth-order part which is about $200p^4$. Samuel [45] has confirmed this estimate. If we bear in mind that "good" estimates [38,39] of the sixth-order part of the electron's moment were low by a factor of ten [36,47], then it is not yet possible to rule out the uneasy thought that the ratio of the coefficient of p^4 to the coefficient of p^3 for the muon may be of the order of $1/p$, or greater. It is hard to see how an independent estimate can be made without the aid of SC.

It is possible to claim that field theory is the property of particle physicists. Certainly the first SC system to compute physical quantities from descriptions of diagrams (Feynman diagrams, in this case) was designed for particle physics [32]. However, there are other subjects (many-body theory, nuclear physics, solid-state physics, model lattice problems in statistical mechanics) which share the property that diagrams in 1:1 relationship with physical quantities (amplitudes) by field-theoretic rules analogous to Feynman rules are the most convenient means of shorthand descriptions of problems. The lesson for people who wish to extend the range of SC applications is obvious. Moreover, there are some very interesting field theories in which the rules are presently ambiguous and possibly incomplete, and in which SC may have a clarifying part to play. In this category, Penrose's twistor field theory [76] may have the greatest relevance to the physics of zero-mass particles.

LISP-like or REDUCE-like systems are flexible but relatively slow and inefficient in the use of storage. Assembly-coded systems such as SCHOONSCHIP are fast and compact, but inflexible. Is there some prospect of a system which will have the virtues of both types? The question can only be answered by future work, but it has already proved possible to build a small system [31] which accepts information in two stages: 1) a description of the properties of unusual factors (e.g. γ_μ) of the computation, which the system then uses to set up a fairly compact representation for these factors in storage, and 2) a statement of the expression to be simplified, and the rules for simplification. More work in this direction should be encouraged.

Finally, I strongly recommend a reading of the three volumes of Knuth's "Art of Computer Programming" which are presently available [18,21,77], as an exposure to techniques useful for SC and to the large number of rewarding problems for mathematical research which arise from SC. There is something for everybody in this field. For those with very large computers, immediate experiments with the largest versions of the various SC systems are possible. For those with smaller but non-trivial computers, there is the possibility of the construction in LISP or similar languages of additional computing features which are not yet found in the large SC systems. As any careful study of LISP will show

[3,15], a set of LISP functions making a program to do A will, when added to a set to do B, produce a program to do A $\cup$ B, provided that the representation of the basic data is the same in each set. This additive property of the best SC languages means that a cottage industry in developing small pieces of a large system is very feasible. Finally, for those with trivial computers or no computers, "thought experiments" on the mathematics of computation and the analysis of algorithms in SC, in the style of Knuth, will continue for the foreseeable future to give both education and entertainment to particle physicists.

REFERENCES

1. W. H. Jefferys, Celest. Mech. 2, 474 (197o).
2. J. A. Campbell, Comp. Phys. Comm. 1, 251 (197o).
3. J. McCarthy et al., "LISP 1.5 Programmer's Manual", MIT Press, Cambridge, Massachusetts (1965).
4. A. C. Hearn, Comm. A. C. M. 14, 511 (1971).
5. A. C. Hearn, in "Computing as a Language of Physics", International Atomic Energy Agency, Vienna (1972), p. 567.
6. W. H. Jefferys, Comm. A. C. M. 14, 538 (1971).
7. D. Barton and J. P. Fitch, Rep. Prog. Phys. 35, 235 (1972).
8. H. J. Kaiser, Nucl. Phys. 43, 62o (1963).
9. H. Strubbe, "Computations with SCHOONSCHIP", Report DD/73/16, Data-Handling Division, CERN, Geneva (1973)

10. M. Veltman, Comp. Phys. Comm. (Suppl.) 3, 75 (1972).

11. G. E. Collins, in "Proceedings of the Second Symposium on Symbolic and Algebraic Manipulation", A.C.M. Headquarters, New York (1971), p. 144.

12. W. H. Jefferys, Celest. Mech. 6, 117 (1972).

13. R. E. Griswold, J. F. Poage and I.P. Polonsky, "The SNOBOL4 Programming Language", Prentice-Hall Inc., Englewood Cliffs, N.J. (1968).

14. J. A. Campbell, in "Computing as a Language of Physics", International Atomic Energy Agency, Vienna (1972), p. 391.

15. C. Weissman, "LISP 1.5 Primer", Dickenson Publishing Co., Belmont, California (1967).

16. D. Lurié, in "Computing as a Language of Physics", International Atomic Energy Agency, Vienna (1972), p. 529.

17. A. C. Hearn, in "Proceedings of the Second Symposium on Symbolic and Algebraic Manipulation", A.C.M. Headquarters, New York (1971), p. 128.

18. D. E. Knuth, "Fundamental Algorithms", Addison-Wesley, Reading, Massachusetts (1968).

19. D. Barton, S. R. Bourne and J. R. Horton, Comp. J. 13, 243 (1970).

20. A. C. Hearn, Bull. Amer. Phys. Soc. 9, 436 (1964).

21. D. E. Knuth, "Seminumerical Algorithms", Addison-Wesley, Reading, Massachusetts (1969).

22. J. A. Campbell, J. Comp. Phys. 2, 412 (1968).

23. R. H. Risch, Trans. Amer. Math. Soc. 139, 167 (1969).

24. J. H. Griesmer and R. D. Jenks, in "Proceedings of the Second Symposium on Symbolic and Algebraic Manipulation", A.C.M. Headquarters, New York (1971),p.42.

25. H. Strubbe, "Internal Mechanism of a SCHOONSCHIP Calculation", Report DD/73/10, Data-Handling Division, CERN, Geneva (1973).

26. W. A. Martin and R. J. Fateman, in "Proceedings of the Second Symposium on Symbolic and Algebraic Manipulation", A.C.M. Headquarters, New York (1971), p.59.

27. J. A. Campbell, Nucl. Phys. B1, 283 (1967); Nucl. Phys. B10, 190 (E) (1969).

28. J. A. Fox and A. C. Hearn, J. Comp. Phys., to be published.

29. R. A. d'Inverno, Comp. J. 12, 124 (1969).

30. A. D. Payne, Comp. Phys. Comm. 4, 100 (1972).

31. J. A. Campbell, W. H. Jefferys, Celest. Mech. 2, 467 (1970).

32. J. A. Campbell, A. C. Hearn, J. Comp. Phys. 5, 280 (1970).

33. Y-S. Tsai and A.C. Hearn, Phys. Rev. 140, B721 (1965).

34. R. W. Brown, J. Smith, Phys. Rev. D3, 207 (1971).

35. M. J. Levine, J. Comp. Phys. 1, 454 (1967).

36. M. J. Levine, J. A. Wright, Phys. Rev. Lett. 26, 1351 (1971).

37. A. Petermann, Helv. Phys. Acta 30, 407 (1957); C.M. Sommerfield, Ann. Phys. (N.Y.) 5, 26 (1958).

38. S. D. Drell, H. Pagels, Phys. Rev. 140, B397 (1965).

39. R. G. Parsons, Phys. Rev. 168, 1562 (1968).

40. J. Aldins, T. Kinoshita, S. J. Brodsky, A. J. Dufner, Phys. Rev. Lett. 23, 441 (1969).

41. J. A. Mignaco, E. Remiddi, Nuov. Cim. 60A, 519 (1969).

42. J. Aldins, S. J. Brodsky, A. J. Dufner, T. Kinoshita, Phys. Rev. D1, 2378 (1970).

43. J. Calmet, M. Perrottet, Phys. Rev. D3, 3101 (1971).

44. S. J. Brodsky, T. Kinoshita, Phys. Rev. D3, 356 (1971).

45. M. A. Samuel, "Estimates of the Eighth Order Corrections to the Anomalous Magnetic Moment of the Muon", Oklahoma State University preprint (1973).

46. B. Lautrup, A. Petermann, E. de Rafael, Phys. Rep. 3C, 193 (1972).

47. T. Kinoshita, P. Cvitanovic, Phys. Rev. Lett. 29, 1534 (1972).

48. J. Gilleland and A. Rich, Phys. Rev. Lett. 23, 1130 (1969).

49. J. Kahane, J. Math. Phys. 9, 1732 (1968).

50. J. S. R. Chisholm, J. Comp. Phys. 8, 1 (1971).

51. A. C. Hearn, Nuov. Cim. 21, 333 (1961).

52. R. B. Clark, B. L. Manny, R. G. Parsons, Ann. Phys. (N.Y.) 69, 522 (1972).

53. J. Calmet, M. Perrottet, J. Comp. Phys. 7, 191 (1971).

54. P. B. James, G. R. North, J. Comp. Phys. 7, 354 (1971).

55. M. Perrottet, in "Computing as a Language of Physics", International Atomic Energy Agency, Vienna (1972), p. 555.

56. T. W. Pratt, D. P. Friedman, Comm. A.C.M. 14, 460 (1971).

57. H. Umezawa, A. Visconti, Nuov. Cim. 1, 1079 (1955).

58. Y. Le Gaillard, A. Visconti, J. Math. Phys. 6, 1774 (1965).

59. J. Soffer, A. Visconti, Nuov. Cim. 38, 817 (1965).

60. L.L. Lewin, "Dilogarithms and Associated Functions", Macdonald, London (1958).

61. K. S. Kölbig, J. A. Mignaco, E. Remiddi, B.I.T. 10, 38 (1970).

62. J. Moses, SIGSAM Bull. A.C.M. nr. 13, p. 14 (1969).

63. M. J. Levine, Nuov. Cim. 48, 67 (1967).

64. A. Petermann, in "Advanced Computing Methods in Theoretical Physics", Centre National de la Recherche Scientifique, Marseille (1971), vol. 2, p. IV. 62.

65. T. Appelquist, S. J. Brodsky, Phys. Rev. Lett. 24, 562 (1970). A further account of independent checks of parts of the results, again by computer, is given by S. J. Brodsky and S. D. Drell, Ann. Rev. Nucl. Sci. 20, 147 (1970). See also ref. 82.

66. J. L. Gammel, M. T. Menzel, Phys. Rev. A7, 858 (1973).

67. C. Hewitt, in "Second International Joint Conference on Artificial Intelligence", British Computer Society, London (1971), p. 167.

68. T. Winograd, "Understanding Natural Language", Edinburgh University Press, Edinburgh (1972).

69. A. Deprit, Celest. Mech. 1, 12 (1969).

70. M. Gell-Mann, D. Horn and J. Weyers, in "Proceedings of the 1967 Heidelberg International Conference on Elementary Particles", North-Holland, Amsterdam (1968), p. 479.

71. P. M. Morse and H. Feshbach, "Methods of Theoretical Physics", McGraw-Hill, New York (1955), vol. 1, p. 657-666.

72. G. S. Chandler, T. Thirunamachandran, J. A. Campbell, J. Chem. Phys. 49, 3640 (1968).

73. R. A. d'Inverno, R. A. Russell-Clark, J. Math. Phys. 12, 1258 (1971).

74. G. W. Gibbons, R. A. Russell-Clark, Phys. Rev. Lett. 30, 398 (1973).

75. B. Lautrup, Phys. Lett. 38B, 408 (1972).

76. R. Penrose, M.A.H. MacCallum, Phys. Rep. 6C, 241 (1973).

77. D. E. Knuth, "Sorting and Searching", Addison-Wesley, Reading, Massachusetts (1973).

78. J. Schwinger, Proc. Nat. Acad. Sci. 37, 452 and 455 (1951).

79. R. P. Feynman, in "The Quantum Theory of Fields", Interscience, London (1961), p. 75-76.

80. H. G. Kahrimanian, "Analytic Differentiation by a Digital Computer", M. A. thesis, Temple University, Philadelphia (1953)

81. J. Moses, Comm. A.C.M. 14, 548 (1971).

82. E. Remiddi, Comp. Phys. Comm. 4, 193 (1972).

83. M. J. Levine, J. Wright, Phys. Rev. D8, 3171 (1973).

84. D. Billi, M. Caffo, E. Remiddi, Lett. Nuov. Cim. 4, 657 (1972).

85. M. J. Levine, R. Roskies, Phys. Rev. Lett. 30, 772 (1973).

Acta Physica Austriaca, Suppl. XIII, 649–663 (1974)

METHODS OF ARTIFICIAL INTELLIGENCE AND COMPUTATIONS IN PHYSICAL SCIENCES

by

J. A. CAMPBELL*)
King's College Research Centre, King's College,
Cambridge, England

Even among its detractors [1], the subject of artificial intelligence (AI) is credited with advances in technique in non-numerical computing. As the value of this type of computing in physics in particular and physical sciences in general is now beyond doubt, it should be useful to see how work in physical sciences and in AI can be related. To begin, it is reasonable to ask two questions:

(q1) What is AI?

(q2) What implications does AI have for physical problems (and vice versa)?

The first question raises the more general query "What is intelligence?". From the point of view of a physical scientist, the important message is that the two are not

---

*) Seminar given at XIII. Internationale Universitätswochen für Kernphysik, Schladming, February 4-15, 1974.

necessarily related. Nevertheless, I shall mention a simple operational measure of "intelligence" below, without taking it too seriously. The name "artificial intelligence" appeared about 15 years ago in answer to the need to find a title for the collection of programming efforts in draughts and chess-playing, translation of languages, pattern-recognition and problem-solving in general. It did not imply the existence of any explicit philosophical justification for the imitation of intelligent behaviour by machines (although there are several very shaky hidden philosophical assumptions which are best avoided). The confusion between AI as presently practised and intelligence probably goes back to influential speculation by Turing [2] about what constitutes intelligent behaviour by a computer. (In this argument, I intend "computer" to be understood as "computer plus program", where the program is the senior partner). Turing's test is that an interrogator communicates (by teleprinter to remove trivial clues) with one human and one computer, asking questions whose answers are designed to reveal evidence of intelligence. If, on the basis of the answers, the interrogator can never tell which is the human and which is the computer, the computer is to be regarded as behaving intelligently. The trouble with this game is that each of us, looking over the interrogator's shoulder, may have a different view about the questions which are needed to establish intelligent behaviour. Thus, if he believes that the computer is intelligent, we have no obligation to agree with him. Moreover, if we are baffled this week about the identity of the computer, we may be able to identify it next week from exactly the same answers after we have had a chance to think about them in more detail. For example, in 1950 it was freely agreed that chess-playing of a good amateur standard would be evidence of intelligence, but subsequent work on chess programs has achieved this standard without any of the latest programmers making any radical claims about intelligent behaviour. As chess and similar exercises of evident use in Turing's test were active programming problems for research in the late 1950s, it probably seemed natural to choose for this new branch of research a name in which the word "intelligence" occurred. More recent research in AI has moved even further away from the aim of direct insights into human intelligence. Subject to the obvious comment that a physical scientist who uses a teleprinter as a scratchpad can amplify his intelligence by calling selectively on AI programs, we may regard intelligence and AI as decoupled topics [20].

If AI is not a semi-philosophical study, then what is it? The question (q1) has returned unchanged. It is possible to escape the philosophical overtones of (q1) by borrowing a definition of intelligence from a dictionary. One possible definition, according to Webster's New Collegiate Dictionary (1956), is: "The power of meeting any situation, especially a novel situation, successfully by proper behaviour adjustments". Because we can define success as precisely as we like in a physical computation (and decide that no other "situation" interests us), any physical program at all which changes itself or its given constants may be viewed as a example of AI. To refine the idea somewhat, there are several classes of programs with the right properties. The definitions of my choice of classes are as follows:

(0) The borderline cases in which the program and constants do not change, but where researchers at some

stage have viewed developments as instances of AI, e.g. symbolic integration [3] via Risch's algorithm [4], and minimax or alpha-beta searches of trees in programs for playing low-level games like draughts [5]. Dirac algebra has come under this heading too, e.g. in the annual report of the Stanford AI Project for 1967. Presumably the examples might have served as ammunition for Turing's test in 1967, but not now.

(1) While not introducing new types of constant, the program may change its own constants, e.g. in improving an evaluation function for chess positions [6]. If "constant" can mean a symbolic list [7] as well as a number, then an example of such a program in physical sciences has been applied to a problem in celestial mechanics [8].

(2) The program may also introduce new types of constant, i.e. it can change the nature of elements of list structure, by adding new types of data or pointers to elements, to express new types of relationship found during a computation.

(3 $-\varepsilon$) (where ε is a small number). The program changes its flow of control during the computation, by some more sophisticated process than a change of constants. (Unlike (1), programs in this class can alter their own flow charts). Longuet-Higgins and Ortony [9] give the simplest possible example of this type; almost any program in PLANNER [10] fits the description also.

(3) Finally, there are the cases of the programs which change themselves, not only by modifying flow charts but by rewriting significant parts of themselves. This class shades into the previous one, as any complex use of PLANNER or its present subsets indicates - it may

require considerable examination of intermediate trace output from such a use to decide whether a program belongs in Class 3 - ε or Class 3.

AI excludes Class 0, probably includes Class 1 (e.g. with current chess programs, although the inclusion is debatable on aesthetic grounds), and certainly includes the others. This is consistent with the dictionary definition given above. An alternative approach is to ask the practitioners of the art what they consider to be examples of programming in AI. There are now two texts [11,12] with AI as part of their titles, and these may be considered as the best oracles on the question of what AI encompasses. Among their cited programs are members of all classes except 0.

It is difficult to make any ambitious use of AI until we have a mental picture of a program in Class 3. For effective self-modification, a program's syntax should be the same as the syntax of its data, so that the program may treat itself as data. Fortunately, programs written in LISP [7] have that property. (PLANNER is also written in LISP). I shall next present in outline a LISP program for self-modification, after preliminary explanations. It will be written in the ALGOL-like form of REDUCE or "New LISP" [13], so that non-specialists may follow it easily, although it is to be understood as existing inside a computer in an equivalent LISP or "Old LISP" form whose list structure is the same as that of LISP data. When the program operates on itself, it is assumed to act on the Old LISP form. The difference between the two forms is discussed in the notes of my Schladming lectures, as well as in ref. 13. To refresh the memory, a line like

ELSE IF CAAR Z = "Z = NIL" THEN ξ_1(CDR Y, CDAR Z)

in New LISP is equivalent to

((EQUAL (CAAR Z) (QUOTE (NULL Z))) (ξ_1(CDR Y)
(CDAR Z))), (el)

so that CAAR of the form (el) is just the expression (EQUAL (CAAR Z) (QUOTE (NULL Z))), and CDAR of (el) is (ξ_1 (CDR Y) (CDAR Z)).

Suppose that our first guess for how to calculate some function of x is expressed in the LISP function ξ_0(X), and that we have a defined strategy contained in ξ_1 to change or repair ξ_0 if it misbehaves. Similarly, let ξ_2 express in LISP a strategy for repairing ξ_1, and so on. Then all ξ_j for $j > 0$ must be functions of two arguments, one to carry information about the type of misbehaviour Y, and one to stand for the part Z of the structure (of the definition of ξ_{j-1}) which is to be repaired. We can so design ξ_0 that Y in the case of misbehaviour is a list whose CAR or first element is an error number, and whose CDR (remainder) bears the detailed information about the error. Without errors, Y is NIL. Let us suppose that an error number up to 100 denotes a need to change only ξ_0, a number between 100 and 199 denotes a "meta-error" requiring ξ_1 to be changed, a number between 200 and 299 requires ξ_3 to change ξ_2, and so on. Then the entire program is the procedure MM defined by:

```
PROCEDURE MM(X)  M(X,NIL,NIL);

PROCEDURE M(X,Y,Z)
```

```
IF Y THEN BEGIN REDEFINE("M",M(X,Y,DEFINITION("M"));
                  M(X,NIL,NIL)  END
ELSE IF Z = NIL THEN (λ  (U); IF NUMBERP CAR U THEN
                  M(X,U,NIL) ELSE PRINT U) ξ0(X)
ELSE IF CAR(Y) < 100 THEN
      IF Z = NIL THEN NIL
      ELSE IF CAAR Z = "Z = NIL" THEN ξ1(CDR Y, CDAR Z)
      ELSE CAR(Z) . M(X,Y,CDR Z)
ELSE IF CAR(Y) < 200 THEN
      IF Z = NIL THEN NIL
      ELSE IF CAAR Z = "CAAR Z = "Z = NIL"" THEN ξ2
                           (CDR Y, CDAR Z)
      ELSE CAR(Z) . M(X,Y,CDR Z)
ELSE IF       ............ξ3....ξ4.............ξq...
ELSE STOP Y;                                          (e2)
```

In (e2), REDEFINE redefines M as the new function which M computes on the basis of X, Y (the error message) and the old definition of M. REDEFINE and DEFINITION are trivial to write in terms of the standard LISP functions DEFINE and GET. U is $\xi_0(X)$. Brackets round single arguments of functions are optional, and are used to improve readability. (e1) now illuminates the behaviour of the "repairing" part of M. The underlined pieces of (e2) are the lowest members of a hierarchy of tests which is simple to generalize. If an error number is 100q or greater, then no strategy for repairs is available and M must stop, communicating Y to us so that we can see what has gone wrong. Otherwise, M prints the desired answer.

(e2) is just a skeleton: the meat resides in the set of ξ_j for any given case. I shall comment on some examples later. However, it shows that even programming in Class 3 is conceptually simple. More than this, it suggests an amusing way to classify degrees of difficulty in AI. Suppose that the number of alternative results available to ξ_i (answers in the case of ξ_0, repairs in the cases of the other functions) for its possible range of arguments is a finite number a_i. Then

$$I_q = \prod_{i=1}^{q} a_i \qquad \text{and} \qquad I_{q+1} = a_0 I_q \qquad \text{(e3)}$$

both measure the amount of AI required to solve a problem with hierarchical ordering of changes of strategy. The I_q of a program is probably a better measure of artificial intelligence than I_{q+1}, because the estimates of a_0 are usually much more vague than those of the other a_i. In descriptions of physical computing problems, it is most often quite easy to estimate a finite a_j for each $j > 0$. People who wish to give greatest weight to the most global order of repairing can replace a_i by $w_i a_i$ in (e3), where w_i is a weight factor: $w_i = i$ is a possible choice. (e3) may be called a party-game device of no philosophical value, but it is still remarkably helpful in discussion, even if only as a stimulus to the better understanding of a problem which comes from attempts to estimate each a_i.

Next, to consider question (q2), what are the recognized AI methods which will possibly be useful to us, and how do the methods in present symbolic programs for physical problems differ from them? A search through refs. 11 and 12 turns up the following methods in AI: construction of implicit tree-structures for problems (e.g. chess) and efficient searching of trees with the use of static or variable evaluation functions (e.g. for good and bad positions on a chessboard), description of a complex problem in a minimal "problem space" and the automatic reduction of the problem to a sum of simpler sub-problems in the space, abstract pattern-matching, and the application of theorem-proving programs to the solution of problems. Seemingly [14], these methods don't have much to do with the greatest successes of symbolic programs in physical sciences. The reason for the lack of applications lies symply in the nature of the physical problems (e.g. Dirac algebra) which have been tackled in the past. In them, the required algorithms are fully determined and not subject to change during a computation. Now that the simple approach based on a choice of simply-described problems has been so successful [14], it is time to try more complex cases: having demonstrated that we can walk, the next move is to attempt to run. In exercises that are complex and interesting enough, I assert that each one of the AI techniques mentioned above has its special place, and that a study of ref. 11 in particular will suggest extra types of physical problem, previously attacked with pencil and paper alone, which are well suited to computations with AI methods. I shall now give three examples of AI-type computations.

Firstly, consider the (m,n) missionaries-and-cannibals problem. A boat which can hold up to n people is to be used to move m missionaries and m cannibals from one bank of a river to the other. At no stage of the process are cannibals allowed to outnumber missionaries on either bank or in the boat. What sequence of loadings of the boat at each crossing of the river solves the problem? We may write each move as a call to a loading function f with two arguments (the numbers of missionaries and cannibals transported in the move). With suitable drafting of the program, the choice of f, equivalent to a solution, reduces to a choice of function-name out of the set of about [7] 80 elementary LISP functions in the (3,2) case. Then $q = 1$ in (e3), and $I_q = 80$. For higher m and n, $q > 1$ and the higher functions $\xi_2, \xi_3 \ldots$ in the analogue of (e2) have the job of changing the representation of the problem used by ξ_0 and ξ_1 in the quoted case, where the (3,2) strategy leads to the undesirable consumption of missionaries by cannibals. The needed changes follow the pattern described by Amarel [15].

Secondly, there is the peculiar but useful computation which takes an incomplete set of Feynman rules for some field theory and the amplitudes corresponding to given Feynman diagrams, and attempts to find the missing rules. For Compton scattering of electrons, it has proved possible to find either a missing vertex rule or a missing fermion-spinor rule by a procedure with the structure of (e2). The spinor case is slightly more impressive, because it leads to even the correct power of 2π in the multiplier. In both examples, $q = 2$, $a_1 = r + \frac{1}{2}r(r - 1)$, where r is the number of Feynman rules given (6 in this case), and $a_2 = t + \frac{1}{2}t(t - 1)$, where t is the number of separate symbolic objects in the amplitude (9: polarization vector, momentum, mass, Dirac-matrix symbol, Greek index, energy, spinor, e and π), so that $I_q = 945$. This is very much an example of work in progress at an elementary level: the eventual object is to apply the method to at least one field theory [16] in which the set of rules for amplitudes in incomplete. Also, it is an example of a classic AI situation: "given a starting set S of data, and a desired answer A, what must be added to S to get A?".

Finally, there is the question of folding of proteins. A protein is formed as a chain of building blocks (e.g. about 10 for pressins, about 100 for varieties of cytochrome c) with standard links, and each block is taken from the basic set of b = 20 amino-acid residues R_i. During formation, the chain folds up into a 3D configuration which is unique for a given sequence of residues. How is the sequence of folds determined? Evidently the links must bend with respect to each other to produce a molecule with a minimum of free energy, in which residues with observed affinities for each other can be brought close to each other while repulsions tend to maintain the straight-line configuration. The AI approach consists of the replacement of the continuous variation of bond angles with a small number of fixed positions p (the data in any case do not reflect precise measurements), and the computation of how the initial structure $R_1(0)$ $R_2(0)R_3(0)$.... is changed in discrete steps to a final $R_1(p_1)R_2(p_2)R_3(p_3)$.... while respecting all the constraints from physics and chemistry (e.g. no increase of free energy beyond an idealized threshold at each step, no geometrically impossible folds, no folds leading to highly unlikely values of van der Waals forces between any pair of residues, favourable treatment for short sub-sequences known from experiments to occur frequently in other proteins). So far the successes of the method are confined to the folding of an 8-residue imitation "protein" with properties (not made available to the program) guaranteeing a specific final structure, and the checking of the self-consistency of some results determined in other ways [17] for triples of residues. It is possible to get a good idea of the problem and the most significant advances made with conventional methods from the papers of Chou and Fasman [18] and Wu and Kabat [19]. For the AI approach so far, $q = 2$ and $I_q \geq 462000$: $a_1 = b^2 + b^3 = 8400$, and a_2 begins at $c + \frac{1}{2}c(c - 1)$, where c is the number of different types of constraint present in the data. The inequality occurs because the program can compute new constraints during the folding process.

I believe that the I_q ratio 80 : 945 : $\geq$ 462000 is actually a good measure of the relative difficulty of these three examples.

(e3) and the hierarchical construction of (e2) hide one large philosophical assumption common to many popular speculations about intelligence, artificial or otherwise. That is, that the basis for intelligence is the breaking-down of problems into the smallest possible pieces of irreducible "atomic" data, then their solution by the application of well-defined rules (ξ_0), rules about rules (ξ_1), rules about rules about rules (ξ_2) etc.. Dreyfus [1] has uncovered this implicit picture of the world in many of the AI researches of the last 15 years. For those who support it, despite the philosophical risks of an infinite regress which it contains, (e3) may be distastefully simple-minded but it is nevertheless the reductio (ad absurdum or not, according to taste) or the hierarchical picture. It also hints that there is some characteristic value (or at least an order of magnitude) of I_q at which "intelligence" begins, even though the value may increase with time (e.g. it may have been below the amateur chess I_q in 1950, but not now). On the other hand, Dreyfus argues that tasks which characterize human intelligence best (e.g. the understanding of ambiguities in natural language) do so just <u>because</u> they cannot be atomized. Hence there is yet another possible suggestion about "intelligent" activity: namely, it may be that kind of activity which resists all attempts to program it in the style of (e2), as Turing's computer resists all attempts to identify it as a computer.

A final word about AI. The examples of AI methods in refs. 11 and 12 have it in common that they perform to their best advantage in problems which are given in terms of a set of initial data (probably incomplete), a set of <u>constraints</u> and a stated goal which may be reached in many possible ways under the constraints. Symbolic computing in physical sciences has by contrast confined itself in the past to problems where the steps are fully prescribed (as in a FORTRAN program) and the statement of the goal is therefore unimportant. I have mentioned three AI-type problems, two in physical sciences, which may be contrasted strongly with traditional problems [14] in particle physics, relativity and celestial mechanics. There is no reason why the scope of AI methods cannot be extended now to many interesting new exercises with applications in the physical sciences.

REFERENCES

1. H. L. Dreyfus, "What Computers can't do", Harper and Row, New York (1972).

2. A. M. Turing, Mind 59, 433 (1950). For a short commentary on some other objections to Turing's test, with references, see P. H. Millar, Mind 82, 595 (1973).

3. J. Moses, Comm. A. C. M. 14, 548 (1971).

4. R. H. Risch, Trans. Amer. Math. Soc. 139, 167 (1969).

5. A. Samuel, IBM J. Res. Develop. 11, 601 (1967).

6. M. M. Botvinnik, "Computers, Chess and Long-Range Planning", Springer-Verlag, New York (1970).

7. J. McCarthy et al., "LISP 1.5 Programmer's Manual", MIT Press, Cambridge, Massachusetts (1965).

8. J. A. Campbell, W. H. Jefferys, Celest. Mech. 2, 467 (1970).

9. H. C. Longuet-Higgins, A. Ortony, in "Machine Intelligence 3" (ed. D. Michie), Edinburgh University Press, Edinburgh (1968), p. 311.

10. C. Hewitt, in "Second International Joint Conference on Artificial Intelligence", British Computer Society, London (1971), p. 167.

11. J. R. Slagle,"Artificial Intelligence", McGraw-Hill, New York (1971).

12. N. J. Nilsson, "Problem-Solving Methods in Artificial Intelligence", McGraw-Hill, New York (1971).

13. D. Lurié, in "Computing as a Language of Physics", International Atomic Energy Authority, Vienna (1972), p. 529; A. C. Hearn, ibid., p. 567.

14. D. Barton, J. P. Fitch, Rep. Prog. Phys. 35, 235 (1972).

15. S. Amarel, in "Machine Intelligence 3" (ed. D. Michie), Edinburgh University Press, Edinburgh (1968), p. 131.

16. R. Penrose, M. A. H. MacCallum, Phys. Rep. 6C, 241 (1973).

17. B. Honig, E. A. Kabat, L. Katz, C. Levinthal and T. T. Wu, J. Mol. Biol. 80, 277 (1973).

18. P. Y. Chou, G. D. Fasman, J. Mol. Biol. 74, 263 (1973).

19. T. T. Wu, E. A. Kabat, J. Mol. Biol. 75, 13 (1973).

20. Global claims for the scope of AI are not now heard outside stories by science-fiction writers; Nilsson's operational definition of AI reflects the new modesty which is abroad in the subject (N.J.Nilsson, in "Third International Joint Conference on Artificial Intelligence", Stanford Research Institute, Menlo Park, California (1973), p.XIII.).

Acta Physica Austriaca, Suppl. XIII, 665–677 (1974)

CORRECTIONS TO LINEAR TRAJECTORIES FROM DUAL LOOP AMPLITUDES

by

A. D. KARPF*)
Institut für Elementarteilchenphysik
Freie Universität Berlin, Germany

GENERAL POLE SERIES

Obvious non-unitarity of dual-resonant amplitudes is the reason for going to higher - order (loop) corrections with appropriate square-root threshold singularities. This accounts for unitarity in a perturbative way [1] which means that upon writing the connected part iR of the S-matrix as a sum over all these loop amplitudes

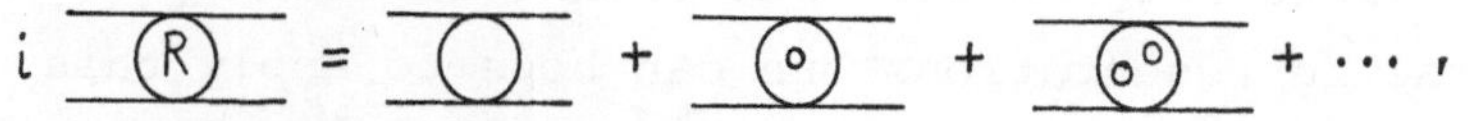

(1)

the unitarity relation [2]

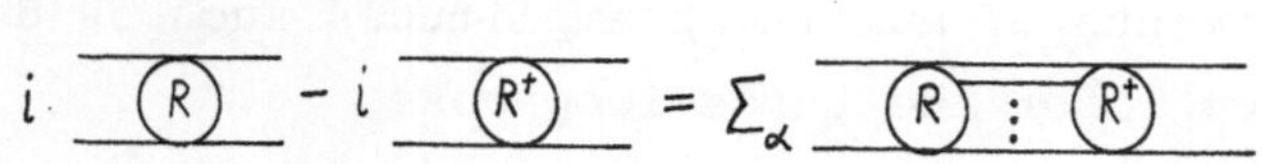

(2)

---

*) Seminar given at XIII. Internationale Universitätswochen für Kernphysik, Schladming, Austria, February 4-15, 1974.

is seen to split up into a perturbative series of relations

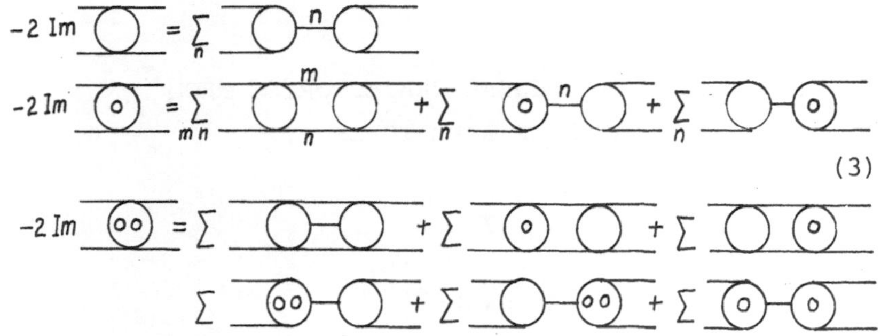
-2 Im
= ∑
n
-2 Im
= ∑
m n
+ ∑
+ ∑
-2 Im
= ∑
+ ∑
+ ∑
∑
+ ∑
+ ∑

(3)

where internal lines represent $\delta_+(q^2-m_n^2) \equiv \theta(q_o)\,\delta(q^2-m_n^2)$, m_n being the masses of the intermediate resonances.

This is as far as unitarity can be proven in the usual formulation of dual models since there is no method of actually performing the sum (1). Yet such a sum can have very interesting implications for the Regge-trajectories, their daughters and the residue functions due to some kinematical regions where the summation is possible approximately.

One may, as one possibility, resort to a very elegant method developed in Feynman-graph theory [4] to sum infinitely many ladder diagrams in order to get the renormalised Regge poles. Due to the similarity of the integral representations one can hope to apply this technique to sum all multipoles α_s^{-k} that arise from different segments of multiloop amplitudes such as depicted in fig. 1 for the three-loop case

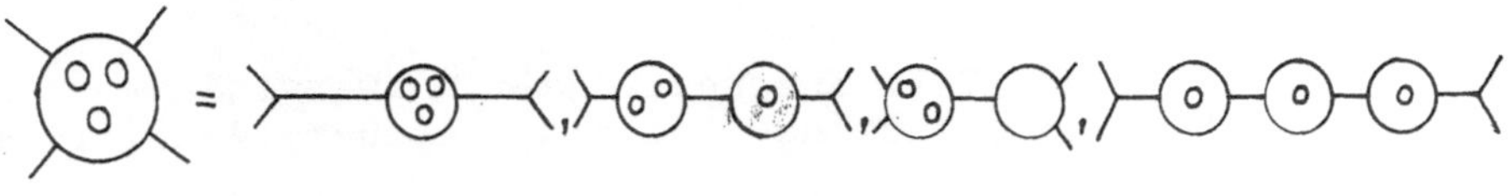

Fig. 1

The various segments are well defined by different integration regions of the integral representation. As in Feynman theory the result should emerge in the form [4]

$$\sum_{N=0}^{\infty} L_N = \sum_{N=0}^{\infty} \sum_{k=1}^{N} \frac{\beta_N^k}{\alpha^k} = G(\alpha) \frac{1}{\alpha - \bar{F}(\alpha)} G(\alpha) \tag{4}$$

with

$$\alpha - \bar{F}(\alpha) = 0$$

defining the (renormalised) positions of the Regge poles. Achieving this result would be of great help to determine the influence of unitary corrections.

ASYMPTOTIC POLE SERIES

While these ambitious results have so far remained beyond reach certain sums of selected contributions of the loop amplitudes have been performed. Kikkawa, Sakita and Virasoro [5] developed for their early model of dual loops a technique to sum the multiple poles in, say, the variable s in the asymptotic region of the variable t. This method was re-applied lateron [6] for the Abelian Integral formulation of dual loops after it had been shown how to factorise these expressions consistently.

In both cases a general N-loop amplitude [7] (e.g. for four external particles)

$$L_N = \int d\tau \; U^{-1} U_o^{-\alpha_o} U_s^{-\alpha_s} U_t^{-\alpha_t} \tag{5}$$

is analysed [6] with respect to the maximum number of possible intermediate poles (as exemplified in diagram e of fig. 1). This is achieved by exhibiting the integration variables $x_1 \dots x_{N+1}$ representing the propagators, and by expanding the remainder in the exponent (note that α_t is assumed large so that no expansion of the exponential is acceptable)

$$L_N = \int_0^1 dx_1 \dots dx_{N+1} (x_1 \dots x_{N+1})^{-\alpha_s - 1} \cdot \int d\bar{\tau}\, \bar{U}^{-1} \bar{U}_o^{-\alpha_o} \bar{U}_s^{-\alpha_s} e^{-\alpha_t x_1 \dots x_{N+1} \bar{E}_t} . \tag{6}$$

Poles in α_s obviously arise from the lower integration boundaries where $x_i^{-\alpha_s - 1}$ can become singular. A further expansion of the barred expressions $\bar{U}$ and $\bar{E}$ in terms of x_i shows that to lowest order these are now constants factorising into terms associated with the blobs of e.g. fig. 1e

$$L_N \approx \int_0^1 dx_1 \dots dx_{N+1} (x_1 \dots x_{N+1})^{-\alpha_s - 1} \cdot \int \prod_{\mu=1}^{N} d\rho_\mu u_\mu^{-1} u_o^{-\alpha_o} u_{s\mu}^{-\alpha_s} e^{-\alpha_t x_1 \dots x_{N+1} F_1 \dots F_N} . \tag{7}$$

A clever substitution [5] of the integration variables x_i allows to perform the integral and yields to lowest order for the N-loop term

$$L_N \approx \Gamma(-\alpha_s)\,\frac{1}{N!}(-\alpha_t)^{\alpha_s}[\Sigma(\alpha_s)\ln(-\alpha_t)]^N \tag{8}$$

whence one derives the sum

$$\sum_{N=0}^{\infty} L_N \approx \Gamma(-\alpha_s)(-\alpha_t)^{\alpha_s + \Sigma(\alpha_s)} \tag{9}$$

with

$$\Sigma(\alpha_s) \equiv \Sigma(\eta)\Big|_{\eta=\alpha_s} = \int d\rho\; u^{-1} u_o^{-\alpha_o} u_s^{-\alpha_s} F^{\eta}\Big|_{\eta=\alpha_s} \;. \tag{10}$$

This therefore introduces a corrected trajectory

$$\alpha_{corr}(s) \equiv \alpha(s) + \Sigma(s)$$

which ought to have proper cut structure while maintaining the usual asymptotic behaviour.

Before going into any further details, however, let us have a look at the approximations used so far. Apart from the x_i-expansion the integration result (9) is also only the first term of the exacter formula [6]

$$\sum_{N=0}^{\infty} L_N \approx \sum_{m=0}^{\infty} \frac{1}{m!}\,\partial_\eta^m$$

$$\{\Gamma(-\eta)(-\alpha_t)^{\Sigma(\eta)}\Sigma^m(\eta)\}_{\eta=\alpha_s}\;(-\alpha_t)^{\alpha_s} \tag{11}$$

which converges if the corrections to the Regge trajectory $\Sigma(\alpha_s)$ and $\partial_\eta \Sigma(\eta=\alpha_s)$ are small. Making use of the higher order corrections available in eq. (11) necessitates, however, a simultaneous use of higher order x_i-expansion terms.

DIRECT POLE SERIES

Another way to derive approximate pole corrections would be a direct sum of poles as used in a more general way in eq. (4). Approximating the procedure there means again to take only graphs of the sort of fig. (1e), hence to expand as in eq. (6). This time, however, all expansions should be performed downstairs so that in the expression (6) $U_t^{-\alpha_t}$ has to be replaced by unity. Since like in eq. (7) all other terms factorise the result of the x_i-integration this time becomes

$$L_N \approx \frac{1}{-\alpha_s} \left(\Sigma(0) \frac{1}{-\alpha_s}\right)^N \tag{12}$$

with

$$\Sigma(0) = \int d\rho \; u^{-1} u_0^{-\alpha_0} u_s^{-\alpha_s} \cdot 1 \tag{13}$$

where we note the replacement of F by unity.

The sum now becomes a geometric series and yields

$$\sum_{N=0}^{\infty} L_N \approx \frac{1}{-\alpha_s} \sum_{N=0}^{\infty} \left(\Sigma(0)\frac{1}{-\alpha_s}\right)^N = \frac{1}{-\alpha_s} \cdot \frac{1}{1-\Sigma(0)/(-\alpha_s)} = \frac{1}{-\alpha_s - \Sigma(0)} \tag{14}$$

which contains the previous correction function $\Sigma(\eta)$ at the pole position $\eta = \alpha_s = 0$. That this holds true for all pole positions can easily be seen by expanding the exponential factor of eq. (8) in the case of finite α_t as

$$L_N = \sum_{m=0}^{\infty} \frac{(-\alpha_t)^m}{m!} \int \prod_{\mu=1}^{N} d\rho_\mu u_\mu^{-1} u_{o\mu}^{-\alpha_o} u_{s\mu}^{-\alpha_s} F_\mu^m \cdot \tag{15}$$

$$\cdot \int_0^1 dx_1 \ldots dx_{N+1} (x_1 \ldots x_{N+1})^{-\alpha_s + m - 1}$$

which upon integration reads

$$L_N = \sum_{m=0}^{\infty} \frac{(-\alpha_t)^m}{m!} \Sigma^N(m) \left(\frac{1}{-\alpha_s+m}\right)^{N+1} .$$

The sum over all loops is again a geometric series which can be performed to yield

$$\sum_{N=1}^{\infty} L_N = \sum_{m=0}^{\infty} \frac{(-\alpha_t)^m}{m!} \frac{1}{-\alpha_s+m-\Sigma(m)} \tag{16}$$

which shows again $\Sigma(m)$ to be the correction of α_s at $\alpha_s = m$, thus shifting the pole to a position $\alpha_s+\Sigma(m)=m$.

The expansion (15) for finite values of α_t is, of course, inconsistent since equivalent powers of x_i have been neglected from other terms. Only for large α_t would the above expansion be formally true although it will certainly not converge.

What can, however, be taken from expression (16) is that it constitutes one of the terms which arise in a more complete and more correct expansion, and in particular that one with the maximum power m of the variable α_t. In a partial wave decomposition of the residue at $\alpha_s = m$ it is therefore the only term which contributes to the leading trajectory which is why this term is so important. As for the lower powers of α_t all sorts of other terms not included in (16) will contribute thus entering into the partial wave projection of the daughter trajectories. This is the reason why these corrections will in general differ from the leading trajectory correction $\Sigma(\alpha_s)$ so that these resonances may perhaps be shifted further to the left as suggested by experimental evidence. No theoretical results are yet available, though.

The fact that no progress has so far been made in this interesting field hinges on the need for more sophisticated methods to perform the higher order pole expansion of multiloop amplitudes and to interpret the resulting functions. Such methods have, however, recently been developed [6,9] so that numerical results for daughter trajectories should be available before long.

LEADING TRAJECTORY CORRECTIONS

So far there are only numerical results for the leading trajectory where it suffices in principle to factorise a one-loop amplitude which supplies all relevant terms. Hence the correction function $\Sigma(\alpha_s)$ has been long known [8] and has recently been reformulated [6] in terms of Abelian Integrals and Differentials. The two tantamount formulations are

$$\Sigma(\alpha_s)=\int_0^1 \frac{dy}{y(1-y)}\int_b^c \frac{d\zeta}{b-\zeta}\,\frac{b-c}{\zeta-c}\, f^{-4}\beta^{-2}\{\frac{y}{1-y}[\zeta_b^c\ a]e^{-\mathrm{Re}\Omega'_{ac}(ac)}\}^{-\alpha_o}$$

$$\{e^{\mathrm{Re}\Omega'_{ac}(ac)}\}^{-\alpha_s}\{(c-a)^2\partial_a\partial_c\,\mathrm{Re}\Omega_{a\infty}(oc)\}^{\eta}\Big|_{\eta=\alpha_s} \tag{17}$$

$$\Sigma(\alpha_s) = \int_0^1\frac{dx_1}{x_1(1-x_1)}\int_0^1\frac{dx_3}{x_3(1-x_3)}f^{-4}\beta^{-2}\{\frac{x_1}{1-x_1}\,\frac{x_3}{1-x_3}T_1'^{-1}\}^{-\alpha_o}\{T_1'\}^{-\alpha_s}$$

$$\{\partial_{x_2}\partial_{x_4}\,T_2\}^{\eta}\Big|_{\eta=\alpha_s,\ x_2=x_4=1} \tag{18}$$

where square brackets are cross-ratios in an obvious notation and

$$f \equiv \prod_{n=0}^{\infty}(1-K^n); \qquad \beta \equiv \ln K; \qquad K \equiv x_1x_3\ . \tag{19}$$

The identity of eqs. (17) and (18) can be checked by using the usual definition of Chan variables [10]

$$x_1 = y; \quad x_2 = 1-y_2; \quad x_3 = [b \, {}^{a}_{c} \, \zeta]; \quad x_4 = (1-y_4) \qquad (20)$$

for the arrangements of fig. 2 where a,b,ζ and c are consecutive Koba-Nielsen variables. It also becomes clear that e.g. x_2 is the dual variable of y_2.

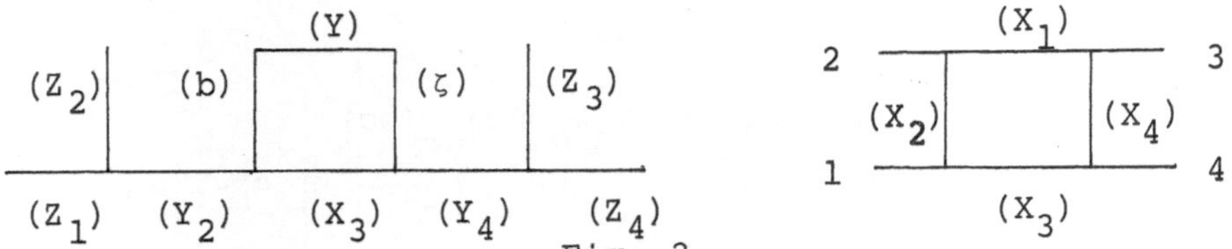

Fig. 2

Furthermore [11]

$$\{T_1'\} = \exp \mathrm{Re}\, \Omega'_{ac}(ac) \Big|_{x_2=x_4=1} = |\psi'(1)/\psi(\tfrac{c}{a})|^{-2} \qquad (21)$$

becomes $$|1/\psi(x_3)|^{-2} \equiv \exp h_1(x_1 x_3) \qquad (22)$$

of ref. 8 if one chooses $\zeta = \infty$, $b = 0$ after having replaced $d\zeta$ by e.g. $dz_4 \to dc$.

The properties of this function $\Sigma(s)$ have been investigated by several authors. As for its discontinuity it was proven in a qualitative discussion [12] that it has the correct behaviour

$$\mathrm{Im}\, \alpha(s) \sim (s-s_k)^{\frac{1}{2} + \alpha(s_k)} \qquad (23)$$

at the trhesholds s_k. Numerical results were only given in arbitrary units. The procedure thereby was to remove the integration over the loop momentum and to use the Cutkosky technique to derive the discontinuity. It would be nice to derive these results directly by using the factorisation properties [9] of the Abelian Integrals.

As for the asymptotic behaviour it can be shown rather easily that for $s \to -\infty$ $\Sigma(s)$ behaves as $(-s)^{1/6}$. The proof that this asymptotic behaviour holds for all directions of the complex s-plane except for parallels to the positive real axis is much more involved, though possible and yields as leading term [13]

$$\Sigma(\alpha_s) \underset{|s|\to\infty}{\to \sim} (-\alpha_t)^{\frac{1}{6}} \ln^{-2}(-\alpha_t)$$
$$\Gamma(-\tfrac{1}{6}) \int_0^1 dv[v(1-v)]^{\alpha_0 - 1} \left(\frac{\sin\pi v}{\pi}\right)^{-2\alpha_0} (2 \sin\pi v)^{\frac{2}{3}} \; . \tag{24}$$

Beyond these two isolated results no complete numerical analysis is available, nor has the real part of $\alpha(s)$ been evaluated although KSV [5] have long ago scetched what they expect to emerge from such a calculation (fig. 3).

Therefore much work has still to be done to discuss the most interesting impacts of the dual loop terms on the trajectory corrections reliably and completely. In particular in view of the many new models [14] and their available one loop corrections [15] it would be helpful to have better numerical results for a phenomenological comparison. For trajectories are the most direct access to pro test unitarity corrections. This becomes an especially important problem with the recent advent of field theoretic dual resonant models [16] which claim to incorporate unitarity automatically.

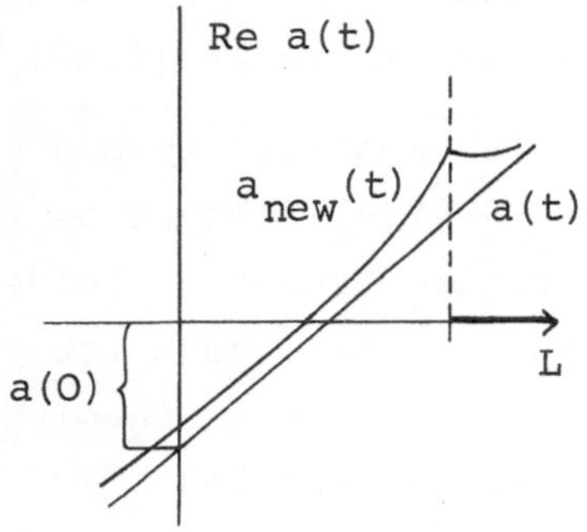

Fig. 3

REFERENCES

1. see e.g. D.J. Gross, A. Neveu, J. Scherk, J. H. Schwarz, Phys. Rev. D2, 697 (1970).

2. R. J. Eden, P. V. Landshoff, D. I. Olive, J. C. Polkinghorne, The Analytic S-Matrix, Cambridge University Press, 1966.

3. see e.g. ref. 2, Chapt. 3.6.

4. K. Kikkawa, B. Sakita, M. A. Virasoro, Phys. Rev. 184, 1701 (1969).

6. A. D. Karpf, H. J. Liehl, Factorisation of Dual Loop Amplitudes II, Nucl. Phys. to be published.

7. C. Lovelace, Phys. Lett. 32B, 703 (1970).
V. Alessandrini, Nuovo Cimento 2A, 321 (1971).
V. Alessandrini, D. Amati, Nuovo Cimento 4A, 793 (1971).

A. D. Karpf, H. J. Liehl, H. F. Schuhmacher, Nucl. Phys. B56, 565 (1973).

8. A. Neveu, J. Scherk, Phys. Rev. D1, 2355 (1970).

9. A. D. Karpf, Factorisation of Dual Loop Amplitudes I, Nucl. Phys. to be published.

10. formula (3.11) and (3.12) of ref. 9 for a one-loop amplitude.

11. see e.g. ref. 8 formula (A.1) and (A.3).

12. H. J. Kaiser, F. Kschluhn, E.Wieczorek, Discontinuity of the Regge Trajectory in the DRM I, Berlin-Zeuthen preprint PHE 71-3.

13. H. Dorn, H. J. Kaiser, Asymptotic Behaviour of the Planar One Loop Correction..., Berlin-Zeuthen preprint PHE 72-15.

14. J. H. Schwarz, Physics Report 8, 275 (1974).

15. L. Brink, D. Olive, Nucl. Phys. B58, 237 (1973).

16. M. Kaku, K. Kikkawa, Field Theory of Relativistic Strings I, City Univ. N. Y. preprint 1974.

Acta Physica Austriaca, Suppl. XIII, 679–709 (1974)

THE N-BODY PROBLEM

by

W. SANDHAS*)
Physikalisches Institut
Universität Bonn
Germany

1. INTRODUCTION

Two years ago, in a lecture given at the Schladming Conference 1972, I reviewed the present status of the quantum mechanical three-body problem [1]. It was the main intention of this survey to illustrate the way in which the properties of the integral equations, studied in this field, are related to basic concepts of multichannel collision theory. Such a consideration makes the typical difficulties of older attempts rather transparent. Moreover, the modern approaches initiated by Faddeev's work [2], are easily understood as natural consequences of general features of the multichannel theory.

A deeper understanding of these equations is, how-

*) Seminar given at XIII. Internationale Universitätswochen für Kernphysik, Schladming, Austria, February 4-15, 1974.

ever, not the only motivation for such an investigation. In addition we may expect that our point of view represents a useful basis for a systematical transition to the general N-body problem. It is the purpose of my talk to show how this can be accomplished [3].

Before going into any details, I emphasize that other proposals for solving the N-body problem exist. Most of them do not show the physically convincing structure of the three-body Faddeev-type equations. Even if mathematically correct, they only generalize some of their special properties. In my opinion, the simple underlying structure which, as emphasized above, should reflect the typical aspects of multichannel theory is usually not pointed out.

2. SOME ASPECTS OF MULTICHANNEL COLLISION THEORY

As mentioned above, we intend to derive N-body integral equations which represent a natural realization of typical aspects of the multichannel collision theory. For the three-body case the basic ideas of such an approach have been explained in our previous lecture [1]. In this section we repeat some considerations, given there, which are of relevance for our present purpose*). This also serves to fix our notation.

---

*) Since the present article generalizes the considerations of Ref. [1], we only refer to this lecture when mentioning further details. A list of original papers is given there.

Two-body scattering

In the two-body (single channel) case we start from the splitting

$$H = H_o + V \tag{2.1}$$

of the Hamiltonian into a kinetic energy part and a potential.

This allows us to introduce eigenfunctions of H_o and H, i.e., plane waves $|\vec{p}\rangle$ describing the relative movement of the colliding particles, and scattering states $|\vec{p}\rangle^{(+)}$ which asymptotically behave like plane waves plus outgoing spherical waves. These states are related by

$$|\vec{p}\rangle^{(+)} = \Omega^{(+)} |\vec{p}\rangle \ , \tag{2.2}$$

via the Møller operator $\Omega^{(+)}$, [4].

The transition probability for a collision process, shown in Fig. 1,

$\vec{p}\,'$  $\vec{p}$

Fig. 1

with relative momenta $\vec{p}$ and $\vec{p}\,'$ before and after the collision, follows from the S-matrix

$$S_{\vec{p}\,',\vec{p}} = {}^{(+)}\langle\vec{p}\,'| \ S \ |\vec{p}\rangle^{(+)} . \tag{2.3}$$

The explicit definition of S by means of Møller operators (see Eq. (6.10) of Ref. 1) is of no interest for our present consideration. What we want to emphasize is, that Eq. (2.3) represents the S-matrix in the Heisenberg picture [5].

For general investigations this representation is useful. Practical approaches, however, are usually based on the interaction picture which follows by inserting Eq. (2.2) in (2.3),

$$S_{\vec{p}',\vec{p}} = \langle \vec{p}'|\ \Omega^{(+)\dagger}\ S\ \Omega^{(+)}\ |\vec{p}\rangle$$

$$= \langle \vec{p}'|\ S_I\ |\vec{p}\rangle\ . \qquad (2.4)$$

Three-body collisions

The difference between both pictures is of particular relevance in the multichannel case. The simplest example of such a situation is the three-body problem. Here processes of the form shown in Fig. 2 occur.

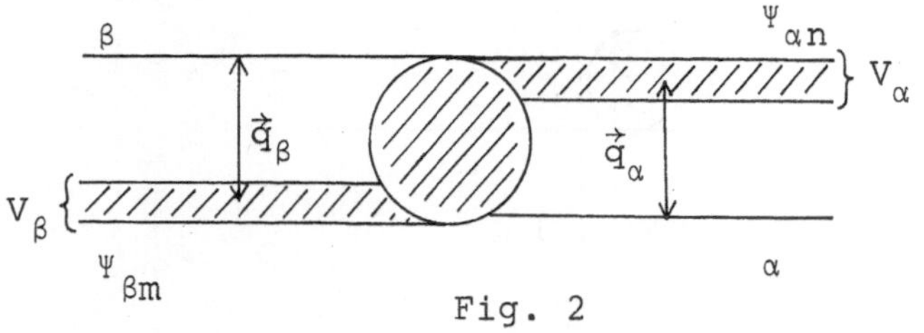

Fig. 2

As in the single channel case, the relative movement of the two colliding fragments is described by a plane wave $|\vec{q}_\alpha\rangle$. Since, however, one of the particles is a bound state $|\psi_{\alpha n}>$, the incoming configuration is now a product[+)],

$$|\Phi_{\alpha n}> = |\psi_{\alpha n}> |\vec{q}_\alpha>. \tag{2.5}$$

Such channel states which replace the plane waves of the two-body case, are eigenfunctions of the channel Hamiltonians[*)]

$$H_\alpha = H_o + V_\alpha \quad . \tag{2.6}$$

The scattering states corresponding to these free asymptotic configurations are introduced in analogy to (2.2), via Møller operators

$$|\Psi_{\alpha n}^{(+)}> = \Omega_\alpha^{(+)} |\Phi_{\alpha n}> \quad . \tag{2.7}$$

They are eigenstates of the total Hamiltonian

$$H = H_o + \sum_{\gamma=1}^{3} V_\gamma \quad . \tag{2.8}$$

It should be emphasized that, due to the existence of several "free" Hamiltonians H_α, a corresponding set of Møller operators $\Omega_\alpha^{(+)}$ has to be introduced (compare definition (4.10) in Ref. 1).

+) The index n collectively denotes the quantum numbers of the subsystem bound states.

*) In our notation V_α is the potential responsible for the binding of the incoming composite particle.

The S-Matrix describing the process shown in Fig. 2, is defined in the Heisenberg picture as a consequent generalization of Eq. (2.3):

$$S_{\beta m,\alpha n} = <\psi^{(+)}_{\beta m}|\; S \;|\Psi^{(+)}_{\alpha n}> \quad . \qquad (2.9)$$

Repeating the procedure which replaces (2.3) by (2.4) we insert (2.7) in (2.9). This yields

$$S_{\beta m,\alpha n} = <\Phi_{\beta m}|\Omega_{\beta}^{(+)\dagger}\; S\Omega_{\alpha}^{(+)}|\Phi_{\alpha m}> = <\Phi_{\beta m}|\; S_{\beta\alpha}|\; \Phi_{\alpha m}> . \qquad (2.10)$$

The S-matrix is now given by operators sandwiched between free (channel) states. Consequently we call this representation, in analogy to the single channel case, the "S-matrix in the interaction picture".

The essential feature of (2.10) is, that a set of S-operators, $S_{\beta\alpha}$, is needed*). In this context we should emphasize that practical approaches, e.g., the ones based on integral equations, require the transition to the interaction picture. The occurrence of a set of S-operators in this picture implies that a system of equations is necessary for the description of multichannel processes.

The N-body case

We have explained the multichannel formalism for the special case of three bodies. When introducing cor-

*) Note that the channel states (2.5), in contrast to the scattering states (2.7), form a non-orthogonal set.

responding channel states and Møller operators, the whole formalism remains valid also in the case $N \geq 4$. But a decisive difference occurs in detail. In contrast to $N = 3$, where one of the colliding particles is elementary, both particles may now be composite. This is demonstrated in Fig. 3, where some possible four-body re-arrangement collisions are shown (with an evident labelling of the channels by σ and ρ).

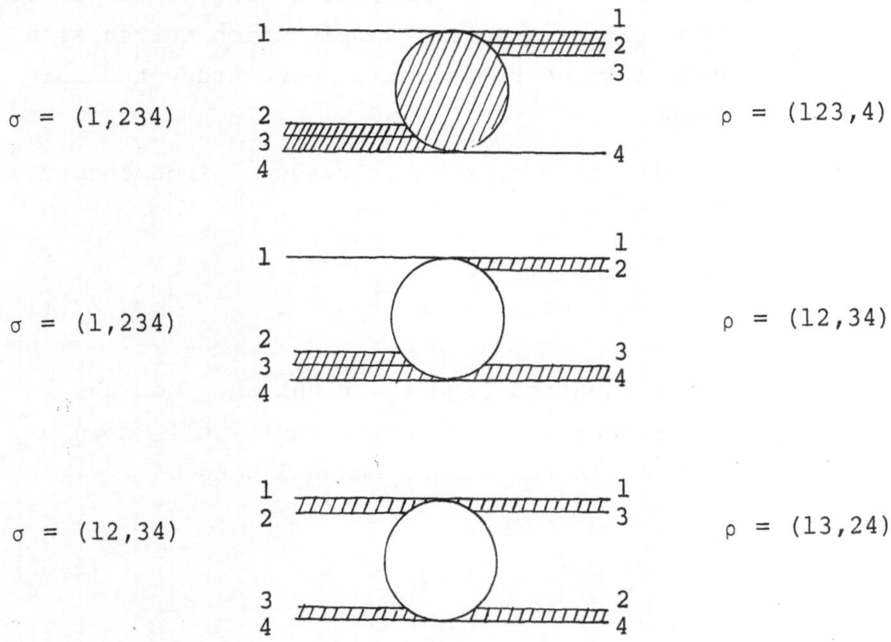

Fig. 3

The consequence of this additional complexity, typical for all collisions with $N \geq 4$, is that the transition from the three-body to the N-body integral equation formalism is not straightforward.

Technically this difference originates from the fact that now the sum of channel potentials, responsible for the binding of the fragments, exceeds the sum of all potentials available.

Unitarity relation

We have emphasized already that in the multichannel case systems of equations are expected when going over to the interaction picture. As an example which serves as a guide to the following investigations, we study the unitarity relation.

In the Heisenberg picture it reads*), independently of the particle number,

$$S^{\dagger}S = S\,S^{\dagger} = 1\ . \tag{2.11}$$

Multiplying this equation from the right and from the left with Møller operators and using their well known properties, we get in the single channel case

$$S_I^{\dagger}\,S_I = S_I\,S_I^{\dagger} = 1\ , \tag{2.12}$$

i.e., an equation identical to (2.11) with the only difference that the original Heisenberg S-operator is replaced by the one of the interaction picture.

In the multichannel case the analogous procedure yields+)

---

*) Eq.(2.11) is valid in the subspace of scattering states.

+) This holds when applied onto channel states.

$$\sum_{\gamma} S^{\dagger}_{\gamma\beta} \, S_{\gamma\alpha} = \sum_{\gamma} S_{\beta\gamma} \, S^{\dagger}_{\alpha\gamma} = \delta_{\beta\alpha} \quad . \tag{2.13}$$

Thus we arrived, as expected, at a system of equations. The decisive point, however, is that it shows the same algebraic structure as its Heisenberg picture analogon, and as equation (2.12) valid in the single channel case.

Consequences

In view of this consideration, we conclude that a systematical N-body generalization of two-body equations should be accomplished by introducing systems of relations which preserve their algebraic structure. In the following we study, in particular, the correct N-body analogue of the single channel Lippmann-Schwinger equation.

Such a concept simplifies considerably the algebraic complexity of the problem. It has, however, an additional advantage. Since the occurring operator valued N-body matrices are introduced as natural generalizations of the corresponding two-body operators, our approach yields automatically their correct physical interpretation.

3. TWO- AND THREE-BODY INTEGRAL EQUATIONS

Single channel processes

It is well known that the transition operator T(z),

related to the S-matrix by

$$S_{\vec{p}'\vec{p}} = \delta(\vec{p}' - \vec{p}) - 2\pi i\ \delta(E'-E) < \vec{p}'\ |T(E+io)|\ \vec{p}>, \quad (3.1)$$

fulfills the Lippmann-Schwinger equation*)

$$T(z) = V + V\ G_o(z)\ T(z)\ . \quad (3.2)$$

Written in momentum space, this operator identity represents an integral equation for the off-shell amplitudes $<\vec{p}'|T(z)|\vec{p}>$. I.e., it defines an extension of the on-shell amplitudes which already determine $S_{\vec{p}'\vec{p}}$. Note that the energy-δ-function in (3.1) implies $p'^2/2\mu = E' = E = p^2/2\mu$.). The structure of Eq. (3.2) is shown in Fig. 4

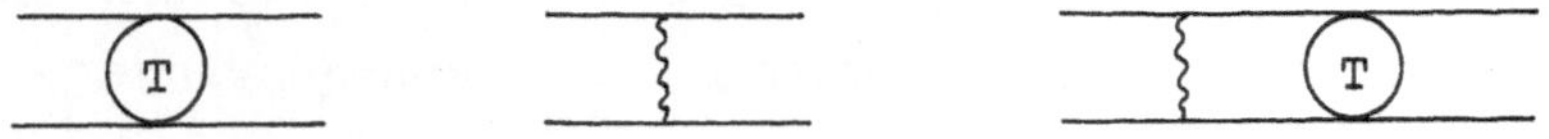

Fig. 4

Three-body collisions

In the three-body-case the analogon of Eq. (3.1) is

$$S_{\beta m,\alpha n}(\vec{q}_\beta{}', \vec{q}_\alpha) = \delta_{\beta\alpha}\delta_{mn}\ \delta(\vec{q}_\beta{}' - \vec{q}_\alpha) \ -$$

---

*) The potential V is introduced in Eq.(2.1). $G_o(z) = (z-H_o)^{-1}$ is the free Green function (resolvent of H_o). For further details see Sec. 7 of Ref. 1.

$$- 2\pi i\, \delta(E'_{\beta m} - E_{\alpha n}) \quad \langle \Phi_{\beta m} | U_{\beta\alpha} \quad (E_{\alpha n} + io) | \Phi_{\alpha n} \rangle \; . \qquad (3.3)$$

Here the $U_{\beta\alpha}$ are suitable off-shell extensions of the on-shell transition operators. They fulfil the Faddeev-type integral equations [6]

$$U_{\beta\alpha} = (1 - \delta_{\beta\alpha})\, G_o^{-1} + \sum_{\gamma \neq \beta} T_\gamma\, G_o\, U_{\gamma\alpha} \; . \qquad (3.4)$$

The kernel of this system of equations is determined by the two-body transition operators T_γ which follow by solving the Lippmann-Schwinger equation (3.2). The additional index γ is necessary to indicate the subsystem where T_γ acts.

Structure of Eq. (3.4)

From our point of view it is essential that Eqs. (3.1) and (3.3) represent the S-matrix in the interaction picture. Consequently, as anticipated by the general arguments given in Sec. 2, the two-body transition operator T is replaced by a set of three-body operators $U_{\beta\alpha}$.

Moreover, writing Eq. (3.4) in the form

$$U_{\beta\alpha} = (1 - \delta_{\beta\alpha}) G_o^{-1} + \sum_{\gamma} (1 - \delta_{\beta\gamma}) G_o^{-1}\; G_o\; T_\gamma\; G_o\; U_{\gamma\alpha} \qquad (3.5)$$

we realize that it can be considered as a matrix equation

$$\mathbf{T} = \mathbf{V} + \mathbf{V}\mathbf{G}_0\mathbf{T} \tag{3.6}$$

with operator valued matrix elements

$$\mathbf{T}_{\beta\alpha} = U_{\beta\alpha} \tag{3.7}$$

$$\mathbf{V}_{\beta\alpha} = (1 - \delta_{\beta\alpha})\, G_0^{-1} \tag{3.8}$$

$$\mathbf{G}_{0,\beta\alpha} = \delta_{\beta\alpha}\, G_0\, T_\alpha\, G_0 \tag{3.9}$$

We have pointed out in the preceding section that a consequent transition to the three-body case should replace the two-body equations by matrix equations of the same algebraic structure. In this sense Eq. (3.6), i.e., the Faddeev-type equations (3.4), represent the correct generalization of the two-body Lippmann-Schwinger equation.

This interpretation is by no means formal.The "potential" (3.8) is evidently the off-shell extension of an exchange potential, and the "free Green function" (3.9) can be regarded as the free propagator of an elementary and a composite particle (compare Sec. 10 of Ref. 1). The general structure of Eq. (3.6) becomes more transparent by giving it in form of diagrams (Fig. 5). This also shows the structural analogy to the single channel case (compare Fig. 4).

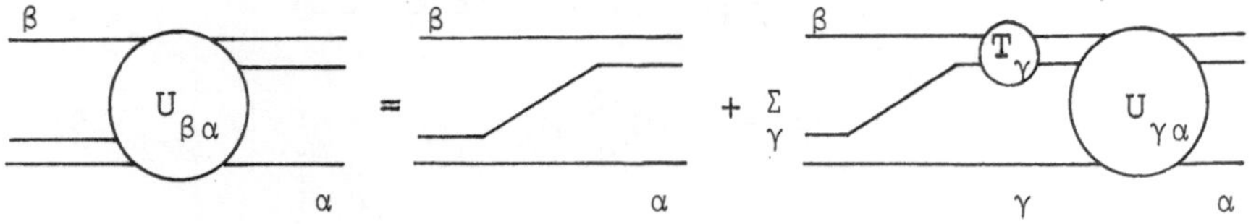

Fig. 5

Uniqueness

First attempts to find the multichannel analogue of the Lippmann-Schwinger equation handled the various channels independently. Such equations are, in contrast to the coupled set of equations introduced by Faddeev, non-unique [7]. The uniqueness of Eqs. (3.4) is a further indication that they represent the correct generalization of the Lippmann-Schwinger equation (3.2) [8].

A general rule

Our interpretation of the Faddeev-theory as a natural generalization of the single channel case, has a useful consequence. Algebraic calculations in the three-body case are considerably simplified by the following rule:

> The whole formal structure of the two-body theory is translated into the three-body case, by replacing the two-body amplitude, the potential and the free Green function by their three-body analogues (3.7), (3.8) and (3.9).

This rule is, of course, a consequence of the general properties discussed in Sec. 2.

Examples

As an application we sketched in Sec. 10 of Ref.[1]

a structurally simple derivation of the three-body unitarity relations. Moreover, replacing in the representation

$$G = G_o + G_o T G_o \tag{3.10}$$

of the two-body resolvent G the operators T and G_o by their three-body analogues (3.7) and (3.9), we define an operator $\mathbf{G}$ which, according to our rule, represents the correct resolvent of the three-body theory. The elements of this matrix operator are (see Sec. 9 of Ref. 1)

$$\mathbf{G}_{\beta\alpha} = G_o \{\delta_{\beta\alpha} V_\alpha + V_\beta G V_\alpha\} G_o = G_o M_{\beta\alpha} G_o . \tag{3.11}$$

I.e., apart from the unessential factors G_o, they are identical to the $M_{\beta\alpha}$ introduced by Faddeev [2].

Since all our basic two- and three-body relations show a one-to-one correspondence, it is trivial that also the analogue of the resolvent equation holds

$$\mathbf{G} = \mathbf{G_o} + \mathbf{G_o} \mathbf{V} \mathbf{G}. \tag{3.12}$$

Inserting here the explicit representations (3.8), (3.9) and (3.11), we arrive at the set of equations

$$M_{\beta\alpha} = \delta_{\beta\alpha} T_\alpha + T_\beta \sum_{\gamma\neq\beta} G_o M_{\gamma\alpha} \tag{3.13}$$

given first in Ref. 2. In other words, these original

Faddeev equations represent the correct generalization of the resolvent equation.

With respect to our general goal, this result has a direct consequence. As emphasized already, we want to perform the transition from three to four bodies in an algebraically equivalent manner. Therefore, as in the three-body case, the kernel of the four-body equations should contain the subsystem transition operators. The above discussion shows that these are the $U_{\beta\alpha}$, given by Eq. (3.4), not the Faddeev operators $M_{\beta\alpha}$.

4. FOUR-BODY EQUATIONS

The interpretation of the Faddeev theory, given in the preceding section, is of advantage already in the three-body case. For our purpose, however, it decisive. It allows us to go over immediately to the four-body problem and then, step by step, even to the general N-body situation.

In this context we recall that the kernel of the three-body equations (3.4) is given by $T_\gamma G_o$. I.e., Eq. (3.4) shows in which way the two-body amplitudes T_γ determine the three-body transition operators $U_{\beta\alpha}$. Applying the same algebra onto our two-body formulation of the three-body theory, we go a step further. In the resulting four-body equations the transition operators are again determined by the subsystem amplitudes. These, however, are now the operators $\mathbf{T}^\tau$ defined by (3.7),

$$\mathbf{T}^{\tau}_{\beta\alpha} = U^{\tau}_{\beta\alpha} \quad . \tag{4.1}$$

As in the three-body case, where an index γ indicates the sub-system in which T_γ acts, we have to label $U_{\beta\alpha}$ by an additional index τ. If, e.g., $\tau = (123,4)$, the $U^{\tau}_{\beta\alpha}$ are the three-body transition operators restricted to the (123)-subsystem.

In view of this discussion it is almost evident that four-body equations of the algebraic form of (3.4) hold*),

$$\mathbf{U}^{\sigma\rho} = (1 - \delta_{\sigma\rho})\, \mathbf{G}_o^{-1} + \sum_{\tau \neq \sigma} \mathbf{T}^{\tau}\mathbf{G}_o\, \mathbf{U}^{\tau\rho} . \tag{4.2}$$

In order to get this result more explicit, we insert the definitions (4.1) and (3.9). This yields

$$U^{\sigma\rho}_{\beta\alpha} = (1 - \delta_{\sigma\rho})\, \delta_{\beta\alpha}\, G_o^{-1}\, T_\alpha^{-1}\, G_o^{-1}$$

$$+ \sum_{\tau\neq\sigma} \sum_{\gamma} U^{\tau}_{\beta\gamma}\, G_o\, T_\gamma\, G_o\, U^{\tau\rho}_{\gamma\alpha} \quad . \tag{4.3}$$

Thus, the kernel of the four-body equations is built up by the three-body and two-body transition amplitudes. The latter ones come into the play via the "free Green function" (3.9).

---

*) The channel indices ρ, σ, τ indicate fragmentations as demonstrated in Fig. 3.

Structure of the kernel

As mentioned above, the subsystem channel index may be $\tau = (123,4)$ labelling the collision of an elementary particle and a three-body bound state. In this case the kernel of Eq. (4.3) is of the form shown in Fig. 6 (for $\beta = 1$ and $\gamma = 3$)

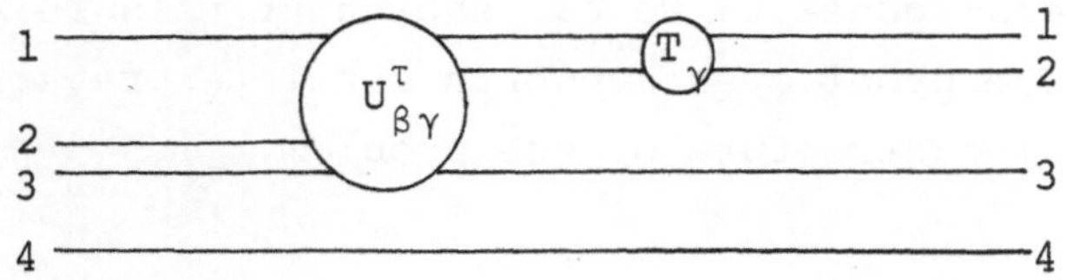

Fig. 6

However, we have emphasized already that in the four-body case also fragmentations of the form $\tau = (12,34)$, corresponding to the collision of two bound states, have to be considered. A more detailed inspection of our derivation shows, indeed, that the sum in (4.3) necessarily runs over such configurations, too. In this case $U^{\tau}_{\beta\alpha}$ is again given by Eq. (3.4), but it has no longer the simple meaning of the transition operator of a three-body subsystem. This part of the kernel of Eq. (4.3) has the form shown in Fig. 7.

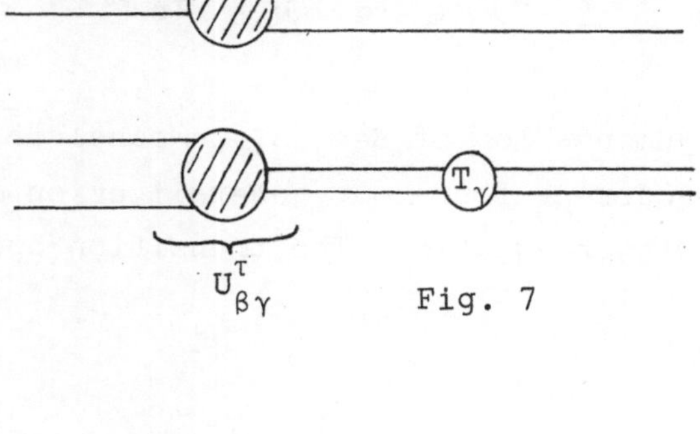

Fig. 7

Comment

Let us finally add a comment on our derivation of the four-body equations (4.3). At first sight it may look like a conjecture, based on structural analogies. In particular, it is not a priori clear that the $U^{\sigma\rho}_{\beta\alpha}$, sandwiched between wave functions, yield the four-body transition amplitudes. However, going into more details of our general concept, we can show that this is really the case. The proof even supports our interpretation of the underlying structure of the problem.

5. N-BODY EQUATIONS

Our method for going over from two- to three- and from three- to four-body problems is, evidently, the beginning of a general scheme which step by step incorporates all subsystem transition operators in the kernel of the N-body integral equations.

In order to make this statement more transparent we give some details of the five-body problem. In analogy to the replacement of Eq. (3.4) by (3.5), we may write our four-body equation (4.2) in the form

$$\mathbf{U}^{\sigma\rho} = (1-\delta_{\sigma\rho})\,\mathbf{G}_o^{-1} + \sum_{\tau} (1-\delta_{\sigma\tau})\,\mathbf{G}_o^{-1}\,\mathbf{G}_o\,\mathbf{T}^{\tau}\,\mathbf{G}_o\,\mathbf{U}^{\tau\rho} \qquad (5.1)$$

Repeating the argumentation of Sec. 3, we conclude that Eq. (4.2) is the correct four-body generalization of the Lippmann-Schwinger equation. The transition operators, generalized potentials and free Green functions are now (compare the definitions (3.7), (3.8) and (3.9))

$$\mathbf{U}^{\sigma\rho} \rightarrow U^{\sigma\rho}_{\beta\alpha} \tag{5.2}$$

$$(1-\delta_{\sigma\rho})\,\mathbf{G}_o^{-1} \rightarrow (1-\delta_{\sigma\rho})\,\delta_{\beta\alpha}\quad G_o^{-1}\,T_\alpha^{-1}\,G_o^{-1} \tag{5.3}$$

$$\delta_{\sigma\rho}\quad \mathbf{G}_o\,\mathbf{T}^{\tau}\,\mathbf{G}_o \rightarrow \delta_{\sigma\rho} G_o T_\beta\; G_o U^{\tau}_{\beta\alpha}\; G_o T_\alpha G_o \,. \tag{5.4}$$

Generalizing the argumentation given in Sec. 4 we conclude that the kernel of the five-body equations is a product of the four-body transition operators (5.2) and the generalized free Green function (5.4). Thus, the five-body kernel has the structure demonstrated in Fig.8.

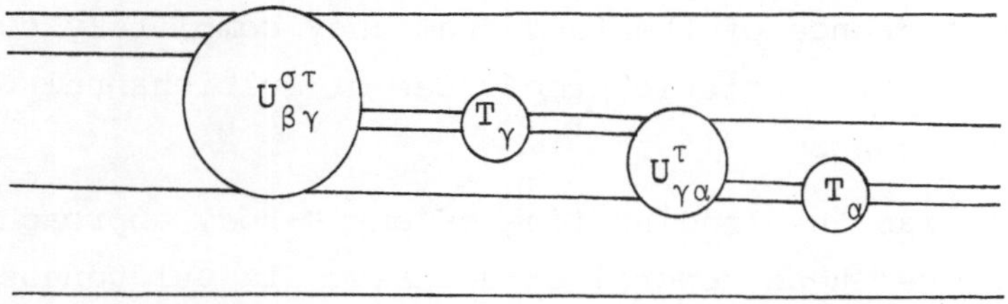

Fig. 8

Note that the two- and three-body transition operators come into the play via the generalized free Green function (5.4). More generally the kernel of the N-body equation contains the (N-1)-body transition operator and a "free Green-function" which is built up by the transition operators of the subsystems with lower particle numbers.

Break-up channels

Up to now we have only discussed the re-arrangement collision of two composite particles. However, it is well known that the three-body Faddeev theory also describes break-up processes. Since our N-body theory represents a consequent generalization of the three-body approach, it can also be extended to the case of transitions into final states with more than two particles.

Uniqueness

As emphasized in Sec. 3 and shown in Ref. 1 the uniqueness of the Faddeev equations can be understood as a consequence of the fact that they completely contain the characteristic properties of multichannel collision theory.

It was the leading idea of our N-body approach to preserve these general structures. Its uniqueness is, therefore, guaranteed.

Moreover, since our N-body theory is algebraically equivalent to the three-body Faddeev equations, the detailed proofs of uniqueness, given in this context [2], can also be taken over to the general case.

6. POLE APPROXIMATION

The uniqueness of the three-body Faddeev equations

is rather often considered as its most important feature. However, from the physical point of view (and also for many practical approaches) the occurrence of the two-body transition operators T_γ in the kernel is as well decisive. Approximating them by their bound state and resonance poles we can immediately study in a simple way the effect of dominant subsystem structures. Moreover such an approximation represents the first step of powerful iteration schemes.

At a pole T_γ has the separable form

$$T^s_\gamma(z) = |\gamma> t_\gamma(z) < \gamma| \,, \tag{6.1}$$

where, in the case of a bound state $|\psi_\gamma>$, the "form factors" are $|\gamma> = V_\gamma |\psi_\gamma>$. Near the binding energy E_γ, the function $t_\gamma(z)$ is of the form

$$t_\gamma(z \sim E_\gamma) \sim \frac{1}{z - E_\gamma} \,. \tag{6.2}$$

Three-body case

In order to simplify our notation we assume that only one pole exists in each channel. Inserting then (6.1) in (3.4) or (3.5) and sandwiching the resulting equation with $<\beta| G_o$ and $G_o |\alpha>$,

$$<\beta| G_o \left\{ U_{\beta\alpha} = (1-\delta_{\beta\alpha}) G_o^{-1} + \sum_\gamma (1-\delta_{\beta\gamma}) G_o^{-1} G_o \underbrace{T_\gamma}_{|\gamma> t_\gamma <\gamma|} G_o U_{\gamma\alpha} \right\} G_o |\alpha>, \tag{6.3}$$

we get the matrix equation

$$\tilde{\mathbf{T}} = \tilde{\mathbf{V}} + \tilde{\mathbf{V}}\, \tilde{\mathbf{G}}_o\, \tilde{\mathbf{T}} \quad . \tag{6.4}$$

The occurring matrix elements are

$$\tilde{\mathbf{T}}_{\beta\alpha} = \langle\beta|\, G_o\, U_{\beta\alpha}\, G_o\, |\alpha\rangle \tag{6.5}$$

$$\tilde{\mathbf{V}}_{\beta\alpha} = (1 - \delta_{\beta\alpha})\, \langle\beta|\, G_o\, |\alpha\rangle \tag{6.6}$$

$$\tilde{\mathbf{G}}_{\mathbf{o},\beta\alpha} = \delta_{\beta\alpha}\, t_\alpha \quad . \tag{6.7}$$

Since in these definitions the original three-body operators are only sandwiched between two-body wave functions, they are still operators in the space of relative momentum states $|\vec{q}_\alpha\rangle$. In other words, by the pole approximation the original three-body problem is reduced to effective two-body equations.

We emphasize that the pole approximation preserves the Lippmann-Schwinger structure of the original three-body equations (3.5)*). This is essential not only with respect to technical details of calculations, but, in particular, for the physical interpretation. Eq. (6.4) has the structure shown in Fig. 9 (compare also Fig. 5).

The half circles, of course, indicate the form factors $|\gamma\rangle$ occurring in Eq. (6.1).

---

*) Compare also the effective two-body operators (6.5), (6.6) and (6.7) with the corresponding three-body operators (3.7), (3.8) and (3.9).

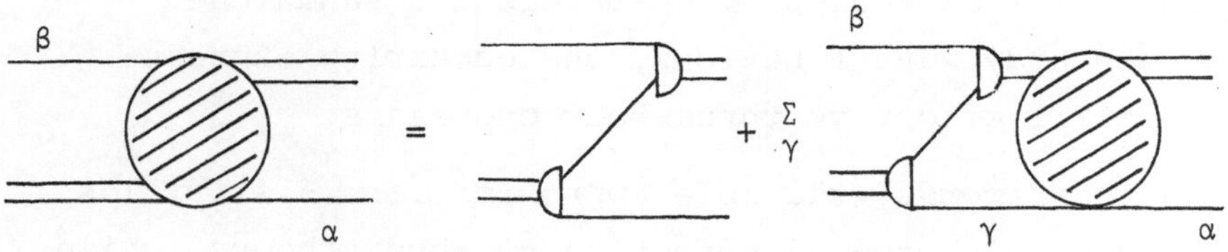

Fig. 9

Four-body case

It was the decisive feature of our four-body equations that in their kernel the subsystem operators occur. This suggests to apply also in this case pole approximations. In a first step we insert again the two-body pole approximation (6.1) in Eq. (4.3). Sandwiching it furthermore between $<\beta|\, G_o$ and $G_o\, |\alpha>$,

$$<\beta|\, G_o \left\{ U^{\sigma\rho}_{\beta\alpha} = (1-\delta_{\sigma\rho})\,\delta_{\beta\alpha}\, G_o^{-1}\, T^{s-1}_{\alpha}\, G_o^{-1} \right. \tag{6.8}$$

$$\left. + \sum_{\tau} \sum_{\gamma} U^{\tau}_{\beta\gamma}\, G_o \underbrace{|\gamma> \quad t_{\gamma} \quad <\gamma|}_{=\, T^{s}_{\gamma}}\, G_o\, U^{\tau\rho}_{\gamma\alpha} \right\} G_o |\alpha>,$$

we find, with the notations (6.5) and (6.7), an equation of the form

$$\tilde{\mathbf{U}}^{\sigma\rho} = (1-\delta_{\sigma\rho})\, \tilde{\mathbf{G}}_o^{-1} + \sum_{\tau} \tilde{\mathbf{T}}^{\tau}\ \tilde{\mathbf{G}}_o\, \tilde{\mathbf{U}}^{\sigma\rho}. \tag{6.9}$$

Since our original four-body operators are sandwiched between two-body wave functions, the occurring expressions are still effective three-body operators.

In other words, the pole approximation (6.1) yields in the three-body case an <u>effective two-body</u> theory, while in the four-body case we are left with <u>effective three-body</u> equations. Moreover, the equations (6.9) have the same form as the three-body equations.

This suggests to proceed as in the genuine three-body case. There we have reduced the dimension of the problem by introducing the pole approximation (6.1) for T_γ. Replacing now its analogue T^τ by pole terms

$$\tilde{\mathbf{T}}^{\tau} \sim |\,\tau > \mathbf{t}_{\tau} < \tau\,| \qquad (6.10)$$

we reduce, in the same way, Eq. (6.9) to an effective two-body equation.

Since, according to (6.5), the operator

$$\tilde{\mathbf{T}}^{\tau}_{\beta\alpha} = <\beta|\; G_o\, U^{\tau}_{\beta\alpha}\, G_o\; |\alpha> \qquad (6.11)$$

is the transition amplitude of the three-body subsystem, expression (6.10) represents a three-body pole approximation.

In other words, our procedure is based, in principle, on a successive pole approximation of the two- and three-body transition operators occurring in the kernel of our four-body equations (4.3):

$$U^{\sigma\rho}_{\beta\alpha} = \ldots\ldots + \sum_{\tau\neq\sigma}\sum_{\gamma} \underbrace{U^{\tau}_{\beta\gamma}}_{|\tau>\mathbf{t}_{\tau}<\tau|}\, G_o \underbrace{T_{\gamma}}_{|\gamma>t_{\gamma}<\gamma|} G_o\, U^{\tau\rho}_{\gamma\alpha} \;. \qquad (6.12)$$

The final effective two-body equations are easily written down. Instead of giving the algebraic details, it is more instructive to represent them in form of diagrams. Generalizing in an evident manner the symbols of Fig. 9, we have the structure demonstrated in Fig. 10.

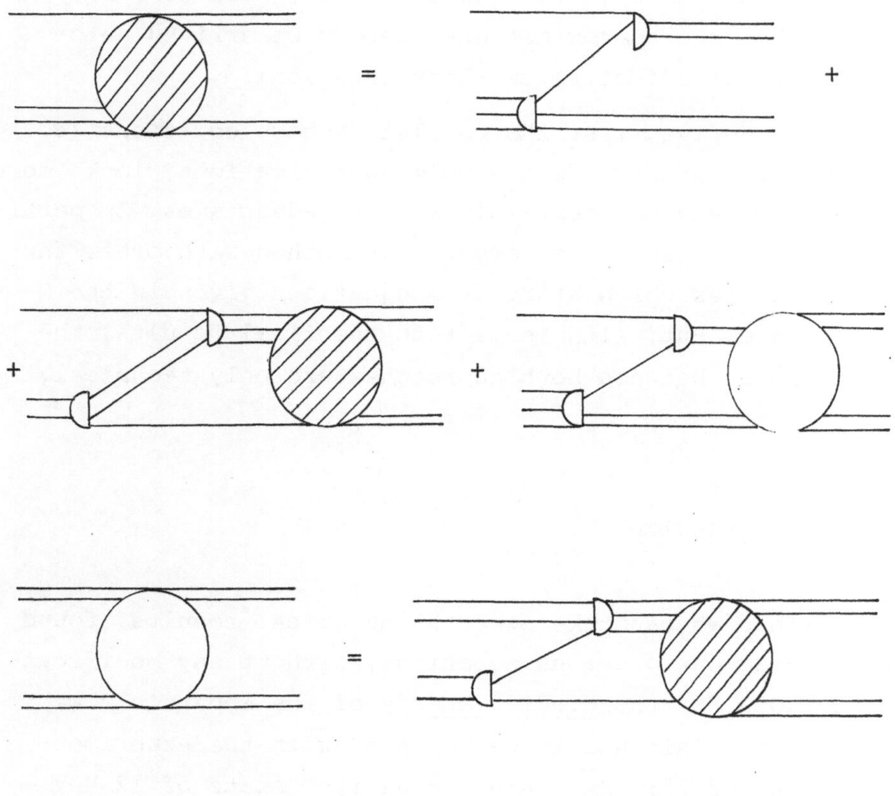

Fig. 10

Previous approach

Starting from the four-body-equations (4.3) and introducing step by step the subsystem pole approximations, we reproduce the effective two-body theory derived by Grassberger and Sandhas some years ago [9].

The general concept of this previous approach is very similar to the one described in Secs. 3 and 4. It was also based on a successive incorporation of the subsystem transition operators. In contrast to our present derivation, we applied, however, in any step simultaneously the pole approximations. Hereby we arrived automatically at effective two-body equations.

To derive first the complete N-body equations (4.3), introducing afterwards the pole approximations, looks more convincing and has certainly several advantages. In particular, it allows us to compare our method with other few-body theories which start from equations given in the N-body space. But, with respect to the final results, the differences between both approaches are only technical.

Numerical results

Thus we can take over the numerical results, found by solving our previous equations, without any modifications [10]. For the binding energy of the α-particle we got 50 MeV. This has to be compared with the experimental value of 28.3 MeV. Also an excited state of 12 MeV is found, while corresponding experimental data are 6.9 or 5.9 MeV. The overbinding is not astonishing, since we used for the form factors in the two-body pole approximation (6.1) the conventional Yamaguchi functions which already in the three-body case yield a remarkable overbinding of the triton.

Starting from the system of effective two-body

equations demonstrated in Fig. 10, we also calculated the deuteron cross section for the process deuteron + deuteron → proton + triton. Instead of solving the equations correctly we applied a simple K-matrix Born approximation to them. None the less the result shown in Fig. 11 is encouraging. Not only the shape of the forward peak but also its absolute value is rather well reproduced (Fig. 11).

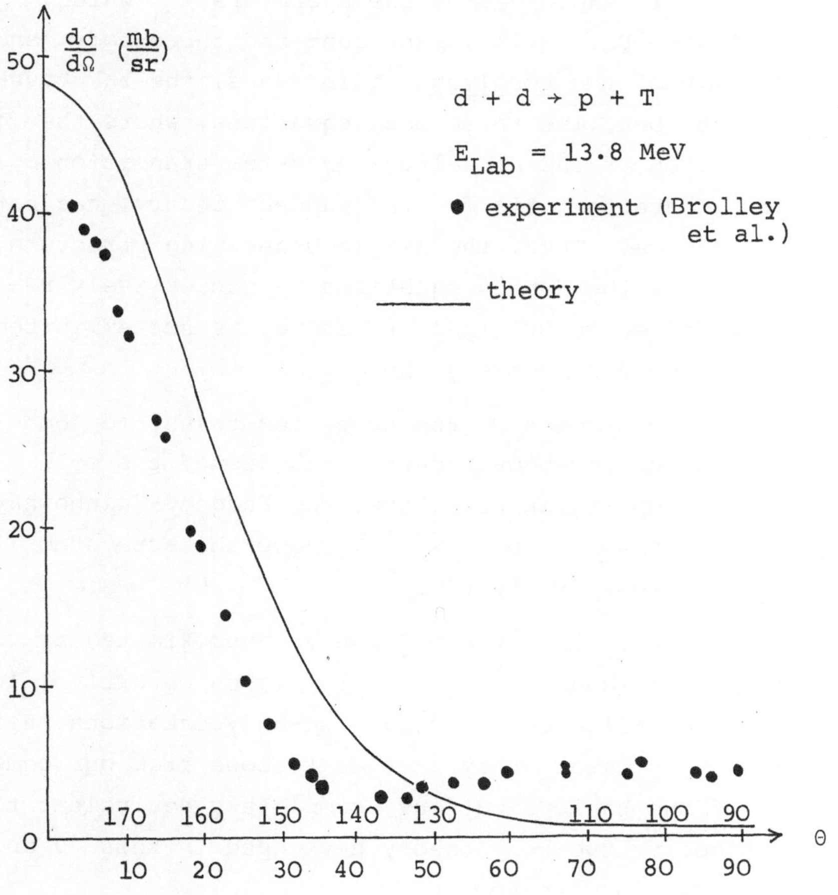

Fig. 11

7. COMPARISON WITH THE FADDEEV-YAKUBOVSKY EQUATIONS

We have emphasized already that our present formulation, in contrast to the previous one, is well suited for comparison with other N-body approaches [11,12].

Comparing, e.g., the kernel of Eq. (4.3) with the one of the Faddeev-Yakubovsky four-body equations [11], we find that they are closely related. However, the latter one is expressed by the operators $M^{\tau}_{\beta\alpha}$ which, according to Eq. (3.11), represent the three-body generalization of the resolvent. This means, the Yakubovsky equations look like three-body equations, where the representation of the kernel by subsystem transition operators, characteristic for the Faddeev theory, has not been performed. Thus, the simple underlying structure which yields the N-body equations by successively repeating Faddeev's original procedure, is not exhibited in the Faddeev-Yakubovsky theory.

The occurrence of the subsystem transition operators in our few-body equations suggested the pole approximations explained above. The Faddeev-Yakubovsky kernel, represented by the $M^{\tau}_{\beta\alpha}$, shows the subsystem structures less explicitly.

None the less, the $M^{\tau}_{\beta\alpha}$ can be approximated by their dominant two- and three-body poles. Then we arrive also in this formalism at effective two-body equations. Being different in detail they are, as follows from our comparison of the basic equations, completely equivalent to the effective two-body theory developed in Refs. 9 and 10 and described in Sec. 6.

This means, attempts to apply pole approximations

onto the Faddeev-Yakubovsky equations only reproduce our previous N-body theory*).

Consequently, recent numerical results, found in this way are almost identical to ours. The binding energy of the α-particle and of an excited state given in Ref. 13, are 52.03 MeV and 12.15 MeV, i.e., rather close to our values of 50 MeV and 12 MeV. No recent calculations of cross sections, based on effective two-body equations, exist which could be compared with our results. But, as explained above, nothing new can be expected if almost identical approximations are applied onto equations which look algebraically different but are, in fact, equivalent. Therefore, progress in this field can only be achieved by starting from improved subsystem approximations, or by solving the effective equations more accurately. Such calculations are in progress.

REFERENCES

1. W. Sandhas, The Three-Body Problems, in Elementary Particle Physics, ed.: P. Urban (Acta Physica Austriaca, Suppl. IX, 57 (1972)).

2. L. D. Faddeev, Mathematical Aspects of the Three-Body Problem in the Quantum Scattering Theory (English translation: Israel Program for Scientific Translation, Jerusalem, 1965).

*) As explained in Ref. 9, the pole approximation is only a starting point of an exact iteration scheme (quasi particle method). I.E., we have not only an approximate but an exact equivalence of both approaches.

3. Compare also: E. O. Alt, P. Grassberger, W. Sandhas, JIRN, E4-6688, Dubna, 1972. See furthermore p. 299 of Few Particle Problems, ed.: I. Slaus et al., North-Holland, 1972 (Proceedings of the Los Angeles Conference 1972).

4. The definition of $\Omega^{(+)}$ by a time limit is given in Eq. (2.9) of Ref. 1.

5. Without going into any details we recall that in the Heisenberg picture the relevant operators have to be sandwiched between eigenstates of the total Hamiltonian. These, however, are in our case the scattering states $|\vec{p}\rangle^{(+)}$.

6. E. O. Alt, P. Grassberger, W. Sandhas, Nucl. Phys. B2, 167 (1967).

7. L. D. Faddeev, Soviet Phys. - JETP 12, 1014 (1961).

8. A detailed investigation, sketched in Ref. 1, shows that the uniqueness of the Faddeev equations is a consequence of the fact that their algebraic structure corresponds exactly to decisive aspects of the multichannel collision theory. The original, more technical proof of uniqueness [2] was based on methods of integral equations theory. From both points of view it is important that the expression (3.8) represents an exchange potential, different from zero only for $\alpha \neq \beta$, or, in other words, that the kernel of the system of equations (3.4) has no diagonal elements.

9. P. Grassberger, W. Sandhas, Nucl. Phys. B2, 181 (1967).

10. E. O. Alt, P. Grassberger, W. Sandhas, Phys. Rev. C1, 85 (1970).

11. O. A. Yakubovsky, Sov. J. Nucl. Phys. 5, 937 (1967).

L. D. Faddeev, Three-Body Problem in Nuclear and Particle Physics, ed.: J. S. C. McKee a.P.M. Rolph (North-Holland, Amsterdam, 1970). Compare also K. Hepp, Helv. Phys. Acta 42, 425 (1969).

12. An incomplete list of further publications is:
S. Weinberg, Phys. Rev., 133, B232 (1964);
L. Rosenberg, Phys. Rev., 140, B217 (1965);
A. N. Mitra, J. Gillespie, R. Sugar, N. Panchapakesan, Phys. Rev., 140, B1336 (1965); V. V. Komarov, A. N. Popova, Nucl. Phys. 69, 253 (1965); Phys. Lett., 28B, 476 (1969);
N. Mishima, Y. Takahashi, Progr. Theor. Phys. 35, 440 (1966);
I. Weyers, Phys. Rev., 145, 1236 (1966); Phys. Rev. 151, 1159 (1966);
R. Omnes, Phys. Rev. 165, 1265 (1968);
I. Sloan, Phys. Rev. C6, 1945 (1972);
G. Bencze, Nucl. Phys. A210, 568 (1973);
A. N. Mitra, Flinders-preprint: FUPH-R-87
E.F. Redish, Saclay-preprint: DPh. T. 74/3.

13. V. F. Kharchenko, Few Particle Problems, ed. I. Slaus et al., North-Holland, 1972 (Proceedings of the Los Angeles Conference 1972).

Acta Physica Austriaca, Suppl. XIII, 711–736 (1974)

ON THE EIKONAL APPROXIMATION IN QUANTUM FIELD THEORY*)

by

K. DIETZ
Physikalisches Institut der Universität Bonn
Bonn, West-Germany

1. INTRODUCTION

In any of the current theories in physics you have to make approximations in calculating experimentally observable quantities: the more complex the system you want to describe the more you have to labour to get reasonable answers. This is obvious for anyone who is involved in theoretical hadron-dynamics or who is tampering with quantum field theory (QFT). If you want to apply this latter theory to high energy reactions, you clearly have to look for an approximation scheme which goes beyond finite-order perturbation theory. One of these schemes is the eikonal approximation [1] which is the subject of this talk.

---

*) Seminar given at XIII. Internationale Universitätswochen für Kernphysik, Schladming, Austria, February 4-15, 1974.

The general idea is that at very high-energies the de Broglie wave-length of the incident particle of a scattering process is small compared with the spatial variations in the target, so that the "geometrical optics"-limit might be reasonable as a first approximation. Indeed, the gross features of proton-proton reactions at ISR energies show that an appreciable part of the cross section has to do with processes that are dominated by "through-going" particles which fragment without changing their initial momentum very much.

Here, we will be exclusively concerned with purely theoretical aspects of the eikonal approximation (EA). In a first section we recapitulate the EA for the Green's function and the scattering matrix of a particle moving in an external field, the relative high-energy behaviour of higher EA-approximants is calculated. It is illustrated that the eikonal asymptotics depend on the dimension of the coupling constant. We then include loops, that is, amplitudes for processes in which the particle virtually disintegrates and reconstitutes itself while interacting arbitrarily many times before, during and after the virtual disintegration. The key to the eikonalisation of these graphs will be a new method of calculating loops in coordinate space. For the purpose of this talk, we think, it will suffice to illustrate our procedure for the case of one loop only. A systematic extension to more, especially overlapping loops will be given in a separate publication.

2. SCATTERING FROM AN EXTERNAL FIELD

To begin with the simplest case, let us consider

the motion of a spinless particle in a scalar external field.

The equation of motion for the Green's function (we demand Feynman boundary conditions),

$$(\Box + m^2 - g\,A(x))\;G(x,y) = \delta(x-y)\;, \qquad (2.1)$$

has an obvious iterative solution which, graphically, reads as follows:

$$G(x,y) = \sum_{n=o}^{\infty}$$

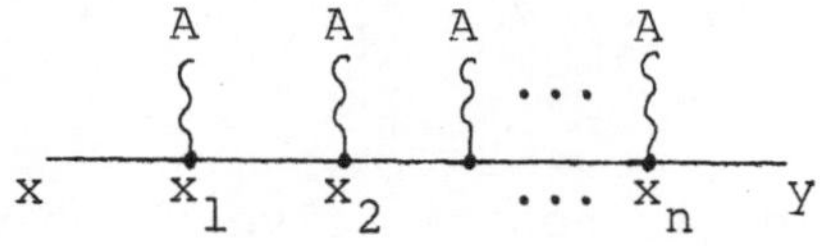

Fig.1. At each point appears a factor g; straight lines represent free propagators of the projectile.

To define an eikonal approximation we introduce a four-momentum q, components of which will eventually tend to asymptotically large values, and a Green's function

$$G_q(x,y) = e^{iq\cdot(x-y)}\;G(x,y)\;. \qquad (2.2)$$

One immediately gets the equation of motion for $G_q(x,y)$

$$(\Box + m^2 - q^2 - 2iq\cdot\partial - g\,A(x))\;G_q(x,y) =$$

$$= \delta(x-y) \qquad (2.3)$$

for which we construct the following iterative solution

$$G_q(x,y) = \sum_{r=o}^{\infty} G_q^{(r)}(x,y)$$

$$(- q^2 - 2iq\cdot\partial - g\, A(x))\, G_q^{(r)}(x,y) = \delta_{ro}\, \delta(x-y) -$$

$$- (\Box + m^2)\, G_q^{(r-1)}(x,y)$$

$$r = 0,\ 1,\ \ldots\ ; \quad G_q^{(-1)} \equiv 0\ . \qquad (2.4)$$

Our problem is now to solve these equations and to work out the relative large-q behaviour of the $G_q^{(r)}(x,y)$; it will turn out that our expansion yields an asymptotic large-q expansion under conditions to be specified.

The essential advantage we achieved by introducing $G_q(x,y)$ and the iteration (2.4) is that the latter equations are ordinary differential equations, the d'Alembertian being treated as a "perturbation" in the envisaged large-q limit.

To make this point explicit we write the derivative in the direction q as

$$2\, q\cdot\partial =: \frac{d}{ds} \qquad (2.5)$$

where s is a parameter along the straight line

$$x(s) = y + 2qs\ . \qquad (2.6)$$

In the first step we solve

$$(- q^2 - 2iq\cdot\partial - g\,A(x))\; G_q^{(o)}(x,y) = \delta(x-y) \qquad (2.7)$$

with an ansatz (we assume that the external field vanishes at infinity)

$$G_q^{(o)}(x,y) = e^{ig\int_o^\infty d\tau[A(x-2\tau q) - A(y-2\tau q)]}\,\bar{G}(x,y) \qquad (2.8)$$

such that

$$(- q^2 - 2iq\cdot\partial)\;\bar{G}(x,y) = \delta(x-y)\;. \qquad (2.9)$$

We write formally

$$\bar{G}(x,y) = i\int_o^\infty ds\; e^{-is(-q^2-2iq.\partial-io)}\,\delta(x-y)$$

and use the fact that exp $(-2sq.\partial)$ performs a translation by -2sq to obtain

$$\bar{G}(x,y) = i\int_o^\infty ds\; e^{-is(-q^2-io)}\,\delta(x-y-2sq)\;. \qquad (2.10)$$

$\bar{G}(x,y)$ is, of course, the first eikonal approximant to the free Feynman propagator $\Delta_F\,(x-y)$.

Inserting into (2.8) we finally have

$$G_q^{(o)}(x,y) = i\int_o^\infty ds\; e^{-is(-q^2-io)}\, e^{ig\int_o^s d\tau A(x-2\tau q)}\,\delta(x-y-2sq)\;. \qquad (2.11)$$

The physical interpretation of these formulae is quite clear: the δ-function in $\bar{G}$ and $G_q^{(o)}$ constraints the propagation from x to y to a classical free-particle motion

$$x = y + 2\,sq$$

where the parameter $\tau = 2\,ms$ has the physical meaning of a proper time. In the presence of an external field this free-particle motion is modified by an eikonal phase.

The transition matrix element for a particle of initial momentum k_i scattered to a final momentum k_f is given by

$$T_{fi} = \lim_{k_i^2 \to m^2} (-k_i^2+m^2) \int d^4x \int d^4y\; e^{ik_f\cdot x}\, e^{-ik_i\cdot y}\, gA(x)G(x,y). \tag{2.12}$$

Its first eikonal approximation is then

$$T_{fi}^{(o)} = \lim_{k_i^2 \to m^2} (-k_i^2+m^2) \int d^4x \int d^4y\; e^{ik_f\cdot y}\, e^{-ik_i\cdot x}\; g\,A(x)\,\cdot$$

$$\cdot\, i\int_0^\infty ds\; e^{-is(-q^2-io)}\; e^{ig\int_0^s d\tau A(x-2\tau q)}\; \delta(x-y-2sq) \tag{2.13}$$

where we have set

$$q = k_i + k_f$$

$$k_f^2 = m^2 \quad .$$

Performing the y-integration and extracting the mass-shell pole by a partial s-integration we obtain the well-known eikonal formula

$$T_{fi}^{(o)} = \int d^4x \; e^{i(k_f-k_i)\cdot x} \; g\,A(x)\; e^{ig\int_o^\infty d\tau A(x-2\tau q)} \tag{2.14}$$

The next eikonal approximant $G_q^{(1)}(x,y)$ is the solution of (r = 1 in eq.(2.4))

$$(-q^2-2iq\cdot\partial-g\,A(x))G_q^{(1)}(x,y) = -(\Box+m^2)G_q^{(o)}(x,y). \tag{2.15}$$

Again, we use the same procedure as above and set

$$G_q^{(1)}(x,y) = e^{ig\int_o^\infty d\tau\, A(x-2\tau q)}\;\tilde{G}(x,y) \tag{2.16}$$

so that

$$(-q^2-2i\partial.q)\;\tilde{G}(x,y) = -e^{-ig\int_o^\infty d\tau\, A(x-2\tau q)}\;(\Box+m^2)G_q^{(o)}(x,y)$$

and

$$\tilde{G}(x,y) = -i\int_o^\infty ds\; e^{-is(-q^2-io)}\; e^{-ig\int_o^\infty d\tau A(x-2(s+\tau)q)}$$

$$\cdot(\Box+m^2)\;G_q^{(o)}(x-2sq,y)\;.$$

A short calculation gives the final answer

$$G_q^{(1)}(x,y) = \int_0^\infty ds\; e^{-is(-q^2-io)}\; e^{ig\int_0^s d\tau A(x-2\tau q)}\; E(s)\,\delta(x-y-2sq) \qquad (2.17)$$

where

$$E(s) = \int_0^s ds'\left((\partial_\mu + (\partial_\mu \phi))^2 + m^2\right)$$

and

$$\phi = \phi(x;s,s') = ig \int_{s'}^{s} d\tau\; A(x-2\tau q) \;.$$

The second EA of the transition matrix is now easily computed

$$T_{fi}^{(1)} = \lim_{k_i^2\to m^2} \;(-k_i^2+m^2)\int d^4x \int d^4y\; e^{ik_f\cdot y}\; e^{-ik_i\cdot x}\; g\,A(x)\;\cdot$$

$$\cdot\int_0^\infty ds\; e^{-is(-q^2-io)}\; e^{ig\int_0^s d\tau A(x-2\tau q)}\; E(s)\,\delta(x-y-2sq)\,.$$

As we did before, we carry out the y-integration and a partial s-integration

$$T_{fi}^{(1)} = \int d^4x\; e^{-ik_i x}\{E(\infty)\; e^{ik_f\cdot x}\}\; e^{ig\int_0^\infty d\tau A(x-2\tau q)}\; g\,A(x)$$

where the curly brackets indicate that E now operates

on the exponential.

More explicitly, we have [2]

$$T_{fi}^{(1)} = \int d^4x \, e^{i(k_f - k_i)\cdot x} \, g\, A(x) \, e^{ig\int_0^\infty d\tau A(x-2\tau q)} \cdot$$

$$\cdot \int_0^\infty ds' \, (\Box \tilde{\phi} + (\partial_\mu \tilde{\phi})^2 + 2ik_f \cdot \partial \tilde{\phi}) \tag{2.18}$$

with

$$\tilde{\phi}(x,s') = \phi(x;\infty,s') \; .$$

We now have to compare the large-q behaviour of $T_{fi}^{(o)}$ and $T_{fi}^{(1)}$ for a suitably specified class of external fields in order to justify our assertion that the eqs. (2.4) generate an asymptotic series for the transition matrix (under certain kinematic constraints which we shall elaborate in the course of the argument).

Writing A(x) as a Fourier-transform allows us to do the s-integrations in (2.18):

Defining

$$A(x) = M \int d^4k \;\; e^{ik\cdot x} \, f(k) \tag{2.19}$$

we get

$$\tilde{\phi}(x,s') = ig \, M \int d^4k \, e^{ik\cdot x} \, f(k) \int_{s'}^{\infty} d\tau \, e^{-2i\tau \, k\cdot q}$$

$$= \frac{gM}{2} \int d^4k \;\; e^{ik\cdot x} \, f(k) \, \frac{e^{-2is'k\cdot q}}{k\cdot q} \; . \tag{2.20}$$

(The mass M is included because of obvious dimensional

reasons; f(k) is assumed to vanish sufficiently rapidly at infinity that the contribution of the upper τ-integration limit vanishes).

This integral is not defined in general unless we ensure the non-vanishing of the denominator $(k\cdot q)^{-1}$.

Since q is chosen to be the time-like vector $q = k_i + k_f$ in eqs. (2.14) and (2.18), we do so by simply demanding time-like support of f(k)

$$f(k) = \delta(k^2 - \mu_o^2)\, \tilde{f}(k)$$
$$\mu_o^2 > 0 \qquad (2.21)$$

(If we include $\mu_o^2 = 0$, we have to extract the well-known infinite phase [3].).

Using eqs. (2.14) and (2.18) we get after inserting (2.20)

$$T_{fi} = T_{fi}^{(o)} + T_{fi}^{(1)} + \dots =$$

$$= \int d^4x\, e^{i(k_f - k_i)\cdot x}\, g\, A(x)\, e^{ig\int_o^\infty d\tau A(x-2\tau q)} \cdot$$

$$\cdot\ (1 - \{\frac{igM}{4} \int d^4k\, e^{ik\cdot x}\, f(k)(-k^2 - 2k_f\cdot k)\, \frac{1}{(k\cdot q)^2}$$

$$- \frac{ig^2M^2}{8} \int d^4k \int d^4k'\, e^{i(k+k')\cdot x}\, \frac{(k\cdot k')f(k)f(k')}{(k\cdot q)(k'\cdot q)(k+k')\cdot q}\}$$

$$+ \dots) . \qquad (2.22)$$

We see by inspection that the expression in the curly

brackets vanishes for sufficiently decreasing f(k) at least as $O((E_i + E_f)^{-1})$, E_i and E_f being the initial and final energies of the projectile, for all scattering angles $\theta\,(|\vec{k}_i||\vec{k}_f| \cos\theta = \vec{k}_i \cdot \vec{k}_f)$.

The notion "at least as" accounts for the fact that we allow for time-dependent external fields and, therefore, in general treat a dissipative system. This means that the energy E_f of the final state of the projectile, occurring in eq. (2.22), can vary quite independently of its initial energy E_i.

This is easily generalised to higher approximants: one infers from eqs. (2.4) (essentially by counting dimensions) that our eikonal iteration generates approximants to the transition amplitude T_{fi}, two consecutive terms of which have the same relative high energy behaviour as $T_{fi}^{(o)}$ and $T_{fi}^{(1)}$.

To be slightly more explicit we assume that f(k) is focussed in a narrow band around an average momentum k. Then the envisaged eikonal asymptotic series can be written as

$$T_{fi} = \sum_{k=o} T_{fi}^{(k)} = \int d^4x\, e^{i(k_f - k_i)\cdot x}\, gA(x)\, e^{ig\int_o^\infty d\tau A(x-2\tau q)} \cdot$$

$$\cdot\ (1 + \sum_{k=1} (\frac{g\ M}{\hat{k}\cdot q})^k F_k(x,q;\ \frac{k_f\cdot\hat{k}}{q\cdot\hat{k}})) \qquad (2.23)$$

$$q = k_i + k_f\ ,$$

where F_k decreases at least as a constant, the part of the q-dependence which results from the finite width of the spectrum f(k) will, in general, be weak (for F_1 see eq. (2.22)). The effective asymptotic expansion parameter, therefore, is

$$\kappa = \frac{g\ M}{\hat{k}\cdot q} \ . \tag{2.24}$$

Substantially different asymptotics is revealed in theories in which the external field couples with a dimensionless coupling constant. Take for example the case of a minimally coupled vector field:

$$((\partial_\mu - \frac{ig}{2} A_\mu)^2 + m^2)\ G(x,y) = \delta(x-y) \tag{2.25}$$

is the equation of motion and the eikonal iteration now is ($G_q(x,y)$ is defined as in (2.2))

$$G_q(x,y) = \sum_{r=o}^{\infty} G_q^{(r)}(x,y)$$

$$(-q^2-2iq.(\partial - \frac{ig}{2} A))\ G_q^{(r)}(x,y) = \delta_{ro}\ (x-y) -$$

$$- ((\partial - \frac{ig}{2} A)^2 + m^2)\ G_q^{(r-1)}(x,y)$$

$$r = 0,1,\ \ldots\ ;\quad G_q^{(-1)} \equiv 0\ . \tag{2.26}$$

Using exactly the same techniques as above we find

$$G_q^{(o)}(x,y) = i\int_o^\infty ds\ e^{-is(-q^2-io)}\ e^{ig\int_o^s d\tau\ q\cdot A(x-2\tau q)}$$

$$\cdot\ \delta(x-y-2sq) \tag{2.27}$$

and

$$G_q^{(1)}(x,y) = \int_0^\infty ds\ e^{-is(-q^2-io)}\ e^{ig\int_0^s d\tau\ q\cdot A(x-2\tau q)}\ \cdot$$

$$\cdot\ \tilde{E}(s)\ \delta(x-y-2sq)$$

$$\tilde{E}(s) = \int_0^s ds'(D^2 + m^2) \tag{2.28}$$

with

$$D_\mu = D_\mu(x,s,s') = \partial_\mu - \frac{ig}{2} A_\mu(x-2sq) + ig\int_{s'}^{s} d\tau q^\rho F_{\mu\rho}(x-2\tau q)$$

$$F_{\mu\nu} = \partial_\mu A_\nu - \partial_\nu A_\mu\ .$$

We now compute the transition matrix (defined as in eq. (2.12)) and find for the first EA

$$T_{fi}^{(o)} = \int d^4x\ e^{i(k_f-k_i)\cdot x}\ g\ q\cdot A(x)\ e^{ig\int_0^\infty d\tau q\cdot A(x-2\tau q)}\ . \tag{2.29}$$

The next term is

$$T_{fi}^{(1)} = \int d^4x\ e^{-ik_i x}\ \{\tilde{E}(\infty)\ e^{ik_f\cdot x}\}\ g\ q\cdot A(x)\ e^{ig\int_0^\infty d\tau q\cdot A(x-2\tau q)}\ . \tag{2.30}$$

As we did before, we write $A_\mu(x)$ as an Fourier integral

$$A_\mu(x) = M\int d^4k\, e^{ik\cdot x} f_\mu(k)$$

and find ($f_\mu(k)$ is again assumed to have a time-like support)

$$T_{fi} = T_{fi}^{(o)} + T_{fi}^{(1)} + \ldots$$

$$= \int d^4x\, e^{i(k_f-k_i)\cdot x} g\, q\cdot A(x)\, e^{ig\int_o^\infty d\tau q.A(x-2\tau q)} \cdot \tag{2.31}$$

$$\cdot\ (1 - \frac{ig\,M}{2}\int d^4k\, e^{ik\cdot x} (\frac{k_f\cdot f}{k\cdot q} - \frac{k\cdot k_f\ q\cdot f}{(k\cdot q)^2}) + \ldots);$$

the dots indicate terms which vanish as $O((E_i + E_f)^{-1})$ for all k_i and k_f and suitably chosen $f_\mu(k)$.

For $f_\mu(k)$ concentrated in a narrow band around $\hat{k}$ it suffices to discuss the high energy behaviour of

$$R(k_i,k_f) = \frac{k_f\cdot f}{\hat{k}\cdot q} - \frac{\hat{k}\cdot k_f\ q\cdot f}{(\hat{k}\cdot q)^2} .$$

Take fixed momentum transfer

$$\Delta = k_i - k_f \quad \text{fixed} ,$$

then

$$R(k_i,k_f) = \frac{k_f\cdot f(\hat{k})}{4k_f\cdot\hat{k}}(\frac{\Delta\cdot\hat{k}}{k_f\cdot\hat{k}} - \frac{\Delta\cdot f(\hat{k})}{k_f\cdot f(\hat{k})} + O(E_i^{-2})) =$$

$$= O(E_i^{-1})\,;\ \Delta \text{ fixed},\ E_i \to \infty .$$

Hence, what we get for the validity of our asymptotic eikonal series, in the case of an external vector field, is a restriction to forward angles; in fact, for Δ fixed, the scattering angle θ has to be [1]

$$\theta \approx O(E_i^{-1}) .$$

Such a restriction does not hold in the case of a scalar external fied as was shown above.

Two final remarks seem to be in place. Firstly, it is easy to see that the sum of graphs depicted in Fig. 1 yield $G_q^{(o)}(x,y)$ (eq. (2.11)) if the free propagators are replaced by their first eikonal approximants $\bar{G}(x,y)$ (eq. (2.10)). Secondly, we want to remind [3] the reader that the Green's function for scattering in an external field A can be used to construct the ladder-approximation (including <u>all</u> crossings) of the 4-point function for projectile-projectile scattering via A-exchange

$$G(x_1,x_2,x_3,x_4) = e^{-\frac{i}{2}\int d^4x\, d^4y\, \frac{\delta}{\delta A_1(x)}\Delta_F(x-y)\frac{\delta}{\delta A_2(y)}} \cdot$$

$$\cdot\, G(x_1,x_2|A_1)\; G(x_3,x_4|A_2)\Big|_{\substack{A_1 = 0\\ A_2 = 0}} \tag{2.32}$$

where we indicated the functional dependence on $A(\mathbf{x})$ of our Green's function $G(x,y)$ explicitly. Eikonalizing the latter leads to an eikonalised sum of ladder graphs.

2. THE ONE-LOOP PROPAGATOR IN AN EXTERNAL FIELD

In this section we are going to outline a method for the eikonalisation of the one-loop two-point Green's function for the motion of a particle in an external field $A(x)$, i.e. the eikonalisation of the following sum of graphs:

$$G_1(x,y) = \Sigma$$

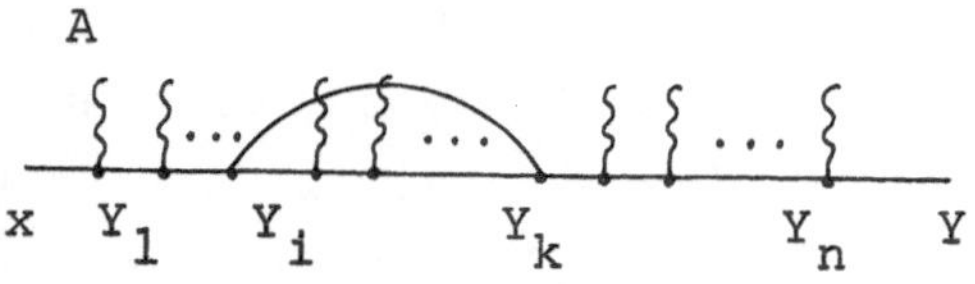

Fig.2. One-loop graphs for the two-point function; the sum runs over all external field insertions.

The essential point of our treatment is a replacement of the Feynman propagators connecting y_i and y_k in Fig. 2 by non-local differential operators. The loop is thus contracted to a multi-external field vertex. In an eikonal limit (to be defined below), this vertex reduces to an effective coupling which will be calculated. Hence, the graphs of Fig. 2 are reduced to the "chain-graphs" of Fig. 1 except that they contain a multi-external field coupling at all possible positions. These new graphs can be eikonalised as described in the previous section.

Following this line of argument we start by writing the graphs of Fig. 2 in the following manner:

First define

$$y_i = x + x_1 + \dots + x_i \qquad i = 1,2 \dots n$$

$$x_{i+1} = y_{i+1} - y_i \tag{2.1}$$

and the m'th order chain-graph connecting the space-time point a and b

$$f_m(a,b) = \int d^4y_1 \dots \int d^4y_m \; \Delta_F(a-y_1)A(y_1)\dots A(y_m)\Delta_F(y_m-b). \tag{2.2}$$

An n'th order contribution to $G_1(x,y)$ can then be expressed as

$$G_1^{(n)}(x,y) = \int d^4y_i \; f_{i-1}(x,y_i)K^{-1}_{x_{i+1}} A(y_{i+1})\dots K^{-1}_{x_{k-1}} A(y_{k-1})\cdot$$

$$\cdot K^{-1}_{x_k} \, i\Delta_F(x_k+\dots+x_{i+1}) f_{n-k}(x+x_1+\dots+x_k,y)$$

$$x_j = 0$$

$$j = i+1,\dots,k. \tag{2.3}$$

where we retained explicitly the notion of the inverse Klein-Gordon operator (with Feynman boundary conditions, of course)

$$K_x = \Box + M^2 \quad . \tag{2.4}$$

These inverse operators are seen to act on products of a Feynman propagator and an external field or a chain-graph. For their evaluation we use the following (formal) short-distance expansion [4]

$$K_{x_1}^{-1} f(x+x_1)\, g(y+x_1) =$$

$$\sum_{\ell=0}^{\infty} \sum_{n,m=0}^{\infty} 2^{\ell} \frac{(-1)^{n+m+\ell}}{M^{2(n+m+1+\ell)}} \binom{n+m+\ell}{n} \binom{m+\ell}{m} [\Box^{n} \partial_{\nu_1} \ldots \partial_{\nu_\ell} f(x)] \cdot$$

$$\cdot\ [\Box^{m} \partial^{\nu_1} \ldots \partial^{\nu_\ell} g(y)] + O(x_1) \tag{2.5}$$

In applying this formula we set e.g.

$$g(y) \equiv F_\lambda(y) = \frac{1}{(2\pi)^4} \int d^4p\, e^{ip\cdot x} (M^2 - p^2 - io)^{\lambda} \tag{2.6}$$

and perform a partial summation to arrive at

$$K_{x_1}^{-1} f(x+x_1)\, F_\lambda(y+x_1) \Big|_{x_1 = 0} =$$

$$\sum_{\ell,n=0}^{\infty} 2^{\ell} (-1)^{n+\ell} \binom{n+\ell}{n} [\Box^{n} \partial_{\nu_1} \ldots \partial_{\nu_\ell} f(x)] \cdot$$

$$\cdot [\partial^{\nu_1} \ldots \partial^{\nu_\ell} F_{\lambda-n-\ell-1}(y)\ ] . \tag{2.7}$$

In deriving this formula we used

$$K_x F_\lambda(x) = F_{\lambda+1}(x) \quad . \tag{2.8}$$

Furthermore, we have to keep in mind the following properties of the distributions $F_\lambda(x)$

$$\partial_\mu F_\lambda(x) = \frac{x_\mu}{2(\lambda+1)} F_{\lambda+1}(x) \tag{2.9}$$

and, of course,

$$F_{-1}(x) = \Delta_F(x) \quad .$$

Having assembled the necessary tools, we apply them to invert the Klein-Gordon operators in eq. (2.3): We consecutively employ eq. (2.7), starting with $K_{x_k}^{-1}$ until we reach $K_{x_{i+1}}^{-1}$. Introducing the following abbreviations

$$c_{n\ell} = 2^\ell \ (-1)^{n+\ell} \ \binom{n+\ell}{n}$$

$$D_{n\ell,\nu} = \Box^n \ \partial_{\nu_1} \ldots \partial_{\nu_\ell} \tag{2.10}$$

$$D_{n\ell}{}^{\nu} = \Box^n \ \partial^{\nu_1} \ldots \partial^{\nu_\ell}$$

$$d_{ki} = \sum_{j=i+1}^{k} (n_j+\ell_j+1)$$

we find

$$G_1^{(n)}(x,y)=\int d^4y_i \; f_{i-1}(x,y_i) \; \Omega(y_i)f_{n-k}(y_i,y)$$

with

$$[N=\{\ell_k,n_k,\ldots,\ \ell_{i+1},n_{i+1}\},\ c_N = c_{n_k\ell_k}\ldots c_{n_{i+1}\ell_{i+1}}]$$

$$\Omega(y_i)f_{n-k}(y_i,y)=\sum_{N=o}^{\infty} c_N[D^{\nu_{i+1}}_{n_{i+1}\ell_{i+1}} A(y_i) \; [D^{\nu_{i+2}}_{n_{i+2}\ell_{i+2}} A(y_i)$$

$$[\ldots D^{\nu_k}_{n_k\ell_k} \; f_{n-k}(y_i,y)]\ldots] \; .$$

$$\cdot[D_{o\ell_{i+1},\nu_{i+1}}\ldots D_{o\ell_k,\nu_k} \; F_{-1-d_{ki}}(o)] \quad (2.11)$$

where it is understood that the differentiations D act to the right on the product within the square-brackets following them.

The second factor in (2.11) is simply a known numerical tensor; in fact we have [5]

$$\partial_{\nu_1}\ldots\partial_{\nu_k} \; F^{(o)}_{-2-j} = \begin{cases} 0 & k=2\ell+1 \\ a_{jk} \; T^{(k)}_{\nu_1\ldots\nu_k} & k=2\ell \end{cases} \qquad \ell=0,1,\ldots$$

$j > 1,\ k \leq j-1$

and ($g_{\mu\nu}$ is the metric tensor)

$$T^{(2\ell)}_{\nu_1 \cdots \nu_{2\ell}} = g_{\nu_1\nu_2} \cdots g_{\nu_{2\ell-1}\nu_{2\ell}} + \ldots \text{ symmetric}$$

$$a_{j\,2\ell} = \frac{-i}{16\pi^2} \frac{(j-\ell-1)!}{M^{2(j-1)}\ 2^{\ell}(j+1)!} \tag{2.12}$$

Hence, the loops in Fig. 2 are represented as differential operators acting on chain graphs and thus are constricted to a coupling at y_i, containing derivatives of infinite order.

To elucidate the validity of such a representation, at least in a qualitative manner, we replace the external field A acting within the loop by plane waves (M is again a dimensional factor)

$$A(y_j) \rightarrow Me^{i\,p_j \cdot y_j} \qquad j=i+1,\ldots k-1\ . \tag{2.13}$$

Then the differentiations in (2.11), contracted by the tensor (2.12), acting on these plane waves, give factors p_t^2 $(p_r \cdot p_s)$, and acting on the chain f they give the momentum P flowing into the loop. So, altogether, we get monomials in $(p_r \cdot p_s)(p_j \cdot P)\ p_t^2 P^2$.

A thorough discussion [4] of the series (2.11) gives the following domain of convergence (the sum runs between any vertices in the loop)

$$-4M^2 < (\Sigma p_r)^2,\ P^2,\ (\Sigma p_r + P)^2 < 4M^2 \qquad (2.14)$$

i.e. the series breaks down at the momentum space singularities of our loop graph, as was to be expected.

To arrive at an eikonal approximation of these loop graphs we propose to consider "soft-particle" emission within the loop, that is, to choose momenta p_r such that

$$p_u \cdot P,\ p_s \cdot p_r,\ p_t^2 << P^2. \qquad (2.15)$$

Under this condition we are able to sum the series (2.11), and, hence, to continue to $^{P^2}/4M^2 > 1$.

Effectively, this means that we approximate each term in the sum (2.11) (taking (2.13) into account)

$$c_N[\ldots [D^{\nu_k}_{n_k \ell_k} \quad f_{n-k}\ (y_i, y)]\ \ldots] \rightarrow$$

$$\hat{c}_N\ e^{i\Sigma p_j \cdot y_i}\ \Box^{\bar{d}}\ f_{n-k}(y_i, y) \qquad (2.16)$$

with

$$\bar{d} = \sum_{j=i+1}^{k} (n_j + {}^{\ell}j/2).$$

The sum can then be evaluated and we obtain

$$\Omega(y_i) \rightarrow \Omega^{\text{Eik}}(y_i)$$

$$\Omega^{Eik}(y_i)=(A(y_i))^r \frac{i}{16\pi^2} \sum_{J=0}^{\infty} \frac{(J+r-1)!(J+r)!}{(2J+r+1)!} \left(-\frac{\Box}{M^2}\right)^J ; \qquad (2.17)$$

$r = k-i-1$ is the number of external fields acting within the loop, the replacement (2.13) is tacitly understood. For $r = 0$, the first term in the sum has to be left out; it is infinite, a fact which can be traced to the logarithmic singularity of the ordinary loop [5].

To get a better grasp of this expression, let us consider the following simplified situation. We assume that the projectile radiates soft particles on all its vertices depicted in Fig. 2 so that

$$\begin{aligned} \text{initial momentum } k_i &\approx \text{momentum } P \text{ entering or leaving the loop} \\ &\approx \text{final momentum } k_f . \end{aligned}$$

In momentum space we then obviously have r-particle vertices with an effective coupling "constant"

$$G_r = g^{r+2} \frac{i}{16\pi^2} \sum_{J=0}^{\infty} \frac{(J+r-1)!(J+r)!}{(2J+r+1)!} \left(\frac{k_i^2}{M^2}\right)^J \qquad (2.18)$$

This sum is convergent for

$$k_i^2 < 4\, M^2 , \qquad (2.19)$$

it can be summed and continued to larger values of k_i^2.

Asymptotically, we obtain

$$G_r = \frac{g^{r+2}}{16\pi^2} \; z^{-1} \; \frac{1}{M^{2r} \; r(r-1)} \; , \; r \geq 2$$

$$G_1 = M^{-2} \; \frac{g^3}{16\pi^2} \; z^{-1} \log z$$

$$z = \frac{k_i^2}{4\,M^2} \gg 1 \; . \qquad (2.20)$$

Physically speaking, $z \gg 1$ corresponds to either a "strongly heated" off-shell projectile, M is then the projectile mass, or an on-shell projectile decaying virtually in two light clusters which radiate r-times during the disintegration, M is then the cluster mass.

By inventing models of this kind for the interpretation of eq. (2.17) we achieved to reduce the loop graphs of Fig. 2 to chain graphs containing multi-external field couplings. The eikonalisation of the latter graphs proceeds in precisely the same way as discussed in the previous section: we replace the free Feynman propagators by the eikonalised ones (eq. (2.10)) and sum all chains. (Another way to pursue is to eikonalise iteratively a generating differential equation; details, as well as an extension to the case of more than one loop, will be published elsewhere.) In this way we have gone the first steps in an approximation scheme, valid in certain kinematical regions, for e.g. the transition matrix which sums all orders in the coupling constant g of graphs characterised by the number of loops.

It is well-known that such a loop-wise summation is determined by an expansion parameter proportional to $\hbar^s$, where s is the number of loops.

ACKNOWLEDGMENT

It is a pleasure to thank Profs. D. Atkinson and K. Meetz for very helpful discussions.

REFERENCES AND FOOTNOTES

1. The number of references to the eikonal method in physics is legion; a few general surveys are:

   R.J. Glauber in Lectures in Theoretical Physics, ed. by W.E. Brittin and L.G. Dunham (Interscience Publishers, Inc. New York, (1959)), Volume I, page 315.

   H.D.I. Abarbanel in "Strong Interaction Physics", International Summer Institute on Theoretical Physics in Kaiserslautern, 1972; Springer 1972.

   H.M. Fried in "Functional Methods and Models in Quantum Field Theory", The MIT Press, Cambridge, Mass., USA, 1972.

2. This formula might be called a relativistic generalisation of the Saxon-Schiff formulae:

   D.S. Saxon, L.I. Schiff, Nuovo Cim. 6, 614 (1957).

   R. Blankenbecler, R.L. Sugar, Phys. Rev. 183, 1387 (1969).

   M. Schlindwein, Diplomarbeit Bonn 1972.

3. H. M. Fried, loc. cit. chapter 5.

4. A further discussion and generalisation of this formula will be published elsewhere.

5. For the simple loop (no external field in the loop) i.e. $k = i + 1$, the term $\ell_k = n_k = 0$ in the sum (2.11) contains an F_{-2} which, in turn, is proportional to the Bessel function K_o; $K_o(o)$ is logarithmically infinite. This is the well-known logarithmic divergence of the loop. For renormalisation we simply replace this first term in the sum (2.11) by a subtraction constant to be suitably chosen.

Acta Physica Austriaca, Suppl. XIII, 737–766 (1974)

REVIEW OF RECENT WORK ON FINITE QUANTUM ELECTRODYNAMICS*) +)

by

M. P. FRY
Institut für Theoretische Physik
der Universität Graz, Austria

According to Pauli [1], Ehrenfest was perhaps the first to note that Dirac's 1927 version of quantum electrodynamics must lead to a divergent electron self-energy. Opinion in the literature since then on the consistency of finite quantum electrodynamics has been conspicuously inconsistent. Notable, in the sense of commanding widespread attention, are the papers of Källén [2], Gell-Mann and Low [3], Landau and collaborators [4], Johnson, Baker, and Willey [5], and Adler [6]. Here we will review in some detail the work of Johnson, Baker, and Willey (JBW) and Adler. This will serve as an intro-

---

*) Seminar given at the XIII. Internationale Universitätswochen für Kernphysik, Schladming, February 4-15, 1974.

+) Supported by "Fonds zur Förderung der wissenschaftlichen Forschung in Österreich".

duction to and motivation for some work to be discussed later on. A very readable account of the assumptions underlying Källén's result that at least one of the renormalization constants must be infinite, together with a summary of the criticism of his result and his reply to it, may be found in his lectures at Schladming in 1965 [7]. Bjorken's summary of the conference proceedings is also relevant [8]. Gell-Mann and Low's work on the small-distance behavior of quantum electrodynamics, although not directly concerned with the full problem of the finiteness of quantum electrodynamics, is mentioned here because of its profound influence on later developments in the subject. An excellent review of their work is given by Wilson [9]. Discussion of the result of Landau and co-workers that the photon wave function renormalization constant Z_3 is infinite may be found in a lecture of Källén [10] and a paper of Kamefuchi [11].

I. BASIC ASSUMPTIONS

It is assumed that quantum electrodynamics by itself, namely, a charged spin-1/2 fermion minimally coupled to the Maxwell field, is a complete theory [12]. The starting point of JBW in their study of the short-distance behavior of quantum electrodynamics is the Schwinger-Dyson equation for the unrenormalized electron propagator S,

$$\frac{1}{S(p)} = \gamma\cdot p + m_o +$$

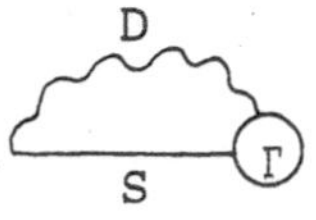

,

where $D_{\mu\nu}$ is the exact unrenormalized photon propagator; Γ_μ is the exact unrenormalized proper electron-photon vertex function, and m_o is the bare electron mass. The idea is to expand Γ_μ in terms of the exact D and S functions

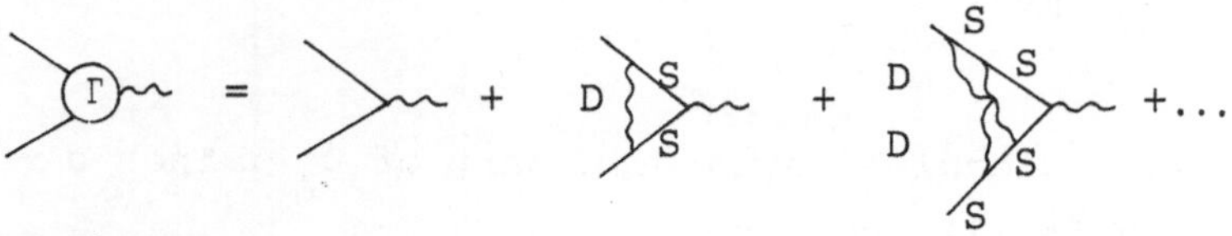

so that

$$\frac{1}{S(p)} = \gamma \cdot p + m_o + \ldots \quad (1)$$

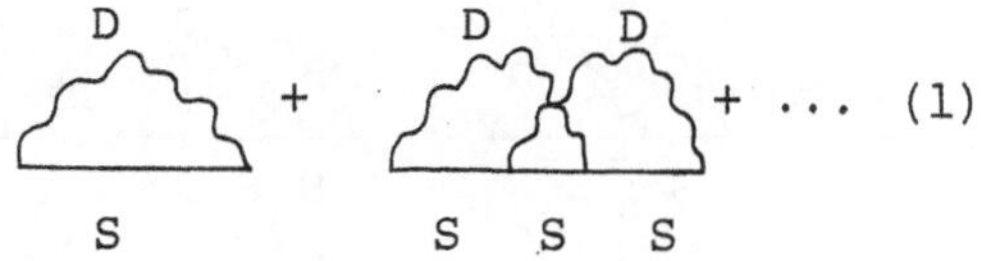

The study of the divergences in quantum electrodynamics is thus centered on the ultraviolet divergences in S and $D_{\mu\nu}$; those in Γ_μ are related to those in S by the gauge covariance of these functions.

Now the above equation for S is terribly complicated and requires some simplification in order to make further progress. Since the primary interest here is the ultraviolet divergences encountered in perturbation theory, and since these divergences are intimately connected with the large spacelike momentum behavior of the propagators in typical self-energy graphs, it seemed reasonable to JBW to replace the exact photon propagator in such graphs by its conjectured asymptotic value. Since $D_{\mu\nu}$ is always sandwiched between two vertices, it always appears multiplied by e_c^2, where e_c is the bare charge. By choosing the gauge of $D_{\mu\nu}$ properly [13], $e_c^2 D_{\mu\nu} = e^2\tilde{D}_{\mu\nu}$, where $\tilde{D}_{\mu\nu}$ is the exact renormalized photon propagator, and e is the physical electron charge. Thus, an assumption about the asymptotic behavior of the product

$e_c^2 \, D_{\mu\nu}$ is equivalent to an assumption about the large spacelike momentum behavior of $\tilde{D}_{\mu\nu}$.

So, for example, suppose that

$$\tilde{D}_{\mu\nu}(q) \underset{q^2\to\infty}{\sim} D^o_{\mu\nu}(q) \quad ,$$

where $D^o_{\mu\nu}$ is the leading asymptotic part of $\tilde{D}_{\mu\nu}$. Then a typical graph in (1) such as

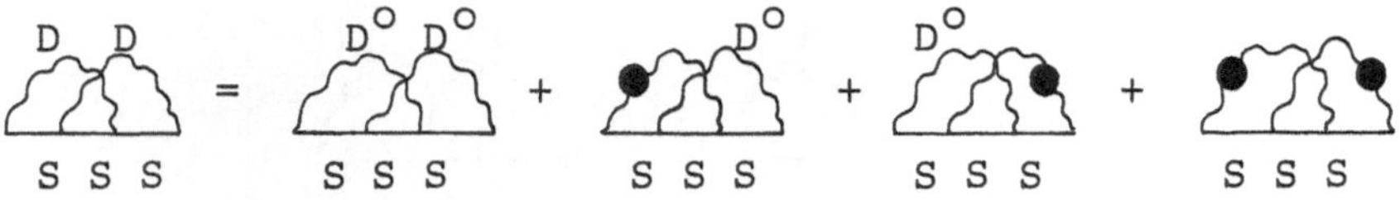

where the shaded blob denotes the nonasymptotic part of $\tilde{D}_{\mu\nu}$, is replaced with the single graph

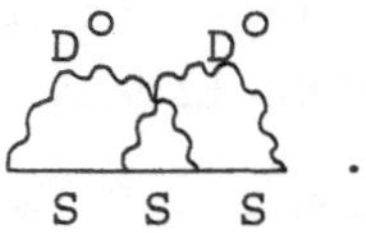

This is obviously a nontrivial step. It amounts to the assumption that the nonasymptotic part of $\tilde{D}_{\mu\nu}$, neglected in each of the graphs in (1), does not contribute to the dominant asymptotic behavior of S(p) for large spacelike values of p^2. Assuming that $\tilde{D}_{\mu\nu}$ has the asymptotic behavior specified below, the nonasymptotic part of $\tilde{D}_{\mu\nu}$ gives a contribution to the right-hand side of (1) that vanishes <u>graph by graph</u> as $p^2 \to \infty$ when the propagator S in the base lines of these graphs is replaced by its renormalized power-series expansion in e^2. So the neglect of the nonasymptotic part of $\tilde{D}_{\mu\nu}$ is at least valid in perturbation theory.

Specifically, it is assumed that the renormalized photon propagator is asymptotically free within a constant factor:

$$\alpha\tilde{D}_{\mu\nu}(q) = (g_{\mu\nu} - \frac{q_\mu q_\nu}{q^2})(\alpha_o + h(q^2/m^2))\frac{1}{q^2} + \text{gauge terms} \quad (2.1)$$

with

$$h(q^2/m^2) \xrightarrow[q^2 \gg m^2]{} 0 \; . \quad (2.2)$$

Here m is the physical electron mass; $\alpha (= e^2/4\pi)$ is the physical fine structure constant, and α_o is a finite, positive constant that Adler calls the asymptotic coupling. For technical reasons α_o is not necessarily the bare fine structure constant $\alpha_c (= e_c^2/4\pi)$. It is further assumed that h vanishes asymptotically with power-law behavior and not like, say,

$$1/\log(q^2/m^2) \quad \text{or} \quad 1/\log\log(q^2/m^2)$$

for $q^2 \gg m^2$. We remark that the asymptotic freedom of $\tilde{D}_{\mu\nu}$ implies that α_c is cutoff-independent.

The assumption that $\tilde{D}_{\mu\nu}$ is asymptotically free is so fundamental that we cannot pass it by entirely without some comment on its physical motivation.

II. QUANTUM ELECTRODYNAMICS AND CRITICAL PHENOMENA

It was assumed that $\tilde{D}_{\mu\nu}(q)$ is asymptotically free and that $h(q^2/m^2)$ vanishes as $q^2 \to \infty$. If so, then the equation

$$\alpha\tilde{D}_{\mu\nu}(q) = \frac{\alpha_0}{q^2} g_{\mu\nu} + \text{gauge terms}$$

is exact for massless electrons. Since this just differs from the free photon propagator by a constant factor, the absorptive part of $\tilde{D}_{\mu\nu}$ must vanish at $m = 0$. But this is just the Fourier transform of the renormalized contracted current correlation function $\langle 0|j_\mu(x)j^\mu(0)|0\rangle$, where j_μ is the electromagnetic current operator. Thus [14],

$$\lim_{m\to 0} \langle 0|j_\mu(x)j^\mu(0)|0\rangle = 0 .$$

For physical electrons, covariance and simple dimensional considerations imply that

$$\langle 0|j_\mu(x)j^\mu(0)|0\rangle = \frac{1}{x^6} f(x^2m^2) ,$$

where f is a dimensionless function which vanishes at $m = 0$. So for massive electrons the requirement that $\tilde{D}_{\mu\nu}$ is asymptotically free can be stated as

$$\lim_{\rho\to 0} \rho^6 \langle 0|j_\mu(\rho x)j^\mu(0)|0\rangle = \frac{1}{x^6} \lim_{\rho\to 0} f(\rho^2x^2m^2)$$

$$= 0 .$$

In perturbation theory this limit diverges as powers of $\log\rho$ beginning in fourth order.

To understand how a vanishing limit might come

about [15], suppose quantum electrodynamics is continued to Euclidean space. Suppose further that a non-covariant ultraviolet cutoff is introduced by slicing this space up into a four-dimensional hypercubic lattice with spacing ℓ $<< \hbar/mc$ and defining the field operators only at the lattice points as illustrated in Fig. 1. Such a

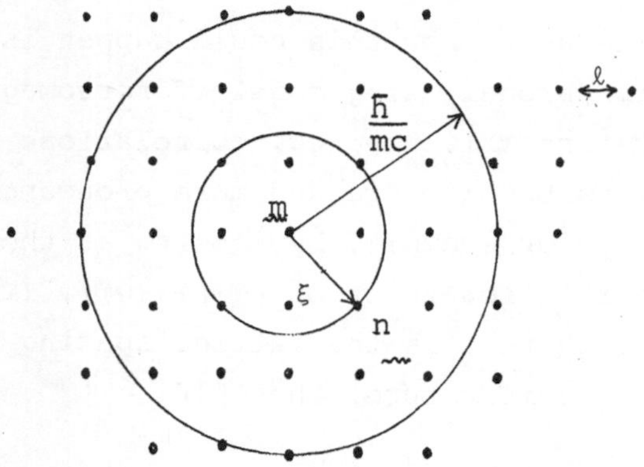

Fig. 1

model of quantum electrodynamics has three fundamental lengths: the lattice spacing ℓ, the electron Compton wavelength $\hbar/mc$, and the distance ξ over which the correlation of a field operator at lattice site $\vec{m}$, say, with a field operator at lattice site $\vec{n}$ is appreciable. Call ξ the correlation length. In this model an asymptotically free photon propagator implies that somehow the correlation length becomes so large that the ground state current correlation function

$$\langle\Omega|j_\mu(\vec{n})j^\mu(\vec{0})|\Omega\rangle \qquad (3)$$

calculated for

$$\ell << |\vec{n}| << \hbar/mc \tag{4}$$

falls off <u>more slowly</u> than $(1/|\vec{n}|)^6$. It is conjectured that this happens when $\xi \gtrsim \hbar/mc$.

A clue as to how this could happen is given by the two-dimensional Ising model of ferromagnetism. As the temperature T is lowered, correlations between lattice sites become more and more pronounced. This is verified, for example, by looking at the spin-spin thermal correlation function $<\sigma(\vec{n})\sigma(\vec{0})>$. If $|\vec{n}| >> \ell$ and $T > T_c$, where ℓ is the lattice spacing and T_c is the critical temperature, then [16]

$$<\sigma(\vec{n})\sigma(\vec{0})> \sim \frac{1}{|\vec{n}|^{1/2}} e^{-|\vec{n}|/\xi(T)} .$$

The correlation length ξ increases as T decreases. At $T = T_c$ the correlation length becomes infinite, signifying a phase transition to the ferromagnetic state, and resulting in a correlation function of infinite spatial extent whose form, for $|\vec{n}|>>\ell$, is [17]

$$<\sigma(\vec{n})\sigma(\vec{0})> \sim \frac{1}{|\vec{n}|^{1/4}} .$$

The important thing to note here is that the spin-spin correlation function falls off very slowly with increasing $|\vec{n}|$ at $T = T_c$, indicating the onset of the long-range spin ordering effects responsible for the

Ising model's broken reflection symmetry for $T < T_c$. The behavior of the Ising model in both two and three dimensions near the critical temperature appears to be determined mainly by the short-range character of the interaction and only to a lesser extent by the details of the Hamiltonian [18].

This being the case, it is tempting to speculate that quantum electrodynamics also exhibits critical behavior. Firstly because the Maxwell and electron fields at each point of space are separate degrees of freedom so that any spatial volume contains an infinite number of degrees of freedom. And secondly because the degrees of freedom interact locally. In place of thermal noise there is quantum noise whose level is fixed by α.

Consider quantum electrodynamics on a lattice again. In analogy with ferromagnetism, we suspect the assumption that (3) falls off more slowly than $(1/|\vec{n}|)^6$ when $|\vec{n}|$ satisfies (4) will be correct if this lattice model of quantum electrodynamics undergoes some kind of phase transition when α is lowered to a critical value α_{crit} so that $\xi = \infty$. This phase transition might take the form of a permanent vacuum electric dipole moment related, perhaps, to Dyson's conjecture [19] of the nonexistence of a stable vacuum state for negative α if quantum electrodynamics is analytic at $\alpha = 0$. The less singular behavior of the current correlation function at distances that are small compared to $\hbar/mc$ may permit a smooth transition to the continuum limit of the lattice when $\alpha = \alpha_{crit}$.

Thus the following possible analogy between ferromagnetism and quantum electrodynamics emerges:

Ferromagnetism	Quantum Electrodynamics
A. Thermal noise	A.′ Quantum noise

B. Noise level fixed by T

C. $<\sigma(\vec{n})\sigma(\vec{0})> \sim (1/|\vec{n}|)^{1/4}$ for $T = T_c$ and $|\vec{n}| >> \ell$.

D. $<\sigma> \neq 0$ for $T < T_c$

B'. Noise level fixed by α

C'. $\lim_{\rho\to 0} \rho^6 <0|j_\mu(\rho x) j^\mu(0)|0> = 0$ for $\alpha = \alpha_{crit}$?

D'. Permanent vacuum electric dipole moment for $\alpha < \alpha_{crit}$

The above analogy is imperfect on one major account: There is no mass in the Ising model. Consequently, one cannot state condition C above in the more suggestive form

$$<\sigma(\vec{n})\sigma(\vec{0})> \sim \frac{1}{|\vec{n}|^{1/4}},$$

for $T = T_c$ and $\ell << |\vec{n}| << \hbar/Mc$, where M is some characteristic mass.

Summarizing, the conjectured nonanalyticity in α of the infinite-momentum limit of $\tilde{D}_{\mu\nu}$ in the sense that

$$\alpha \lim_{q^2\to\infty} q^2 D_{\mu\nu}(q) = g_{\mu\nu}\alpha_o + \text{gauge terms if } \alpha = \alpha_{crit},$$

$$= \infty \text{ if } \alpha > \alpha_{crit},$$

is suggested by the known behavior of other systems with many degrees of freedom within a correlation length. For we know that the behavior of such systems - of which quantum electrodynamics is an example <u>par excellence</u> - is characterized by a nonanalytic dependence on their input parameters. The points of nonanalyticity mark the transition to states with qualitatively different properties as, for example, the breaking of the rotational invariance of a lattice of spins at its critical temperature or the onset of turbulance in a fluid at a critical Reynold's number.

These very qualitative remarks can be misleading. In particular, if the renormalized photon propagator is asymptotically free, then Eq. (2) indicates that α_o, not α, is the coupling constant governing the small-distance behavior of quantum electrodynamics. So the condition that $\tilde{D}_{\mu\nu}$ be asymptotically free may instead place a constraint on α_o rather than α. More will be said about this in Sec. V.

III. BEHAVIOR OF $S(p)$ FOR LARGE p^2 AND THE VALUE OF m_o

If $\alpha_c D_{\mu\nu}$ is now replaced in Eq. (1) by its conjectured asymptotic form

$$\alpha_c D_{\mu\nu} = \alpha \tilde{D}_{\mu\nu} \rightarrow (g_{\mu\nu} + G \frac{q_\mu q_\nu}{q^2}) \frac{\alpha_o}{q^2} \equiv \alpha_o D^o_{\mu\nu},$$

where G specifies the gauge, one gets an equation for S alone:

$$\frac{1}{S(p)} = \gamma \cdot p + m_o +$$

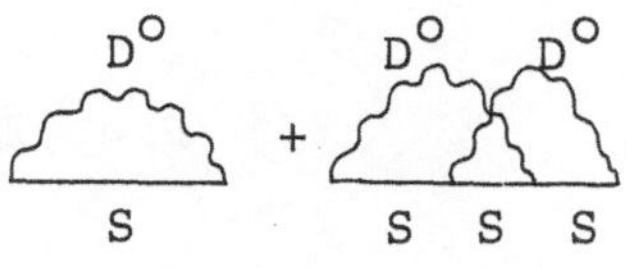

$$+ \cdots (5)$$

In this equation the asymptotic coupling α_o replaces α_c. The claim is that this simplified equation contains all the ultraviolet divergences associated with Z_2 and m_o in the exact Schwinger-Dyson equation for S and that it gives the correct asymptotic behavior of $S(p)$ for large spacelike values of p^2 - all this provided $\hat{D}_{\mu\nu}$ is asymptotically free and h vanishes with power - law behavior. In order to remove the ultraviolet divergence due to m_0, a subtraction is made at $p^2 = -m^2$:

$$\frac{1}{S(p)} = \gamma \cdot p + m + \overset{D\,\circ}{\diagdown}_{S} + \overset{D\,\circ\,D\,\circ}{\diagdown}_{S\ S\ S} + \cdots - \left(\overset{D\,\circ}{\diagdown}_{S} + \overset{D\,\circ\,D\,\circ}{\diagdown}_{S\ S\ S} + \cdots \right)_{\gamma \cdot p = -\mathrm{I}} \quad (6)$$

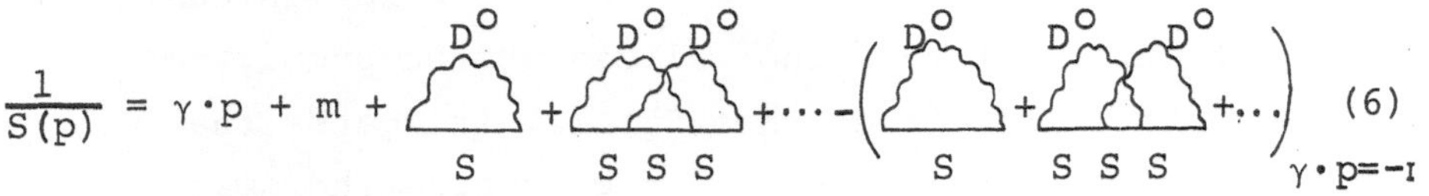

The crucial step that enables Johnson and Baker to solve this equation in the region $p^2 \gg m^2$ is the following theorem proved by JBW [20]:

Suppose the solution of (6) has the asymptotic behavior

$$S(p) \underset{p^2 \gg m^2}{\sim} \frac{\text{const.}}{\gamma \cdot p}$$

to order α_0^n. Then the off-mass shell e^+-e^- scattering amplitude of order α_0^{n+1} (Fig. 2)

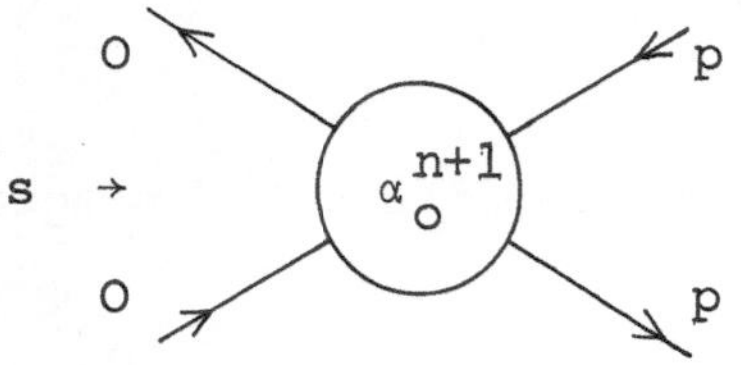

Fig. 2

calculated for $p^2 > 0$ and by omitting all photon self-energy parts and all one- and two-particle s-channel intermediate states, remains finite in the limit $m = 0$.

This is perhaps the central theorem of JBW's work; without it many of their results would be false or at least would have to be rederived following some other route. The intimate connection between a finite theory of quantum electrodynamics and the insensitivity of the above e^+-e^- scattering amplitude to the electron mass has been repeatedly stressed by JBW.

Using renormalization group techniques, which the above theorem partly justifies, Baker and Johnson [21] then go on to show that the solution of Eq. (6) has the scaling form

$$S(p) \underset{p^2 >> m^2}{\sim} (\text{const.})\left[\frac{1}{\gamma \cdot p} + (\text{const.}) \frac{m}{p^2}\left(\frac{m^2}{p^2}\right)^{\varepsilon} \right] \qquad (7)$$

in a fixed gauge provided $-1/2<\varepsilon<1$. The parameter ε and the gauge G are functions of α_o alone and are known only in lowest order:

$$\varepsilon = \frac{3\alpha_o}{4\pi} + \frac{3\alpha_o^2}{32\pi^2} + \ldots$$

$$G = -1 + \frac{3\alpha_o}{4\pi} + \ldots$$

Even though G is only known explicitly to order α_o, JBW [20] have proved by induction that a gauge can always be found that renders Z_2 cutoff-independent in each order of α_o provided $\tilde{D}_{\mu\nu}$ has the properties listed in Sec. II.

The result (7) can also be obtained from the Callen-Symanzik equation for S, as shown by Adler and Bardeen [22]. No constraint is placed on ε in this approach. In both cases it is assumed that terms that vanish in each order of α_0 as $p^2 >> m^2$ do not asymptotically dominate (7) when summed up. If it turns out that $\varepsilon < -1/2$, then the mass term in (7) dominates, and one begins to doubt if the neglect of vanishing mass-dependent terms in the derivation of the Callen-Symanzik equation for S is legitimate.

Now let us see what implication these results have for the electron bare mass m_0. Suppose $\varepsilon > 0$. If the asymptotic solution (7) is substituted into (5), then by a simple scaling argument applied to each self-energy graph one gets

$$S(p) \underset{p^2>>m^2}{\sim} (\text{const.})\left[\frac{1}{\gamma\cdot p} + \frac{(\text{const.})m_0}{p^2}\right]. \qquad (8)$$

Comparing the two asymptotic expressions (7) and (8) for S(p), one sees immediately that they are consistent only if

$$m_0 = 0 .$$

In this case all of the electron's mass is electromagnetic [23]. The value of m cannot be calculated, since the theory has no scale of length when $m_0 = 0$. For $\varepsilon \leq 0$, the right-hand side of (5) diverges term by term. Whether this implied divergence in m_0 is real or is simply due to an invalid expansion of Γ_μ in terms of the exact S and D functions remains an open problem.

IV. THERE ARE NO GOLDSTONE BOSONS IN QUANTUM ELECTRODYNAMICS

The case $m_o = 0$ with $m > 0$ (as represented by the asymptotic solution (7) for $S(p)$ with $\varepsilon > 0$) appears to be a chiral symmetry breaking solution of quantum electrodynamics and hence raises the danger of unwanted Goldstone bosons. There are none. To see this, consider the Ward identity for the unrenormalized proper vertex Γ_5^μ associated with the axial-vector current $j_{5\mu}$:

$$q_\mu \Gamma_5^\mu(p+q,p) = \Gamma_5(p+q,p) + S^{-1}(p+q)\gamma_5 + \gamma_5 S^{-1}(p) \ . \qquad (9)$$

The proper pseudoscalar vertex Γ_5 corresponds to the pseudo-scalar density j_5. In lowest order $\Gamma_5 = 0$ when $m_o = 0$. For $m \neq 0$, the last two terms in (9) give a finite contribution at $q_\mu = 0$, and hence Γ_5^μ will be singular at $q_\mu = 0$ if the <u>exact</u> vertex $\Gamma_5(p,p)$ also vanishes at $m_o = 0$. Such a singularity would suggest the presence of a zero-mass pseudoscalar particle associated with the breaking of the formal γ_5 invariance of quantum electrodynamics when $m_o = 0$. But if the integral equation for $\Gamma_5(p,p)$ has a nonvanishing solution for $m_o = 0$ satisfying the constraint

$$\Gamma_5(p,p) + S^{-1}(p)\gamma_5 + \gamma_5 S^{-1}(p) = 0, \qquad (10)$$

the would-be Goldstone boson is eliminated. This is precisely what happens after summing <u>all</u> orders of perturbation theory [24]. The formal chiral invariance of quantum electrodynamics when $m_o = 0$ is dynamically broken. Of course, one also has the option of choosing the trivial solution $\Gamma_5 = 0$ when $m_o = 0$, in which case there are indeed Goldstone bosons [25]. Quantum electrodynamics selects the nontrivial solution (10). A solution for S with $m > 0$ also breaks the formal scale invariance of quantum electrodynamics when $m_o = 0$. It is conjectured [26] that anomalies in the divergence of the dilation current, which are known to be present in perturbation theory [27], will prevent the occurance of potential scalar Goldstone bosons.

V. BEHAVIOR OF $\tilde{D}_{\mu\nu}(q)$ FOR LARGE q^2

To complete the short-distance study of quantum electrodynamics, the consistency of the initial assumption of an asymptotically free photon propagator must be checked. We will briefly describe the considerable progress that has been made on this subject.

The dynamics of $\tilde{D}_{\mu\nu}$ is governed by the renormalized photon proper self-energy Π_c:

$$\alpha\tilde{D}_{\mu\nu}(q) = (g_{\mu\nu} - \frac{q_\mu q_\nu}{q^2}) \frac{1}{q^2} \frac{\alpha}{1+\alpha\Pi_c(q^2)} + \text{gauge terms}$$

where $\Pi_c(q^2) = \Pi(q^2) - \Pi(0)$, and $\Pi(q^2)$ can be calculated in perturbation theory from a sum of graphs of the form

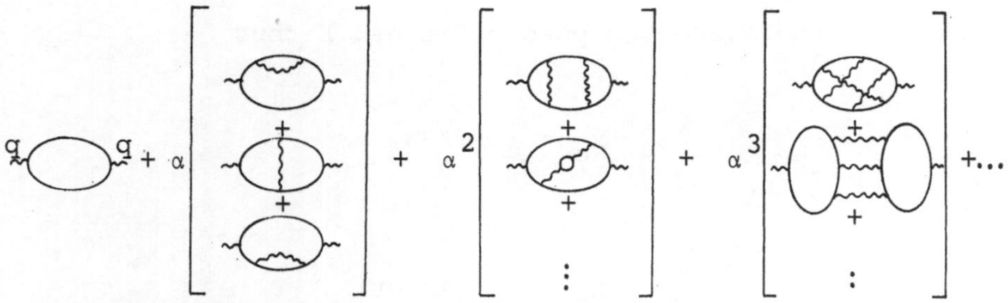

In perturbation theory

$$\Pi_c(q^2) \underset{q^2>>m^2}{\sim} C_o(\alpha)+C_1(\alpha)\log(q^2/m^2)+C_2(\alpha)\log^2(q^2/m^2)+\ldots \tag{11}$$

This sum must converge to a finite constant for consistency.

The following remarkable result has been obtained by Adler [28]: Take all one-loop graphs in the series for $\Pi_c(q^2)$, substitute an arbitrary coupling constant x for α, and sum them to obtain

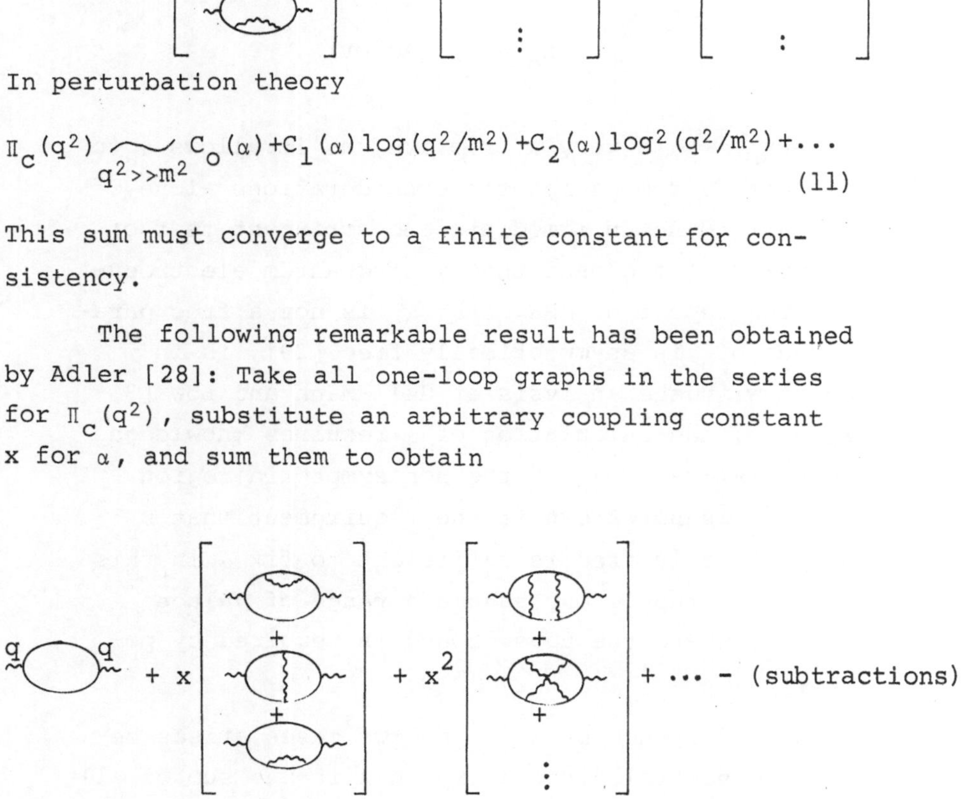

$$\underset{q^2>>m^2}{\sim} (g_{\mu\nu}q^2-q_\mu q_\nu)[F(x)\ \log\ (q^2/m^2) + \text{const.}]. \tag{12}$$

Then a <u>necessary condition</u> for an asymptotically free photon propagator is that F(x) vanish with a positive infinite-order zero at $x = x_o$:

$$F(x_o) = 0^\infty .$$

Moreover, there are two possibilities: Either

$$x_o = \alpha_o \quad \text{(JBW)}$$

or

$$x_o = \alpha \quad \text{(Adler)} \ .$$

The case $x_o = \alpha$ is very attractive, as it allows α to be calculated from asymptotic considerations alone, thereby affording an immediate and stringent test of the adequacy of a closed theory of quantum electrodynamics. The result of JBW, that α_o is not a free parameter when $\tilde{D}_{\mu\nu}$ is asymptotically free [29], is in agreement with the analysis of Gell-Mann and Low [3]. For $x_o = \alpha_o$, the calculation of α requires knowledge of the behavior of $\tilde{D}_{\mu\nu}$ in the nonasymptotic region $q^2 \lesssim m^2$. It is not known if the requirement that $\tilde{D}_{\mu\nu}$ be asymptotically free is sufficient to fix α in this case. If not, then α may assume a range of values $0 < \alpha < \alpha_o$, where the upper bound is required by positivity.

The difference between the two cases arises because the essential singularity in x in the sum of all one-loop photon self-energy parts (if present) will also be present in the full perturbation series for Π_c when calculated with coupling constant x. Consequently, different prescriptions for summing the series can give inequivalent results. Adler's condition emerges by first summing all one-loop graphs, then all two-loop graphs, and so on. The result of JBW is obtained by first summing up all graphs for Π_c that differ only by photon self-energy insertions and then summing the remaining graphs.

The summation procedures are inequivalent, since by positivity $\alpha_o > \alpha$, and hence the point $x = \alpha_o$ lies outside the radius of convergence of loopwise summation. Both procedures may be correct, in which case they simply define different theories [30].

The distinction between $x_o = \alpha$ and $x_o = \alpha_o$ shows up quite dramatically in the rate that the nonasymptotic part $h(q^2/m^2)$ of $\tilde{D}_{\mu\nu}(q)$ vanishes with increasing q^2. Its expected behavior for the two cases is qualitatively sketched in Fig. 3. Thus, for the case $x_o = \alpha$ we expect the radiative correction logarithms in $\tilde{D}_{\mu\nu}$ to disappear much sooner than in the case $x_o = \alpha_o$ [31,32].

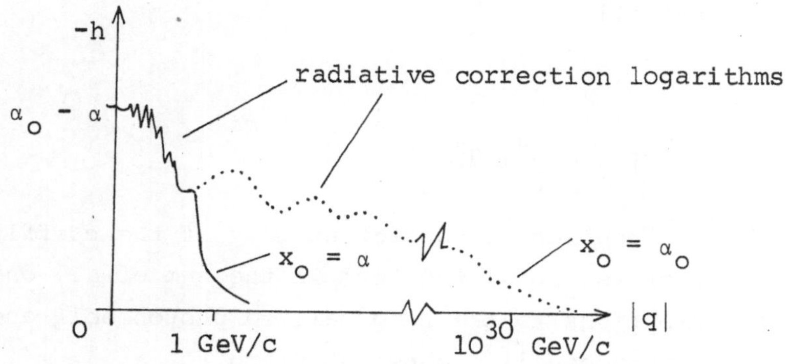

Fig. 3

This immediately suggests that the physical solution of quantum electrodynamics can be discovered from accurate elastic electron and electron-positron scattering experiments for large values of q^2 [32]. The difficulty is that even at the 1 to 2% level of accuracy for a colliding beam experiment with c.m. energies ranging from 1.2 GeV to 3.0 GeV, the data can be fitted very well by cross sections calculated only to order α^3 [33].

So, as far as the photon propagator is concerned, such experiments are so far only sensitive to the large q^2 behavior of the lowest-order contribution to the photon self-energy. Since the probing of higher-order contributions at presently available machine energies will require accuracies of better than 0.1%, this route to the discovery of how quantum electrodynamics sums itself up is out of the question for some time to come.

VI. OVERVIEW

Suppose the exact renormalized photon propagator is asymptotically free:

$$\alpha \tilde{D}_{\mu\nu}(q) \underset{q^2>>m^2}{\sim} g_{\mu\nu} \frac{\alpha_o}{q^2} + \text{gauge terms} .$$

Then this assumption is consistent only if the coefficient $F(x)$ of the logarithm term in the sum of all one-loop contributions to the renormalized photon self-energy, calculated with coupling constant x and for $q^2 >> m^2$, has an infinite-order zero for a positive value x_o of x: $F(x_o) = 0^{\infty}$. Secondly, α_c, the bare fine structure constant, is cutoff-independent and can, in principle, be calculated in terms of α_o. There are at least two possibilities: $x_o = \alpha$ and $x_o = \alpha_o$.

For the case $x_o = \alpha$, Z_2 $(= Z_1)$ is cutoff-independent in an appropriate gauge. In this gauge the unrenormalized electron propagator has the behavior

$$S(p) \underset{p^2>>m^2}{\sim} \frac{\text{const.}}{\gamma \cdot p} ,$$

for $\varepsilon(\alpha_o) > -1/2$. If $\varepsilon(\alpha_o) > 0$, the electron bare mass m_o is zero, and the physical electron mass is arbitrary. There are no Goldstone bosons when $m > 0$.

For the case $x_o = \alpha_o$, $h(q^2/m^2)$ decreases more slowly than any power of q^2/m^2, and the JBW analysis of Z_2, m_o, and the asymptotic behavior of S fails. Except for the positivity requirement $\alpha < \alpha_o$, the value of α may be left undetermined. A preliminary analysis of this case has been made by Crewther, et al. [34] using the Callan-Symanzik equation for S. The reader is referred to their paper for details.

Finally, the condition $F(x_o) = 0^\infty$ does not guarantee that $\tilde{D}_{\mu\nu}$ is asymptotically free. A sufficient condition for this when $x_o = \alpha$ is that all the c_i's in Eq. (11) for $i \geq 1$ separately vanish at the value of α calculated from $F(\alpha) = 0^\infty$. For the case $x_o = \alpha_o$, a sufficient condition is that the coefficient c_1 in Eq. (8) with the value of α_o substituted for α satisfy $c_1(\alpha_o) = 0^\infty$ [35].

Until all the assumptions [36] underlying these detailed results are checked out, and quantities such as F and ε are calculated, all that can honestly be said is that a finite, closed theory of quantum electrodynamics cannot be ruled out. Even if all is well, the basic incompleteness of such a theory is manifest by its inability to predict the electron mass should $x_o = \alpha$ be the physical solution.

VII. CALCULATION OF F

Obviously it would be very desirable to know the

eigenvalue function F(x). So far the most ambitious attempts to determine its analytic properties have exploited the invariance of quantum electrodynamics under the full conformal group in the zero electron mass limit when photon self-energy insertions are omitted and the photon propagator is restricted to a spcial gauge [37]. In particular, Adler [38] has considered a model of massless electrodynamics in which electrons are treated exactly, but with the short-distance singularity of the free photon propagator cut off. It is found that the sum of <u>all</u> graphs contributing to F(x) <u>cannot</u> develop an infinite-order zero as x approaches a positive value x_o from below. So, if F(x) does have such an infinite-order zero, it must be sensitive to the small-distance behavior of the free-photon propagator.

Let us now briefly describe an alternative approach [39] to the calculation of F that may prove useful. One important simplification is that F can be calculated in a model of quantum electrodynamics without vacuum polarization. Specifically, the one-loop diagrams in Eq. (12) can be obtained from the Schwinger-Dyson equation for the photon self-energy,

$$(g_{\mu\nu}q^2 - q_\mu q_\nu)\Pi(q^2) =$$

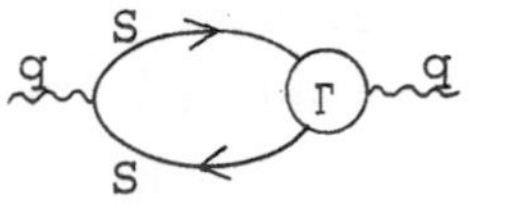

, (13)

where S and Γ_μ are calculated without closed electron loops. Thus, graphs like

are excluded from S and Γ_μ.

Already we are encouraged because the exact equation for S in this model takes the form of a very simple looking functional differential equation [40]

$$[\gamma^\mu(\frac{1}{i}\partial_\mu - e\int D^o_{\mu\nu}(x-y)J^\nu(y) + ie\frac{\delta}{\delta J^\mu(x)}) + m_o]S(x,x'|J) = \delta(x-x'), \tag{14}$$

where J_μ is an arbitrary external current; $D^o_{\mu\nu}$ is the free photon propagator in an arbitrary gauge; m_o is the electron bare mass, and e is some coupling constant. Yet it takes but a moment to realize that its outward simplicity is deceptive. For knowledge of the functional dependence of S on J_μ allows one to calculate all processes (without vacuum polarization) of the type $e^+ + e^- \to n\gamma$ simply by functionally differentiating S n-times. So S[J]contains an enormous amount of information that is superfluous for our purpose. We are also led to suspect that a nonperturbative solution of (14) is out of the question. Yet one is reluctant to abandon it altogether, since it is a very compact expression of the field dynamics: It is equivalent to an infinite coupled system of equations for all Green's functions with two external charged legs. If Eq. (14) could be used as the starting point for ridding S of its functional dependence on J_μ without explicitly introducing higher-point Green's functions, then perhaps there would be some justified hope for finding an explicit expression for S in the absence of closed electron loops.

This has in fact been accomplished [41]. Starting

from Eq. (14), it has been shown that the problem of calculating S in the limit $J_\mu = 0$ reduces to the problem of calculating the motion of an electron in the path-dependent potential

$$A_\mu(x) = -e^2\int_{x_o}^{x} d\xi^\nu D^o_{\mu\nu}(x-\xi),$$

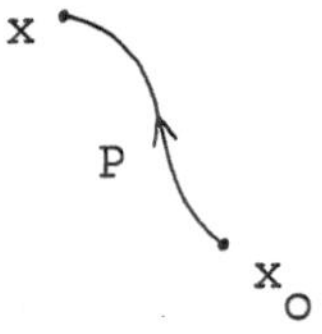

where the point x_o and the path P joining x_o with x are both left arbitrary. Specifically, the <u>exact</u> unrenormalized electron propagator S in the absence of closed electron loops is obtainable from the path-dependent propagator S(P) calculated from

$$[\gamma^\mu(\frac{1}{i}\partial_\mu+e^2\int_{x_o}^{x} d\xi^\nu D^o_{\mu\nu}(x-\xi))+m_o]S(x,x'|P)=\delta(x-x'), \qquad (15)$$

by shrinking the path P to zero by letting x_o approach x along P:

$$S(x-x') = \lim_{x_o\to x} S(x,x'|P). \qquad (16)$$

The divergences associated with m_o and Z_2 can be suppressed by introducing an ultraviolet cutoff in $D^o_{\mu\nu}$. We note that Eqs. (15) and (16) with $m_o = 0$ are an <u>exact</u> description of massless electrons provided the exact renormalized photon propagator $\tilde{D}_{\mu\nu}$ is asymptotically free. They are exact, since the approximation of neglecting closed electron loops in the massive electron case is exact for massless electrons when $\tilde{D}_{\mu\nu}$ is asymptotically free [14]. In this case, e^2 in (15) is replaced with $4\pi\alpha_o$, where α_o is the asymptotic coupling defined by Eq. (2). We also note that S(P) is gauge-dependent, which makes it essentially different from the path-dependent but gauge-invariant electron propagator introduced some years ago by Mandelstam [42].

A major difficulty that has so far prevented a nonperturbative solution of (15) is that one cannot first choose a convenient path, do the line integral, then solve for S(P). Instead, one must first solve for S(P) keeping the line integral, then shrink it to zero according to (16). This being the case, the equation for S(P) has a kind of nonlocality built into it, since the derivative of S(P) at x depends not only on information at x, but also on the path joining x to a point x_o that may be far removed from x.

It has also been possible to obtain an exact equation for the once-amputated unrenormalized electron-photon vertex in the absence of closed electron loops:

$$[\gamma^{\alpha}(\frac{1}{i}\partial_{\alpha}+e^2\int_{x_o}^{x} d\eta^{\beta}D^{o}_{\alpha\beta}(x-\eta))+m_o]V^{\mu}(x,y,\xi|P)=\delta(x-y)\int_{x_o}^{x}d\eta^{\mu}\delta(\xi-\eta), \tag{17}$$

where

$$iS\Gamma^{\mu}S = \lim_{x_o\to x} V^{\mu}(P) .$$

Hence, knowledge of the path-dependent vertex $V^{\mu}(P)$ allows one to calculate F from the Schwinger-Dyson equation (13).

What has been shown is that a system with an infinite number of degrees of freedom can be simulated by a path-dependent potential. Whether this is just a mathematical trick or has some physical significance is not clear. We suspect the former, since the procedure used to derive (15) and (17) does not work outside Abelian gauge theories, or even within them when closed electron loops are present. What is clear is that the full exploitation of this approach to the calculation of F is very closely linked to our imperfect knowledge of the dynamics of an electron in a path-dependent potential.

REFERENCES

1. W. Pauli, Naturwiss. 21, 841 (1933);
   R. Jost, in Aspects of Quantum Theory, edited by A. Salam and E. P. Wigner (Cambridge Univ. Press, Cambridge, England, 1972), p. 73.

2. G. Källén, K. Dan. Vidensk. Selsk. Mat.-Fys. Medd. 27, No. 12 (1953);
   CERN Report No. CERN/T/GK/3, 1955 (unpublished);
   in Proc. of the CERN Symposium on High Energy Accelerators and Pion Physics (CERN, Geneva, 1956), Vol. 2, p. 187;
   Handbuch der Physik V/1 (1958);
   Acta Phys. Austr., Suppl. II, 133 (1965).

3. M. Gell-Mann, F. E. Low, Phys. Rev. 95, 1300 (1954).

4. L. D. Landau, in Niels Bohr and the Development of Physics, edited by W. Pauli (McGraw-Hill, New York, 1955), and references cited therein.

5. K. Johnson, M. Baker, R. Willey, Phys. Rev. Letters 11, 518 (1963).
This preliminary account of their work is further developed in K. Johnson, M. Baker, R. Willey, Phys. Rev. 136, B1111 (1964);
K. Johnson, R. Willey, M. Baker, ibid. 163, 1699 (1967)
Zh. Eksp. Teor. Fiz. 52, 318 (1967) (Sov. Phys.-JETP 25, 205 (1967));
M. Baker, K. Johnson, Phys. Rev. 183, 1292 (1969); Phys. Rev. D3, 2516 (1971); 3, 2541 (1971). This work, as well as that of S. L. Adler (Phys. Rev. D5, 3021 (1972); 7, 1948 (E) (1973)) is reviewed in K. Johnson and M. Baker, Phys. Rev. D8, 1110 (1973).

6. S. L. Adler, Phys. Rev. D5, 3021 (1972); 6, 3445 (1972); 7, 1948(E) (1973); 7, 3821(E) (1973); 8, 2400 (1973).

7. G. Källén, Acta Phys. Austr., Suppl. II, 133 (1965).

8. J. D. Bjorken, Acta Phys. Austr., Suppl. II, 239 (1965)

9. K. Wilson, Phys. Rev. D3, 1818 (1971), Sec. II and the Appendix.

10. G. Källén, in Proc. of the CERN Symposium on High Energy Accelerators and Pion Physics (CERN, Geneva, 1956), Vol. 2, p.187.

11. S. Kamefuchi, K. Dan. Vidensk. Selsk. Mat.-Fys. Medd. 31, No. 6 (1957).

12. Therefore, this eliminates from consideration here open theories of the sort proposed by T. D. Lee and G. C. Wick, Phys. Rev. D2, 1o33 (1970);
A. Salam, J. Strathdee, Nuovo Cimento Letters 4, 101 (1970);
C. J. Isham, A. Salam, J. Strathdee, Phys. Rev. D3, 18o5 (1971); 5, 2548 (1972).

13. S. L. Adler, Phys. Rev. D5, 3021 (1972), Sec. 2.

14. M. Baker, K. Johnson, Phys. Rev. D3, 2541 (1971).

15. Much of the work in this section was suggested by K. Wilson and J. Kogut, Phys. Reports (to be published), especially Secs. I, II, and X, and by K. Johnson, talk delivered at the University of Maryland (1972) (unpublished).

16. L. P. Kadanoff, Nuovo Cimento 44B, 276 (1966).

17. T. T. Wu, Phys. Rev. 149, 380 (1966).

18. See, for example, H. E. Stanley, Introduction to Phase Transitions and Critical Phenomena (Clarendon Press, Oxford, England, 1971), Ch. 8 and references cited therein.

19. F. J. Dyson, Phys. Rev. 85, 631 (1952).

20. K. Johnson, R. Willey, M. Baker, Phys. Rev. 163, 1699 (1967), Sec. IV and Appendix A.

21. M. Baker, K. Johnson, Phys. Rev. D3, 2516 (1971).

22. S. L. Adler, W. A. Bardeen, Phys. Rev. D4, 3045 (1971); 6, 734(E) (1972).

23. See Ref. 21, Sec. VI.

24. K. Johnson, in 9th Latin American School of Physics, Santiago, Chile, 1967, edited by I. Saavedra (Benjamin, New York, 1968), Sec. 8; M. Baker, K. Johnson, Phys. Rev. D3, 2516 (1971), Sec. VII. We note that Eq. (10) implies that $\partial^\mu j_{5\mu} \neq 0$. This would seem to imply that the triangle anomaly would be sufficient to avoid potential Goldstone bosons. However, since it gives no contribution to the Ward identity (9) at $q_\mu = 0$, it cannot

prevent $\Gamma_5^{\mu}(p + q,p)$ from developing a singularity at $q_{\mu} = 0$.

25. H. Pagels, Phys. Rev. Letters 28, 1482 (1972); Phys. Rev. D7, 3689 (1973).

26. R. Jackiw, K. Johnson, Phys. Rev. D8, 2386 (1973), Sec. II. D. The argument at the beginning of this section has immediate application to quantum electrodynamics.

27. K. Wilson, Phys. Rev. D2, 1478 (1970); S. Coleman, R. Jackiw, Ann. Phys. (N.Y.) 67, 552 (1971).

28. S. L. Adler, Phys. Rev. D5, 3021 (1972); 7, 1948(E) (1973).

29. K. Johnson, R. Willey, M. Baker, Phys. Rev. 163, 1699 (1967); M. Baker, K. Johnson, Phys. Rev. D3, 2541 (1971).

30. Unitarity may place a constraint on the possible summation procedures. See R. J. Crewther, S. S. Shei, and T. M. Yan, Phys. Rev. D8, 3396 (1973), Footnote 45.

31. See Ref. 28, Secs. 3.B. and 4.

32. K. Johnson, M. Baker, Phys. Rev. D8, 1110 (1973), Sec. V.

33. M. Bernardini et al., Phys. Letters 45B, 510 (1973); V. Alles-Borelli et al., Nuovo Cimento 7A, 345 (1972).

34. R. J. Crewther, S. S. Shei, T. M. Yan, Phys. Rev. D8, 3396 (1973), Sec. III. See also the remarks of S. L. Adler, Phys. Rev. D5, 3021 (1972), end of Sec. 3.C.

35. M. Baker, K. Johnson, Phys. Rev. 183, 1292 (1969); S. L. Adler, Phys. Rev. D5, 3021 (1972), Sec. 3.A. The function $c_1(\alpha_o)$ is JBW's $f(\alpha_o)$.

36. One technical assumption made throughout the work of JBW and Adler is that terms that vanish asymptotically in each order of perturbation theory do not, when summed up, become asymptotically dominant, as briefly discussed in Sec. I. G. Marques and C.H.Woo (Phys. Rev. (to be published)) have illustrated with a simple model the failure of this assumption when the physical coupling constant coincides with one of the zeros of the Callen-Symanzik function β. This is the situation in Adler's case: $F(\alpha) = 0^{\infty}$ implies $\beta(\alpha) = 0^{\infty}$ (S.L. Adler, Phys. Rev. D5, 3021 (1972), Sec. 4). The need to check this assumption has also been stressed by K. Symanzik, Comm. Math. Phys. 23, 49 (1971).

37. R. A. Abdellatif, Ph. D. thesis, University of Washington, 197o (unpublished); S. L. Adler D6, 3445 (1972); 7, 3821(E) (1973);
H.J. Schnitzer, Phys. Rev. D8, 385 (1973).
N. Christ, Phys. Rev. D9, 946 (1974).

38. S. L. Adler, Phys. Rev. D8, 2400 (1973).

39. For a third approach, see K. Johnson and M. Baker, Phys. Rev. D8, 1110 (1973), Sec. V.

40. The use of Eq. (14) as the starting point for the calculation of F was first suggested by M. Baker and K. Johnson, Phys. Rev. D3, 2541 (1971).

41. M. P. Fry, Acta Phys. Austr. (to be published).

42. S. Mandelstam, Ann. Phys. (N.Y.) 19, 1 (1962); Phys. Rev. 175, 1580 (1968);
A. Q. Sarker, Ann. Phys. (N.Y.) 24, 19 (1963).

Acta Physica Austriaca, Suppl. XIII, 767–773 (1974)

RECENT PROGRESS IN PARTICLE PHYSICS*)

by

William R. FRAZER+)
Department of Physics
University of California, San Diego
La Jolla, California 92037

Although I am honored to have been asked to give the final summary talk of this conference, I must say right away that it is impossible to summarize such a wide range of topics as we have discussed in the past two weeks. The particles discussed ranged from point-like partons, through our familiar mesons and baryons, to the other extreme of the particles Ehlers was concerned with - which turned out to be small rocks or even golf balls! Therefore I will not try to summarize, but will instead give you some personal reflections on the theme of this conference, progress in particle physics.

Every year there is progress in particle physics, but I have the feeling that we are now in an exciting

---

*) Summary Talk given at Schladming Winterschool, Schladming, Austria, 1974.

+) Work supported in part by the United States Atomic Energy Commission.

era of rapid progress. Last night I tried to write down the most exciting developments in particle physics within my memory. Here's what occurred to me

1957 - Parity nonconservation, dispersion relations
1958 - V-A theory, Mandelstam representation
1960 - Resonances
1961 - Eight-fold way, current algebra, Regge poles, Froissart bound
1962 - Two neutrinos
1963 -
1964 - CP violation
1965 -
1966 -
1967 - Duality
1968 - Scaling in deep-inelastic electron scattering
1969 -
1970 -
1971 -
1972 - Short-range correlations and scaling in multiparticle hadronic reactions
1973 - Neutral currents (?), successes of parton model

If any of you listening to or reading this talk were to make up such a list, it would probably not be identical to mine, and I have probably forgotten some items I would like to add if you were to remind me of them. But the point I want to make is that there was a relatively dull period of almost a decade. At the Irvine conference in 1971 there was a panel discussion at which J. D. Jackson pointed out that at that conference we were discussing the same things as at the Aix-en-Provence conference a decade earlier!

At that time many of us were feeling a lack of excitement with particle physics. I am reminded of a conversation I once had with Carl Eckart, concerning how it felt to be working during the 1920's and 1930's, when quantum mechanics was being developed. I said, "It must have been very exciting." "No," he replied, "I grew up during that period, and to me it was the norm. It's just been rather dull ever since."

Now I think we have entered another period of exciting advances. The most recent has been the observation of neutral weak currents. Before these two weeks, I was quite skeptical. In fact, I was surprised at the degree of confidence I found among you here in the validity of the observations. It seems that the confidence level varies inversely with the distance from C.E.R.N.! The hadronic events require sizeable Monte Carlo corrections, and as for the leptonic event, one is reluctant to believe in such an important effect on the basis of one event. The Ω^- is a counter-example, but this neutral-current event is much less impressive: I can reproduce it by dragging my pen carelessly across the overhead projector:

Now, however, a very important observation has been reported at this conference: a second such event! That is enough, I think, to allow us to celebrate the discovery (meanwhile anxiously awaiting final results from N.A.L. experiments).

The discovery of neutral currents is a significant new datum about the weak interaction - an interaction about which precious few data exist. Although it will often be discussed in the framework of gauge field theories, we should remember that its fundamental importance remains, even if those theories should fail.

Now let's turn to gauge field theories, where there is much excitement. At this conference we have celebrated the marriage of our beloved quantum electrodynamics to that ill-mannered youth, weak-interaction field theory. If fear that it is, to use some American slang, a shotgun wedding, and I will be surprised if it lasts more than a couple of years.

In trying to correct the notorious bad behavior of weak interactions, we proposed an interaction (the W-boson) which, while somewhat improving the behavior of the weak interaction, unfortunately got Q.E.D. into trouble. (I refer, of course, to the divergences in the electrodynamics of charged vector bosons.) At this point some clever physicists saw that the problems of both Q.E.D. and weak interaction could be solved by marrying them!

Perhaps they are compatible, but they are so different! One conserves parity, while the other squanders it. One has infinite range, while the range of the other is at present experimentally indistinguishable from zero. Is such a union plausible?

Let me now drop this metaphor before it gets out of control, and discuss this union of Q.E.D. and weak interactions in more precise language. From the S-matrix point of view of Llewellyn-Smith and Bell, one can see the problem by considering the process $e^+ + e^- \to W^+ + W^-$ in lowest order:

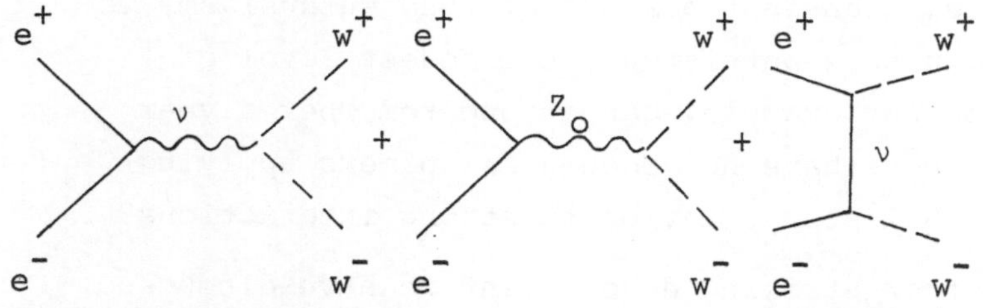

These diagrams individually have unacceptable high-energy behavior, but the new theory arranges their couplings so that cancellations occur, leaving the sum with acceptable behavior. It is not necessary, however elegant it may seem, to put the two theories together. One could arrange for them to cancel separately, at the cost of introducing more new particles - theorists in this field have already overcome their aversion to introducing new particles of unobservably high mass. Thus there is no compelling reason to unite the theories, and your opinion is as good as mine concerning how elegant and natural it is to do so.

Nevertheless, gauge field theories are remarkable, and even if they do not live up to all expectations, I think they are here to stay. I have noticed that we theorists usually retain some affection for an old theory, even if it has failed to live up to our original fond expectations. Most theories follow a curve of significance vs. time which is similar to the following:

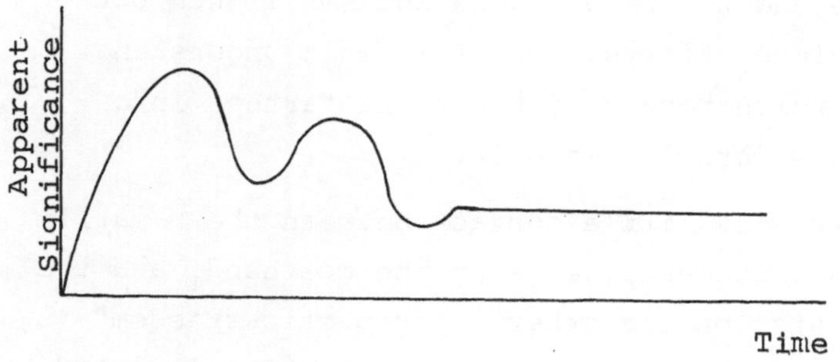

After a few transient periods of fond enthusiasm, we settle down to a persistent, but modest level of affection. For example, can anyone remember a year in which there have not been a few papers applying the Bethe-Salpeter equation to strong interactions?

Another striking development we have discussed is the quark-parton model. The successes of this daring model are indeed spectacular, (although its difficulty with e^+e^- annihilation cannot be overlooked). The model is at first sight so outrageous! At a parton conference at N.A.L. last year, Feynman gave as an opening motto, "Angels rush in where fools fear to tread". I could not resist composing my own distortion of Einstein's remark, "God is not malicious, but he <u>is</u> sophisticated!". Nevertheless, the success of this daring, apparently naive, model should challenge and stimulate all of us.

Having insulted everyone else, I must not leave out my own specialty, strong interactions. Here the most notable development is the veritable glut of data given us by the ISR and by NAL. A few years ago, we theorists would have only developed indigestion if fed so much data! But now, equipped with powerful new enzymes (short-range correlations, scaling, Mueller-Regge...) we confidently digest it. Unfortunately, the digestion is incomplete and the result does not smell much better than our previous efforts. We still don't understand the pomeron, which recedes farther and farther into asymptopia as we crawl toward it.

There is very little contact between the S-matrix theorists and phenomenologists on the one hand, and the "fundamentalists" on the other. "Asymptotic freedom" is a very ingenious theory, but it has not yet calculated one number! A story told me by Geoffrey Chew is apropos here: At a Solvay Conference Chew had the opportunity to ask Heisenberg about his attitude toward S-matrix theory. (Heisenberg, after first proposing S-matrix theory, turned toward the search for a fundamental field theory.) Heisenberg answered, to paraphrase, that both approaches had the same goal. One started from the inside and tried to work out toward contact with experiment, whereas the other started from the outside and tried to work toward the inside. At this conference, we have seen successes of both approaches, but I cannot even hazard a guess as to whether and when we shall see them meet.